ASTRONAUTICAL RESEARCH 1971

ASTRONAUTICAL RESEARCH
1971

PROCEEDINGS OF THE 22ND CONGRESS OF THE
INTERNATIONAL ASTRONAUTICAL FEDERATION
BRUSSELS, 20–25 SEPTEMBER 1971

Editor-in-chief

L. G. NAPOLITANO

Editors

P. CONTENSOU and W. F. HILTON

D. REIDEL PUBLISHING COMPANY
DORDRECHT-HOLLAND/BOSTON-U.S.A.

Library of Congress Catalog Card Number 72–92536

ISBN-13: 978-94-010-2561-4   e-ISBN-13: 978-94-010-2559-1
DOI: 10.1007/ 978-94-010-2559-1

Published by D. Reidel Publishing Company,
P.O. Box 17, Dordrecht, Holland

Sold and distributed in the U.S.A., Canada, and Mexico
by D. Reidel Publishing Company, Inc.
306 Dartmouth Street, Boston,
Mass. 02116, U.S.A.

PREFACE

The International Astronautical Federation is the only professional society in the field of aerospace engineering and Sciences which brings together specialists of all countries interested in the exploration and peaceful exploitation of space.

At its annual Congresses a large number of invited and/or carefully selected contributed papers are presented which cover a wide variety of topics and are distributed over a number of sessions, each one being organized by two leading scientists who later chair the session itself.

Each year the selection of specific topics to be dealt with is dictated either by significant new progress achieved in some sectors or by new developments and trends which are liable to influence substantially the objectives toward which space research and/or application of space technology will be oriented in the immediate future.

A second rigorous screening, performed with the help of the Session Chairmen and carried out according to the same criteria identifies finally the papers which are published in the Proceedings.

The outcome of all this is reliable and authoritative information as to the actual status and future trends of space activities, both from the research point of view and from the point of view of utilization and/or application.

These indications are all the more valuable since they offer a global, panoramic view which includes the often different perspectives of the several nations and/or groups which have any activity in Space, from those which are heavily involved to those which are moderately involved and all the way down to those which are just starting to get interested in Space.

For all these reasons, Astronautical Research 1971 should constitute an invaluable and irreplaceable reference tool for all who are concerned, at any level, with the present status and the immediate future trends of space activities.

The book, which consists of four parts, appropriately and meaningfully begins with an overview of International Cooperation in Astronautics, a topic which formed the subject of the Third I.A.F. Invited Lecture, delivered at the Congress by M. J.-P. Causse.

The first part of the Proceedings contains papers addressing basic problems and consists of three sections devoted, respectively, to Astrodynamics, Fluid-Mechanics Aspects of Space Flight and Bioastronautics.

The second part groups the papers concerned with the Engineering and/or Management Aspects of Space Technology. Specific topics considered are: Earth-to-Orbit and Orbit-to-Orbit Space Transportation; Propulsion; Structure and Materials; Telemetering and Data Management.

The third part is devoted to the utilization and/or applications of Space Technology.

Of particular and timely importance is the section dealing with Earth Resources

Satellites. It contains discussion of the latest developments of remote sensors of the type used on the Earth Resource Technology Satellite (ERTS) (which will certainly be in orbit by the time the Proceedings appear) and up-dated and valuable assessments of two problems of extreme relevance for their practical implications: global monitoring from satellites and utilization of satellites in the management of human environment.

The other two sections of the third part are concerned with Scientific Spacecraft and problems related to Education, particular emphasis being given, in the latter section, to the application of space benefits to Education and to safety in youth rocket experiments.

The fourth part should have an appeal of its own since it is devoted to the International Conference of graduating and/or post-graduate aerospace students which, for the first time in the world, the International Astronautical Federation has convened.

The papers included in this fourth part testify that the upcoming younger generation, throughout the world, has a continuing interest in aerospace studies and is ample evidence of the validity of its creative production.

In closing, I wish to thank all those who have helped me in editing these Proceedings: the Session Chairmen for their valuable suggestions and advice, the Language Editors, Dr P. Contensou and Dr W. Hilton, for carrying out their task with enthusiasm, promptness and efficiency; the Secretariat in Paris for an extremely valid cooperation and a much appreciated coordinating action; the managing editor of the Reidel Publishing Company for a patient and competent handling of the many problems associated with the publication of international Proceedings. To the authors, I am sure, will go the appreciation of the readers.

LUIGI G. NAPOLITANO
Editor-in-Chief

TABLE OF CONTENTS

LA COOPERATION INTERNATIONALE DANS LE DOMAINE
DE L'ASTRONAUTIQUE

M. JEAN-PIERRE CAUSSE

Secrétaire Général Adjoint, Directeur des Activités Futures du CECLES/ELDO

Je suis très sensible au redoutable honneur qui m'est fait de m'adresser aujourd'hui, lors de sa séance d'ouverture, au Congrès de la Fédération Internationale d'Astronautique après que les deux années précédentes deux éminentes personnalités, l'une américaine, l'autre russe, aient présenté des contributions majeures à la connaissance des projets spatiaux.

Il me paraît tout à fait symptomatique qu'en demandant à un européen de prendre la parole à son tour, votre président lui ait suggéré de traiter de la coopération internationale dans le domaine de l'astronautique. Nous autres européens vivons en effet sans doute plus intensément que d'autres, et ce depuis bientôt dix ans, l'expérience quotidienne de la collaboration internationale avec ses alternances de crises et d'espoirs.

Mais lorsque le Président Jaumotte s'est adressé à moi, ma première réaction a été de penser qu'il y avait en Europe de nombreuses autres personnalités plus qualifiées que moi pour traiter de ce sujet passionnant mais ô combien difficile. En réfléchissant il m'est cependant apparu que mon expérience personnelle était tout de même un peu particulière et que je pourrais essayer de vous communiquer quelques réflexions qu'elle a pu m'inspirer. Je ne parle ici, est-il besoin de le dire, qu'en mon nom purement personnel.

Ma première participation directe à l'aventure spatiale, dès 1958, remonte en effet à l'époque où je travaillais à titre privé aux États-Unis. Rentré en France pour y diriger le programme de satellites au Centre National d'Études Spatiales, je repris sans tarder le chemin de l'Amérique, cette fois-ci à la tête d'une délégation de mon pays pour négocier avec la NASA ce qui devait devenir le projet 'FR-1', premier satellite scientifique français, qui fut lancé par une fusée américaine en 1965.

Quelques années plus tard je me trouvais désigné comme responsable du côté français du projet de satellite 'Roseau', projet de collaboration avec l'Académie des Sciences de l'Union Soviétique dans lequel je fus l'interlocuteur de l'académicien Georges Babakin, prématurément disparu cet été, au souvenir duquel je tiens ici à rendre hommage.

A la même époque il me fallait porter aussi sur les fonds baptismaux le projet franco-allemand 'Symphonie', autre exemple de coopération bi-latérale. Depuis quatre ans c'est aux programmes européens que je consacre l'essentiel de mon activité et c'est à ce titre que je me suit trouvé à la tête d'une délégation européenne essayant de préparer la participation du vieux continent au programme post-Apollo de la NASA.

C'est dire que j'ai vu les choses 'par l'intérieur' mais je suis un ingénieur et non un

L. G. Napolitano et al. (eds.), Astronautical Research 1971, 1–10. All Rights Reserved.
Copyright © 1973 by D. Reidel Publishing Company, Dordrecht-Holland.

spécialiste des relations internationales. Je ne vous présenterai donc pas une théorie exhaustive de la coopération et me bornerai à une série de remarques. J'insisterai, en particulier, sur les expériences européennes qui, à condition de savoir en tirer des leçons, me paraissent contenir une série d'enseignements de grande signification pour toute entreprise de ce genre.

La Fédération Astronautique Internationale est l'un des organismes où la coopération spatiale a pris naissance et s'est le mieux développée. C'est donc un cadre de choix pour réfléchir sur les problèmes qu'elle pose.

En tout cas il ne me paraît guère nécessaire de démontrer l'intérêt de la collaboration internationale dans le domaine de l'espace. L'espace n'appartient à personne et transcende les frontières nationales. Cette collaboration est même *inévitable* dans beaucoup de cas. Elle apparaît en fait *indispensable* pour les projets les plus vastes qui deviennent aujourd'hui techniquement possibles, mais dont l'ampleur même semble dépasser les plus grandes puissances spatiales. Les dernières années, les derniers mois même, ont vu des progrès sensibles dont nous ne pouvons que nous rejouir dans les relations entre les deux grands où la rivalité initiale fait place aujourd'hui à des possibilités de collaboration. Je pense que cette tendance ne pourra que se développer dans les années à venir.

Si les grands acceptent de collaborer entre eux, le nombre de pays désirant avoir des activités spatiales et capables de les exercer augmente également sans cesse cependant que les gouvernements cherchent, pour des raisons financières, à éviter des doubles emplois et à tirer le meilleur parti des sommes consacrées à l'espace. Tous ces facteurs me semblent favorables à une extension importante de la collaboration internationale dans les années qui viennent. Encore faut-il que celle-ci soit fructueuse et apporte effectivement ce que l'on attend d'elle.

Sans entrer dans les détails de l'évolution de l'attitude des différents gouvernements sur la valeur de la coopération internationale dans le domaine des activités spatiales, on peut dire qu'elle est aujourd'hui très largement reconnue et encouragée par les plus hautes autorités. Le Traité de l'Espace en 1967 a marqué une étape importante dans cette évolution. Les déclarations du Président des États-Unis, l'offre américaine de participation à post-Apollo, le succès apparent des premières réunions américano-soviétiques sur le rendez-vous et l'arrimage de cabines spatiales, constituent autant d'indications que sur le principe même tout le monde est aujourd'hui d'accord.

Il faut bien reconnaître cependant que la coopération ne s'est pas développée également dans tous les domaines des activités spatiales. C'est certainement la recherche scientifique qui s'est trouvée le milieu le plus favorable à son développement, sans doute parce qu'il y avait moins de barrières là qu'ailleurs, mais aussi parce qu'il existe de longue date une tradition de coopération internationale dans le domaine de la science. Il faut aussi souligner le rôle important des unions scientifiques ou d'organisations comme le COSPAR et la F.A.I. Cette coopération a d'ailleurs pris des formes diverses et nombreuses: en premier lieu le lancement par les pays disposant de lanceurs de satellites construits par d'autres nations. Les États-Unis ont, vous le savez, ouvert la voie dans ce domaine et de nombreux pays ont bénéficié de ces facilités.

L'Union Soviétique a fait de même avec plusieurs pays de l'Est et avait proposé à la France, comme je l'ai mentionné plus haut, de mettre sur une orbite très excentrique de la terre, grâce à une puissante fusée, le satellite 'Roseau' pour l'étude de la magnétosphère. Ce projet a été abandonné à la suite de difficultés budgétaires du côté français mais la voie était ouverte et ce fut une grande déception pour nous tous de ne pas mener à terme ce projet. Ce ne furent pas, et de loin, les problèmes techniques qui furent les plus difficiles. Je me souviens de certaines séances de travail lors de nos premières réunions où nous nous connaissions à peine et où nous explorions avec prudence nos interfaces techniques : l'équipe française décrivait la solution proposée pour tel ou tel problème et pendant que l'interprète traduisait je voyais s'illuminer le visage de mon interlocuteur, le Professeur Babakin, qui terminait dans un grand éclat de rire. Il fallait alors attendre l'inévitable traduction en retour, qui était "Tous les ingénieurs sont les mêmes. Aux mêmes problèmes nous trouvons la même solution. Nous sommes d'accord". Cette coopération continue d'ailleurs, mais sur un plan moins ambitieux. De petits satellites technologiques construits en France doivent être lancés par des fusées soviétiques, mais comme passagers auxiliaires de la charge principale. Je pense que le premier d'entre eux doit être lancé incessamment.

En France, à notre tour, nous avons pu faire bénéficier d'un tir d'essai du Diamant B les chercheurs de la République Fédérale d'Allemagne. Ce fut le satellite scientifique 'Dial', programme conclu, je crois, à la satisfaction de tous.

En sens inverse une équipe italienne a déjà réussi la mise sur orbite d'un satellite américain depuis sa plate-forme équatoriale 'San Marco', et d'autres essais du même genre sont prévus.

Encore plus répandue est l'adjonction à l'intérieur d'un satellite de dispositifs expérimentaux réalisés par d'autres. Dans ce cas il n'est plus besoin de disposer des compétences complètes de technologie spatiale nécessaires à la construction d'un satellite, mais seulement de celles requises par l'expérience. Les difficultés cependant ne sont pas négligeables car nombreux sont les problèmes d'interface avec les systèmes toujours complexes d'un satellite étranger.

En ce sens, et pour parler des projets que je connais le mieux, il faut saluer les succès que sont, par exemple, l'envoi, à bord d'une sonde soviétique Mars, de l'expérience française 'Stéreo', qui permet l'étude des émissions radioélectriques du soleil ou le dépôt sur la lune de l'engin Lunakhod porteur d'un réflecteur laser construit également en France. Sans oublier, bien sûr, la série impressionnante d'expériences européennes embarquées sur les satellites observatoires américains 'OSO' et 'OGO'.

Enfin je mentionnerai une dernière forme de coopération bilatérale, celle de l'exploitation des résultats dont l'exemple le plus frappant est sans doute la distribution à de nombreux laboratoires de tous les pays d'échantillons du sol lunaire rapportés par les expéditions américaines et soviétique.

Ces exemples d'accords entre deux pays ont pour eux l'avantage de l'extrême souplesse qui permet de s'adapter au sujet précis, aux conditions des participants, etc. Ainsi les États-Unis indiquent-ils dans leur rapport aux Nations-Unies qu'ils ont signé deux cent cinquante accords avec trente-cinq pays.

Certains accords bi-latéraux peuvent d'ailleurs servir de tremplin à une coopération plus large. Ainsi le programme de satellites ionosphériques 'Alouette', commencé sous la forme d'un satellite canadien lancé par une fusée américaine et dont le succès technique aussi bien que scientifique fut remarquable, se transforma en un véritable programme international ('ISIS') de l'étude de l'ionosphère avec de nombreux participants.

Enfin en Europe, mais j'y reviendrai plus loin en détail, la coopération scientifique spatiale prit la forme d'une organisation institutionnalisée, le CERS/ESRO, cependant que d'autres formes, aussi bien nationales que bi-latérales ou multi-latérales, de recherche scientifique continuaient à exister.

Si la coopération dans le domaine scientifique ne pose pas de problèmes sérieux, il n'en a pas été de même dans le domaine des applications. C'est seulement cet été qu'a pu être mis sur pied l'accord définitif sur les télécommunications spatiales entre les 80 pays d'INTELSAT. Si techniquement le système donne satisfaction à ses utilisateurs, et s'il a permis un développement spectaculaire des télécommunications à l'échelle mondiale, il n'en reste pas moins que sur le plan de son organisation le moins que l'on puisse en dire est qu'il ne suscite pas l'enthousiasme général. Malgré une évolution sensible par rapport aux accords intérimaires, l'accord final tend à geler la situation telle qu'elle était au début des années 60 où, dans le monde occidental, seuls les États-Unis pouvaient construire et lancer des satellites de télécommunications. En particulier, le désir des Européens de devenir des partenaires, non seulement pour l'utilisation mais aussi pour la construction du secteur spatial, n'a pas été pleinement satisfait. Il représente peut-être, il est vrai, une de ces motivations divergentes qui rendent si difficile de trouver un but commun à une entreprise commune.

En revanche, dans ce contexte marqué d'amertume, on ne peut que se réjouir du projet de satellite pour le contrôle de la navigation aérienne que l'Europe et les États-Unis envisagent aujourd'hui de *développer* en commun. La genèse de ce projet est intéressante. C'est le mérite des organisations de recherche spatiale, principalement le CERS/ESRO et le Centre National d'Études Spatiales français, d'avoir fait connaître les possibilités des satellites dans ce domaine aux organisations qui assurent la responsabilité du trafic aérien dans les pays européens. Celles-ci ont reconnu l'intérêt de cette méthode et c'est en commun qu'elles ont fait les études nécessaires aboutissant, en particulier, à une bonne solution technique, condition sine qua non à mes yeux d'un succès réel à long terme. C'est donc en *partenaires* bien préparés et unis que les Européens ont pu s'adresser aux Américains et que, devant la perspective de systèmes concurrents, perspective absurde pour les utilisateurs, un accord de coopération est actuellement en train de se réaliser. De nombreuses questions restent encore à régler (sur les modes de financement en particulier), mais il est permis d'espérer aujourd'hui qu'un système véritablement conçu, construit, géré et utilisé en commun, guidera un jour les avions de toutes les compagnies aériennes, quelles qu'elles soient, sur les océans du globe.

En fait, pour chaque nouvelle forme de satellite d'application le même problème se pose, avec plus ou moins d'acuité. Si, pour la météorologie, domaine qui en lui-même fait déjà l'objet d'une coopération internationale satisfaisante, il n'y a pas de difficultés sérieuses, de nouveaux types de satellites font beaucoup parler d'eux. Il s'agit de la télévision directe et des ressources terrestres. Il est intéressant d'observer que dans les deux cas les implications politiques de ces progrès techniques, dont nous savons bien qu'ils sont réalisables à court terme, sont telles que la plupart des pays considérés appellent de leurs voeux un contrôle international *efficace* de ces activités.

Si l'on considère le problème de la coopération dans son aspect le plus général, l'expérience permet de dégager un certain nombre de règles: En premier lieu, pour être véritablement efficace et aboutir à des résultats concrets, la coopération devrait tout au moins s'inscrire dans une politique d'ensemble de la science et de la technologie entre les mêmes participants. Ces politiques n'existent malheureusement pas encore, ni au niveau mondial, ni même au niveau régional. Elles n'existent même pas toujours au niveau national. En deuxième lieu il me semble bien qu'il ne faut pas trop espérer qu'une coopération dans le domaine spatial puisse être très en avance sur les relations d'ensemble entre les dits participants. Certes, par sa nature même la coopération dans le domaine spatial peut servir de pionnier mais elle n'ira pas très loin tant que les relations techniques et industrielles et surtout politiques entre les mêmes participants n'auront pas atteint un niveau adéquat. C'est peut-être le rôle et l'honneur de nous autres, spécialistes de l'astronautique, de pousser à ce mouvement, car nous savons qu'il est indispensable dans notre science, mais nous devons aussi être réalistes et ne pas tenter l'impossible.

Nous devrions aussi rappeler que notre activité n'est possible que dans le cadre d'une planification à long terme, car les politiques nationales ou coopératives sont soumises à des fluctuations nombreuses. Les 'intégrer' sur un grand nombre des cas devrait tendre à faire disparaître les à-coups, mais il faut aussi se garder des tendances générales, des modes. Aujourd'hui on assiste à une récession générale de l'intérêt dans l'exploration de l'espace et à une tendance à se limiter aux activités qui offrent le plus de retours économiques rapides et sûrs. S'il faut sans doute aujourd'hui marquer un temps d'arrêt pour exploiter l'acquis technique et technologique de la première décennie d'activité spatiale, il faut aussi penser à l'avenir.

En tout cas il est possible de rappeler que quelle que soit la forme de la coopération spatiale internationale, pour être vraiment fructueuse elle devra être efficace et donc obéir à certaines règles de gestion qui sont d'ores et déjà bien connues. La première de ces règles est que les discussions sur les objectifs à atteindre doivent être préalables à l'adoption d'un programme et que ces objectifs doivent être véritablement communs et non représenter la simple addition des objectifs particuliers des différents participants. Il sera aussi nécessaire de faire des études préalables et approfondies des bénéfices escomptés par chaque partie, de voir si ceux-ci sont compatibles entre eux, faute de quoi certains se considéreront comme lésés. Enfin il est indispensable de disposer d'une définition précise des projets du point de vue technique mais aussi du point de vue opérationnel; un projet réussi sans utilisateur ne peut conduire qu'à des regrets.

C'est le rôle délicat des organisations responsables du développement spatial de savoir aller suffisamment de l'avant pour démontrer les possibilités nouvelles de l'espace à des utilisateurs qui ne sont pas toujours conscients de ses possibilités et qui ne sont pas toujours organisés pour cela ou à d'autres qui hésitent à s'engager, mais il leur faut naturellement aussi savoir s'arrêter et éventuellement se réorienter.

Il me semble que l'on peut ainsi dire que la coopération a un double aspect : elle joue un rôle dynamique, c'est-à-dire qu'elle seule permet de réaliser certaines grandes choses. C'est ce rôle, il faut bien le dire, qui est le plus proche de nos préoccupations à nous spécialistes de l'astronautique. Mais aussi la coopération joue un rôle régulateur en ce sens qu'elle conduit à fixer des limites à ces activités, à leur donner un cadre et des lois. Dans ce sens la tendance est claire, on s'oriente de plus en plus vers une définition de la règlementation sur le plan véritablement mondial à travers les Nations-Unies ou les organisations qui en dépendent comme l'Union Internationale des Télécommunications.

Passons maintenant, si vous le voulez bien, à la coopération spatiale dans le cadre européen. J'essaierai de distinguer d'abord quelles en sont les caractéristiques propres puis je tenterai d'indiquer quelles leçons me paraissent pouvoir être tirées de cette expérience.

L'Europe s'est lancée très tôt dans la coopération spatiale internationale pour toute une série de raisons : clairement aucun pays européen ne dispose des ressources suffisantes pour agir seul. Même aujourd'hui, en additionnant la totalité des sommes consacrées à l'espace en Europe, projets nationaux ou projets européens, cela ne représente encore qu'un dixième de ce qu'y consacrent les États-Unis. C'était un facteur 20 il y a seulement quelques années. C'est-à-dire que pour la plupart des pays d'Europe il n'est pratiquement pas possible d'avoir un programme spatial autonome et que la collaboration européenne constitue en fait leur seule forme possible d'action dans ce domaine.

Enfin, et surtout à l'époque où les organisations spatiales européennes ont été fondées, cette coopération était considérée comme un élément d'intégration dans le cadre d'une politique scientifique et technologique commune.

La création du CERS/ESRO, orienté vers la recherche scientifique dans le domaine spatial, doit certainement beaucoup au brillant succès du CERN qui avait montré à la fin des années 50 que lorsque les Européens étaient capables de s'unir ils pouvaient se hisser au niveau des Américains, voire les dépasser, même dans un domaine très ambitieux.

Quant au CECLES/ELDO, créé pour construire un puissant lanceur européen, on pouvait considérer qu'il représentait un modèle de coopération technologique, plusieurs pays mettant en pool leurs connaissances dans le domaine de la propulsion sans avoir à aller trop loin dans l'intégration de leurs techniques propres, en se chargeant chacun d'un étage ou d'un élément bien défini de la fusée européenne.

Si l'on cherche à tracer un bilan de cette coopération européenne dans l'espace on trouve naturellement des aspects très positifs, mais malgré tout il faut bien reconnaître que la situation n'est pas totalement satisfaisante.

Au premier plan de ces aspects positifs je place l'investissement humain qui a été réalisé. Il ne faut pas sous-estimer en effet les difficultés qu'il y a eu à rassembler, à faire travailler en commun, des gens d'horizons, de cultures, de langues différents.

Il a fallu aussi établir des règles de gestion acceptables pour tous les pays, tout en respectant les principes sages du management. J'estime que dans ce domaine des progrès décisifs ont été réalisés. Certes on voit assez bien aujourd'hui que diverses erreurs auraient pu être évitées, en particulier que les succès auraient pu être atteints plus tôt, ce qui eut été important. Car aujourd'hui où nous trouvons en présence d'une tendance générale pour le ralentissement de l'effort spatial celle-ci se combine malencontreusement pour nous avec le désenchantement des gouvernements lié à nos difficultés propres. En tout cas ces dix années de travail en commun comptent.

Sur le plan des réalisations on peut également parler de bilan très positif: le CERS/ESRO a déjà construit et fait lancer quatre satellites scientifiques importants dont le fonctionnement a donné toute satisfaction. Trois autres satellites sont en cours de fabrication et doivent être lancés dans les prochains mois. Du côté du CECLES/ELDO, si le succès complet, c'est-à-dire la mise en orbite, n'a pas encore été atteint, il y a de bonnes raisons de penser qu'il est à notre portée. Le prochain tir, le premier dans la configuration à quatre étages, dite EUROPA II, est prévu pour le mois de novembre 1971, c'est-à-dire dans deux mois, depuis la base équatoriale que le CECLES/ELDO a construite en Guyane Française. Là encore, il ne faut pas que pour le besoin de l'autocritique nous oubliions le fait qu'il s'agit d'une fusée très puissante et que le progrès accompli est déjà considérable.

Parallèlement les travaux technologiques nécessaires à la préparation de la génération suivante de fusées, utilisant en particulier la propulsion par hydrogène liquide, se déroulent de façon très satisfaisante.

Si nous en venons maintenant aux aspects négatifs, il faut bien reconnaître que les activités spatiales européennes souffrent gravement de l'absence à peu près complète de politique générale. Les deux organisations sont nées dans des conditions différentes avec des partenaires différents, des objectifs différents. L'utilisation de l'espace à des fins économiques et l'apparition des satellites d'application devaient poser le problème: un programme de lanceurs européens a-t-il une signification en dehors d'une référence au programme de satellites européens? Dès lors le besoin évident de coordonner les organisations est apparu. C'est en 1966 qu'a été créée la Conférence Spatiale Européenne qui devait se réunir en principe une fois par an au niveau ministériel pour prendre les grandes décisions de principe sur l'ensemble des programmes et assurer ainsi leur harmonisation. Des efforts considérables ont été déployés depuis cette date, mais il existe encore de profondes divergences et je vois mal à l'heure actuelle comment et quand les points de vue pourront être rapprochés.

Il y a divergence sur les motifs, sur les objectifs, sur le contenu même du programme. Par motifs je veux dire: "Que veut-on faire dans l'espace? Quelles sont les raisons pour lesquelles l'Europe doit se donner un programme spatial?" Nous voyons s'affronter fréquemment les positions de ceux qui sont motivés par le désir d'obtenir rapidement un lanceur utilisable et celles de ceux qui considèrent comme également important de développer leur technologie, de faire progresser l'industrie, voire même de favoriser l'intégration de l'Europe.

Divergences sur les objectifs? S'il est maintenant généralement admis que le CERS/ESRO, tout en continuant un programme de recherche scientifique doit avoir également une activité dans le domaine des satellites d'application, je ne vois pas de signe d'amélioration dans l'irritante et interminable querelle des lanceurs: "L'Europe doit-elle disposer de ses propres lanceurs?" Pour ma part la réponse paraît bien évidemment devoir être affirmative. En effet, les ambitions que l'Europe est en droit de cultiver pour l'avenir dans le domaine spatial ne peuvent être pleinement réalisées que si elle dispose de tous les moyens nécessaires. Si posséder des lanceurs était hors d'atteinte, il faudrait évidemment se contenter d'autres solutions, mais aujourd'hui, pratiquement, le lanceur est réalisé. Ne pas s'en servir quand il sera au point, après que son développement ait coûté 650 millions de dollars, ma paraît une absurdité. C'est pourtant ce à quoi semblent se résoudre plusieurs de nos états membres.

Dans ces conditions il est bien difficile de se mettre d'accord sur le contenu des programmes, sur le niveau de ressources à affecter à chaque type d'activité, et jusqu'à un certain point on peut qualifier de dangereuse la situation actuelle d'incertitude sur les programmes futurs des deux organisations.

Peut-être plus grave encore est l'absence d'harmonisation entre les programmes nationaux et les programmes coopératifs réalisés par les organisations CERS/ESRO et CECLES/ELDO. Il est certain qu'au début les programmes nationaux, qui n'avaient pas à souffrir des contraintes propres à la vie internationale, ont pu se développer plus rapidement et atteindre plus aisément le succès.

Suivant l'exemple des premiers et observant simultanément l'extrême difficulté qu'il y avait à définir entre partenaires européens des objectifs véritablement communs, certains pays qui n'avaient pas de programme national en ont entrepris un. La situation est caractérisée aujourd'hui par la coexistence d'une série de programmes nationaux, entretenant d'ailleurs entre eux certaines relations bi-latérales, ou des relations bi-latérales extérieures, la plupart du temps avec les États-Unis, et le programme communautaire européen.

On a souvent défendu la thèse que ces activités, nationales constituent un support solide pour l'activité communautaire. Je n'en suis pas convaincu aujourd'hui. En particulier la limitation des resources disponibles pour l'activité spatiale rend inévitablement ces projets concurrents des projets européens. Certains d'entre eux sont de taille et d'ambition considérables et pèsent lourdement dans la balance au moment des décisions budgétaires. D'autres sont très modestes et même souscritiques. Cette famille nombreuse et mal nourrie ne me paraît pas être en bien meilleure santé que celle des projets européens.

Certes, dans la structure politique actuelle de l'Europe, et dans sa structure industrielle, cette situation est sans doute aujourd'hui inévitable. Elle permet d'atteindre certains objectifs à court terme, mais à plus longue échéance je crois qu'il est absolument nécessaire d'évoluer dans un sens ou dans un autre vers une organisation plus rationnelle.

Une première solution consisterait à laisser une certaine autonomie d'exécution des programmes aux états membres quitte à renforcer la coordination au niveau de leur adoption. Ceci conduirait à dé-européeniser certaines activités, par exemple en confiant les établissements européens aux états sur le territoire duquel ils sont placés. On confierait à l'échelon européen:

– une activité de coordination des programmes qui seraient exécutés en principe à la demande et pour le compte de la communauté par tel ou tel établissement;

– la réalisation de certains projets d'une taille vraiment trop grande pour un établissement isolé tels que les lanceurs lourds et les très gros satellites.

Mais puisque aujourd'hui la tendance politique en Europe est à l'extension de la communauté et partant à sa consolidation, pourquoi ne deviendrait-il pas possible maintenant de créer bientôt une véritable communauté spatiale européenne en fusionnant réellement toutes les activités spatiales quelle que soit leur forme actuelle? C'est, me semble-t-il, le seul moyen de rendre véritablement efficace l'effort entrepris et aujourd'hui encore très incertain.

Pourquoi ne pas commencer par réunir les deux organisations spatiales existantes en une seule, comme le principe en avait été décidé ici même à Bruxelles il y a plus d'un an mais jamais exécuté? La Conférence Spatiale Européenne avait alors prévu une série de mesures qui, sans même attendre un changement de la structure juridique, rendaient possible sur le plan des hommes et pratiquement sans délai, d'assurer la cohérence des deux programmes.

Le stade suivant pourrait être de mettre en pool les établissements nationaux, tout au moins ceux exclusivement consacrés à l'espace, ainsi sans doute que les projets déjà entrepris. Ce serait une mise en commun effective des moyens et des hommes sous une autorité unique qui devrait bien évidemment être internationale. En sens inverse ce renforcement de l'exécutif unique par la mise en commun des moyens et des hommes me paraît pouvoir rendre possible une certaine diversité dans la participation des états membres suivant leurs désirs ou leurs possibilités et la nature des objectifs qu'ils entendent poursuivre. Cette formule 'à la carte' a déjà été étudiée dans le cadre actuel du CERS/ESRO et du CECLES/ELDO; les mécanismes juridiques en ont été explorés et ne paraissent pas poser de difficultés fondamentales.

Une telle évolution des activités spatiales européennes dans un sens communautaire pose toute une série de problèmes qui devraient également être résolus. Sur le plan technologique ceci exige que la compétition passe au niveau européen alors qu'elle s'exerce aujourd'hui au niveau national.

Bien souvent la coopération européen actuelle est synonyme de refus de concurrence. Pour des raisons de politique nationale chaque pays tend à développer sa propre technologie et ensuite cherche à l'imposer à la communauté. Certes il ne faut surtout pas renoncer au puissant moteur de progrès qui est la concurrence. La question est seulement de savoir si cette concurrence doit s'exercer entre groupes ou firmes nationaux. Il me semble que poser la question c'est la résoudre.

Il faudrait pour cela obtenir une réforme des structures industrielles pour faire disparaître, là aussi, les réflexes nationalistes dûs à la coïncidence des intérêts des firmes avec les frontières géographiques. En attendant les sociétés européennes dont nous avons tellement besoin, il nous faut continuer à encourager les regroupements au sein de consortiums qui les préfigurent.

Je pense que tout cela est possible sinon dans l'immédiat tout au moins dans un avenir proche et que les activités spatiales pourraient ouvrir la voie.

Les difficultés que j'ai évoquées sur l'absence de politique générale pèsent également d'un certain poids dans la question de la participation européenne au programme américain post-Apollo que je voudrais brièvement évoquer pour terminer.

Il y a bientôt deux ans que le Dr Paine alors Administrateur de la NASA, a fait cette offre généreuse et passionnante à l'Europe, ainsi d'ailleurs qu'à d'autres pays du monde. La réaction européenne dans son ensemble a été très favorable. Pour la première fois se présentait la possibilité de participer vraiment au développement en non pas seulement à l'utilisation d'un grand système, et quel système! Dès le début d'ailleurs le gouvernement américain a marqué sa préférence pour que les Européens agissent collectivement plutôt que par un ensemble d'accords bi-latéraux. Tout naturellement c'est donc dans le cadre de la Conférence Spatiale Européenne qu'ont été menées les discussions avec les États-Unis, ainsi que les principales études entreprises en Europe pour mieux comprendre le système et éventuellement se préparer à y participer.

Il est aujourd'hui difficile de dire ce que sera l'avenir de cette participation. Tout d'abord pour des raisons purement techniques les thèmes de la participation éventuelle ne sont pas encore définis. D'ailleurs, du côté de la NASA le projet lui-même a évolué sensiblement depuis les premiers contacts. Du côté européen l'intérêt se porte surtout sur le développement du nouveau système de transport spatial plutôt que sur celui de la station orbitale. Mais aucun élément précis n'a encore été retenu.

De même des précisions sont nécessaires sur le niveau des engagements financiers à souscrire par l'Europe, sur les méthodes de gestion qui seront adoptées, les conditions d'accès aux informations, etc.

Mais d'autre part le projet a des aspects politiques qui restent encore à clarifier. Très important est celui de la disponibilité des lanceurs américains classiques et du nouveau système de transport spatial dont la solution conditionne la décision des Européens sur leur propre programme. Il a été en effect assez rapidement admis que l'effort financier nécessaire pour une participation significative à post-Apollo demandait que l'on renonce à certaines activités purement européennes, par exemple au développement d'une nouvelle fusée EUROPA III. C'est aux gouvernements de la Conférence Spatiale Européenne de prendre une décision dans les mois qui viennent. N'est-il pas permis de penser que si l'Europe avait déjà résolu ses problèmes internes elle aurait été en meilleure position pour envisager cette participation? En ce sens la proposition américaine est venue peut-être un peu trop tôt.

Mais si nos gouvernements prennent une position positive ce devrait être avec notre partenaire américain une forme complètement nouvelle de coopération internationale adaptée à la place que devraient tenir et que tiendront, j'en suis convaincu, les activités spatiales dans les prochaines décennies.

PART I

BASIC PROBLEMS

A. ASTRODYNAMICS

REVUE GÉNÉRALE DES MÉTHODES AUTOMATIQUES DE CALCUL DE THÉORIES ANALYTIQUES DE SATELLITES ARTIFICIELS

J. KOVALEVSKY

*Groupe de Recherches de Géodésie Spatiale
et Bureau des Longitudes, Paris*

1. Introduction

Au fur et à mesure que la précision des observations des satellites artificiels a augmenté et que le réseau de stations de poursuite s'est densifié, la nécessité de représenter numériquement avec précision l'orbite probable du satellite est devenue de plus en plus pressante.

Pour parvenir à ce but, deux types de problèmes se posent:

(1) Avoir une représentation aussi correcte que possible des forces en présence (modèle Terre).

(2) Avoir une solution aussi exacte que possible des équations du mouvement.

Nous ne considérons, dans cet exposé, que le second problème. Il peut en effet être résolu en principe indépendamment du premier, soit en adoptant pour les paramètres les meilleures valeurs numériques connues à l'époque, soit en leur donnant des valeurs indéterminées exprimées de façon littérale.

De nombreuses variantes existent pour résoudre ce problème. Elles sont basées sur l'emploi plus ou moins simultané des deux techniques fondamentales: l'intégration numérique et la théorie analytique littérale.

Dans son exposé, Mr B. Moynot décrira une méthode basée sur l'intégration numérique, pour obtenir l'orbite d'un satellite à partir des observations.

Mais cette méthode peut être coûteuse et on peut avoir intérêt à exprimer tout ou partie des perturbations causées par les forces de gravitation sous forme analytique et non pas en intégrant pas à pas des équations dont le second membre est complexe, mais dont la solution est connue par ailleurs.

Il est facile en effet d'ajouter aux résultats de l'intégration numérique les petites perturbations calculées analytiquement et qui modifient trop peu les coordonnées du corps en mouvement pour que cela réagisse de façon appréciable sur les forces calculées. C'est le cas des effets d'un grand nombre d'harmoniques tesséraux.

Une autre méthode, d'emploi plus délicat, est toutefois à conseiller pour les perturbations luni-solaires ou pour les termes empiriques du frottement atmosphérique: on tient compte des perturbations calculées analytiquement pour le calcul des forces dans l'intégration numérique, mais on n'ajoute leur effet aux coordonnées qu'après l'intégration.

L. G. Napolitano et al. (eds.), Astronautical Research 1971, 15–25. All Rights Reserved.
Copyright © 1973 by D. Reidel Publishing Company, Dordrecht-Holland.

Enfin, on peut, comme cela est fait systématiquement au Smithsonian Astrophysical Observatory (où se font les programmes les plus importants de géodésie dynamique par satellites), utiliser à tous les stades des théories analytiques et déterminer les éléments orbitaux et les paramètres inconnus et utilisant des équations aux variations également écrites une fois pour toutes sous forme littérale.

Pour préparer les expressions qui servent à ces programmes, il n'est plus possible de les calculer à la main, comme l'ont fait il y a dix ans Brouwer [5], Kozai [20] et bien d'autres. La précision requise est désormais de l'ordre du décimètre, puisque les observations laser se font dans certaines stations avec une erreur moyenne de 40 cm. Dans quelques années, après le lancement de Cannon Ball ou de satellites analogues, lorsque la précision du laser aura été réduite à quelques centimètres, ou encore lorsqu'on fera de la poursuite de satellites par satellites, c'est le centimètre que les théories devront assurer. La même précision sera aussi exigée de la théorie du mouvement de la Lune, lorsque les lasers-Lune centimétriques auront commencé à observer.

Le nombre de termes qu'il faut maintenant conserver est tel qu'il est indispensable de les faire calculer par un ordinateur. Ceci a été tenté – avec succès – par plusieurs équipes, surtout aux États-Unis et en France. L'objet de cette revue est de faire une présentation comparative de ces tentatives.

En fait, il y a dans la théorie du mouvement d'un satellite artificiel, deux problèmes très différents que nous considérons indépendamment:

(1) *Le problème principal:* calcul des termes dépendant uniquement du facteur d'ellipticité géopotentielle J_2, mais à l'ordre 3, sinon 4 de ce paramètre. Il faut programmer des théories d'ordre élevé, ce qui exige de faire de nombreuses opérations algébriques sur des expressions littérales.

(2) *Le calcul des perturbations dues aux harmoniques tesséraux.* Ces quantités ne sont utiles qu'au premier ordre des paramètres. Les opérations à effectuer sont simples, mais il faut les répéter pour un grand nombre d'expressions semblables: par exemple la Standard Earth 1970 (Gaposchkin et Lambeck [13]) fait intervenir 296 coefficients d'harmoniques tesséraux et on peut estimer à 400 ou 500 le nombre qu'il faudra déterminer pour construire les prochains modèles de Terre.

2. Le Problème Principal des Satellites Artificiels

Dans ce problème, la fonction perturbatrice par rapport au potentiel newtonnien est:

$$\mathscr{R} = \mu J_2 \, \frac{R^2}{a^3} \left[(\tfrac{1}{2} - \tfrac{3}{4} \sin^2 i) \, \frac{a^3}{r_3} + \tfrac{3}{4} \sin^2 i \, \frac{a^3}{r_3} \, \cos (2\omega + 2v) \right]. \tag{1}$$

Dans cette équation, comme dans la suite de l'exposé, nous utiliserons les notations suivantes:

μ constante géocentrique de la gravitation;
R rayon équatorial terrestre;
J_2 facteur d'ellipticité géopotentielle;
a demi grand axe (éléments osculateurs);
e excentricité;
i inclinaison;
Ω ascension droite du noeud;
ω argument du périgée;
M anomalie moyenne;
r rayon vecteur;
v anomalie vraie.

Une des façons d'écrire la solution du problème principal est classiquement (voir par exemple, Kovalevsky [18]) la suivante:
(1) Tout élément métrique σ (a, e ou i) a la forme:

$$\sigma = \sigma_0 + \sum_{i=1}^{\infty} \sum_{j=0}^{\infty} \sum_{k=-\infty}^{+\infty} J_2^i A_{ijk}(a_0, e_0, i_0) \cos(j\overline{M} + 2k\overline{\omega}), \qquad (2)$$

où les quantités a_0, e_0, i_0 sont les termes constants des développements correspondants,

où
$$\overline{M} = M_0 + n_M(t - t_0),$$
$$\overline{\omega} = \omega_0 + n_\omega(t - t_0),$$

et où n_M, n_ω ainsi que les coefficients A_{ijk} sont des fonctions de a_0, e_0 et i_0.
(2) De même, tout élément angulaire θ (Ω, ω, M) a la forme:

$$\theta = \theta_0 + n_0(t - t_0) + \sum_{i=1}^{\infty} \sum_{j=0}^{\infty} \sum_{k=-\infty}^{+\infty} J_2^i \beta_{ijk}(a, e_0, i_0) \sin(j\overline{M} + 2k\overline{\omega}). \quad (3)$$

En fait, pour des besoins pratiques, seul un nombre fini de termes suffira. On connait, par certaines propriétés des séries, comme les propriétés de l'Alembert, des relations d'inégalité entre les coefficients entiers i, j et k, et l'ordre en e_0 (on admet e_0 pas trop grand). En effet, on développe A, B et n en série entière de e, avec en général, des coefficients dépendant de $\eta = \sqrt{1 - e_0^2}$.
Il s'ensuit dès lors que ces expressions sont des fractions rationelles de a_0, e_0, $\sqrt{1 - e_0^2}$, $\cos i_0$ et $\sin i_0$ (ou d'autres quantités qui en sont des fonctions simples).
Pour résoudre les équations du mouvement d'un satellite artificiel, quelle que soit la méthode employée (résolution des équations de Lagrange par approximations successives, méthode de Von Zeipel, méthode des séries des Lie, etc....), on est ramené à effectuer un certain nombre d'opérations simples sur des séries de la forme (2) ou (3).

Ces opérations sont:
- Addition ou multiplication de séries,
- Intégration par rapport au temps,
- Dérivation par rapport à des paramètres.

Toutes les autres opérations: calcul de racines carrées ou d'inverses, fonctions sinus ou cosinus, se ramènent à une suite d'additions et de multiplications en utilisant soit des développements formels (type développements de Mac Laurin) soit des formules d'approximations successives ou de récurrence. Certaines conditions sur l'ordre de grandeur du terme constant doivent cependant être vérifiées.

De telles formules sont en général faciles à établir. On en trouvera quelques exemples chez plusieurs auteurs, en particulier Kovalevsky [17] ou Rom [21].

Donnons un exemple: soit à calculer $S = \Sigma^{-1/2}$.

Supposons que l'on ait une première approximation S_0 de S, et posons

$$S = S_0 + \Delta S.$$

On a

$$\Sigma = \frac{1}{(S_0 + \Delta S)^2}$$

et, en développant, puis en négligeant ΔS^2, on obtient:

$$\Delta S = \tfrac{1}{2}(S_0 - S_0^3 \Sigma). \tag{4}$$

On peut poser $S_1 = S_0 + \Delta S$ et appliquer de nouveau la formule (4).

Ainsi, pour construire la théorie du mouvement d'un satellite artificiel à l'aide d'un ordinateur, il faut avant tout disposer d'un programme permettant de faire effectuer ces opérations élémentaires sur des expressions (2) et (3).

Il y a toutefois une condition supplémentaire: ces programmes doivent pouvoir traiter des expressions très longues. A titre d'exemple, les perturbations en J_2^2 écrits sous la forme (2) ou (3) après avoir développé les coefficients en séries entières jusqu'à l'ordre 5 en e_0, comprennent environ 400 monômes. Développer à l'ordre 10 et pousser les calculs en J_2^3, multiplierait ce nombre par 20. Par ailleurs, on voudra aussi appliquer ces programmes à des problèmes plus ambitieux, comme par exemple la théorie du mouvement de la Lune. Pour cela le nombre de termes sera cent ou mille fois plus grand encore.

Un grand nombre de programmes de calcul algébrique ont été mis au point par divers centres de recherche de mathématiques appliquées et certains sont disponibles (voir, par exemple, Davis [8]). Ils sont tous d'un caractère très général et permettent d'effectuer un grand nombre d'opérations inutiles en Mécanique Céleste. Cette généralité a pour conséquence une grande lourdeur de programmation – d'où lenteur du calcul et encombrement des programmes. Ce dernier défaut est redhibitoire car il empêche de travailler sur des expressions aussi longues que celles qui sont communes en Mécanique Céleste. L'expérience de Sconzo et Valenzuela qui ont appliqué FORMAC pour le calcul des termes en J_2^2 en témoigne. L'importance des expressions était telle que le calcul n'a pu être entièrement achevé que sur l'ordinateur IBM 360-91.

Cette impuissance des grands programmes généraux à résoudre ce type de problème n'est pas étonnante. Une des principales raisons en est que le rassemblement des termes semblables (qui se présentent en grand nombre dans les multiplications de séries) ne se fait qu'à la fin du calcul d'une expression et non pendant le calcul. Il faut donc toujours réserver une mémoire bien plus grande que la dimension du résultat. Les divers autres groupes de recherche qui s'étaient proposés de travailler sur ce sujet ont donc préféré écrire une série de programmes capables de traiter plus particulièrement le type de séries propres à la Mécanique Céleste et ayant la forme suivante:

$$\sum_{i_1 \ldots i_n} R_{i_1 \ldots i_n}(x_1, x_2 \ldots x_p) \frac{\sin}{\cos}(i_1 y_1 + i_2 y_2 \ldots + i_n y_n) \tag{5}$$

où R est soit un polynôme de la forme

$$\sum_{j_1 \ldots j_p} A_{j_1 \ldots j_p} x_1^{j_1} x_2^{j_2} \ldots x_p^{j_p}$$

soit un rapport de deux polynômes de ce type.

Chaque terme est ainsi défini par un nombre A (en virgule flottante ou sous forme rationnelle) et par deux ensembles de nombres entiers:
– l'argument $Y = (i_1, i_2 \ldots i_n)$
– la caractéristique $X = (j_1, j_2 \ldots j_p)$.
Il faut y ajouter des indicateurs concernant le type de série (sin ou cos) et la place du monôme (en numérateur ou dénominateur).

La façon dont ces séries sont placées en machine varie d'un programme à un autre. Diverses techniques de mises en tableau ou sous forme de listes chaînées ont été adoptées. Elles sont décrites par les auteurs et j'en ai donné par ailleurs une description comparée (Kovalevsky [19]). Les solutions adoptées conditionnent la structure des programmes eux-mêmes.

Trois ensembles de programmes, avec dans chaque cas plusieurs variantes, ont été appliqués au problème principal du mouvement d'un satellite artificiel. Nous ne citerons qu'à titre de référence d'autres ensembles de programmes comme ceux de Barton [1], de Broucke et Garthwaite [4] ou de Jefferys [15] qui n'ont pas été utilisés dans ce but, bien que leur conception l'aurait permis. Nous n'analyserons donc pas ces travaux.

(1) *Au Bureau des Longitudes*, un ensemble de programmes a été développé pour permettre les calculs de mécanique céleste dans les théories du mouvement de la Lune et des planètes sur une petite machine, la Gamma 30 S Bull. Ces programmes sont tous basés sur le principe de la représentation des polynômes sous force d'un

tableau des coefficients à structure prédéterminée. Une des variante de cet ensemble de programmes a été utilisée par P. Bretagnon pour calculer les termes du second ordre en J_2, par la méthode des équations de Lagrange, en substituant la théorie du premier ordre dans les seconds membres et en intégrant de nouveau.

Les expressions obtenues vont être publiées incessamment (Bretagnon [3]). Elles ont été comparées avec une intégration numérique par Berger qui a confirmé que les termes à courte période représentent l'orbite à 10 cm près. Notons que les termes séculaires du second ordre obtenus sont en désaccord avec ceux de Brouwer [5]. Cette différence provient d'une définition un peu différente des constantes d'intégration dans la méthode de Von Zeipel. Mais pour maintenir cette précision sur un temps de l'ordre du mois, il est indispensable de calculer les termes à longue période en J_2^3.

(2) *Au Smithsonian Astrophysical Observatory*, Hall et Cherniack [14] ont construit un ensemble de programmes SPASM (Smithsonian Package for Algebra and Symbolic Mathematics) dont l'utilisation est calquée sur le symbolisme de FORTRAN. Ce système est d'un caractère plus général que le précédent, mais néanmoins il a été spécialement écrit pour traiter des séries trigonométriques et des polynômes à plusieurs variables. Les expressions sont introduites en machine sous forme de listes chaînées.

Récemment, avec ces programmes, les équipes du S.A.O. ont calculé la théorie du mouvement jusqu'au 3ème ordre en J_2 par la méthode de Von Zeipel. La comparaison de la théorie avec une intégration numérique a montré que les expressions obtenues sont exactes à 5 cm près (Gaposchkin *et al.* [12]).

Par ailleurs, les perturbations luni-solaires sur un satellite artificiel ont aussi été calculées grâce à ces programmes par Cherniack et Gaposchkin [7]. Les résultats de cette théorie sont désormais employés dans les études de géodésie dynamique effectuées par ce groupe.

(3) *Aux Boeing Scientific Research Laboratories*, un important ensemble de programmes d'algèbre sur des séries trigonométriques à coefficients polynômiaux du type (5) a été écrit et mis au point par Rom. Il s'agit du Mechanized Algebric Operation (MAO) écrit pour une IBM 360/44 (Rom [21]) et conçu essentiellement pour construire une théorie littérale du mouvement de la Lune. L'ensemble de sous-programmes constituant ce système est utilisé par le système FORTRAN. Les termes ne sont effectivement stockés dans la machine que s'ils ne sont pas nuls. Ils sont placés de manière séquentielle dans un espace dont les dimensions sont déterminées par le programme lui-même.

Ce programme a été appliqué par Deprit et Rom [10] au problème principal du mouvement d'un satellite artificiel. La méthode employée est celle des transformées de Lie mise au point par Deprit [9]. Cette méthode est basée sur un emploi de transformations canoniques que l'on peut construire explicitement ainsi que leurs inverses sans passer par les pénibles inversions de séries nécessitées par l'emploi de fonctions génératrices mixtes comme dans la méthode de Von Zeipel. Enfin, les auteurs ont employé plutôt les variables de Delannoy au lieu des variables elliptiques afin de conserver les avantages de la canonicité, tout en exprimant les résultats finaux à l'aide de variables non singulières aux excentricités nulles.

Les calculs ont été menés au troisième ordre en J_2 pour les termes à courte période et au quatrième ordre pour les termes à longue période. Les coefficients des expressions finales ont 12 chiffres significatifs. La comparaison avec des intégrations numériques a montré que cette théorie permettait de répondre à tous les besoins actuels ou même prévisibles dans un proche avenir.

Ainsi, les différences entre les expressions algébriques et une intégration numérique ne dépassent-elles pas 20 cm après 210 jours d'intégration pour un satellite du type ANNA 1.B et 2,4 m après 350 jours pour le satellite RELAY II.

Notons encore que le programme MAO a été récemment amélioré pour permettre d'introduire des dénominateurs dans les coefficients algébriques (Rom [22]). Les autres principes généraux de MAO ont été conservés dans cet 'Echeloned Series Processor' (ESP) qui, toutefois, n'a pas été utilisé pour la théorie du mouvement de satellites artificiels.

Ainsi nous avons en présence trois théories, construites par trois méthodes analytiques différentes, à l'aide de programmes algébriques différents. Les séries littérales ainsi obtenues ont été comparées à des intégrations numériques avec des résultats analogues.

Les théories de l'ordre 2 pour les courtes périodes et les longues périodes assurent une exactitude de 5 à 10 cm sur une dizaine de jours, peut être un mois (S.A.O. et Bureau des Longitudes). L'addition des termes du 3è et du 4è ordre (Boeing) donne une précision de l'ordre du mètre en un an, c'est à dire en fait une précision équivalente.

Une extrapolation de l'étude numérique de Berger [2] montre que l'ordre de grandeur des termes à courte et à longues périodes du 3é ordre, pour des excentricités assez faibles (0, 10) est effectivement de l'ordre du décimètre. Il est certain que le calcul de Deprit et Rom, à moins d'erreur improbable, a produit ces termes et on peut s'étonner que la comparaison avec l'intégration numérique ne soit pas meilleure. Toutefois, l'examen des courbes donnant les résidus, en particulier l'aspect systématique des variations de a, $e \sin \omega$, $e \cos \omega$ croissant en amplitude avec le temps indique avec certitude qu'il s'agit d'erreurs d'accumulation des arrondis ou de troncature dans l'intégration numérique.

On peut donc conclure que, pour les besoins présents, les théories complètes du second ordre sont suffisantes. Pour les besoins futurs, lorsqu'il faudra assurer la précision du centimètre, il apparait avec une bonne certitude qu'il suffira de pousser les calculs à l'ordre suivant en J_2.

3. Perturbations dues aux Harmoniques Tesseraux

La seule représentation du potentiel terrestre qui soit commode pour le calcul analytique est celle qui donnée par le développement en harmoniques sphériques du potentiel. La fonction perturbatrice s'écrit:

$$\mathcal{R}_T = \frac{\mu \varepsilon}{r} \sum_{n=2}^{\infty} \sum_{q=1}^{n} \frac{R^n}{r} P_{nq}(\sin \varphi) J_{nq} \cos q(\lambda - \lambda_{nq}), \tag{6}$$

où

$$P_{nq}(x) = (1 - x^2)^{q/2} \frac{\mathrm{d}^q}{\mathrm{d}x^q} P_n(x); \quad \eta = \sqrt{1 - e^2};$$

$\varepsilon = \pm 1$ selon les valeurs des indices, $P_n(x)$ est le polynôme de Legendre d'ordre n; r, λ et φ sont les coordonnées sphériques géocentriques des satellites; J_{nq} et λ_{nq} sont les coefficients et les phases des harmoniques sphériques et sont des nombres sans dimension définissant le modèle de potentiel terrestre adopté.

D'une façon générale, les coefficients J_{nq} sont d'une dimension telle que le potentiel résultant est de l'ordre de 10^{-6} à 10^{-8} fois le potentiel newtonien pour les valeurs de n qui sont utiles ($n < 20$). Il en résulte qu'un calcul de perturbations du premier ordre par rapport à ces paramètres est suffisant. Pour cela, il suffit de substituer dans les seconds membres des équations de Lagrange écrits avec la fonction perturbatrice (6), la solution simple

$$\left.\begin{aligned} a &= a_0, & \bar{\Omega} &= \Omega_0 + n_\Omega(t - t_0), \\ e &= e_0, & \bar{\omega} &= \omega_0 + n_\omega(t - t_0), \\ i &= i_0, & \bar{M} &= M_0 + n_M(t - t_0). \end{aligned}\right\} \tag{7}$$

obtenue en ne conservant que les termes séculaires de la solution du problème principal.

Pour cela, il faut exprimer (6) en fonction des éléments elliptiques. Plusieurs développements de ce type ont été publiés par Kaula [16], Gaposchkin [11], Challe et Laclaverie [6], etc. . . . .

Ces auteurs ont démontré, avec plus ou moins de calculs, que $\mathcal{R}_T$ se développe sous la forme suivante:

$$\mathcal{R}_T = \sum_{n=2}^{\infty} \sum_{q=1}^{n} \frac{\mu R^n J_{nq}}{a^{n+1}} \sum_{p=-\infty}^{+\infty} \sum_{k=-n}^{+n}$$
$$\times A_{nq}^k(i) B_{np}^k(e) \begin{smallmatrix} \sin \\ \cos \end{smallmatrix}(k\omega + q\Omega - qT - q_{nq}^\lambda + pM), \tag{8}$$

où T est le temps sidéral de Greenwich.

Ce qui importe dans cette formule (7), c'est qu'il y a séparation entre l'inclinaison et l'excentricité.

Les quantités $B_{np}^k(e)$ ne sont autres que les coefficients de Hansen que l'on rencontre dans le développement en série de Fourier des quantités relatives au problème des deux corps (voir par exemple Tisserand [23]).

On trouve ainsi:

$$B_{np}^k(e) = \frac{1}{2\pi} \int_0^{2\pi} \left(\frac{a}{r}\right)^n \exp \sqrt{-1}(kv + pM)\, \mathrm{d}M$$

où v est l'anomalie vraie.

On peut exprimer ces coefficients sous forme de développement explicite de l'excentricité. Il est plus utile de connaître des formules de récurrence qui permettent de les calculer tous à partir des premiers. Il est aussi nécessaire d'avoir ces relations pour les dérivées qui s'introduisent dans les seconds membres des équations différentielles. Challe et Laclaverie utilisent les relations suivantes:

(a) *Relations générales*

$$
\left.
\begin{aligned}
B_{n,\,p}^{k}(e) &= \frac{1}{\eta^2}\left[B_{n-1,\,p}^{k}(e) + \frac{e}{2}\left(B_{n-1,\,p}^{k+1}(e) + B_{n-1,\,p}^{k-1}(e)\right)\right] \\[2mm]
B_{n,\,p}^{k}(e) &= \frac{1}{p\eta}\left[kB_{n+1,\,p}^{k}(e) + (n+k)\frac{e}{2}B_{n+1,\,p}^{k+1}(e)\right. \\[2mm]
&\qquad\qquad\qquad\qquad\left. - (n-k)\frac{e}{2}B_{n+1,\,p}^{k-1}(e)\right] \\[2mm]
\frac{d}{de}B_{n,\,p}^{k}(e) &= \frac{n-1}{e}[B_{n+1,\,p}^{k}(e) - B_{n,\,p}^{k}(e)] + \frac{p}{2\eta}[B_{n-1,\,p}^{k+1}(e) \\[2mm]
&\qquad - B_{n-1,\,p}^{k-1}(e)] + \frac{k}{2\eta^2}[B_{n,\,p}^{k}(e) - B_{n,\,p}^{k-1}(e)]
\end{aligned}
\right\} \quad (9)
$$

(b) *Relations à n constant*

$$
\left.
\begin{aligned}
4\left[\eta^3 - k\left(1 + \frac{e^2}{2}\right)\right]B_{n,\,p}^{k}(e) &= e^2(k-n)B_{n,\,p}^{k-2}(e) \\[2mm]
+ 2e(2k-n)B_{n,\,p}^{k-1}(e) &+ 2e(2k+n)B_{n,\,p}^{k+1}(e) + e^2(k+n)B_{n,\,p}^{k+2}(e). \\[2mm]
\frac{d}{de}B_{n,\,p}^{k}(e) &= \frac{1}{\eta^2}\left[\left(\frac{ne}{2} - \frac{p\eta^3}{e} + k\left(\frac{e}{2} + \frac{1}{e}\right)\right)B_{n,\,p}^{k}(e)\right. \\[2mm]
&\qquad\left. + (n+2k)B_{n,\,p}^{k+1}(e) + (n+k)\frac{e}{2}B_{n,\,p}^{k+2}(e)\right], \\[2mm]
\frac{d}{de}B_{n,\,p}^{k}(e) &= \frac{1}{\eta^2}\left[\left(\frac{ne}{2} + \frac{p\eta^3}{e} - k\left(\frac{e}{2} + \frac{1}{e}\right)\right)B_{n,\,p}^{k}(e)\right. \\[2mm]
&\qquad\left. + (n-2k)B_{n,\,p}^{k-1}(e) + (n-k)\frac{e}{2}B_{n,\,p}^{k-2}(e)\right].
\end{aligned}
\right\} \quad (10)
$$

Les quantités $A_{nq}^{k}(i)$ sont des fonctions, de i seulement dont ces mêmes auteurs ont donné aussi les relations de récurrence ci-dessous où l'on a posé

$$s = \sin i; \qquad c = \cos i$$

$$\frac{2n + 1}{2} \, s \cdot [A_{n,q}^k(i) - A_{n,q}^{k+2}(i)] = (n + q)A_{n-1,q}^{k+1}(i) - (n - q + 1)A_{n+1,q}^{k+1}(i).$$

Relations à n et k variables

$$A_{n,n}^k(i) = \frac{n(2n - 1)}{n + k} \, (1 + c)A_{n-1,n-1}^{k+1}(i),$$

$$A_{n,n}^k(i) = \frac{n(2n - 1)}{n - k} \, (1 - c)A_{n-1,n-1}^{k+1}(i);$$

Relation à n constant

$$A_{n,n}^k(i) = \frac{n - k + 2}{n + k} \cdot \frac{1 + c}{1 - c} \cdot A_{n,n}^{k-2}(i);$$

Relation à k constant

$$A_{n,n}^k(i) = \frac{n(n - 1)(2n - 1)(2n - 3)}{(n - k)(n + k)} \, s^2 \cdot A_{n-2,n-1}^k(i). \qquad (11)$$

Les relations entre dérivées par rapport à i s'en déduisent aisément par simple dérivation.

Les quantités ainsi calculées permettent d'exprimer analytiquement toutes les perturbations du premier ordre dues aux harmoniques tesséraux.

Pour en avoir des évaluations numériques, il suffit de connaître la valeur des éléments moyens (8) et les substituer dans ces équations. Or, on peut programmer les relations (9) à (11) et, par conséquent, programmer les valeurs numériques des coefficients de ces perturbations pour chaque ensemble d'indices (k, p, q) c'est à dire pour chaque argument.

Un test peut être introduit pour ne retenir que les coefficients significatifs et l'ordinateur construit finalement des séries à coefficients numériques de la forme

$$\sum_k \sum_q \sum_p \left[\sum_n K_{kqpn}(a_0, e, i_0, J_{nq}, \lambda_{nq}) \right] \begin{matrix} \cos \\ \sin \end{matrix} (k\bar{\omega} + q\bar{\Omega} - qT + p\bar{M}). \qquad (12)$$

On peut introduire ces séries dans les programmes de calcul en position du satellite pour tout instant t lié aux arguments par les relations (7).

C'est le principe du programme de calcul utilisé au G.R.G.S. à Meudon et à Brétigny. Le programme du Smithsonian Astrophysical Observatory est d'un principe analogue. Toutefois, des formules directes pour le calcul des A_{nq}^k et B_{np}^k y sont utilisées au lieu de formules de récurrence.

Quoiqu'il en soit, dans les deux cas, c'est l'ordinateur qui construit les expressions analytiques nécessaires au calcul. Il n'y a aucune limitation ni dans le nombre d'harmoniques que l'on peut conserver, ni dans l'ordre des développements en k, q et surtout p. La précision théorique peut donc être aussi grande que l'on veut tant que le principe même d'une théorie du premier ordre est suffisant. Ceci est vrai tant qu'il n'y a pas résonance serrée, c'est à dire tant que le diviseur qui se produit lors de l'intégration n'est pas trop petit

$$D = |kn_\omega + qn_\Omega - qv + pn_M| > \varepsilon$$

où v est la pulsation correspondant à la rotation de la Terre.

Ce type de théorie ne convient donc pas pour les satellites géostationnaires. Jusqu'à présent, même pour certains satellites à résonance assez forte comme D1.D, ces calculs assurent une précision amplement suffisante, eu égard en particulier aux incertitudes même du modèle.

Conclusion

Nous voyons donc que, quelque soit la nature du problème analytique qui se pose, il existe maintenant des méthodes qui permettent d'établir les expressions littérales pour calculer les perturbations d'un satellite artificiel. Les expressions obtenues sont d'ores et déjà utilisées dans les calculs et ont montré leur équivalence sinon leur supériorité sur les lourdes méthodes basées uniquement sur l'intégration numérique.

Bibliographie

[1] Barton, D., *Computer J.* **9** (1967), 340.
[2] Berger, X., *Bull. G.R.G.S.*, Paris, 1971, No. 2.
[3] Bretagnon, P., *Bull. G.R.G.S.*, Paris, 1971, No. 2.
[4] Broucke, R. et Garthwaite, K., *Celest. Mech.* **1** (1969), 271.
[5] Brouwer, D., *Astron. J.* **64** (1959), 378.
[6] Challe, A. et Laclaverie, J. J., *Astron. Astrophys.* **3** (1969), 15.
[7] Cherniack, J. et Gaposchkin, E. M., in B. Morando (ed.), *Dynamics of Satellites*, Springer Verlag, 1970, p. 36.
[8] Davis, M. S., *Astron. J.* **73** (1968), 195.
[9] Deprit, A., *Celest. Mech.* **1** (1969), 12.
[10] Deprit, A. et Rom, A., *Celest. Mech.* **2** (1970), 166.
[11] Gaposchkin, E. M., Smithsonian Astrophysical Observatory Special Report, No. 200, 1966, p. 156.
[12] Gaposchkin, E. M., Cherniack, J., Briggs, R. E. et Benima, B., Communication au Congrès de l'American Geophysical Union, Washington, avril 1971.
[13] Gaposchkin, E. M. et Lambeck, K., Smithsonian Astrophysical Observatory Special Report, No. 315, 1970.
[14] Hall, N. M. et Cherniack, J. R., Smithsonian Astrophysical Observatory Special Report No. 291, 1969.
[15] Jefferys, W. H., *Celest. Mech.* **2** (1970), 474.
[16] Kaula, W. M., in *Theory of Satellite Geodesy*, Blaisdell Publ., Waltham, Mass., 1966.
[17] Kovalevsky, J., *Bull. Astron.* **23** (1959), 1.
[18] Kovalevsky, J., *Introduction to Celestial Mechanics*, Reidel Publ. Co., Dordrecht, Holland, 1967, p. 88.
[19] Kovalevsky, J., *Astron. J.* **73** (1968), 203.
[20] Kozai, Y., *Astron. J.* **67** (1962), 446.
[21] Rom, A., *Celest. Mech.* **1** (1970), 301.
[22] Rom, A., *Celest. Mech.* **3** (1971), 331.
[23] Tisserand, F., *Traité de Mécanique Céleste*, réed. Gauthier Villars, Paris, Vol. 1, 1960, p. 249.

CARACTÉRISTIQUES D'UN PROGRAMME DE CORRECTION DIFFÉRENTIELLE EN GÉODÉSIE DYNAMIQUE

BERNARD MOYNOT

Groupe de Recherches de Géodésie Spatiale, Centre National d'Études Spatiales,
B.P. No. 4, 91-Brétigny-sur-Orge, France

1. But

Le nouveau programme de correction différentielle qui est en préparation au Centre National d'Études Spatiales, aura pour but essentiel la redétermination précise et simultanée des éléments suivants:
– les coordonnées des stations d'observation rapportées au centre de gravité de la Terre,
– les coefficients du potentiel terrestre, à partir des mouvements des satellites artificiels.
Ce but est maintenant mis à notre portée par la campagne ISAGEX grâce à la réunion des 3 conditions:
– disposer d'une famille d'orbites de satellites bien diversifiées, surtout en inclinaison,
– disposer d'un nombre important de stations réparties sur toute la Terre,
– disposer de mesures précises et suffisamment diversifiées.

2. Principe

Le principe va consister à ajuster par la méthode des moindres carrés un jeu de paramètres qui se subdivise en 2 parties:
– *Les paramètres externes*: apportant une contribution commune à tous les satellites et toutes les mesures: ce sont essentiellement les éléments que nous nous proposons de redeterminer: les coordonnées des stations et les coefficients du potentiel.
– *Les paramètres internes*: propres à chaque satellite, à chaque arc d'orbite traité: ce sont essentiellement les éléments orbitaux de ces satellites aux instants de début des arcs traités. De tels paramètres seront à éliminer au cours du traitement.
Le traitement sera divisé essentiellement en 5 étapes:
Étape A: Traitement des observations regroupées en lots distincts, chaque lot étant constitué des observations faites par toutes les stations sur un seul satellite pendant une période de temps donnée qui sera de l'ordre d'une quinzaine de jours, cette étape fournit un jeu d'équations linéarisées au voisinage d'une orbite de référence et que l'on notera matriciellement:

$$AX = B.$$

Vu sa taille, le stockage des éléments de ce système est impossible: aussi, doit-il être immédiatement converti en un système d'équations normales noté: $CX = D$

L. G. Napolitano et al. (eds.), Astronautical Research 1971, 27–35. All Rights Reserved.

où :

$$C = A^T \prod A \quad \text{et} \quad D = A^T \prod B.$$

Étape B: Elimination, pour chaque arc traité, de ces paramètres internes, ce qui donne un jeu d'équations dites externes, et noté:

$$C^* X_E = D^*$$

ne contenant plus que les paramètres externes.

Étape C: Mélange des jeux d'équations externes: cela donne un jeu résultant ayant la même structure et mieux conditionné.

Étape D: Résolution au moins partielle du jeu résultant, cela donne un jeu de corrections à apporter aux paramètres externes.

Dans cette étape, il est prévu la possibilité de figer certains des paramètres inconnus à des valeurs données au cas où le système à résoudre se révèlerait encore mal conditionné: cela donnera des résultats provisoires en attendant de rajouter des nouvelles équations de condition obtenues à partir de nouvelles observations. De toutes façons, l'examen de la matrice 1er membre, pourra fournir les combinaisons linéaires des paramètres qui sont les mieux déterminées.

Étape E: C'est une étape de vérification: on restitue les paramètres internes pour chaque arc traité et on redétermine les résidus (o. – c.). La comparaison des résidus de l'orbite améliorée à ceux de l'orbite de référence utilisée dans l'étape A sera le seul moyen d'apprécier les améliorations effectivement obtenues par l'ensemble du traitement, les équations de condition individuelles n'ayant pas pu être conservées.

Parmi ces 5 étapes, l'étape A sera, de beaucoup la plus coûteuse en ordinateur, pour plusieurs raisons. Elle devra être répétée pour chaque arc à traiter. Le calcul de la matrice 1er membre nécessitera le calcul des dérivées partielles du vecteur-station-satellite par rapport aux paramètres à ajuster. Les éléments orbitaux initiaux et les coefficient du potentiel intervenant par l'extrémité satellite, les dérivées partielles correspondantes s'obtiendront au prix de l'intégration numérique de nombreux systèmes décalés ou linéarisés. L'impossibilité de stocker les matrices A et B ne permettra plus de revenir sur les mesures: il est donc capital que les mesures fournies à l'étape A soient soigneusement filtrées et pondérées. De plus, la linéarisation sera d'autant plus valable que l'orbite de référence sera plus proche de la réalité. Ces 2 impératifs peuvent être satisfaits simultanément à l'aide du programme de correction différentielle actuellement opérationnel qui, au cours d'une étape préliminaire, fournit une orbite de référence avec des résidus de l'ordre de 10 m, en faisant appel à une méthode semi-dynamique. Une mesure mauvaise se manifeste immédiatement à cette étape par un résidu anormalement élevé, ce qui permet de l'éliminer. Par contre, la pondération pourra poser des problèmes, surtout si on a mélangé des mesures de types différents.

Des nombreux problèmes qui se sont posés lors de la mise en oeuvre d'une méthode dynamique, nous ne retiendrons dans la suite que ceux d'origine mathématique. Nous examinerons plus spécialement la méthode d'intégration numérique et la choix des repères à adopter.

A. AMÉLIORATION DE LA MÉTHODE D'INTÉGRATION NUMÉRIQUE

Pour bien situer le problème, indiquons d'abord les systèmes différentiels à intégrer : on aura :

- l'orbite de référence,
- 6 orbites décalées résultant des décalages de base sur les 6 paramètres orbitaux initiaux,
- 2 systèmes linéarisés pour ajuster 2 coefficients de proportionnalité que nous aurons libérés dans les expressions des forces de frottement atmosphérique et de pression de radiation,
- plusieurs centaines de systèmes linéarisés résultant des décalages de base sur les coefficients du potentiel terrestre (que nous serons obligés de tronquer à un certain ordre).

Considérons en premier lieu, les 7 systèmes restituant l'orbite de référence et les 6 orbites décalées. L'orbite de référence doit être restituée avec la meilleure précision

TABLEAU I

Point de départ	e	Δa après 5 tours (mètres)	ΔM après 5 tours (radians)
Périgée	0.8	$> 2 \times 10^7$ (v. absol.)	plus. fois (2π)
	0.6	-2.15	-0.79×10^{-3}
	0.4	-2.12×10^{-4}	-0.50×10^{-6}
	0.2	-1.05×10^{-6}	-0.90×10^{-9}
	0	-0.33×10^{-8}	-0.52×10^{-13}
Apogée	0.2	-1.05×10^{-6}	1.08×10^{-13}
	0.4	-2.09×10^{-4}	0.91×10^{-10}
	0.6	-0.65×10^{-2}	-2.41×10^{-9}
	0.8	-598	1.88×10^{-5}

possible, et les 6 orbites décalées devront être restituées exactement par la même méthode que l'orbite de référence, pour que les écarts entre ces orbites restent significatifs.

La méthode d'intégration numérique adoptée est la méthode de Cowell.

Cette méthode, comme toute autre, est entachée d'erreurs de tronçature. Ces erreurs de tronçature sont de 2 sortes :

- Les erreurs permanentes ou de croisière se manifestant à chaque pas d'intégration du fait que les formules ne sont pas rigoureuses. De ce fait, les trajectoires de croisière restituées ne coincident jamais exactement avec les trajectoires réelles : elles en sont d'autant plus voisines que le pas d'intégration est plus petit.

- Les erreurs de démarrage qui affectent le début de la trajectoire : celle-ci ne tend asymptotiquement vers une trajectoire de croisière qu'au bout d'un certain nombre de pas d'intégration.

La méthode de Cowell appliquée à l'intégration du mouvement Képlérien a donné les résultats suivants:

– On se donne un astre central ponctuel ayant la masse de la Terre, et une période de révolution de 8000 s: le demi grand axe de l'orbite est de l'ordre de 8600 km. On se fixe un pas d'intégration constant de 40 s, ce qui correspond à $\frac{1}{200}$ d'orbite. On donne à l'excentricité plusieurs valeurs de 0 à +0.8, et on choisit le point de départ au périgée ou à l'apogée. Les écarts sur le $\frac{1}{2}$ grand axe et l'anomalie moyenne sont les suivants (Tableau I).

Ce tableau montre une propagation des erreurs de tronçature d'autant plus importante que l'orbite est plus excentrique et que le point de départ est plus près du périgée. Elle devient catastrophique pour des excentricités atteignant 0.8, surtout si on part du périgée. Nous allons remédier à ces erreurs de tronçature en 2 étapes:

1. Changement de variable indépendante

Il est naturel de chercher un moyen de restituer une orbite excentrique dans les mêmes conditions qu'une orbite circulaire. Or, les seules formules introduisant une erreur de tronçature sont les formules de l'algorithme de Cowell: elles sont toutes linéaires, donc insensibles à une affinité. Ce n'est donc pas l'ellipticité de l'orbite qui est gênante, mais la répartition des points calculés sur cette orbite: ils sont trop espacés au périgée, trop concentrés à l'apogée. Pour les rendre équidistants, à l'affinité près, le moyen le plus simple est de prendre comme variable indépendante, non plus le temps, mais l'anomalie excentrique, ou mieux, la quantité suivante qui lui est proportionnelle:

$$\sigma = \int \frac{1}{r} \, dt$$

où r est la distance géocentrique. Cette dernière quantité à l'avantage de permettre une transformation simple de l'équation du mouvement perturbé:

$$\frac{d^2\mathbf{r}}{dt^2} = \mathbf{F}$$

cette équation est remplacée par le système:

$$\frac{d^2\mathbf{r}}{d\sigma_2} = \frac{1}{r^2} \left(\mathbf{r} \, \frac{d\mathbf{r}}{d\sigma} \right) \frac{d\mathbf{r}}{d\sigma} + r^2 \mathbf{F}$$

$$\frac{dt}{d\sigma} = r$$

l'équation supplémentaire permettant de récupérer l'instant associé à une position donnée. Il est à remarquer qu'elle est du 1er ordre ce qui ne présente aucun inconvénient: l'algorithme de Cowell s'adapte très facilement à l'intégration d'un système mixte composé d'équations du 1er et du 2ème ordre.

Reprenons l'exemple précédent: les résultats obtenus par cette méthode améliorée sont les suivants; à nombre de pas égal (200 pas) par orbite (Tableau II).

TABLEAU II

Point de départ	e	Δa après 5 tours (mètres)	ΔM après 5 tours (radians)
Périgée	0.8	0.21×10^{-7}	-0.82×10^{-11}
	0.6	0.61×10^{-7}	-1.79×10^{-12}
	0.4	1.06×10^{-7}	-0.52×10^{-12}
	0.2	1.34×10^{-7}	-1.76×10^{-13}
———	0	1.33×10^{-7}	-0.61×10^{-13}
Apogée	0.2	1.27×10^{-7}	-1.89×10^{-14}
	0.4	1.21×10^{-7}	-0.37×10^{-14}
	0.6	1.16×10^{-7}	-1.32×10^{-15}
	0.8	1.15×10^{-7}	0.50×10^{-14}

Ce tableau montre, comme on s'y attendait, l'amélioration spectaculaire de la précision dans le cas des orbites excentriques, sans aucune augmentation du nombre de pas d'intégration.

2. Modification de l'algorithme de Cowell

Ayant choisi la variable indépendante, on peut améliorer davantage la restitution de l'orbite de référence en modifiant l'algorithme de Cowell, de façon à le rendre rigoureux pour le mouvement Képlérien: le mouvement réel en étant très proche, ne serait plus entaché que d'erreurs de troncature résiduelles dues uniquement à ces perturbations.

La solution Képlérienne elliptique s'explicite rigoureusement de façon simple:

$$\mathbf{r} = \mathbf{a}_1 (\cos \nu_0 \sigma - e) + \mathbf{a}_2 \sqrt{1 - e^2} \sin \nu_0 \sigma$$

$$t = t_0 + a \cdot \left(\sigma - \frac{e}{\nu_0} \sin \nu_0 \sigma \right)$$

où les 2 vecteurs a_1 et a_2, constants, sont orthogonaux et ont pour module le demi grand axe a, e représente l'excentricité, ν_0 est la quantité $\sqrt{GM/a}$ et où t_0 est l'instant de passage au périgée.

Le problème se ramène au suivant:

– rendre l'algorithme de Cowell rigoureux pour toute solution dont chaque composante est de la forme:

$$x = x_0 + x_1 \cos \nu_0 \sigma + x_2 \sin \nu_0 \sigma$$

admettant une partie constante et un terme sinusoidal de période égale à $T_0 = 2\pi/\nu_0$ que l'on peut se fixer à l'avance (cette valeur sera donc déterminée au départ de l'intégration numérique en fonction des éléments orbitaux initiaux du satellite).

Pour ramener le temps à une solution de cette forme, on calculera en réalité:

$\tilde{t} = t - a\sigma$ en intégrant l'équation:

$$\frac{d\tilde{t}}{d\sigma} = r - a.$$

Rappelons brièvement le principe de l'algorithme de Cowell:

– il consiste à calculer de proche en proche les éléments de 5 tableaux:

– le tableau des coordonnées noté X_n ($n = -3, 2, \ldots$),

– le tableau des dérivées 1ères noté X_n',

– le tableau des dérivées 2èmes noté X_n'',

– 2 tableaux auxiliaires notés $\delta_{n-(1/2)}^{-1}$ et δ_n^{-2},

où n est le n° du point associé à la valeur $\sigma_n = \sigma_0 + nh$ de la variable indépendante σ (h est le pas d'intégration).

Son ordre a été fixé à la valeur 8: on conserve donc en mémoire les valeurs de ces éléments pour 8 valeurs consécutives de n.

Les formules utilisées sont les suivantes:

$$X_{n+p} = h^2 \left[\delta_{n+p}^{-2} + \sum_{k=-3}^{3} \alpha_{pk} X_{n+k}'' \right], \tag{1}$$

$$X_{n+p}' = h \left[\delta_{n+p-(1/2)}^{-1} + \sum_{k=-3}^{3} \beta_{pk} X_{n+k}'' \right], \tag{2}$$

$$X_{n+p}'' = \mathscr{F}(\sigma_{n+p}, X_{n+p}, X_{n+p}'), \tag{3}$$

$$X_{n+p}'' = \delta_{n+p+(1/2)}^{-1} - \delta_{n+p-}^{-1}, \tag{4}$$

$$\delta_{n+p-(1/2)}^{-1} = \delta_{n+p}^{-2} - \delta_{n+p-1}^{-2}. \tag{5}$$

Dans les formules (1) et (2), les α_{pk} et les β_{pk} représentent 2 matrices de coefficients indépendantes du système différentiel à intégrer, donc précalculées une fois pour toutes.

Le démarrage s'effectue en faisant $n = 0$ dans ces formules et en utilisant pour p les valeurs entières de -3 à $+3$; les progression par pas s'effectue en faisant $p = 4$ (séquence 'prédictor'), puis en incrémentant n de 1 unité et en utilisant les valeurs de p de -3 à $+3$ (séquence 'corrector').

Or: seules les formules (1) et (2) introduisent une erreur de troncature.

Il est facile de l'éliminer rigoureusement (pour les solutions particulières précisées plus haut) en rajoutant aux 2èmes membres de ces formules, les termes correctifs suivants:

$$h^2 c_p X_{n+p*}'' - h\rho_p X_{n+p}'$$

$$hc_p' X_{n+p*}'' - \rho_p' X_{n+p}'$$

ou: $p^* = p$ pour $-3 \leqslant p \leqslant 3$; $p^* = 3$ pour $p = 4$ (cette restriction étant due à la non connaissance de X_{n+4}'' au pas n°n).

Il est facile de montrer que les coefficients c_p, c'_p, ρ_p, ρ'_p sont, pour les solutions particulières envisagées, des constantes qui ne dépendent que du rapport h/T_0: ces constantes peuvent être calculées une fois pour toutes au démarrage de l'intégration numérique, une fois donné le pas d'intégration.

L'allongement des formules (1) et (2) apparaîtra en pratique insignifiant à côté de la complexité de l'expression des 2èmes membres des systèmes à intégrer.

Cette amélioration permet donc, à peu de frais, à pas égal d'améliorer considérablement la précision, ou, à précision égale d'élargir le pas de calcul et donc du même coup, de réduire le temps de calcul et les erreurs d'arrondi. Ce sont les forces perturbatrices qui, seules, imposeront une limite supérieure au pas d'intégration.

3. Intégration des Systèmes Auxiliaires Donnant Accès aux Effets des Écarts sur les Coefficients du Potentiel

Les solutions de ces systèmes étant composées de termes sinusoïdaux mélangés de périodes diverses, il est exclu de chercher à éliminer les erreurs de tronçature.

Ces systèmes étant les plus nombreux (plusieurs centaines), le souci sera au contraire de trouver l'algorithme le plus rapide possible, en espérant qu'il aura la précision suffisante.

Explicitons ces systèmes: ils sont obtenus par linéarisation du système restituant l'orbite de référence que nous écrivons sous la forme suivante au 1er ordre (6 composantes):

$$\frac{dX}{dt} = F(t, X)$$

en gardant provisoirement la variable temps t pour la clarté de l'exposé. Le problème sera un peu différent avec la variable σ.

Les systèmes linéarisés s'écrivent:

$$\frac{d\delta X_{lms}}{dt} = M(t)\delta X_{lms} + \delta y_{lms}(t)$$

où:

$$M(t) = \frac{\delta F}{\delta X}$$

où δy_{lms} est une fonction vectorielle dépendant du temps par l'intermédiaire de la position de référence, mettant en jeu les polynômes de Legendre et les déplacements élémentaires δC_{lm} où δS_{lm}; où le 3e indice prend la valeur 0 pour un C_{lm}; 1 pour un S_{lm}.

Or: la partie homogène associée est la même pour tous ces systèmes.

L'intégration de ces centaines de systèmes se ramène donc:

– à la détermination de la solution générale du seul système:

$$\frac{d\delta X}{dt} = M(t)\delta X$$

lequel problème est équivalent à la détermination de l'orbite de référence, et des 6 orbites décalées,

– à des centaines de quadrature vectorielles:

$$\delta\mathbf{X}_{lms}(t) = \mathbf{K}(t_0, t)\int_{t_0}^{t} \mathbf{K}^{-1}(t_0, u)\delta y_{lms}(u)\,\mathrm{d}u$$

$K(t_0, t)$ étant la matrice résolvante du système homogène ci-dessus.

Le choix de la variable indépendante σ pour l'orbite de référence et les orbites décalées va évidemment se répercuter sur le calcul des δX_{lms}.

On intégrera en réalité:

$$\delta\mathbf{X}_{lms} = \mathbf{K}(t_0, t(\sigma))\int_{\sigma_0}^{\sigma} \mathbf{K}^{-1}(t_0, t(\sigma'))\delta y_{lms}(t(\sigma'))r(\sigma')\,\mathrm{d}\sigma'.$$

Une seule remarque à faire, concernant le calcul de la matrice $K(t_0, t(\sigma))$: cette matrice ne peut s'obtenir que pas à pas à partir de l'orbite de référence et des écarts entre cette orbite et les 6 orbites décalées, ces écarts étant corrigés des écarts entre les temps respectifs restitués par les orbites décalées.

Aussi, les δX_{lms} seront intégrés par la méthode de Cowell adaptée aux systèmes du 1er ordre. Or, les inconnues δX_{lms} ne figurent pas dans les seconds membres: le calcul de leurs dérivées à σ donné peut donc se faire du premier coup: la séquence d'intégration sur un pas se réduit donc à la séquence 'prédictor', d'où un gain de temps considérable par rapport à l'algorithme complet.

4. Choix des Repères

Abordons maintenant le problème du choix des repères.

L'intégration numérique des mouvements des satellites s'effectue dans un repère céleste inertiel ou quasi-inertiel.

Les coordonnées des stations d'observation et les coefficients du potentiel sont déterminés dans un repère terrestre qui tourne avec la Terre.

Le fait de travailler dans 2 repères différents nécessite de déterminer avec précision la matrice de passage. Cette matrice sera simplifiée si on décide de choisir le 3ème axe commun. L'axe le plus simple qui s'impose alors, est l'axe des pôles instantané. Le repère céleste $OXYZ$ est alors quasi-inertiel: il est animé d'un mouvement résiduel déterminé par la précession et la nutation. Or: une précision métrique de 10 cm à la surface de la Terre, correspond à une précision angulaire de l'ordre de $0''003$: il est donc nécessaire de tenir compte de ce mouvement par l'introduction, dans les seconds membres des équations du mouvement, des termes d'entraînement et de Coriolis.

Le repère terrestre $Oxyz$ n'est pas rigoureusement fixe par rapport à la Terre: il est animé d'un mouvement correspondant au mouvement du pôle. Ce mouvement se répercutera: d'une part, sur les coordonnées des stations; d'autre part, sur les coefficients du potentiel. Dans ce dernier cas, il est facile de montrer que le seul effet appréciable est l'apparition de coefficients C_{21} et S_{21}.

Enfin, le matrice de passage, réduite à une simple matrice de rotation, sera déterminée par le temps sidéral vrai.

COUPLED LIBRATIONAL DYNAMICS AND ATTITUDE
CONTROL OF SATELLITES IN PRESENCE OF
SOLAR RADIATION PRESSURE*

V. J. MODI** and K. KUMAR‡

The University of British Columbia, Vancouver, Canada

Abstract. The paper studies the coupled librational dynamics of gravity oriented cylindrical satellites under the influence of the solar radiation pressure. The investigation suggests substantial destabilizing influence of the solar pressure on the attitude behaviour of a satellite. The possibility of utilizing this minute force to advantage for achieving general three-dimensional librational damping and attitude control is explored. The concept offers a great degree of versatility permitting the satellite to attain any general spatial configuration.

1. Introduction

The precise orientation of a satellite with respect to the Earth is of considerable importance in many space applications such as telecommunications, meteorology, military missions, etc. Of the numerous methods proposed for the station-keeping, gravity gradient stabilization has gained much attention [1–8 *et al.*]. It represents an attractive passive stabilization technique in terms of conceptual simplicity and satellite life-time.

The gravity restoring torques being rather weak, it is important to ascertain the influence of perturbing environmental forces on the satellite response and stability. Of course, there is always a possibility of utilizing these environmental forces to advantage.

As pointed out by Roberson [9] and Holl [10], the major external disturbance at high altitudes appears to be due to the solar radiations. More recently, a study conducted by Modi *et al.* [11–13] showed that from the dynamical considerations, the solar radiations indeed constitute the only significant force above 6000 mile altitude.

This paper studies the coupled librational dynamics of cylindrical satellites, orbiting in the ecliptic plane, under the influence of the solar radiation pressure. The development of a general semi-passive controller capable of not only damping the librational motion but also adjusting its attitude in orbit is described. The precise pointing accuracy together with low power consumption makes the system quite attractive for the next generation of communications satellites and space stations of the future.

* The investigation was supported by the National Research Council and the Defence Research Board grants A2181 and 9551-18, respectively.
** Professor.
‡ Graduate Research Assistant.

L. G. Napolitano et al. (eds.), Astronautical Research 1971, 37–52. All Rights Reserved.
Copyright ©1973 by D. Reidel Publishing Company, Dordrecht-Holland.

Nomenclature

a, L	radius and length of satellite, respectively
c	solar parameter, $(2/3)\ R^3SAl\left\{1 + \rho - \tau + \dfrac{1 - \rho - \tau}{2}\right\}$
c_ψ	$R^3SA_p l_1(1+\rho-\tau)/(\mu c'I)$
c_β	$R^3SA_p l_2(1+\rho-\tau)/(\mu c'I)$
c_λ	$R^3SA_p m_1(1+\rho-\tau)/(\mu c'I)$
c'	velocity of light
c_{max}	physical limit on c_ψ, c_β and c_λ as imposed by the controller design
e	eccentricity of orbit
g	aspect ratio, $\dfrac{3\pi a}{4L}\left\{(1 - \rho - \tau)\Big/\left(1 + \rho - \tau + \dfrac{1-\rho-\tau}{2}\right)\right\}$
l	distance of geometric centre of satellite from the centre of mass
l_1, m_1	distances of geometric centre of first set of controller plates (G_1, G_1') from satellite mass centre along z_2 and y_2 axes, respectively, Figure 6
l_2	distance of geometric centre of second set of controller plates (G_2, G_2') from satellite mass centre along z_2 axis, Figure 6
p	constant depending upon material reflectivity and transmissivity, $(1-\rho-\tau)/\{2(1+\rho-\tau)\}$
$\bar{u}_r, \bar{u}_\sigma$	unit vectors in x–y plane normal and parallel to the representative element, Figure 1
$\bar{u}_z$	unit vector along z axis, Figure 1
x, y, z	principal body coordinates with origin at the mass centre s and the z axis pointing along the axis of symmetry, Figure 1
$\left.\begin{array}{l} x_0, y_0, z_0 \\ x_1, y_1, z_1 \\ x_2, y_2, z_2 \end{array}\right\}$	intermediate body coordinates with origin at the centre of mass during the modified Eulerian rotations ψ, β, λ, respectively, Figure 1
x_p, y_p, z_p	rectangular coordinate axes with x_p normal to the plates (G_2, G_2'), Figure 6
A	diametral cross-sectional area, $2aL$
A_p	total area of each set of controller plates
I_{ti}	mass moment of inertia about i axis, $I_{xx}=I_{yy}=I$
K_i	inertia parameter, $1-(I_{zz}/I)$
M_1, M_2, M_3	the components of the total radiation moment along $s1$, $s2$ and $s3$, respectively, $M_3=0$
$Q_\psi, Q_\beta, Q_\lambda$	generalized forces
R	radius of the orbit
S	solar constant
θ	satellite position angle as measured from the perigee, Figure 1
μ	gravitational field constant
μ_i, ν_i	proportionality constants in controller characteristic relation; $i=\psi, \beta, \lambda$
γ_i	position control angle in controller characteristic relation; $i=\psi, \beta, \lambda$
ρ, τ	reflectivity and transmissivity of material, respectively
σ	position angle of the representative element, Figure 1
σ_0	$(\pi/2)+\tan^{-1}[-\sin(\theta+\psi-\varphi)/\{\sin\beta\cos(\theta+\psi-\varphi)\}]$
φ	solar aspect angle
ψ	rotation in the orbital plane, Figure 1
β	rotation across the orbital plane, Figure 1
λ	rotation about the axis of symmetry, Figure 1

Dots and primes over ψ, β, λ denote differentiation with respect to time and θ, respectively; subscript 0 refers to initial conditions.

2. Formulation of the Problem

Consider a cylindrical satellite orbiting in the ecliptic plane about the centre of force O and executing coupled librational motion (Figure 1). Let x, y, z represent the principal body axes of the satellite. The Eulerian angles ψ, β and λ specify the instantaneous satellite configuration.

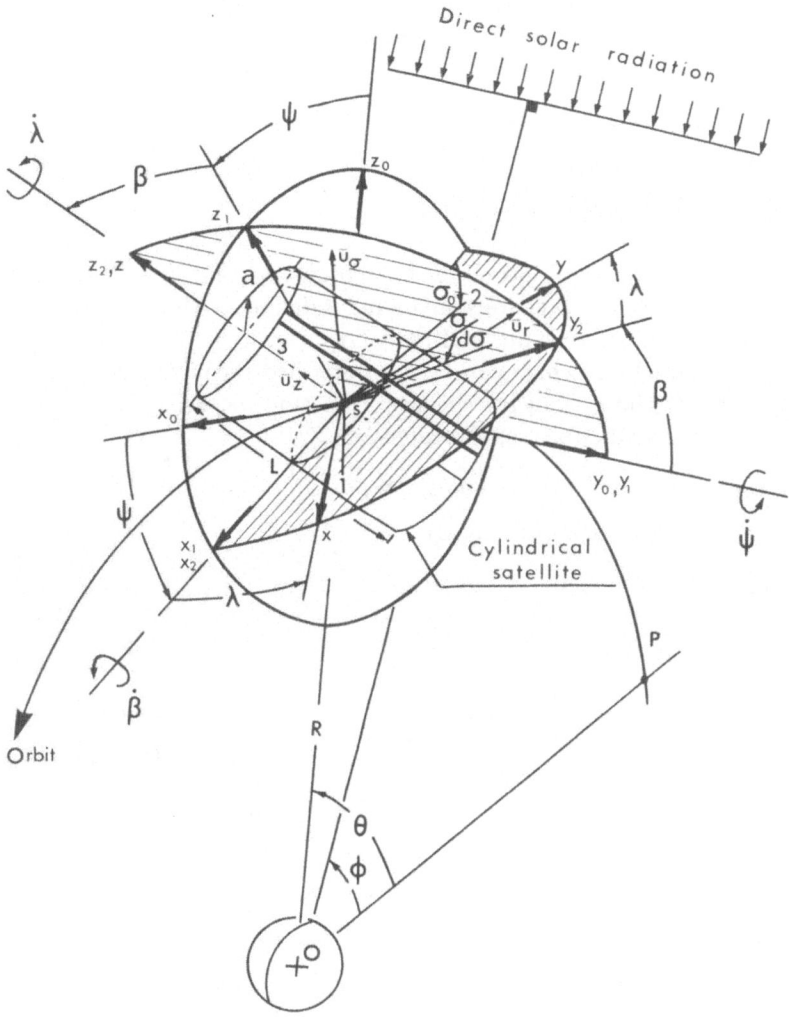

Fig. 1. Geometry of satellite motion.

Using Lagrangian formulation, the governing equations of motion in ψ, β and λ degrees of freedom can be written as:

$$I(\ddot{\theta} + \ddot{\psi}) \cos^2 \beta - 2I(\dot{\theta} + \dot{\psi}) \sin \beta \cos \beta \cdot \dot{\beta}$$

$$+ \frac{3\mu I}{R^3} K_i \sin \psi \cos \psi \cos^2 \beta \tag{1a}$$

$$- I_{zz} \frac{d}{dt} [\{\lambda - (\dot{\theta} + \dot{\psi}) \sin \beta\} \sin \beta] = Q_\psi$$

$$I\ddot{\beta} + I \left\{ (\dot{\theta} + \dot{\psi})^2 + \frac{3\mu}{R^3} K_i \cos^2 \psi \right\} \sin \beta \cos \beta$$

$$+ I_{zz}(\dot{\theta} + \dot{\psi}) \cos \beta \{\lambda - (\dot{\theta} + \dot{\psi}) \sin \beta\} = Q_\beta \tag{1b}$$

$$I_{zz} \frac{d}{dt} \{\lambda - (\dot{\theta} + \dot{\psi}) \sin \beta\} = Q_\lambda \tag{1c}$$

where the generalized forces $Q_i (i = \lambda, \beta, \psi)$, representing the solar radiation effect, are lengthy complicated expressions involving satellite orientation and geometry, material properties, solar constant, and the velocity of light. Without going into the details of their evaluation, which can be accomplished through the principle of virtual work, it can be shown that for specular reflection the *exact* expressions for the generalized forces take the form as indicated:

$$Q_\psi = (M_1 \cos \sigma_0 + M_2 \sin \sigma_0) \cos \beta$$

$$Q_\beta = (M_1 \sin \sigma_0 - M_2 \cos \sigma_0) \tag{2}$$

$$Q_\lambda = 0$$

where

$$M_1 = \frac{SAl}{2c'} \left[(1 + \rho - \tau)\{\sin^2(\theta + \psi - \varphi)(1 + \tfrac{1}{3} \cos 2\sigma_0) \right.$$

$$+ \sin^2 \beta \cos^2 (\theta + \psi - \varphi)(1 - \tfrac{1}{3} \cos 2\sigma_0)$$

$$+ \tfrac{1}{3} \sin \beta \sin 2(\theta + \psi - \varphi) \sin 2\sigma_0\}$$

$$+ \frac{(1 - \rho - \tau)}{2} \{\tfrac{4}{3} \sin^2 (\theta + \psi - \varphi) \cos 2\sigma_0$$

$$- \tfrac{4}{3} \sin^2 \beta \cos^2 (\theta + \psi - \varphi) \cos 2\sigma_0$$

$$\left. + \tfrac{4}{3} \sin \beta \sin 2(\theta + \psi - \varphi) \sin 2\sigma_0\} \right]$$

$$+ \frac{\pi SaA}{4c'} \cdot \frac{2l}{L} (1 - \rho - \tau) \cos \beta |\cos (\theta + \psi - \varphi)|$$

$$\times \{\sin (\theta + \psi - \varphi) \cos \sigma_0 + \sin \sigma_0 \cos (\theta + \psi - \varphi) \sin \beta\}$$

$$M_2 = \frac{SAl}{2c'}\left[\frac{2}{3}(1 + \rho - \tau)\{\sin 2\sigma_0 \sin^2(\theta + \psi - \varphi)\right.$$
$$- \sin 2\sigma_0 \sin^2\beta \cos^2(\theta + \psi - \varphi)$$
$$- \cos 2\sigma_0 \sin\beta \sin 2(\theta + \psi - \varphi)\}$$

$$+ \frac{2}{3}\frac{(1 - \rho - \tau)}{2}\{\sin 2\sigma_0 \sin^2(\theta + \psi - \varphi)$$
$$- \sin 2\sigma_0 \sin^2\beta \cos^2(\theta + \psi - \varphi)$$
$$\left. - \cos 2\sigma_0 \sin\beta \sin 2(\theta + \psi - \varphi)\right]$$

$$+ \frac{\pi SaA}{4c'}\cdot\frac{2l}{L}(1 - \rho - \tau)\cos\beta|\cos(\theta + \psi - \varphi)|$$
$$\times \{\sin(\theta + \psi - \varphi)\sin\sigma_0 - \cos\sigma_0 \sin\beta \cos(\theta + \psi - \varphi)\}$$

Q_λ being zero, $\lambda - (\theta + \psi)\sin\beta$ becomes a constant of the system. For a non-spinning satellite this constant must vanish leading to a considerable simplification of the equations of motion which now take the form:

$$I(\ddot{\theta} + \ddot{\psi})\cos^2\beta - 2I(\dot{\theta} + \dot{\psi})\dot{\beta}\sin\beta\cos\beta + \frac{3\mu I}{R^3}K_i \sin\psi\cos\psi\cos^2\beta = Q_\psi$$

$$I\ddot{\beta} + I\left[(\dot{\theta} + \dot{\psi})^2 + \frac{3\mu}{R^3}K_i\cos^2\psi\right]\sin\beta\cos\beta = Q_\beta$$

$$(3)$$

Recognizing that:

$$\dot{\psi} = \psi'\dot{\theta}$$
$$\dot{\beta} = \beta'\dot{\theta}$$
$$\ddot{\psi} = \psi''\dot{\theta}^2 + \psi'\ddot{\theta}$$
$$\ddot{\beta} = \beta''\dot{\theta}^2 + \beta'\ddot{\theta}$$

and for a circular orbit

$$\frac{\mu}{R^3} = \dot{\theta}^2 = \text{constant}, \quad \text{and} \quad \ddot{\theta} = 0$$

the equations of librational motion can be written as:

$$\psi'' - 2(1 + \psi')\beta'\tan\beta + 3K_i \sin\psi\cos\psi$$
$$= \frac{R^3}{\mu I}[M_1 \cos\sigma_0 + M_2 \sin\sigma_0]\sec\beta = F_\psi$$

$$\beta'' + \{(1 + \psi')^2 + 3K_i\cos^2\psi\}\sin\beta\cos\beta$$
$$= \frac{R^3}{\mu I}[M_1 \sin\sigma_0 - M_2 \cos\sigma_0] = F_\beta$$

$$\lambda' - (1 + \psi')\sin\beta = 0. \tag{4}$$

where

$$
\begin{aligned}
F_\psi = \frac{c}{\cos \beta} \Bigg[\frac{\cos \sigma_0}{1 + p} & \{\tfrac{3}{4} \sin^2 (\theta + \psi - \varphi)(1 + \tfrac{1}{3} \cos 2\sigma_0) \\
& + \tfrac{3}{4} \sin^2 \beta \cos^2 (\theta + \psi - \varphi)(1 - \tfrac{1}{3} \cos 2\sigma_0) \\
& + \tfrac{1}{4} \sin \beta \sin 2(\theta + \psi - \varphi) \sin 2\sigma_0 \\
& + p(\sin^2 (\theta + \psi - \varphi) \cos 2\sigma_0 \\
& - \sin^2 \beta \cos^2 (\theta + \psi - \varphi) \cos 2\sigma_0 \\
& + \sin \beta \sin 2(\theta + \psi - \varphi) \sin 2\sigma_0)\} \\
& + g \cos \sigma_0 \cos \beta |\cos (\theta + \psi - \varphi)| \{\sin (\theta + \psi - \varphi) \cos \sigma_0 \\
& \hspace{6em} + \sin \sigma_0 \cos (\theta + \psi - \varphi) \sin \beta\} \\
& + \sin \sigma_0 \{\tfrac{1}{2} \sin 2\sigma_0 \sin^2 (\theta + \psi - \varphi) \\
& - \tfrac{1}{2} \sin 2\sigma_0 \sin^2 \beta \cos^2 (\theta + \psi - \varphi) \\
& - \tfrac{1}{2} \cos 2\sigma_0 \sin \beta \sin 2 (\theta + \psi - \varphi) \\
& + g \cos \beta |\cos (\theta + \psi - \varphi)| (\sin (\theta + \psi - \varphi) \sin \sigma_0 \\
& - \cos \sigma_0 \cos (\theta + \psi - \varphi) \sin \beta)\} \Bigg]
\end{aligned}
$$

$$
= F_\psi(c, K_i, g, p, \theta, \varphi, \psi, \beta)
$$

$$
\begin{aligned}
F_\beta = c \Bigg[\frac{\sin \sigma_0}{(1 + p)} & \{\tfrac{3}{4} \sin^2 (\theta + \psi - \varphi)(1 + \tfrac{1}{3} \cos 2\sigma_0) \\
& + \tfrac{3}{4} \sin^2 \beta \cos^2 (\theta + \psi - \varphi)(1 - \tfrac{1}{3} \cos 2\sigma_0) \\
& + \tfrac{1}{4} \sin \beta \sin 2(\theta + \psi - \varphi) \sin 2\sigma_0 \\
& + p(\sin^2 (\theta + \psi - \varphi) \cos 2\sigma_0 \\
& - \sin^2 \beta \cos^2 (\theta + \psi - \varphi) \cos 2\sigma_0 \\
& + \sin \beta \sin 2(\theta + \psi - \varphi) \sin 2\sigma_0)\} \\
& + g \sin \sigma_0 \cos \beta |\cos (\theta + \psi - \varphi)| \{\sin (\theta + \psi - \varphi) \cos \sigma_0 \\
& \hspace{6em} + \sin \sigma_0 \cos (\theta + \psi - \varphi) \sin \beta\} \\
& - \cos \sigma_0 \{\tfrac{1}{2} \sin 2\sigma_0 \sin^2 (\theta + \psi - \varphi) \\
& - \tfrac{1}{2} \sin 2\sigma_0 \sin^2 \beta \cos^2 (\theta + \psi - \varphi) \\
& - \tfrac{1}{2} \cos 2\sigma_0 \sin \beta \sin 2(\theta + \psi - \varphi) \\
& + g \cos \beta |\cos (\theta + \psi - \varphi)| (\sin (\theta + \psi - \varphi) \sin \sigma_0 \\
& - \cos \sigma_0 \cos (\theta + \psi - \varphi) \sin \beta)\} \Bigg]
\end{aligned}
$$

$$
= F_\beta(c, K_i, g, p, \theta, \varphi, \psi, \beta)
$$

3. System Response

The destabilizing influence of the solar radiation pressure and other system parameters can be vividly illustrated by considering just a simple case of planar librations as indicated in Figure 2. A need for a systematic study of the satellite dynamics in such a hostile environment is quite evident.

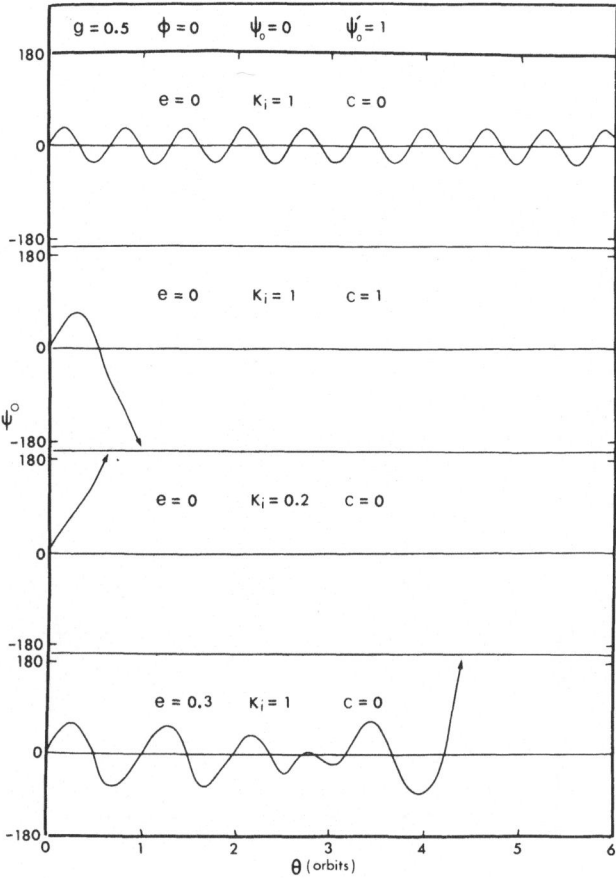

Fig. 2.   Instability induced by solar radiations, satellite inertia and orbital eccentricity.

In the present case the system does not possess any known closed form solution. In such a situation, numerical techniques were used to advantage to evaluate the librational performance of the satellite as affected by the design parameters K_i, c and g. Numerical integration was performed using the Adams-Bashforth predictor corrector technique in conjunction with the Runge-Kutta starter taking a suitable step-size of $3°$.

Figure 3 shows the librational response of the satellite in two typical situations. A comparison of the response of the satellite for two different values of solar parameter c clearly points out a substantial detrimental effect of the solar radiations on the planar degree of freedom. On the other hand, its effect on the cross-plane motion is relatively less significant.

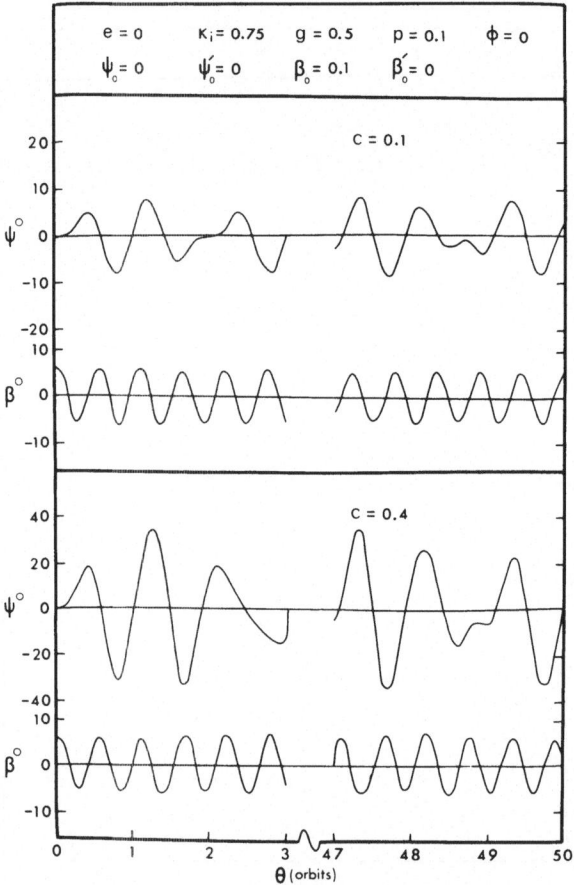

Fig. 3.   Satellite response showing the effect of solar parameter.

To better understand the dynamical behaviour of the satellite over a wide range of system parameters and initial conditions, numerous response plots were obtained as functions of c, K_i, g, φ, and initial disturbances. The resulting information was condensed in the form of system plots. Only a few of the representative plots, sufficient to establish trends, are presented here.

Figure 4a exhibits the effect of solar parameter on the maximum amplitude of motion and the average period expressed as a fraction of the orbital period, for ψ

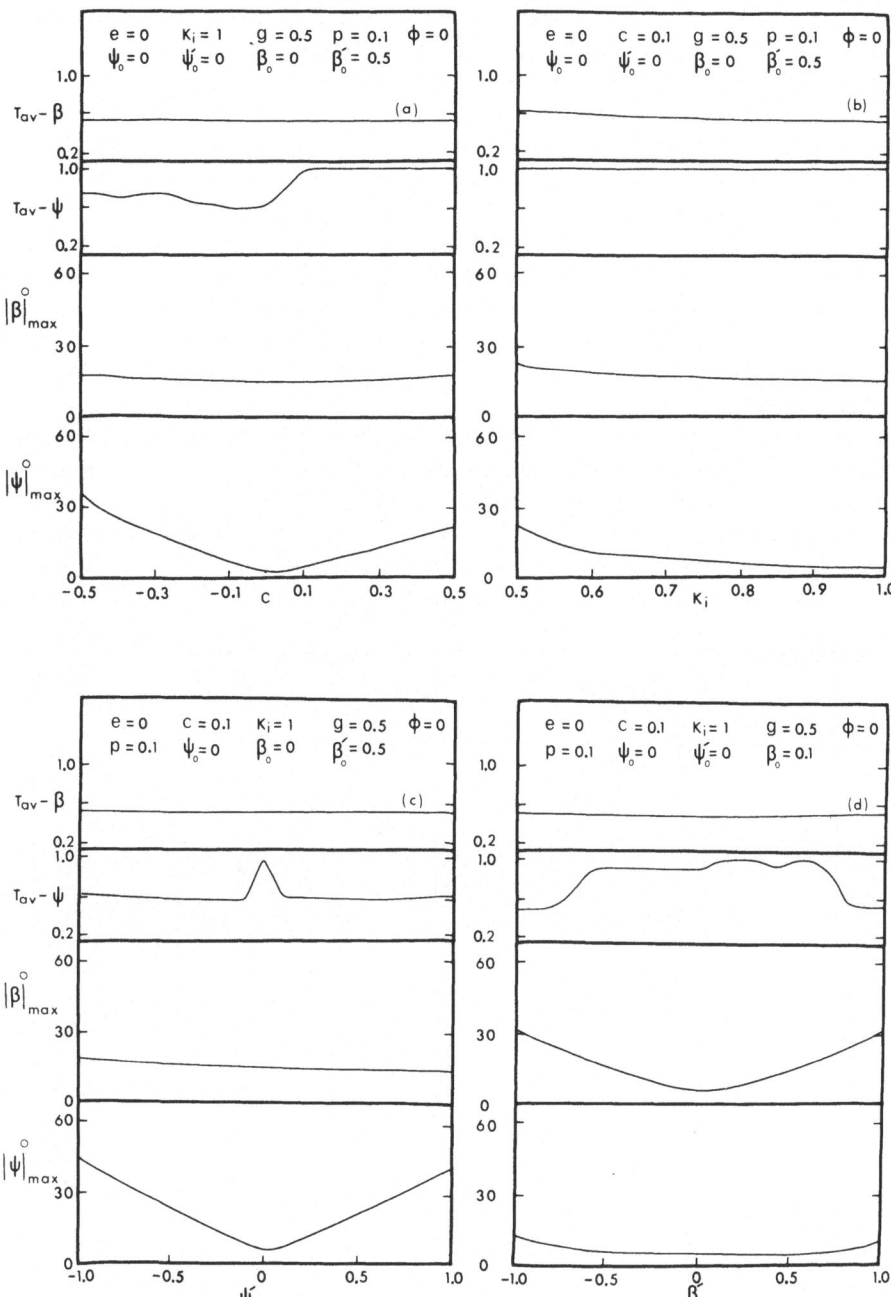

Figs. 4a–d. System plots showing the maximum amplitude and average period in ψ and β degrees of freedom as affected by: (a) solar parameter c; (b) inertia parameter K_i. (c) Initial disturbance ψ_0'; (d) initial disturbance β_0'.

and β degrees of freedom. It establishes the importance of solar parameter c, affected by the area distribution, in the design of a gravity oriented system.

The influence of inertia parameter is indicated in Figure 4b. It may be observed that the gravity gradient torque continues to be effective in limiting the librational amplitude even in the presence of the solar radiation pressure. Furthermore, both the solar parameter as well as the inertia parameter affect the average period of librations in the two degrees of freedom.

The aspect ratio did not appear to have any significant influence on the librational amplitude or period of the motion. The same is true for the effect of the solar aspect angle which was found to influence mainly the period of the in-plane motion. A system plot showing the effect of φ would be useful in predicting the long range performance of the satellite. This is of particular importance in the case of communications satellites which have a life span of 3 to 5 yr.

Figures 4c and 4d point out the limitation of the gravity gradient stabilization in the presence of the solar radiations when the satellite is subjected to severe disturbances. As the equations of motion are strongly coupled, a large disturbance across the orbital plane substantially affects the in-plane motion, however, the converse is not true. This is apparent from the fact that when $\beta_0 = \beta_0' = 0$, $F_\beta = 0$ and the equilibrium configuration remains unaffected by the in-plane motion.

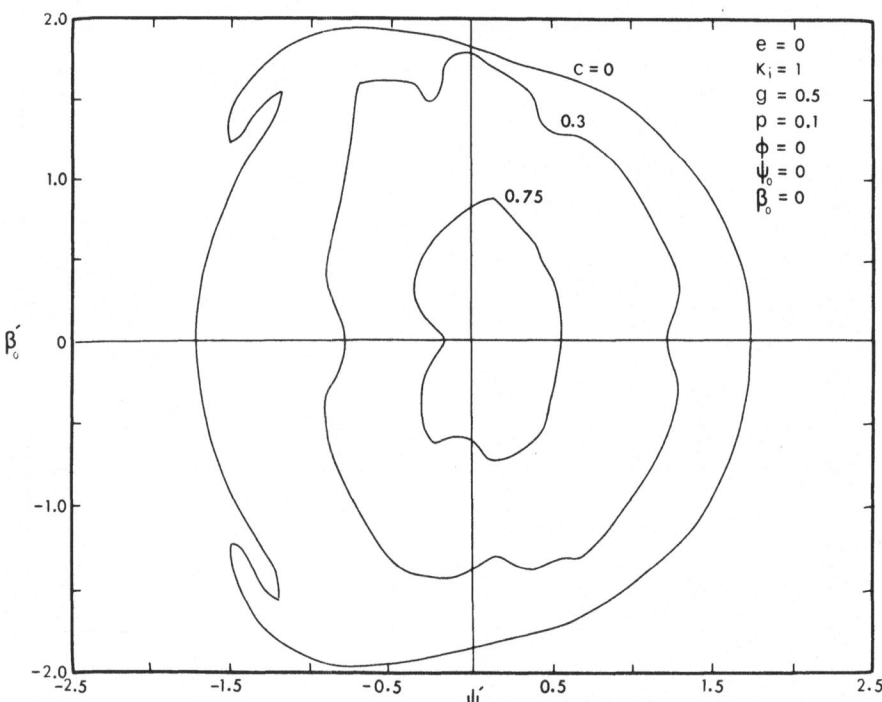

Fig. 5.   Stability plot showing the effect of solar parameter on allowable impulsive disturbances for stable operation.

4. Librational Stability

The stability bounds for the general motion were established by analysing the librational response, over 40–50 orbits, to the systematically varied initial conditions ψ_0' and β_0', for a range of the solar parameter values. The vast amount of information thus gathered is condensed in the form of stability plots which indicate allowable impulsive disturbances $(\psi_0 = \beta_0 = 0)$ for non-tumbling motion (Figure 5). It may be noted that an increase in the value of the solar parameter, in general, leads to a reduction in the region of stability. This, once again, emphasizes the importance of the solar radiation pressure, signified by c, in obtaining the design data useful for gravity oriented satellites.

5. Attitude Control

The satellite response, obtained over a wide range of initial conditions and physical parameters showed a substantial influence of the solar radiation pressure on the librational dynamics of a gravity gradient system. In general, the effect was destabilizing. However, the results suggest that through a judicious control of the generalized forces Q_ψ, Q_β and Q_λ, normally involving the solar parameter, the radiations may be able to provide an effective damping torque to attain a desired attitude. Hence the feasibility of using the solar pressure in achieving general spatial attitude control was investigated.

The proposed controller model is shown in Figure 6. It consists of two sets of plates $G_1 G_1'$ and $G_2 G_2'$ made of light, rigid and highly reflective material (e.g. aluminized Mylar membrane) and is intended to govern the forces Q_ψ, Q_β and Q_λ by regulating the motion of these plates for general attitude control. The angular position of these sets is kept stationary with respect to $x_2 y_2 z_2$ axes, themselves being fixed at 45° $(\lambda - \alpha = 45°$, Figure 6) to each other. The motion of the plates $G_1 G_1'$ w.r.t. the satellite body is controlled along the z_2 axis (l_1) as well as the y_2 axis (m_1) whereas the $G_2 G_2'$ is free to move along the z_2 axis (l_1). The equation of motion for such a system can be derived on the lines as indicated earlier. The governing relations in the three degrees of freedom are found to have the form:

$$\psi'' - 2(1 + \psi')\beta' \tan \beta - \frac{\sin \beta}{\cos^2 \beta} (1 - K_i) \frac{d}{d\theta} \{\lambda' - (1 + \psi') \sin \beta\}$$

$$- (1 - K_i) \frac{\beta'}{\cos \beta} \{\lambda' - (1 + \psi') \sin \beta\}$$

$$= \left[\left\{ \frac{c_\psi \cos \beta + c_\lambda \sin \beta}{\cos^2 \beta} \right\} \sin (\theta + \psi - \varphi) |\sin (\theta + \psi - \varphi)| \right.$$

$$+ \frac{c_\beta}{\sqrt{8} \cos \beta} \{(1 + 2p) \sin (\theta + \psi - \varphi)$$

$$- (1 - 2p) \sin \beta \cos (\theta + \psi - \varphi)\}$$

$$\left. \times |\sin (\theta + \psi - \varphi) - \sin \beta \cos(\theta + \psi - \varphi)| \right] \tag{5a}$$

$$\beta'' + \{(1 + \psi')^2 + 3K_i \cos^2 \psi\} \sin \beta \cos \beta + (1 - K_i)$$

$$\times (1 + \psi') \cos \beta \{\lambda' - (1 + \psi') \sin \beta\}$$

$$= \left[2p\{c_\psi \sin \beta \cos(\theta + \psi - \varphi) \right.$$

$$- c_\lambda \cos \beta \cos(\theta + \psi - \varphi)\}|\sin(\theta + \psi - \varphi)|$$

$$+ \frac{c_\beta}{\sqrt{8}} \{-(1 - 2p) \sin(\theta + \psi - \varphi)$$

$$+ (1 + 2p) \sin \beta \cos(\theta + \psi - \varphi)\}$$

$$\left. \times |\sin(\theta + \psi - \varphi) - \sin \beta \cos(\theta + \psi - \varphi)| \right] \qquad (5b)$$

$$\lambda'' - \psi'' \sin \beta - (1 + \psi')\beta' \cos \beta$$

$$= - \{1/(1 - K_i)\}c_\lambda \sin(\theta + \psi - \varphi)|\sin(\theta + \psi - \varphi)| \qquad (5c)$$

where

$$c_\psi = \frac{R^3 S A_p l_1 (1 + \rho - \tau)}{\mu c' I}$$

$$c_\beta = \frac{R^3 S A_p l_2 (1 + \rho - \tau)}{\mu c' I}$$

$$c_\lambda = \frac{R^3 S A_p m_1 (1 + \rho - \tau)}{\mu c' I}$$

As the main objective here is to establish the feasibility of the solar radiation pressure controller it has been assumed, for simplicity, that the satellite mass centre lies at its geometric centre or very close to it so that the main satellite body does not contribute to the expressions for the generalized forces.

The movements of the plates have been controlled according to the following relations:

$$c_\psi = \begin{cases} -[\mu_\psi \psi' + \nu_\psi(\psi - \gamma_\psi)], & 2k\pi < (\theta + \psi - \varphi) \leqslant (2k + 1)\pi \\ +[\mu_\psi \psi' + \nu_\psi(\psi - \gamma_\psi)], & (2k + 1)\pi < (\theta + \psi - \varphi) \leqslant (2k + 2)\pi \end{cases}$$

$$c_\beta = \begin{cases} +[\mu_\beta \beta' + \nu_\beta(\beta - \gamma_\beta)], & 2k\pi < (\theta + \psi - \varphi) \leqslant (2k + 1)\pi \\ -[\mu_\beta \beta' + \nu_\beta(\beta - \gamma_\beta)], & (2k + 1)\pi < (\theta + \psi - \varphi) \leqslant (2k + 2)\pi \end{cases}$$

$$c_\lambda = \begin{cases} +[\mu_\lambda \lambda' + \nu_\lambda(\lambda - \gamma_\lambda)], & 2k\pi < (\theta + \psi - \varphi) \leqslant (2k + 1)\pi \\ -[\mu_\lambda \lambda' + \nu_\lambda(\lambda - \gamma_\lambda)], & (2k + 1)\pi < (\theta + \psi - \varphi) \leqslant (2k + 2)\pi \end{cases}$$

with $|c_\psi|, |c_\beta|, |c_\lambda| < c_{max}$, k being an integer. Here, μ_ψ, μ_β, μ_λ, ν_ψ, ν_β and ν_λ represent the gains in the controller characteristic relations while γ_ψ, γ_β and γ_λ denote the position control angles. The scheme essentially involves a generalization of the planar attitude control system [14].

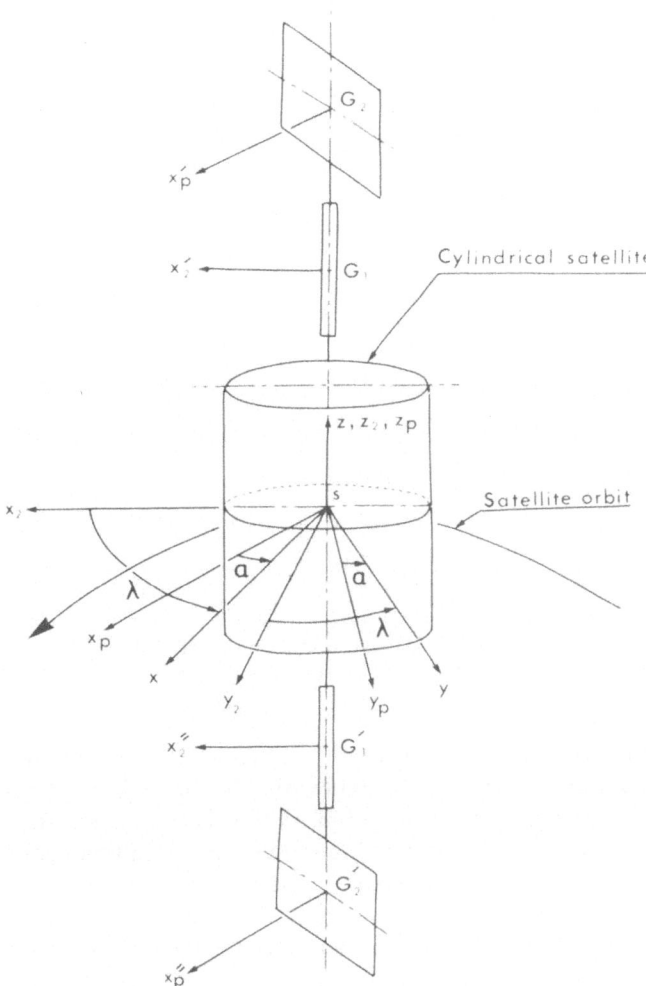

Fig. 6.   Geometry of proposed solar controller configuration.

Figure 7 shows a few typical situations in which the proposed solar controller attempts to achieve the desired spatial orientation. It may be pointed out that the satellite can be oriented along the local vertical as well as the local horizontal, normally an unstable configuration for gravity gradient satellites. Further, it is possible to stabilize the satellite in any general intermediate configuration by choosing suitable values of the position control parameters γ_ψ, γ_β and γ_λ. The amplitude of limit cycle resulting in the case of the intermediate attitude control can be minimized to $\pm 2°$ by an appropriate choice of the control parameters.

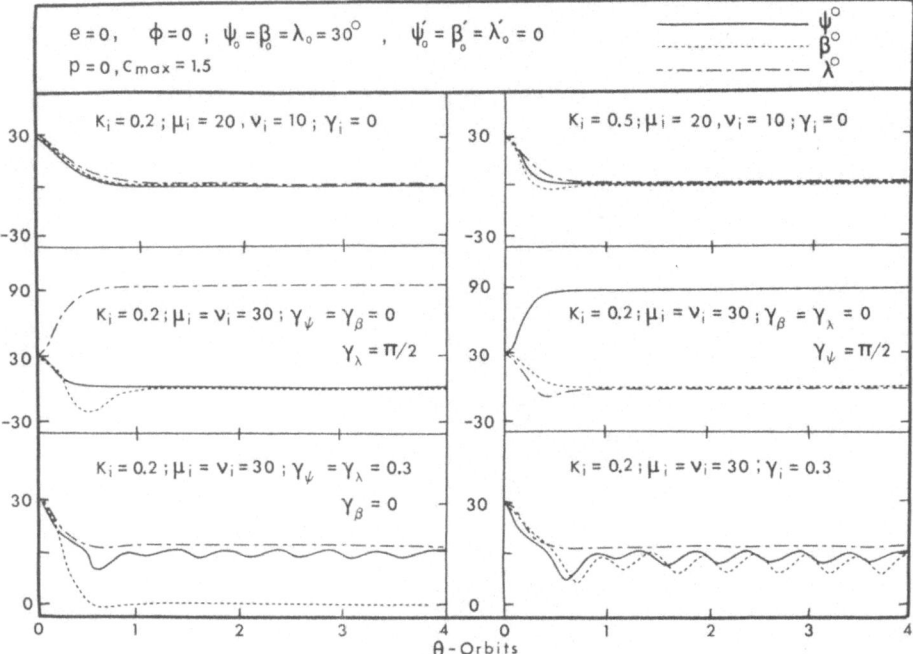

Fig. 7. Typical examples of satellite response showing the general 3-dimensional attitude control using the proposed solar controller.

The concept presents an exciting possibility of controlling communications satellites and space stations of the future (Figure 8). The size of the plates required is rather modest; so is the power consumption of the servo-system. The semi-passive character of the system promises increased life time with overall reduction in the cost. The ability of the controller to change the orientation of a space vehicle in orbit would impart it a great degree of versatility. A communications satellite can now improve upon its zone of usefulness and a skylab crew can undertake diverse scientific missions.

6. Concluding Remarks

The important features of the analysis and the conclusions based on them may be summarized as follows:

(i) The solar radiation pressure considerably affects the librational dynamics of a satellite. For a non-zero c, the satellite always executes librational motion. Hence neglecting radiation forces when predicting the attitude behaviour of gravity gradient satellites could result in gross errors.

(ii) The solar radiation pressure has substantial influence on the ψ degree of freedom and this effect is generally detrimental to the satellite performance. On the other hand, its effect on the cross-plane motion is relatively less significant.

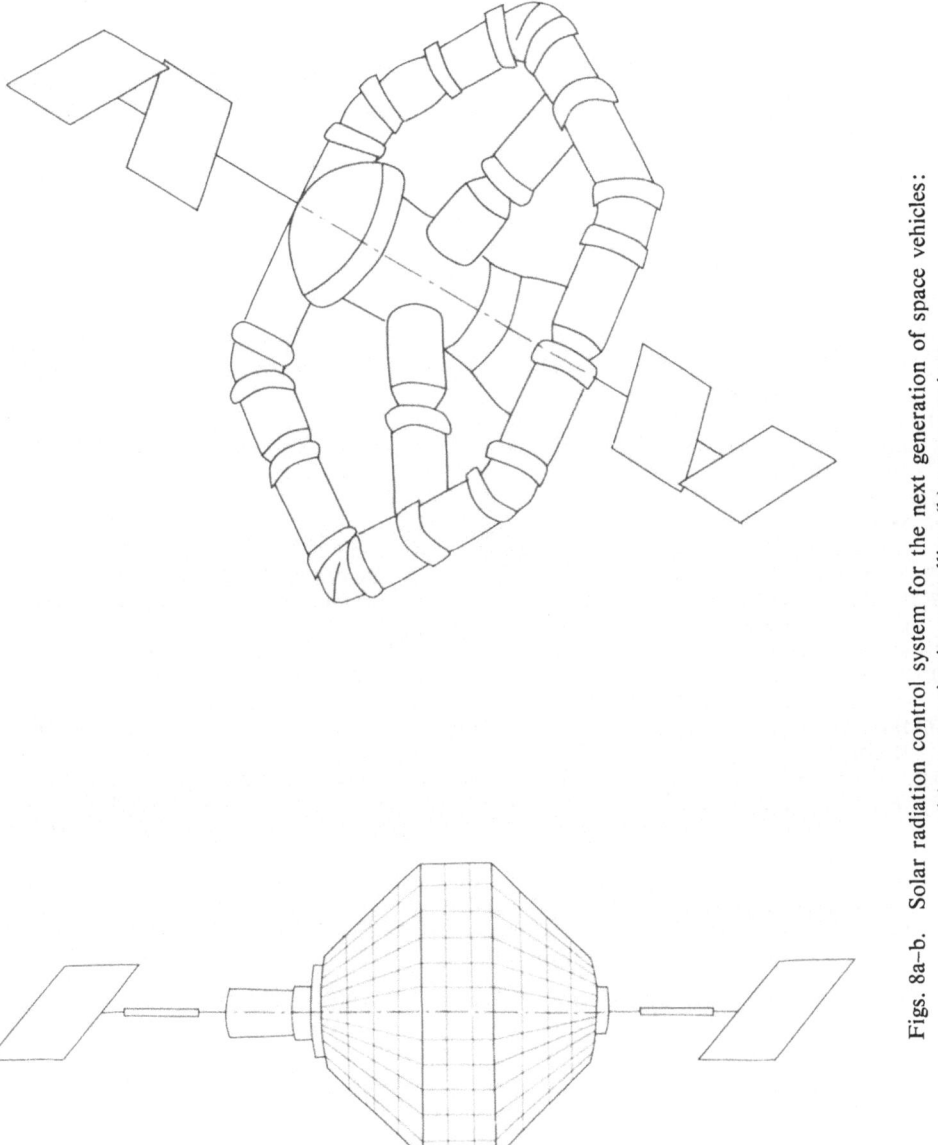

Figs. 8a–b. Solar radiation control system for the next generation of space vehicles: (a) communications satellite; (b) space station.

(iii) The solar parameter c substantially affects the stability of the motion and hence merits equal consideration with the inertia parameter in the satellite design.

(iv) The radiation pressure controller promises to be quite effective in damping librational motion even in the presence of large disturbances. Furthermore, it presents a possibility of a controlled change of the satellite attitude in orbit.

(v) The semi-passive character of the controller promises a substantial increase in life-span. It appears to be well suited for controlling the next generation of spacecrafts.

References

[1] Klemperer, W. B., *ARS J.* **30** (1960), 123.
[2] Beletskiy, V. V., in M. Roy (ed.), *Dynamics of Satellites*, Academic Press, New York, 1963, p. 219.
[3] Zlatousov, V. A., Okhotsimsky, D. E., Sarghev, V. A., and Torsheveskey, A. P., in H. Görtler (ed.), *Proceedings of XI International Congress of Applied Mechanics*, Springer-Verlag, Berlin, 1964, p. 436.
[4] Brereton, R. C. and Modi, V. J., in M. Lunc (ed.), *Proceedings of the XVII International Astronautical Federation Congress*, Gordon and Breach Science Publishers, Inc., New York, 1967, p. 179.
[5] Kane, T. R., *AIAA J.* **3** (1966), 726.
[6] Modi, V. J. and Brereton, R. C., in M. Lunc (ed.), *Proceedings of the XVIII International Astronautical Federation Congress*, Pergamon Press, London, 1966, p. 109.
[7] Modi, V. J. and Shrivastava, S. K., *Trans. Canadian Aeronaut. Space Inst.* 4 (1971), 32.
[8] Modi, V. J. and Shrivastava, S. K., 'On Coupled Librational Dynamics of Non-Autonomous Gravity Stabilized Systems in Presence of Atmosphere', *AIAA preprint* No. 71-89, AIAA 9th Aerospace Sciences Meeting, New York, Jan. 1971, also *AIAA J.* (in press).
[9] Roberson, R. E., in F. Hecht (ed.), *Proceedings of the VIII International Astronautical Congress*, Wein, Springer-Verlag, Berlin, 1958, p. 317.
[10] Hall, H. B., 'The Effects of Radiation Force on Satellite of Convex Shape', *NASA TN* D-604, 1961.
[11] Flanagan, R. C. and Modi, V. J., *Trans. Canadian Aeronaut. Space Inst.* 3 (1970), 147.
[12] Modi, V. J. and Flanagan, R. C., 'Effects of Environmental Forces on the Attitude Dynamics of a Gravity Oriented Satellite: Part I – High Altitude Orbits', *Aeronaut. J.*, the Royal Aeronautical Society (in press).
[13] Modi, V. J. and Kumar, K., in G. M. L. Gladwell (ed.), *Proceedings of the Symposium on Computer Aided Engineering*, May 1971, p. 359.
[14] Modi, V. J. and Kumar, K., in G. Partel (ed.), *Proceedings of III International Conference of Space Technology*, Rome, May 1971 (in press).

INERTIAL FORCE FIELD PATTERNS DUE TO
NUTATIONAL MOTION OF SPINNING SATELLITES

CHIKARA MURAKAMI and YOSHIAKI OHKAMI

National Aerospace Laboratory, 1880 Jindaijimachi, Chofu, Tokyo, Japan

Abstract. The motion of nutation dampers on spin-stabilized asymmetric spacecraft (including dual-spin satellites) is analyzed in general form. Up to now, many papers on this problem have been presented, but as far as the authors know, there is no paper that treats the effects of mounting position and direction of the damper motion.

In this paper, inertial force fields generated by nutational motion of the satellite are investigated and thereby, the force with nutation frequency is evaluated for an arbitrary position of the satellite. It is shown that, without any constraint on damper mounting, the equation of motion of a single-degree-of-freedom damper which consists of mass, spring and dashpot is reduced to

$$\ddot{u} + 2\zeta\omega_n\dot{u} + \omega_n^2 u = -\boldsymbol{\lambda} \cdot \mathbf{A}$$
$$\mathbf{A} = \dot{\boldsymbol{\omega}} \times \boldsymbol{\beta} + \boldsymbol{\omega} \times (\boldsymbol{\omega} \times \boldsymbol{\beta}) \tag{1}$$
$$\boldsymbol{\beta} = \boldsymbol{\beta}_0 + u\boldsymbol{\lambda}$$

where ζ, ω_n = damping ratio and undamped natural frequency of the damper itself, respectively
   u = damper mass displacement
   $\boldsymbol{\omega}$ = angular velocity vector of the satellite
   $\boldsymbol{\beta}$ = position vector of the damper mass
   $\boldsymbol{\lambda}$ = unit vector of the direction of motion of the damper.

The vector $\mathbf{A}$ at an arbitrarily fixed point $\boldsymbol{\beta}_0$ may be readily obtained by approximately solving Euler equations of the satellite motion (letting the damper mass and displacement equal zero, and coning angle be small). Then the quantity $-\boldsymbol{\lambda} \cdot \mathbf{A}$, the inertial force component along the direction $\boldsymbol{\lambda}$, is obtained and the examination of the components of $\mathbf{A}$ makes it possible to find the most effective direction of motion that will maximize the amplitude of the oscillating term of $\boldsymbol{\lambda} \cdot \mathbf{A}$. As the damper mass moves, the force term $\boldsymbol{\lambda} \cdot \mathbf{A}$ may contain the displacement u, which would yield a parametric excitation, but in most cases it can be neglected because u is small.

Two patterns of the inertial force field are given; the first one is for a dual-spin satellite with an asymmetric stator and a symmetric rotor, and the second is for an asymmetric single-spin satellite. These patterns are very useful not only for damper mounting but also for nutation sensor mounting. Of course, these patterns also indicate the oscillating stress field of the nutating bodies. Furthermore, combining the above results and the energy sink approximation method, nutation damping rate is easily evaluated in terms of ζ, ω_n, moments of inertia, spinning rate and initial nutation angle for arbitrary mounting of the damper.

It is also shown that the procedure and results presented above, are applicable to other types of passive dampers (pendulum damper etc.) with slight modification. For the pendulum type, u becomes angular displacement, and for the fluid-filled loop damper, $\omega_n^2 = 0$ in Equation (1). Especially, for the fluid-filled loop damper, total inertial force along the loop is

$$- \oint \mathbf{A} \cdot \boldsymbol{\lambda} \, du = - \int_s (\nabla \times \mathbf{A}) \cdot \mathbf{n} \, ds = -2 \int_s \dot{\boldsymbol{\omega}} \cdot \mathbf{n} \, ds,$$

where s is the surface area enclosed by the loop, and $\mathbf{n}$ is the normal unit vector of the surface and fixed to the body. In case of planar loop damper, $\mathbf{n}$ becomes constant, therefore

$$- \oint \mathbf{A} \cdot \boldsymbol{\lambda} \, du = -2(\dot{\boldsymbol{\omega}} \cdot \mathbf{n})s.$$

In order to maximize the total inertial force, $\mathbf{n}$ should be made parallel to the direction to which $\dot{\boldsymbol{\omega}}$

L. G. Napolitano et al. (eds.), Astronautical Research 1971, 53–62. All Rights Reserved.
Copyright ©1973 by D. Reidel Publishing Company, Dordrecht-Holland.

has the maximum amplitude. In other words, the optimal direction of **n** coincides with y axis for dual-spin satellites, and with the x axis for single-spin satellites, where x and y axes are lateral principal axes, assuming that the moment of inertia about x axis is larger than that about y axis.

Some numerical examples based on the above analysis are illustrated, and compared favorably with direct digital simulation.

1. Introduction

Many papers have been presented that are concerned with the nutation damper for spin-stabilized spacecraft. Within the limits of the authors' knowledge, however, the damper motion has been analyzed only for a particular mounting position specified before analysis, and the comparative merits of the way of mounting have not been discussed.

In this paper, the equation of motion of a single-degree-of-freedom nutation damper is given for an arbitrary mounting, and inertial force fields generated by nutational motion of asymmetric satellites are investigated. Two patterns of such inertial force fields are given for dual- and single-spin satellites. The effects of mounting position of the damper and its direction of motion are also discussed with some examples.

2. Equations of Motion

Nomenclature

H angular momentum vector of the total system
I inertia dyadic of the stator $= I_x\mathbf{ii} + I_y\mathbf{jj} + I_z\mathbf{kk}$
I' inertia dyadic of the rotor $= I_\zeta(\mathbf{i'i'} + \mathbf{j'j'}) + I\mathbf{kk}$
c viscous friction coefficient
k spring constant
m reduced [1] bob mass
u damper bob displacement
$\boldsymbol{\beta}$ position vector of the damper bob $= x\mathbf{i} + y\mathbf{j} + z\mathbf{k}$
$\boldsymbol{\beta}_0$ neutral position vector of the damper bob
θ_m maximum coning angle
$\boldsymbol{\lambda}$ unit vector of the direction of the damper motion
$\boldsymbol{\omega}$ angular velocity vector of the stator $= \omega_x\mathbf{i} + \omega_y\mathbf{j} + \omega_z\mathbf{k}$
$\boldsymbol{\omega'}$ angular velocity vector of the rotor $= \omega_\xi\mathbf{i'} + \omega_\eta\mathbf{j'} + \omega_\zeta\mathbf{k}$
Subscripts d and s denote dual- and single-spin, respectively.

Figure 1 shows a dual-spin satellite mounting a single-degree-of-freedom nutation damper on the stator. x, y and z are the coordinate axes which coincide with the principal axes of inertia of the stator, and ξ, η and ζ are of the rotor. z and ζ are common. The origin of the coordinates is the mass center of the satellite. It is assumed that the stator is asymmetric, while the rotor is axi-symmetric.

I' and $\boldsymbol{\omega'}$ can be expressed by the unit vectors of the stator axes,

$$\mathbf{I'} = I_\zeta(\mathbf{ii} + \mathbf{jj}) + I\mathbf{kk}$$

$$\boldsymbol{\omega'} = \omega_x\mathbf{i} + \omega_y\mathbf{j} + \omega_\zeta\mathbf{k}.$$

If there is no disturbance,

$$\frac{d\mathbf{H}}{dt} = \frac{d}{dt}\left(\mathbf{I}\cdot\boldsymbol{\omega} + \mathbf{I'}\cdot\boldsymbol{\omega'} + m\boldsymbol{\beta}\times\frac{d\boldsymbol{\beta}}{dt}\right) = 0. \tag{1}$$

On the other hand, the damper motion is given by

$$\left(m \, \frac{d^2 \boldsymbol{\beta}}{dt^2} \right) \cdot \boldsymbol{\lambda} = -(c\dot{u} + ku), \tag{2}$$

where d/dt means time differentiation with respect to the inertial coordinate axes, while dot means with respect to the body axes. As is well known,

$$\frac{d^2 \boldsymbol{\beta}}{dt^2} = \ddot{\boldsymbol{\beta}} + 2(\boldsymbol{\omega} \times \dot{\boldsymbol{\beta}}) + \dot{\boldsymbol{\omega}} \times \boldsymbol{\beta} + \boldsymbol{\omega} \times (\boldsymbol{\omega} \times \boldsymbol{\beta}). \tag{3}$$

Here, we define a vector $\mathbf{A}$, by Equation (4).

$$\mathbf{A} \equiv \dot{\boldsymbol{\omega}} \times \boldsymbol{\beta} + \boldsymbol{\omega} \times (\boldsymbol{\omega} \times \boldsymbol{\beta}). \tag{4}$$

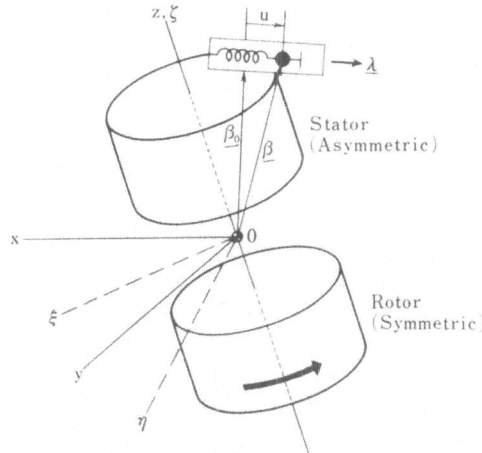

Fig. 1.   Dual-spin system.

Substituting $\mathbf{A}$, $\ddot{\boldsymbol{\beta}} = \ddot{u}\boldsymbol{\lambda}$ and $\dot{\boldsymbol{\beta}} = \dot{u}\boldsymbol{\lambda}$ into Equation (3), we have

$$\ddot{u} + 2\zeta\omega_n\dot{u} + \omega_n^2 u = -\boldsymbol{\lambda} \cdot \mathbf{A},$$

where

$$2\zeta\omega_n = c/m \quad \text{and} \quad \omega_n^2 = k/m. \tag{5}$$

In order to solve Equation (5), it is necessary to determine the vector $\mathbf{A}$. Generally, m is very small comparing with the satellite mass, therefore the Euler Equation (1) can be solved approximately. Since the stator is controlled to be non-spinning, $\omega_z = 0$. Then Equation (1) becomes

$$(I_x + I_\xi)\dot{\omega}_x + I_\xi\omega_\xi\omega_y = 0$$

$$(I_y + I_\xi)\dot{\omega}_y - I_\xi\omega_\xi\omega_x = 0 \tag{6}$$

$$I_\xi\dot{\omega}_\xi - (I_x - I_y)\omega_x\omega_y = 0.$$

If nutation is very small, that is, $|\omega_x|$, $|\omega_y| \ll |\omega_\zeta|$, then

$$\omega_x = P_d \sin \Omega_d t$$
$$\omega_y = - Q_d \cos \Omega_d t \tag{7}$$
$$\omega_\zeta = R_d \text{ (spin, rate, constant)},$$

where

$$\Omega_d = \sqrt{\frac{I_\zeta}{I_x + I_\zeta} \cdot \frac{I_\zeta}{I_y + I_\zeta}} \cdot R_d,$$
$$P_d = \sqrt{\frac{I_y + I_\zeta}{I_x + I_\zeta}} \cdot \Omega_d \theta_m \tag{8}$$
$$Q_d \cdot = \Omega_d \theta_m.$$

Similarly, for single-spin,

$$\omega_x = P_s \sin \Omega_s t$$
$$\omega_y = - Q_s \cos \Omega_s t \tag{9}$$
$$\omega_z = R_s \text{ (spin rate, constant)},$$

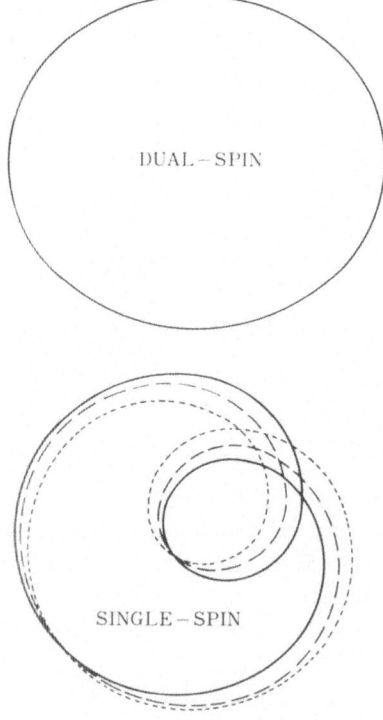

Fig. 2. Loci of spin axis.

$$\Omega_s = \sqrt{\left(\frac{I_z}{I_x} - 1\right) \cdot \left(\frac{I_z}{I_y} - 1\right)} \cdot R_s$$

$$I_z > I_x \geq I_y (I_\zeta, I_\eta, I_\zeta = 0)$$

$$P_s = \frac{I_z}{I_x} \cdot \frac{1}{\sqrt{(I_z/I_x - 1) \cdot (I_z/I_y - 1)}} \cdot \Omega_s \theta_m$$

$$Q_s = \frac{I_z}{I_z - I_y} \cdot \Omega_s \theta_m.$$

(10)

Actually, θ_m changes since m is very small but not zero. But θ_m can be regarded as constant because it changes very slowly.

From Equation (8) and Equation (10), it is evident that Ω_d is quite different from Ω_s. Ω_d never goes to zero. On the other hand, Ω_s approaches zero when I_x or I_y approaches I_z, and it may cause inefficiency for the nutation damper and the instability of the satellite. It is interesting that the locus of the spin axis top is a single ellipse for dual-spin, while in the case of single-spin, the locus is very complex as shown in Figure 2.

3. Inertial Force Field

As the solution has been obtained for the angular velocity, $\boldsymbol{\omega}$, it is possible to estimate the force field vector, $\mathbf{A}$. Substituting $\boldsymbol{\omega}$ and $\boldsymbol{\beta}$ into Equation (4), $\mathbf{A}$ is given by a function of the position on the satellite. Since $\omega_z = 0$ for dual-spin,

$$\mathbf{A_d} = \begin{Bmatrix} z\dot{\omega}_y \\ -z\dot{\omega}_x \\ y\dot{\omega}_x - x\dot{\omega}_y \end{Bmatrix} + \begin{Bmatrix} (y\omega_x - x\omega_y)\omega_y \\ (x\omega_y - y\omega_x)\omega_x \\ -z(\omega_x^2 + \omega_y^2) \end{Bmatrix}$$

$$\simeq \begin{Bmatrix} z\dot{\omega}_y \\ -z\dot{\omega}_x \\ y\dot{\omega}_x - x\dot{\omega}_y \end{Bmatrix}, \quad (\omega_x^2, \omega_y^2, \omega_x\omega_y \ll \dot{\omega}_x, \dot{\omega}_y)$$

(11)

Differentiating Equation (7) and substituting it into Equation (11),

$$\mathbf{A_d} = \begin{Bmatrix} zQ_d\Omega_d \sin \Omega_d t \\ -zP_d\Omega_d \cos \Omega_d t \\ yP_d\Omega_d \cos \Omega_d t - xQ_d\Omega_d \sin \Omega_d t \end{Bmatrix}$$

(12)

Similarly, for single-spin,

$$A_s = \begin{Bmatrix} z(R_s P_s + Q_s\Omega_s) \sin \Omega_s t \\ -z(R_s Q_s + P_s \Omega_s) \cos \Omega_s t \\ x(R_s P_s - Q_s\Omega_s) \sin \Omega_s t + y(-R_s Q_s + P_s \Omega_s) \cos \Omega_s t \end{Bmatrix} + R_s^2 \begin{Bmatrix} -x \\ -y \\ 0 \end{Bmatrix}$$

(13)

4. Inertial Force Field Patterns

From Equations (12), (13), it is seen that A is a sinusoidal function with angular velocity of Ω, but that it contains a constant centrifugal force term for single-spin. Of course, A is an approximate solution in which the second order of terms of ω_x and ω_y are neglected. Except for the constant centrifugal force term, both A_d and A_s are rotating vectors with the angular velocity of Ω. Inspecting A_d, it is found that this vector rotates on a plane fixed to the stator, and the plane is normal to the position vector β. The first term of A_s is a rotating vector, too, but in this case the plane is not always normal to β. It should be noted that the magnitude of A_s as well as A_d is proportional to the magnitude of β. If we discuss the inertial force field on any position of the satellite, it is convenient to make the magnitude of β constant, or to estimate the vector A on a spherical surface whose center coincides with the satellite mass center.

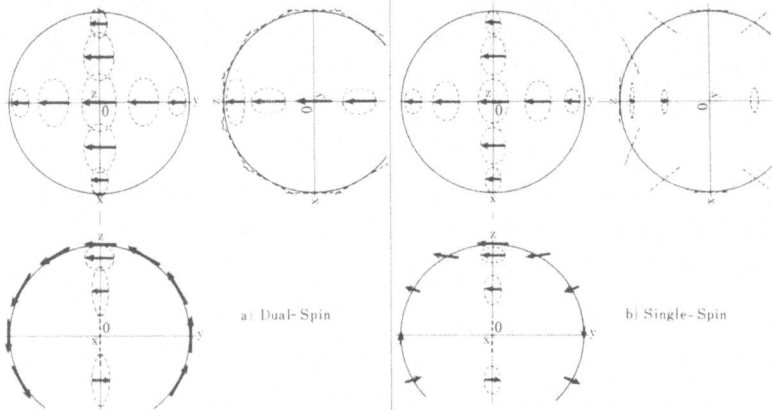

Fig. 3. Examples of inertial force field pattern.

An example of A_d is shown in Figure 3a, where $I_\zeta/(I_x+I_\xi)=1.06$ and $I_\zeta/(I_\zeta+I_\xi)=2.684$, and A_s in Figure 3b where $I_z/I_x=1.06$ and $I_z/I_y=2.684$. Figure 3 consists of the plane views or spin axis top side views, the front views or x axis side views, and the side views or y axis side views. These arrows indicate A_d or A_s vectors at an initial instant, and they all rotate anti-clockwise when seen from the top. Of course, the magnitudes of these vectors are not always constant. Their heads and tails trace ellipses when they rotate, and are shown in the figure by the dotted lines. In other words, the magnitude of the vector A is the length of the chord passing through the ellipse center. Accordingly, the oscillating force is proportional to the chord length. For convenience, we will regard the sphere as the Earth, and use such terms as pole, equator and meridian.

At first, we will investigate the dual-spin case. Each ellipse plane is tangent to the sphere. The largest ellipses are on the poles, whereas on the equator the ellipses become straight lines. Generally, as the position approaches the equator, the ellipse becomes slender. As is evident from the figure, the force component tangent to a meridian does not depend on the position along the meridian. On the pole, the directions of major and minor axes of the ellipse are parallel to the larger and smaller lateral principal axes of the satellite, respectively. Consequently, the single-degree-of-freedom nutation damper should be mounted on the meridian containing the larger lateral principal axis with the direction of motion tangent to the meridian.

In the case of single-spin (Figure 3b), the constant centrifugal force term is omitted. The ellipse planes are not tangent to the sphere except on the poles and the equator.

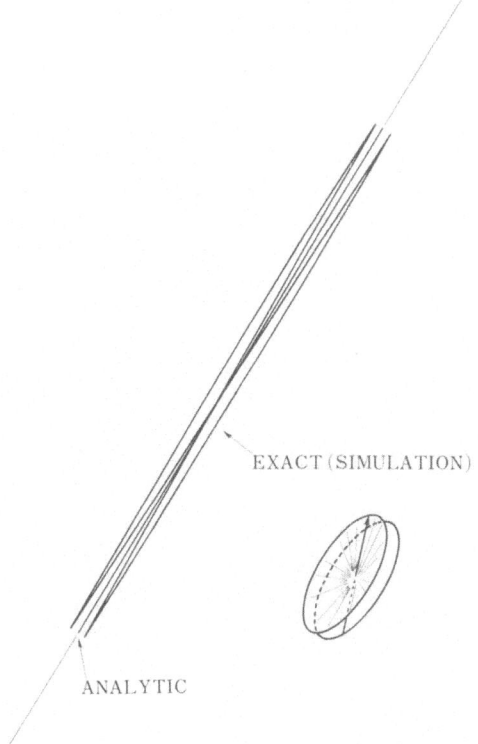

EXACT (SIMULATION)

ANALYTIC

Fig. 4. Locus of an inertial force vector.

On the poles, the ellipses are the largest, while on the equator, they become straight lines. Generally, as the position approaches the equator, the ellipses become slender, as in the case of dual-spin, and smaller.

There are two main differences between A_d and A_s; the first is that the ellipse planes are tangent to the sphere for dual-spin, while they are not always tangent to the sphere for single-spin. The second difference is as follows. Major or minor axes of the ellipses have the same length on the meridian containing them for dual-spin, whereas for single-spin, they become shorter as the positions approach the equator. Therefore, in the case of single-spin, it is more effective for a single-degree-of-freedom nutation damper to be mounted on the poles on any other position of the same distance from the satellite mass center. The most effective directions are easily seen from the figure. Except for the poles and the equator, a biased spring is required because of the constant centrifugal force.

As is already mentioned, A_d and A_s (Equations (12) and (13)) are approximate solutions, resulting that they rotate in each plane. However, since the second order terms of ω_x and ω_y are neglected, they do not always become planes. Actually, they draw circular cones as shown in Figure 4, which are so flat that they can be regarded as planes. That means the above approximation is valid.

5. Examples

For single-spin case, the poles are the most effective positions. But if the poles cannot be used for damper, and if the satellite is cylindrical, then the top or the base plane has the same effect as the poles, because both x and y components of the first term in Equation (13) do not depend on x and y but only on z. And these positions are more effective than the equator as is evident from Equation (13) by comparing the amplitudes of the components, provided that the diameter is equal to the height and $I_z > I_x$ (see Figure 5).

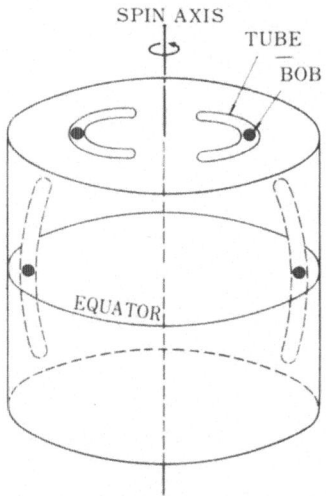

Fig. 5.   Examples of damper mounting for a cylindrical satellite.

Figure 6 illustrates damping effects in case of a cylindrical axi-symmetric single-spin satellite, with a height equal to the diameter. The damper is mounted on the surface of the satellite. Three directions of mounting are considered. The first (A) is the direction of the maximum inertial force. The second (B) is perpendicular to the spin axis and to the radial direction. And the third (C) is parallel to the spin axis.

The maximum inertial force direction defined by the angle θ_3 is a function of mounting angle position θ_1, which is derived from Equation (13) letting $P_s = Q_s$ and $y = 0$.

$$\tan \theta_3 \equiv \frac{x \text{ component of } \mathbf{A_s}}{z \text{ component of } \mathbf{A_s}} = \frac{I_z/I_x}{2 - (I_z/I_x)} \cdot \cot \theta_1. \tag{14}$$

The ordinate of the graph expresses the nutation damping rate normalized by one at the pole. Curves mean the analytic solutions using the energy sink method. Small circles are the results obtained by the direct digital simulation. We can see favorable agreement between them.

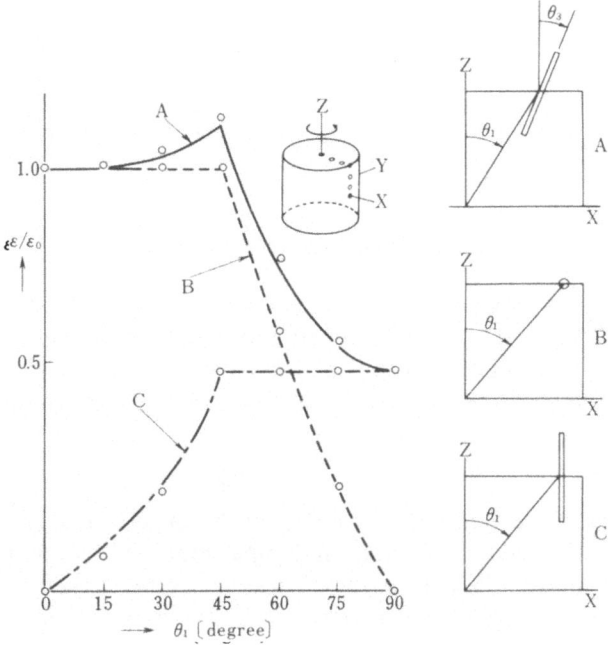

Fig. 6. Comparison of nutation damping rates.

It is shown that the concept of the inertial field force is applicable to a fluid-filled loop damper. If the mean cross-sectional flow model can be applied, the inertial force acting along the loop can be expressed by the following line integration:

$$- \oint \mathbf{A} \cdot \mathbf{\lambda} \, du.$$

By Stokes' theorem, a line integral can be converted into a surface integral,

$$- \oint \mathbf{A} \cdot \mathbf{\lambda} \, du = - \int_s (\mathbf{\nabla} \times \mathbf{A}) \cdot n \, ds = -2 \int_s \dot{\omega} \cdot \mathbf{n} \, ds, \tag{15}$$

where s is the surface area enclosed by the loop and $\mathbf{n}$ is the normal unit vector of the surface (see Figure 7). Therefore, for the case of plane loop damper, the vector $\mathbf{n}$

should be made parallel to the direction to which $\dot{\omega}$ has the maximum amplitude. In other words, the optimal direction of **n** coincides with y axis for dual-spin, whereas x axis for single-spin, assuming that the moment of inertia about x axis is larger than that about y axis. From Equation (15), we can see that the oscillating force acting

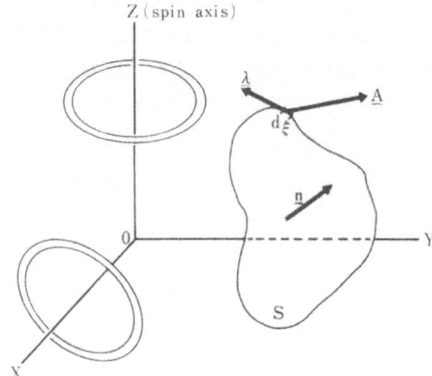

Fig. 7.   Fluid-filled nutation damper.

along the loop is proportional to the area enclosed by the loop, and only Euler's force or angular acceleration term ($\dot{\omega} \times \boldsymbol{\beta}$) can contribute to the nutation damping effect for both single and dual-spin cases. Accordingly, fluid-filled loop damper whose normal unit vector is parallel to the spin axis has little damping effect.

6. Conclusion

From the inertial force field patterns given by the present analysis, it is possible to evaluate the effectiveness of the nutation damper mounted on any position and with any direction of motion. Of course, these patterns can be applied to a two-degree-of-freedom nutation damper. These patterns will be very useful not only for damper mounting but also for nutation sensor mounting.

Reference

[1] Fang, B. T. *AIAA J.* **3** (1965), 1540–1542.

PLANAR LIBRATIONAL MOTION OF A
GRAVITY-GRADIENT SATELLITE DURING DEPLOYMENT

VIRENDER PURI* and PETER M. BAINUM**

Dept. of Mechanical Engineering, Howard University, Washington, D.C. 20001, U.S.A.

Abstract. Two existing analytical methods of predicting the planar librational motion during deployment of a gravity-gradient satellite with extendible booms are studied. The first method by Watson is based on the assumption that during the relatively short time of deployment the integrated effects of gravity-gradient torques can be neglected. The second method by Liu and Mitchell assumes that the librational motion is limited to small amplitudes. The series solution approach of Liu and Mitchell has been adapted here to include a corrected higher order term. Both analytical solutions are applied to predict librational deployment motion of (1) the DODGE satellite during a particular 'dead-beat' capture maneuver, and (2) a Dumbbell satellite initially stabilized with respect to the local vertical. The solutions obtained by both analytical methods are compared with the results of numerical integration of the exact deployment equation.

Nomenclature

A_0, B_0, A_1, B_1	Constants appearing in the zeroth order and the first order series solutions respectively and determined from the initial conditions.
$C_0, C_1, \ldots, C_s$	Coefficients appearing in the series solutions for $r=0$
$D_0, D_1, \ldots, D_s$	Coefficients appearing in the series solutions for $r=-1$
F_g	Force due to gravity on satellite end masses acting toward the center of the Earth
I	Moment of inertia of the central satellite body
$\mathbf{I}$	Moment of inertia tensor
I_i, I_f	Initial and final moments of inertia of the satellite about an axis perpendicular to the orbital plane
I_x, I_y, I_z	Principal moments of inertia about spacecraft body axes
K	Arbitrary constant
K'	Constant determined from initial conditions of the first (energy) integral; $K' = \omega_0^2 (I_x - I_z)/I_y$
$L, L(t)$	Length of the satellite boom at time t
L_0	Length of the satellite boom before initiation of extension
$\mathbf{L}_c$	Angular momentum vector with respect to the center of mass of the system
R_c	Distance from center of the Earth to the satellite center of mass
T	Dimensionless time; $T = \omega_0 t + \beta$
$\mathbf{T}_g$	A vector quantity representing the gravitational torque
$U_0(t), V_0(T),$	Functions to be determined in the variation of parameter solution of the non-
$U_1(T), V_1(T)$	homogeneous parts of the zeroth order and the first order deployment equations
V	Uniform extension rate of the booms; $V = \dot{L}$
c'	Constant; $c' = (I_i/2mV^2)^{1/2}$
k	Radius of gyration of the central satellite body
k_1, k_2	Constants to be determined in the variation of parameter technique
m	Boom end mass
n	Any integral number; $n = 1, 2, 3 \ldots$
r	Root of indicial equation
s	Any integral numbers; $s = 1, 2, 3 \ldots$

* Graduate student.
** Associate Professor of Aerospace Engineering.

L. G. Napolitano et al. (eds.), Astronautical Research 1971, 63–80. All Rights Reserved.

t	Time
t_f	Final time
y_1, y_2, y_3, y_4	Periodic functions of T appearing in the variation of parameter technique

GREEK

α	Dimensionless variable; $\alpha^2 = I\beta^2/(2mL_0^2)$
β	Dimensionless variable; $\beta = L_0\omega_0/V$
φ	Libration angle in the orbit plane (Pitch angle)
$\varphi_0, \varphi_1, \varphi_2$	Contributions to the libration angle by the zeroth, first and second order approximations respectively
ω	Angular velocity vector
ω_0	Magnitude of orbital angular velocity vector
ω_1, ω_f	Magnitude of initial and final angular velocities respectively
ω_y	Component of the inertial angular velocity normal to the orbital plane

SUPERSCRIPT

Indicates differentiation with respect to t
Indicates differentiation with respect to T

MATHEMATICAL SYMBOLS (SERIES)

$$\text{Ci}(\ ) = \ln(\ ) + \sum_{n=1}^{\infty} (-1)^n \frac{(\ )^{2n}}{2n \cdot 2n!}$$

$$\text{Si}(\ ) = \sum_{n=0}^{\infty} (-1)^n \frac{(\ )^{2n+1}}{(2n+1) \cdot (2n+1)!}$$

1. Introduction

It has been shown by Klemperer and Baker [1], and Beletskii [2] that by mutual action of Newtonian attraction of the particles of a rigid body toward the center of gravitation and by centrifugal forces which cause the motion of the center of mass of the body to follow an orbit, librational (oscillatory) motion exists about the position of relative rotational equilibrium. This principal has been applied to gravity-gradient stabilization of satellite missions, which require that a sensor or antenna point toward the Earth. The Transit Research and Attitude Control Satellite (TRAAC) [3], and more recently the Department of Defense Gravity Experiment (DODGE) satellite [3, 4] and the Radio Astronomy Explorer (RAE) satellite [5] have used gravity forces for stabilization. Due to payload restrictions, these satellites employ end masses attached to lightweight booms to achieve the required moment of inertia ratio for gravity-gradient stabilization. Many gravity-gradient satellites employ motorized extendible and retractable booms. These satellites have the capability of extending two opposing booms at the same time or a single boom by itself.

Of interest in this investigation is a comparison of existing analytical techniques that can be used in studying the attitude dynamics during the actual deployment of the extendible booms. The librational motion of a gravity-gradient satellite can be predicted by numerical integration of exact deployment equations. It can also be predicted by two existing analytical methods, both of which assume that the librations occur only within the orbital plane. The first method by Watson [6] is based on the assumption that during the relatively short time of deployment, gravity-gradient torques can be neglected. The second method by Liu and Mitchell [7] assumes that during deployment the librational angle of the satellite with respect to the local vertical, the line extending from the center of the Earth to the satellite mass center, is small. Liu and Mitchell developed an approximate series solution to the inplane deployment equation of motion. Their approach is extended here to include a corrected higher order term.

These techniques can also be applied in studying the attitude motion during the initial capture with respect to the local vertical. When a gravity-gradient satellite is injected into orbit, it has, in general, some inertial angular rate about each of its principal axes. Most of this residual rate, relative to the local vertical, can be removed by different techniques – e.g., an enhanced magnetic damping system [4] which utilizes the interaction of on-board magnetic dipoles with the ambient magnetic field of the Earth, or by transferring the momentum from the spacecraft body to a pair of masses attached to cables wrapped around it [8]. Upon completion of this initial 'de-spin' maneuver, the spacecraft is still free to rotate, primarily about an axis normal to the orbital plane at a smaller inertial angular rate. This remaining librational motion can further be removed in accordance with the principal of conservation of angular momentum by further extending booms at the time of local vertical crossing in such a way that a subsequent increase in the moment of inertia will reduce the angular velocity. This particular maneuver is referred to as the 'dead-beat' capture technique [4], and can also be analyzed by application of the existing analytical methods considered here.

2. Development of Equations of Planar Librational Motion

A. WATSON'S METHOD

In his development of the equation of librational motion, Watson assumes that a gravity-gradient stabilized satellite can be represented by a single rigid body with variable inertia, and that the motion is in the orbital plane [6]. The dynamics of boom extension is further approximated with the assumptions that the time integral of gravitational torques is negligible during extension of the booms, and that the satellite's inertial angular velocity remains fixed in direction.

Newton's second law as applied to rotational motion states that the time rate of change of the angular momentum vector relative to the center of mass equals the resultant torque about the center of mass [9]. The general equation of rotational motion of a satellite in a gravity-field is obtained by relating the time derivatives of the angular momentum vector in both the body and space axes systems. In vector notation [9],

$$\left(\frac{d\mathbf{L}_c}{dt}\right)_{\text{body}} + \boldsymbol{\omega} \times \mathbf{L}_c = \left(\frac{d\mathbf{L}_c}{dt}\right)_{\text{space}} = \mathbf{T}_g. \tag{1}$$

In the above equation, $\mathbf{L}_c$ is the angular momentum vector with respect to the center of mass of system, $\boldsymbol{\omega}$ is the angular velocity vector, and $\mathbf{T}_g$ is the gravitational torque; since $\mathbf{L}_c = \mathbf{I} \cdot \boldsymbol{\omega}$, where $\mathbf{I}$ is the moment of inertia tensor, Equation (1) can be reduced to:

$$\frac{d}{dt}(\mathbf{I} \cdot \boldsymbol{\omega}) + \boldsymbol{\omega} \times (\mathbf{I} \cdot \boldsymbol{\omega}) = \mathbf{T}_g. \tag{2}$$

Neglecting the time integral of the gravitational torque $\int \mathbf{T_g}\, dt$ and assuming that the satellite's inertial angular velocity vector $\boldsymbol{\omega}$ remains fixed in direction during the time of deployment, Equation (2) yields:

$$\frac{d}{dt}\,(\mathbf{I} \cdot \boldsymbol{\omega}) = 0; \quad \text{or} \quad (\mathbf{I} \cdot \boldsymbol{\omega})_i = (\mathbf{I} \cdot \boldsymbol{\omega})_f.$$

In scalar notation, since the libration is assumed to be in the orbital plane,

$$\omega_f = \frac{\omega_i}{(I_f/I_i)}, \tag{3}$$

where ω_i is the magnitude of initial angular velocity, ω_f the magnitude of final angular velocity vector, and I_f/I_i is the ratio of the final moment of inertia of the fully expanded satellite to the initial moment of inertia of the satellite, where both moments of inertia are taken about an axis normal to the orbital plane and passing through the center of mass.

The initial and final moments of inertia are related by,

$$I_f = I_i + 2mL^2 \tag{4}$$

and for the case of a uniform extension rate, V, the length of each boom is represented by,

$$L = Vt, \quad 0 \le t \le t_f \tag{5}$$

after substituting (4) and (5) into (3) and denoting ω_f by $d\varphi_t/dt$, we obtain:

$$\frac{d\varphi_t}{dt} = \frac{\omega_i}{(I_i + 2mV^2t^2)/I_i}$$

This can be integrated to yield:

$$\varphi_t = \omega_i c' \tan^{-1}(t_f/c'), \tag{6}$$

where

$$c' = (I_i/2mV^2)^{1/2}$$

and t_f is the time at which deployment ceases temporarily or permanently. After the boom extension, the satellite will be oriented at some angle φ with respect to the local vertical expressed in terms of φ_t by:

$$\varphi = \varphi_t + \varphi(0) - \omega_0 t_f, \tag{7}$$

where the satellite axis moves at an angle $\omega_0 t_f$ during boom extension, ω_0 is the orbital angular velocity (constant for circular orbits), and $\varphi(0)$ is some initial offset angle.

Watson's final expression, developed here, is useful for estimating the satellite orientation at any time during the deployment maneuver. In the following sections Watson's method will be applied to some physical examples and compared with another analytical method.

B. ADAPTATION OF LIU AND MITCHELL'S METHOD

The equation of planar librational motion [2] for a gravity-gradient satellite, with constant principal moments of inertia in a circular orbit (see Figure 1) is given by,

$$\ddot{\varphi} + 3\omega_0^2 \left(\frac{I_x - I_z}{I_y} \right) \sin \varphi \cos \varphi = 0, \tag{8}$$

where ω_0 is the angular velocity of the satellite's center of mass in its circular orbit, I_x and I_z are the principal moments of inertia about perpendicular body axes in the plane of the orbit, and I_y is the principal moment of inertia about an axis normal to the orbital plane. The origin of the principal body axes system is at the center of mass of the satellite. In deriving Equation (8), it is assumed that the square of the ratio of the length of the satellite boom to the circular orbital radius, $(L/R_c)^2$ is much less than one.

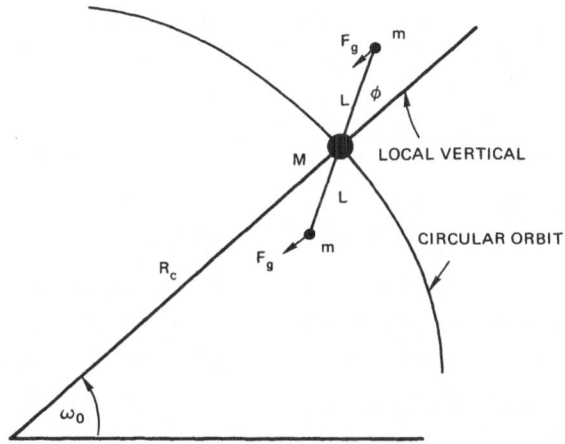

Fig. 1.   Planar librational motion of a satellite.

The equation of planar librational motion for a gravity-gradient satellite with variable principal moments of inertia can be expressed in the following form [7]:

$$\frac{d}{dt} [I_y(\dot{\varphi} + \omega_0)] + 3\omega_0^2(I_x - I_z) \cos \varphi \sin \varphi = 0. \tag{9}$$

For two equal point masses of mass, m, connected by rigid booms of length, L, to the center of a central body with moment of inertia, I,

$$I_x = I_y = I + 2mL^2 \tag{10}$$

$$I_z = I = Mk^2, \tag{11}$$

where M is the mass of the central body and k is its radius of gyration. It is assumed that the deployment takes place uniformly with a constant velocity, V, i.e.,

$$L(t) = L_0 + Vt, \quad 0 \le t \le t_f. \tag{12}$$

Consider the non-dimensional variables:

$$T = \omega_0 t + \beta \tag{13}$$

$$\beta = L_0\omega_0/V; \qquad \alpha^2 = I\beta^2/2mL_0^2. \tag{14}$$

From Equation (13),

$$\frac{d^2(\ )}{dt^2} = \omega_0^2 \frac{d^2(\ )}{dT^2} \tag{15}$$

Substitution of (10), (11) and (15), into (9) yields:

$$\omega_0\varphi'' + \frac{4mLV}{(I + 2mL^2)} \varphi' + \frac{6\omega_0 mL^2}{(I + 2mL^2)} \sin \varphi \cos \varphi + \frac{4mLV}{(I + 2mL^2)} = 0, \tag{16}$$

where the prime denotes derivatives with respect to T.

In terms of α^2, β, and T Equation (16) can be represented by,

$$\varphi'' + \frac{2T}{(\alpha^2 + T^2)} \varphi' + \frac{3T^2}{(\alpha^2 + T^2)} \sin \varphi \cos \varphi = -\frac{2T}{(\alpha^2 + T^2)}. \tag{17}$$

For φ very small, $\sin \varphi \approx \varphi$ and $\cos \varphi \approx 1$, Equation (17) becomes,

$$\varphi'' + \frac{2T}{(\alpha^2 + T^2)} \varphi' + \frac{3T^2}{(\alpha^2 + T^2)} \varphi = -\frac{2T}{(\alpha^2 + T^2)}. \tag{18}$$

For a typical satellite in Earth orbit, L_0 and ω_0 are small; for V within practical limits, α^2 is seen to be very small. In view of the linearity of Equation (18) and the magnitude of α^2, a solution in the form of a power series in α^2 can be assumed [7].

Let

$$\varphi = \varphi_0 + \alpha^2\varphi_1 + \alpha^4\varphi_2 + \cdots \tag{19}$$

Then,

$$\varphi' = \varphi_0' + \alpha^2\varphi_1' + \alpha^4\varphi_2' + \cdots \tag{20}$$

and

$$\varphi'' = \varphi_0' + \alpha^2\varphi_1'' + \alpha^4\varphi_2'' + \cdots \tag{21}$$

The binomial expression for $\left(1 + \dfrac{\alpha^2}{T^2}\right)^{-1}$ is $\left(1 - \dfrac{\alpha^2}{T^2} + \dfrac{\alpha^4}{T^4} \cdots\right)$ \tag{22}

Substituting (19), (20), (21) and (22) into (18) and rearranging,

$$\varphi_0'' + \frac{2}{T}\left(1 - \frac{\alpha^2}{T^2} + \frac{\alpha^4}{T_4}\right)\varphi_0' + 3\left(1 - \frac{\alpha^2}{T_2} + \frac{\alpha^4}{T_4}\right)\varphi_0$$

$$+ \alpha^2\left\{\varphi_1'' + \frac{2}{T}\left(1 - \frac{\alpha^2}{T^2} + \frac{\alpha^4}{T^4}\right)\varphi_1' + 3\left(1 - \frac{\alpha^2}{T^2} + \frac{\alpha^4}{T^4}\right)\varphi_1\right\}$$

$$+ \alpha^4\left\{\varphi_2'' + \frac{2}{T}\left(1 - \frac{\alpha^2}{T^2} + \frac{\alpha^4}{T^4}\right)\varphi_2'' + 3\left(1 - \frac{\alpha^2}{T^2} + \frac{\alpha^4}{T^4}\right)\varphi_2\right\}$$

$$= -\frac{2}{T}\left(1 - \frac{\alpha^2}{T^2} + \frac{\alpha^4}{T^4}\right). \tag{23}$$

Simplification of Equation (23) yields the following differential equations to different orders of magnitude in φ:

Zeroth order differential equation:

$$\varphi_0'' + \frac{2}{T}\,\varphi_0' + 3\varphi_0 = -\frac{2}{T}. \tag{24}$$

First order differential equation:

$$\varphi_1'' + \frac{2}{T}\,\varphi_1' + 3\varphi_1 = -\frac{1}{T^2}\,\varphi_0''. \tag{25}$$

Equations (24) and (25) given above were compared with the zeroth order differential equation (12) and the first order differential equation, (13) with $n=1$, both given in reference [7]. While the zeroth order differential equation given in this paper and reference [7] agreed, discrepancies were observed between the right hand sides of the first order differential equation (25) above, and Equation (13) in reference [7], reproduced below:

$$\varphi_n'' + \frac{2}{T}\,\varphi_n' + 3\varphi_n = \varphi_{n-1}'', \quad n \ge 1. \tag{26}$$

For $n=1$, Equation (26) becomes,

$$\varphi_1'' + \frac{2}{T}\,\varphi_1' + 3\varphi_1 = \varphi_0'' \tag{27}$$

Upon comparison of Equation (25) with Equation (27), it should be noted that the right hand side of (25) has a factor of $-1/T^2$. After further investigation [10], Equation (25) given above was verified to be the correct first order differential equation.

Next, the solutions to Equations (24) and (25) were investigated. The solution to the zeroth order differential equation (24), presented in the next section, agreed with the solution given in reference [7]. But, Equation (15) of reference [7], the solution to the first order differential equation, satisfies neither Equation (25), developed in this study, nor the first order differential equation (27), obtained from reference [7]. Equation (15) in reference [7], with appropriate modifications, can be shown to satisfy the following differential equation [11]:

$$\varphi_1'' + \frac{2}{T}\,\varphi_1' + 3\varphi_1 = -\frac{4}{3T^3} + \frac{2}{T} - \varphi_0''$$

C. ANALYTIC SOLUTION

c1. *Series Solution to the Zeroth Order Deployment Equation*

Linear ordinary differential equations with variable coefficients and of order greater than one can, under suitable conditions, be solved by using power series [12]. Examine the homogeneous part of the zeroth order differential equation (24), repeated here:

$$\varphi_0'' + \frac{2}{T}\,\varphi_0' + 3\varphi_0 = 0. \tag{28}$$

This is a linear differential equation with a variable coefficient and a regular singular point at $T=0$. By using the method of Frobenius [13] the solution to this differential equation can be represented by a power series in φ_0, expanded about the origin,

$$\varphi_0 = T^r \sum_{s=0}^{\infty} C_s T^s. \tag{29}$$

Since there is no other singular point, the radius of convergence of this series solution extends up to infinity in T [12, 13].

Substituting (29), φ_0', and φ_0'', into Equation (28) and rearranging, we get,

$$\sum_{s=0}^{\infty} \{r(r-1) + 2rs + s(s-1) + 2(r+s)\}C_s T^{s+r-2}$$

$$+ \sum_{s=0}^{\infty} 3C_s T^{s+r} = 0. \tag{30}$$

The indicial equation corresponding to Equation (28) is obtained by equating to zero the coefficients of the terms in the smallest power in T and assuming $C_0 \neq 0$. In this case the smallest power of T, obtained by setting $s=0$, is T^{r-2}. Then, from the coefficient of this term:

$$[r(r+1)]C_0 = 0.$$

Since it has already been assumed that $C_0 \neq 0$,

$$r(r+1) = 0. \tag{31}$$

Equation (31) is the indicial equation corresponding to Equation (28) with roots $r=0; -1$, differing by an integer.

Solution for $r=0$

Consider Equation (30) for $s=1$. By equating the sum of the coefficients of T^{r-1} to zero we obtain,

$$(r^2 + 3r + 2)C_1 = 0$$

Since $r=0$, C_1 must also be equal to zero.

In the second series of Equation (30) by letting $s=s-2$ and rearranging, we obtain:

$$C_s = -\frac{3C_{s-2}}{r(r+1) + 2rs + s(s+1)}.$$

Again for $r=0$

$$C_s = -\frac{3C_{s-2}}{s(s+1)}$$

Substitution for $C_0, C_1, C_2, \ldots, C_s$ into Equation (29) and simplification yields:

$$[\varphi_0]_I = \frac{C_0}{\sqrt{3}\,T} \left[\sqrt{3}\,T - \frac{(\sqrt{3}\,T)^2}{3!} + \frac{(\sqrt{3}\,T)^5}{5!} \cdots \right] = \frac{C_0}{\sqrt{3}} \frac{\sin \sqrt{3}\,T}{T}$$

with the subscript, I, corresponding to the root $r=0$.

Solution for r = −1

Since the roots of the indicial equation differ by an integer, the second independent solution is of the form, [13]

$$[\varphi_0]_{II} = K[\varphi_0]_I \ln T + T^r \sum_{s=0}^{\infty} D_s T^s \tag{32}$$

with the subscript, *II*, corresponding to root $r = −1$. By substituting for $[\varphi_0]_{II}$, $[\varphi_0]'_{II}$, and $[\varphi_0]''_{II}$ into Equation (28), it can be shown that $K = 0$. Then,

$$\sum_{s=0}^{\infty} (s + r)(s + r + 1)D_s T^{s+r-2} + 3 \sum_{s=0}^{\infty} D_s T^{s+r} = 0.$$

When $r = −1$, for $s = 0$, $D_0 \neq 0$; and for $s = −1$, $D_1 \neq 0$, then it can be shown that:

$$D_s = \frac{-3D_{s-2}}{s(s-1)} \tag{33}$$

Substitution of (33) into (32) results in

$$[\varphi_0]_{II} = \frac{D_0}{T}\left[1 - \frac{(\sqrt{3}\,T)^2}{2!} + \frac{(\sqrt{3}\,T)^4}{4!} \cdots\right]$$
$$+ \frac{D_1}{\sqrt{3}\,T}\left[\sqrt{3}\,T - \frac{(\sqrt{3}\,T)^3}{3!} + \frac{(\sqrt{3}\,T)^5}{5!} \cdots\right]$$

$[\varphi_0]_{II}$ can also be expressed

$$[\varphi_0]_{II} = D_0 \frac{\cos \sqrt{3}\,T}{T} + D_1 \frac{\sin \sqrt{3}\,T}{\sqrt{3}\,T}.$$

The complete solution to Equation (28) becomes,

$$\varphi_0 = [\varphi_0]_I + [\varphi_0]_{II}$$

or

$$\varphi_0 = A_0 \frac{\cos \sqrt{3}\,T}{T} + B_0 \frac{\sin \sqrt{3}\,T}{T},$$

where $A_0 = D_0$, $B_0 = (C_0 + D_1)/\sqrt{3}$ and can be determined from the initial conditions [11].

c2. *Solution to Non-homogeneous Parts of the Zeroth Order Differential Equation*

Again consider the complete zeroth order differential equation (24)

$$\varphi_0'' + \frac{2}{T}\varphi_0' + 3\varphi_0 = -\frac{2}{T}.$$

This is a non-homogeneous, linear, second order differential equation. A particular solution of this differential equation can be obtained by using the method of variation of parameters [13]. From the corresponding solution to the homogeneous equation, it can be said that the general solution is of the form,

$$\varphi_0(T)_h = k_1 y_1(T) + k_2 y_2(T),$$

where

$$y_1(T) = A_0 \frac{\cos \sqrt{3}\, T}{T} \quad \text{and} \quad y_2(T) = B_0 \frac{\sin \sqrt{3}\, T}{T}.$$

The method consists in replacing k_1 and k_2 by functions $U_0(T)$ and $V_0(T)$ to be determined such that the resulting function,

$$\varphi_0(T)_p = U_0(T) y_1(T) + V_0(T) y_2(T) \tag{34}$$

is a particular solution of the differential equation (24). Upon the differentiation with respect to T, Equation (34) becomes,

$$\varphi_0'(T)_p = U_0'(T) y_1(T) + U_0(T) y_1'(T) + V_0'(T) y_2(T) + V_0(T) y_2'(T). \tag{35}$$

The functions $U_0(T)$ and $V_0(T)$ can be selected such that,

$$U_0'(T) y_1(T) + V_0'(T) y_2(T) = 0. \tag{36}$$

Equations (35) and (36) can be solved by using Cramer's rule for $U_0'(T)$ and $V_0'(T)$,

$$U_0'(T) = \frac{2}{\sqrt{3}\, A_0} \sin \sqrt{3}\, T \tag{37}$$

and

$$V_0'(T) = -\frac{2}{\sqrt{3}\, B_0} \cos \sqrt{3}\, T. \tag{38}$$

Integration of Equations (37) and (38) yields:

$$U_0(T) = -\frac{2}{3 A_0} \cos \sqrt{3}\, T$$

and

$$V_0(T) = -\frac{2}{3 B_0} \sin \sqrt{3}\, T.$$

Then (34) becomes,

$$\varphi_0(T)_p = -\frac{2}{3T} (\cos^2 \sqrt{3}\, T + \sin^2 \sqrt{3}\, T) = -\frac{2}{3T}$$

The general solution of the zeroth order differential equation is:

$$\varphi_0 = \varphi_0(T)_h + \varphi_0(T)_p$$

or

$$\varphi_0 = \frac{1}{T} \left(A_0 \cos \sqrt{3}\, T + B_0 \sin \sqrt{3}\, T - \frac{2}{3} \right). \tag{39}$$

Notice the factor $1/T$ occurring in all terms of Equation (39). It can be seen, in retrospect, that with the transformation $\varphi_0 = T\varphi_0$, Equation (24) can be reduced to the form,

$$\Phi_0'' + 3\Phi_0 = -2 . \tag{40}$$

The homogeneous part of Equation (40) contains constant coefficients, the solution to which can be obtained by employing standard methods.

c3. *Series Solution to the First Order Deployment Equation*

The first order differential equation (25) repeated here,

$$\varphi_1'' + \frac{2}{T} \varphi_1' + 3\varphi_1 = -\frac{1}{T^2} \varphi_0'',$$

can be solved in a fashion similar to the zeroth order differential equation. The homogeneous solution, can be obtained by the method of Frobenius with the result:

$$\varphi_1(T)_h = A_1 \frac{\cos \sqrt{3}\,T}{T} + B_1 \frac{\cos \sqrt{3}\,T}{T}, \tag{41}$$

where A_1 and B_1 are constants to be determined from the initial conditions [11]. Substituting φ_0'' from Equation (39) into the right side of Equation (25) we obtain:

$$\varphi_1'' + \frac{2}{T} \varphi_1' + 3\varphi_1 = -\frac{1}{T^2} \left\{ \frac{2}{T^3} \left(-A_0 \cos \sqrt{3}\,T + B_0 \sin \sqrt{3}\,T - \frac{2}{3} \right) \right.$$
$$- \frac{2\sqrt{3}}{T^2} (A_0 \sin \sqrt{3}\,T + B_0 \cos \sqrt{3}\,T) - \frac{3}{T}$$
$$\left. \times (A_0 \cos \sqrt{3}\,T + B_0 \sin \sqrt{3}\,T) \right\} .$$

Using the variation of parameter technique we assume,

$$\varphi_1(T)_p = U_1(T) y_3(T) + V_1(T) y_4(T) \tag{42}$$

where

$$y_3(T) = A_1 \frac{\cos \sqrt{3}\,T}{T}, \quad y_4(T) = B_1 \frac{\sin \sqrt{3}\,T}{T}$$

and $U_1(T)$ and $V_1(T)$ are functions to be determined. By a method similar to that used in solving the zeroth order differential equation, we find,

$$U_1(T) = -\frac{1}{\sqrt{3}A_1} \int_\beta^T \left\{ -\frac{2}{T^4} \left(A_0 \cos \sqrt{3}\,T + B_0 \sin \sqrt{3}\,T - \frac{2}{3} \right) \right.$$
$$+ \frac{2\sqrt{3}}{T^3} (-A_0 \sin \sqrt{3}\,T + B_0 \cos \sqrt{3}\,T)$$
$$\left. + \frac{3}{T^2} (A_0 \cos \sqrt{3}\,T + B_0 \sin \sqrt{3}\,T) \right\}$$
$$\times \sin \sqrt{3}\,T \, \mathrm{d}T \tag{43}$$

and

$$
V_1(T) = \frac{1}{\sqrt{3}\,B_1} \int_\beta^T \left\{ -\frac{2}{T^4} \left(A_0 \cos \sqrt{3}\,T + B_0 \sin \sqrt{3}\,T - \frac{2}{3} \right) \right.
$$

$$
+ \frac{2\sqrt{3}}{T^3} \left(-A_0 \sin \sqrt{3}\,T + B_0 \cos \sqrt{3}\,T \right)
$$

$$
\left. + \frac{3}{T^2} \left(A_0 \cos \sqrt{3}\,T + B_0 \sin \sqrt{3}\,T \right) \right\}
$$

$$
\times \cos \sqrt{3}\,T \; \mathrm{d}T . \quad (44)
$$

On substitution of the results of integration, i.e., Equations (43) and (44) into Equation (42) the particular solution for φ_1 is obtained. The general solution for φ_1 is represented by a combination of Equations (41) and (42), i.e.,

$$
\varphi_1 = \varphi_1(T)_h + \varphi_1(T)_p .
$$

c4. *Solution Corresponding to the First Order Correction*

The approximate series solution to Equation (17) given by,

$$
\varphi = \varphi_0 + \alpha^2 \varphi_1
$$

can be written as,

$$
\varphi = \{\varphi_0(T)_h + \varphi_0(T)_p\} + \alpha^2 \{\varphi_1(T)_h + \varphi_1(T)_p\}. \quad (45)
$$

On expanding, Equation (45) becomes,

$$
\varphi = \frac{1}{T} \left(A_0 \cos \sqrt{3}\,T + B_0 \sin \sqrt{3}\,T - \frac{2}{3} \right)
$$

$$
+ \alpha^2 \left\{ \frac{1}{T} \left(A_1 \cos \sqrt{3}\,T + B_1 \sin \sqrt{3}\,T \right) \right\}
$$

$$
+ \alpha^2 \left[A_0 \left\{ -\frac{\sin 2\sqrt{3}\,T}{3\sqrt{3}\,T^3} + \frac{1}{6} \frac{\cos 2\sqrt{3}\,T}{T^2} \right. \right.
$$

$$
\left. + \frac{1}{2\sqrt{3}} \frac{\sin 2\sqrt{3}\,T}{T} - \frac{1}{2T^2} - Ci(2\sqrt{3}\,T) \right\}
$$

$$
+ B_0 \left\{ -\frac{1}{3\sqrt{3}\,T^3} + \frac{\cos 2\sqrt{3}\,T}{3\sqrt{3}\,T^3} + \frac{1}{6} \frac{\sin 2\sqrt{3}\,T}{T^2} \right.
$$

$$
\left. - \frac{1}{2\sqrt{3}} \frac{\cos 2\sqrt{3}\,T}{T} + \frac{\sqrt{3}}{2T} - Si(2\sqrt{3}\,T) \right\}
$$

$$
+ \frac{2}{3} \left\{ \frac{2}{3\sqrt{3}} \frac{\sin \sqrt{3}\,T}{T^3} + \frac{1}{3} \frac{\cos \sqrt{3}\,T}{T^2} - \frac{1}{\sqrt{3}} \frac{\sin \sqrt{3}\,T}{T} \right.
$$

$$
\left. \left. + Ci(\sqrt{3}\,T) \right\} \right]_\beta^T \frac{\cos \sqrt{3}\,T}{T}
$$

$$+\alpha^2 \left[A_0 \left\{ \frac{1}{3\sqrt{3}\,T^3} + \frac{\cos 2\sqrt{3}\,T}{3\sqrt{3}\,T^3} + \frac{1}{6}\frac{\sin 2\sqrt{3}\,T}{T^2} \right. \right.$$

$$\left. - \frac{\cos 2\sqrt{3}\,T}{2\sqrt{3}\,T} - \frac{\sqrt{3}}{2T} - Si(2\sqrt{3}\,T) \right\}$$

$$+ B_0 \left\{ \frac{\sin 2\sqrt{3}\,T}{3\sqrt{3}\,T^3} - \frac{1}{6}\frac{\cos 2\sqrt{3}\,T}{T^2} - \frac{1}{2\sqrt{3}}\frac{\sin 2\sqrt{3}\,T}{T} \right.$$

$$\left. - \frac{1}{2T^2} + Ci(2\sqrt{3}\,T) \right\}$$

$$- \frac{2}{3} \left\{ \frac{2}{3\sqrt{3}}\frac{\cos\sqrt{3}\,T}{T^3} - \frac{1}{3}\frac{\sin\sqrt{3}\,T}{T^2} - \frac{1}{\sqrt{3}}\frac{\cos\sqrt{3}\,T}{T} \right.$$

$$\left. \left. - Si(\sqrt{3}\,T) \right\} \right]_\beta^T \frac{\sin\sqrt{3}\,T}{T}. \tag{46}$$

Thus (46) is the approximate series solution to Equation (17) in the form, $\varphi_0 + \alpha^2\varphi_1$, which is correct to order α^2 within the limitation imposed on this solution, due to the small angle approximation, i.e., $\alpha^2\varphi_1 > \varphi^3/3!$. This is because, for small φ, $\varphi^3/3!$ and succeeding terms in the series expansion for $\sin\varphi$ have been neglected. The radius of convergence of this series solution extends from $T=0$ to $T=\pm\infty$.

3. Application of Equations of Librational Motion

A. NUMERICAL ANALYSES

It will be of interest to apply and analyze the solutions of the deployment librational equations of motion for some specific examples. For this purpose, Watson's equation (7), the adapted Liu and Mitchell's solution (46), and the exact deployment equation (17), were programmed for use on the IBM 1130 Computer with a memory capacity of 16 000 words. Due to the small magnitude of the quantities involved, extended precision was used in the computations for the adapted Liu and Mitchell's solution and for the numerical integration of the exact deployment equation.

B. DEAD BEAT CAPTURE MANEUVER

The first example considered here is of the DODGE [4] satellite during a 'dead-beat' [3, 4] capture deployment maneuver. The moment of inertia variation required for stabilization of the satellite is achieved by using extendible and retractable booms. When the satellite is injected into orbit, it has, in general, some angular rate about each of its principal axes. This initial spin rate is removed by employing the enhanced magnetic damping system, described earlier. At the completion of the post injection initial de-spin maneuver, the Y axis of the satellite is oriented along the local magnetic field direction, approximately perpendicular to the equatorial orbital plane [4]. The satellite is then free to rotate about its Y axis at approximately 10 revolutions per orbit, Figure 2A. When the satellite crosses the vertical with correct side downward, $\pm Z$

booms are erected to give one half the final value of the pitch moment of inertia (about the Y axis) I_y, as shown in Figure 2B. The satellite will then swing out from the vertical to some maximum angle, Figure 2C, of about 35.4°, and will return one half of a pitch period later to the local vertical [3], Figure 2D. From the first (energy) integral [3, 11] and the initial conditions of $\varphi(0) = 35.4°$ and $\varphi'(0) = 0$, at the time of the next local vertical crossing: $\dot{\varphi} = \pm 1.0 \, \omega_0 \, ; \varphi' = |1.0| \, ;$ where $\omega_0 = 1.0$ revolutions per orbit, the sign depends on the direction of crossing the vertical. If it is in the same sense as the orbital momentum vector, then $\varphi' = 1.0$ r.p.o. and $\omega_y = 1.0 + \varphi' = 2.0$ r.p.o. as shown in Figure 2D. The next step is to extend $\pm Z$ booms to give the final moment of inertia (ideally twice that indicated in Figure 2B), Figure 2E. Applying the principal

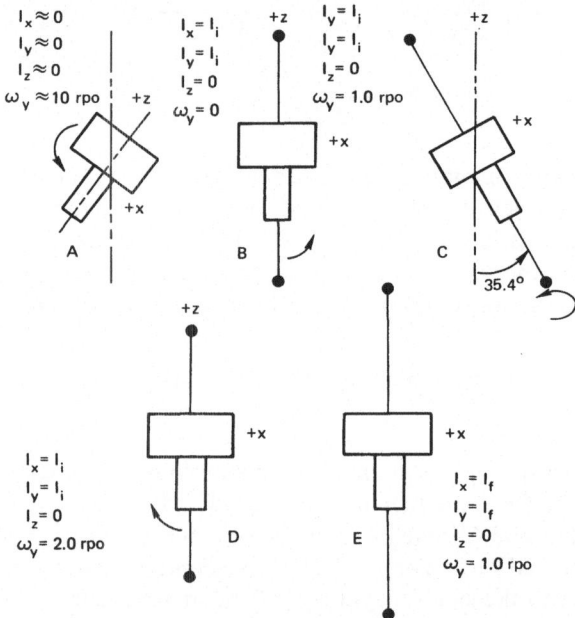

Fig. 2.   Dead-beat deployment maneuver. The DODGE satellite.

of the conservation of angular momentum, it can be seen that the ideal final values are: $\omega_y = 1.0$ r.p.o. and $\varphi' = 0$. Because the booms can never be extended instantaneously, i.e., in zero time, the analytical and numerical solutions described earlier were used to study the librational motion during the actual 'finite-time' deployment.

Figure 3 illustrates the pitch displacement, φ, for a DODGE-type satellite during the 'dead-beat' maneuver as predicted by the analytical methods and numerical integration. The following initial conditions and DODGE satellite parameters were used [3].

$$\varphi(0) = 0, \quad \varphi'(0) = 1.0, \quad L_0 = 95 \text{ ft}, \quad V = 0.25 \text{ fps}$$
Central mass = 13.500 slug, $m = 0.155$ slug, $\omega_0 = 0.79 \times 10^{-4}$ rad s^{-1}.

These parameters give a value of $\alpha^2 = 0.25 \times 10^{-7}$ rad s^{-1}.

It can be seen that Watson's method approximates more closely the exact numerical solution than the adapted method of Liu and Mitchell. The curve representing the zeroth order approximation to the series solution lies to the right of the curve predicted by the exact solution, whereas, the curve obtained by using the first order correction, Equation (46), lies to the left, showing the tendency of the series solution to converge to the exact solution. The ratio of the final to initial moments of inertia for zero angular rate from the plot of the exact numerical solution is seen to be 2.97, indicating the effect of (1) the action of the gravitational torques, and (2) the finite time required for deployment. The librational angle corresponding to this moment of inertia ratio of 2.97 is indicated to be 0.544°.

In order to evaluate the effect of an initial offset angle on the librational motion, $\varphi(0)$ was given values of $-2.86°$, $+2.86°$, and $5.73°$ respectively; all other parameters

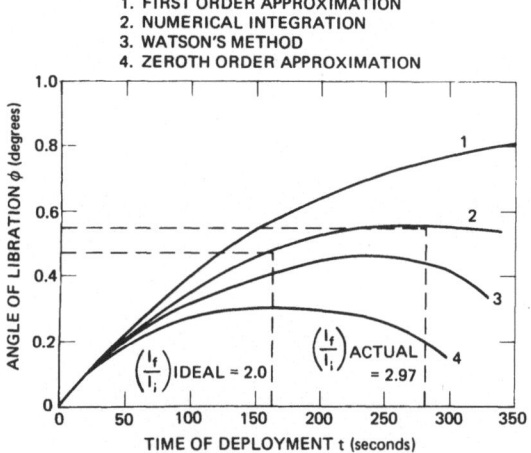

1. FIRST ORDER APPROXIMATION
2. NUMERICAL INTEGRATION
3. WATSON'S METHOD
4. ZEROTH ORDER APPROXIMATION

Fig. 3.   Capture of the DODGE satellite.
$\varphi(0)=0$, $\varphi'(0)=1.0$, $V=0.25$ fps.

indicated for Figure 3 were used. For the first two cases a shifting of the origin of all curves predicting the librational motion was noted [11]. Therefore, a possible way to reduce the small residual libration angle, would be to begin the deployment at a small negative offset angle. Figure 4, the response for the initial conditions of $\varphi(0) = +5.73°$ and $\varphi'(0) = 1.0$, shows a shift in the curves obtained by using numerical integration and Watson's method, whereas the solution given by both the zeroth order and the first order series approximations, indicate the satellite to be librating opposite to the sense of the orbital angular velocity. This divergence of the series solution can be explained by (1) the initial assumption on the magnitude of φ and (2) comparison of the magnitudes of $\alpha^2\varphi_1$ and $\varphi^3/3!$. For this particular set of initial conditions at the response time interval shown, $\alpha^2\varphi_1 < \varphi^3/3!$, whereas for meaningful results with the series solution, the reverse should be true.

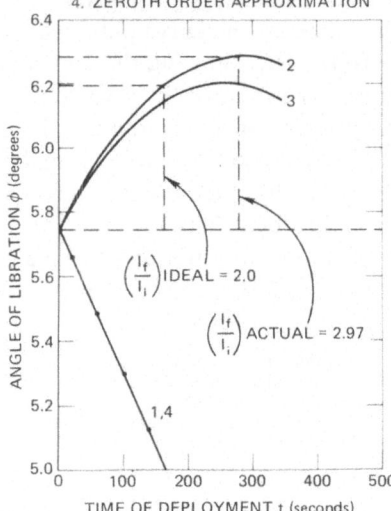

1. FIRST ORDER APPROXIMATION
2. NUMERICAL INTEGRATION
3. WATSON'S METHOD
4. ZEROTH ORDER APPROXIMATION
ANGLE OF LIBRATION φ (degrees)
TIME OF DEPLOYMENT t (seconds)
(If/Ii) IDEAL = 2.0
(If/Ii) ACTUAL = 2.97

Fig. 4. Capture of the DODGE satellite. $\varphi(0) = +5.73°$, $\varphi' = 1.0$, $V = 0.25$ fps.

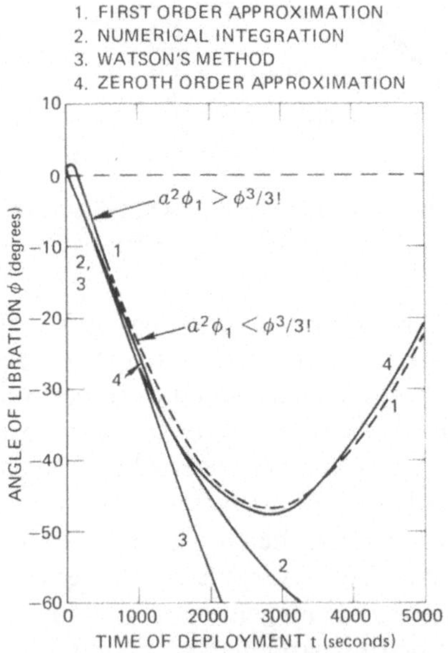

1. FIRST ORDER APPROXIMATION
2. NUMERICAL INTEGRATION
3. WATSON'S METHOD
4. ZEROTH ORDER APPROXIMATION
ANGLE OF LIBRATION φ (degrees)
TIME OF DEPLOYMENT t (seconds)
a²φ₁ > φ³/3!
a²φ₁ < φ³/3!

Fig. 5. Development of a Dumbbell satellite. $\varphi(0) = 0$, $\varphi'(0) = 0$, $V = 50$ cm s^{-1}.

C. DEPLOYMENT OF A DUMBBELL SATELLITE

In this section the deployment of a dumbbell satellite over a long period of time is examined. The satellite is assumed to be initially stabilized and aligned along or near the local vertical. Liu and Mitchell [7] have indicated that deployment over a long period of time can induce tumbling. The purpose here is to verify the previous results of Liu and Mitchell and to compare the results of their adapted method with those obtained by using Watson's method.

Figure 5 shows the deployment of a dumbbell satellite as predicted by the analytical methods and numerical integration. The following initial conditions and satellite parameters were used:

$$\varphi(0) = 0, \quad \varphi'(0) = 0, \quad L_0 = 30 \text{ cm}, \quad V = 50 \text{ cm s}^{-1}$$
$$\text{Central mass} = 500 \text{ kg}, \quad m = 1 \text{ kg} \quad \text{and} \quad \omega_0 = 0.50 \times 10^{-3} \text{ rad s}^{-1}$$

corresponding to a near-Earth satellite. These parameters give a value for $\alpha^2 = 0.625 \times 10^{-4}$.

It can be seen that for the parameters used for the dumbbell satellite, the responses of Watson's method and the zeroth order approximation are similar to that of the exact numerical solution, for the first several hundred seconds of deployment. As the deployment continues the two curves corresponding to the zeroth order approximation and Watson's solution start diverging from the results predicted by numerical integration. Unlike the case for the deadbeat capture maneuver the first order series approximation does not show a better convergence to the exact solution than does the zeroth order solution. The convergence of the series solution appears to be quite sensitive to the selection of the satellite and orbital parameters yielding different values of α^2, as well as to changes in the initial conditions which result in different values of A_0, B_0, A_1, and B_1 [11]. For these cases convergence might be improved by including the terms $\alpha^4 \varphi_2$ in the approximate solution [viz. Equation (19)].

The response predicted by the zeroth and the first order series approximations in Figure 5 is similar to the motion depicted in Figure 5 of reference [7]. It should be noted that for the longer deployment time, both analytical solutions fail to converge.

4. Conclusions and Results

From the development of the equation of planar librational motion of a satellite during deployment and its solutions, coupled with the numerical analysis, the following comments can be made:

(1) With regard to the 'dead-beat' capture maneuver of the DODGE satellite it can be said:

    (a) Watson's method gives a better prediction of the librational motion during the short deployment time associated with this maneuver, than the results obtained with the series solution adaptation of Liu and Mitchell.

    (b) The first order series solution fails to converge for all cases when $\alpha^2 \varphi_1 < \varphi^3/3!$.

(c) In general, for finite time deployment of the booms, a larger final to initial moment of inertia ratio is required to achieve gravity-gradient stabilization than predicted by theory.

(2) With regard to the deployment of a dumbbell satellite initially stabilized with respect to the local vertical:

(a) For short deployment time, Watson's method predicts the librational motion more accurately than the series solutions.

(b) Deployment over very large time intervals will result in large residual libration angles.

References

[1] Klemperer, W. B. and Baker, Jr., R. M., *Proceedings of the VIIth International Astronautical Congress, Rome, September, 1956.*

[2] Beletskii, V. V., 'The Libration of a Satellite', NASA TTF-10, May, 1960 (Translation).

[3] Bainum, P. M., 'On the Motion and Stability of a Multiple Connected Gravity-Gradient Satellite with Passive Damping', Ph.D. dissertation, Catholic University, Washington, D.C., 1966; also the Johns Hopkins University-Applied Physics Laboratory Technical Report TG-872, January, 1967.

[4] Fischell, Robert E., 'A Graviety-Gradient Satellite at Synchronous Altitude', Second IFAC Symposium on Automatic Control in Space, Vienna, Austria, Sept. 1967.

[5] Dow, P. C., Scammell, F. H., Murray, F. T., Carlson, N. A., and Buck, J. H., in Proceedings of the AIAA/JACC Guidance and Control Conference, Seattle, August 15–17, 1966, published by AIAA, New York, N.Y., p. 285.

[6] Watson, D. M., *Proceedings of the Symposium on Passive Gravity-Gradient Stabilization,* NASA-Ames Research Center, NASA-SP-107, 1966, p. 227.

[7] Liu, Han-Shou and Mitchell, Thomas P., 'The Structural and Librational Dynamics of a Satellite Deploying Flexible Booms or Antennas', AIAA 5th Aerospace Sciences Meeting, New York, N.Y., Jan. 23–26, 1967, Paper No. 67–43.

[8] Smola, J. F., 'Momentum Transfer as a Means of Despinning a Rotating Spacecraft', The Johns Hopkins University, Applied Physics Laboratory, Technical Report TG-885, January 1967.

[9] Goldstein, Herbert, *Classical Mechanics,* Addison-Wesley Publishing Company, Inc., Reading, Mass., 1965.

[10] Private Conversation with Dr Han-Shou Liu, NASA Goodard Space Flight Center, Greenbelt, Maryland, October 1969.

[11] Puri, Virender, 'Planar Librational Motion of a Gravity-Gradient Satellite During Deployment', Master's dissertation, Howard University, Washington, D.C. Jan. 1971.

[12] Rainville, Earl D., *Elementary Differential Equations,* The Macmillan Co., New York, 1965.

[13] Kreyzig, Erwin, *Advanced Engineering Mathematics,* John Wiley and Sons, New York, 1967.

TIME-OPTIMAL SEMI-ACTIVE ATTITUDE CONTROL
FOR THE PITCH MOTION OF A SATELLITE

EBERHARD P. HOFER

Dept. of Mechanical Engineering, University of Stuttgart,* Germany

Abstract. In the present paper it is shown that both gravitational and aerodynamic torques can be used successfully for the attitude control of a satellite. This concept is applied to control the pitch motion of an Earth satellite in circular orbit. Moreover, for small pitch angles the problem of a time-optimal alignment with the local vertical is considered. This problem leads to a bilinear optimization problem and the solution is based on the maximum principle. A complete optimal synthesis is given. For a technical realization deployable rods with tip masses and rotatable panels are proposed. Since no jets and no fly wheels are needed for the attitude control process the notation 'semi-active' is used.

1. Introduction

The influence of gravitational and aerodynamic torques on the attitude dynamics of low orbiting Earth satellites at altitudes up to 700 km may be of the same order. Especially, satellites with large appendages, such as solar arrays, present large areas to the wind and, consequently, considerable aerodynamic torques arise. Here, such a satellite model is considered.

In the literature both types of torques mentioned above are discussed; gravitational torques are investigated in detail [1]. For satellites with complex geometry different mathematical flow models have been applied for calculating the aerodynamic force and moment coefficients [2]. The problem of stability of the free motion of a rigid satellite influenced by both gravitational and aerodynamic torques has been considered [3]. Furthermore, it has been shown that gravitational torques can be used for the attitude control of satellites [4]. By variation of the satellite's mass distribution such a control has been realized.

The present paper shows that aerodynamic torques can also be successfully used for the attitude control of satellites. Control by aerodynamic torques is possible by changing the relative position of panels with respect to the satellite's main body. It appears that the combination of a controlled change of the mass distribution with a controlled positioning of the panels, i.e. a controlled utilization of both gravitational and aerodynamic torques, leads to a promising attitude control system.

The pitch attitude control of a satellite is considered in detail below.

2. Satellite Model

As mentioned above, the proposed control program requires changes of the mass distribution of the satellite. Consequently, the satellite can no longer be considered as a rigid body and this leads to great mathematical difficulties.

* 7000 Stuttgart 1, Keplerstr. 17, Germany.

L. G. Napolitano et al. (eds.), *Astronautical Research 1971*, 81–88. *All Rights Reserved.*

An acceptable mathematical model which overcomes these difficulties can be constructed if the methods of changing the mass distribution are restricted such that the
principal axes of the satellite and the principal moment of inertia with respect to the
pitch axis remain fixed; such a satellite model is shown in Figure 1. The body-fixed
principal axes of the satellite are denoted by x, y, z. The four tip masses m can move
along the x and y axes between the indicated extreme positions. The x and y axes lie
in the orbital plane which is given by the flight direction (ξ axis) and the direction
of the local vertical (η axis). It is obvious that the motion of the tip masses can be

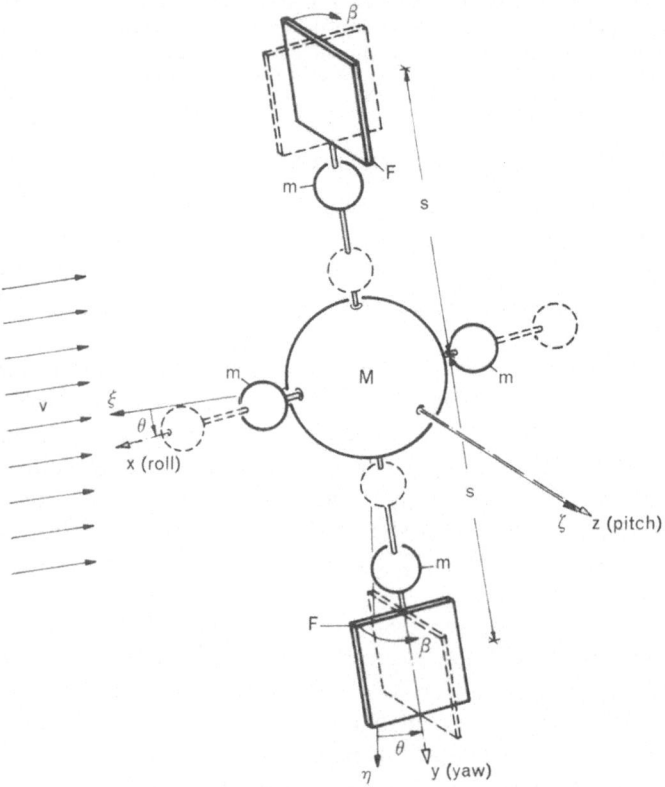

Fig. 1.   Satellite model.

controlled such that the principal axes of the satellite do not change. The two panels
F are rotatable about the y axis. For the positioning of the panels opposite turns are
necessary to avoid any influence on the roll and yaw motions by rotating the panels.
The resulting change of the moment of inertia with respect to the pitch axis can be
compensated by varying the distance s of the panels.

The satellite's principal moments of inertia are denoted by A, B, C. Assuming a
stable satellite, the stability condition

$$C(t) > A(t) > B(t) \tag{1}$$

is fulfilled for all t where t is the time normalized with respect to the orbital angular velocity. Further, the mass distribution of the proposed satellite model satisfies the conditions

$$C(t) = C_0 = \text{const} \tag{2}$$

$$0 < I_{\min} \leqslant A(t) - B(t) \leqslant I_{\max} < 1 \tag{3}$$

for all t.

Since the main body M of the satellite and also the tip masses m are spheres, these masses do not contribute to the aerodynamic torque about the center of mass. On the other hand, changing the position of the panels does not essentially influence the gravitational torque.

3. Statement of the Problem

Based on a Newtonian aerodynamic flow model [5] and considering (2) the equation for the pitch motion of the satellite in Figure 1 is

$$C_0 \ddot{\theta}(t) = -3[A(t) - B(t)] \sin \theta(t) \cos \theta(t)$$
$$- \tfrac{1}{2} \rho v^2 sF'(\beta(t)) c_N(\theta(t)) \cos \theta(t) |\cos \theta(t)| . \tag{4}$$

The pitch angle is denoted by θ and the dots stand for the derivative with respect to time t. The symbols of the second term on the right-hand side of (4) – this term describes the aerodynamic torque – are defined as follows: ρ is the atmospheric density, v is the velocity of the incident flow, c_N is the non-dimensional positive aerodynamic force coefficient, F' denotes the scalar difference of the projected areas of the two panels normal to the incident flow, and β is the control angle of the panel position.

The question of an alignment of the satellite with the local vertical is of practical importance. Moreover, for small deviations from the local vertical the problem of aligning in minimum time is of special interest. Thus, the following optimization problem may be specified: Determine the control program for changing the tip masses and for positioning the two panels such that for small pitch angles θ the satellite is time-optimally aligned with the local vertical.

Linearizing (4) with respect to θ the equation for the pitch motion reads as

$$C_0 \ddot{\theta}(t) = -3[A(t) - B(t)]\theta(t) - \tfrac{1}{2} \rho v^2 sF'(\beta(t)) c_N(\theta(t)) \tag{5}$$

with

$$F'(\beta(t)) = F[|\cos \beta(t)| - |\sin \beta(t)|], \tag{6}$$

$$c_N(\theta(t)) = c_{N0} = \text{const} . \tag{7}$$

For small angles θ relationship (6) follows directly from the geometry of the two panels. According to the used Newtonian flow model and the linearization with respect to θ Equation (7) is justified also [3].

The boundary conditions and the performance index are given by

$$\theta(0) = \theta_0, \quad \dot\theta(0) = \dot\theta_0, \tag{8}$$

$$\theta(T) = \dot\theta(T) = 0, \tag{9}$$

and

$$T \overset{!}{=} \min, \tag{10}$$

where T denotes the duration for the alignment of the satellite. Then, by (5) to (10) the optimization problem is given.

4. Solution of the Problem

Introducing the state variables

$$x_1(t) = \theta(t), \qquad x_2(t) = \dot\theta(t) \tag{11}$$

and defining the control variables

$$u_1(t) = 3\,\frac{A(t) - B(t)}{C_0}, \tag{12}$$

$$u_2(t) = \frac{\rho v^2 s c_{N0}}{2C_0}\,F'(\beta(t)), \tag{13}$$

where u_1 and u_2 describe the positions of the tip masses and the two panels, respectively, Equation (5) reads as the bilinear system [6]

$$\begin{aligned} \dot x_1 &= x_2 \\ \dot x_2 &= -u_1 x_1 - u_2. \end{aligned} \tag{14}$$

Due to the restrictions (3) and (6) for u_1 and u_2 we can state the following constraints

$$0 < u_{1m} \leqslant u_1(t) \leqslant u_{1M} < 1, \tag{15}$$

$$|u_2(t)| \leqslant u_{20}. \tag{16}$$

Considering (11) the boundary conditions (8), (9) are transformed into

$$x_1(0) = x_{10}, \qquad x_2(0) = x_{20}, \tag{17}$$

$$x_1(T) = x_2(T) = 0. \tag{18}$$

For the solution of the bilinear optimization problem [7] given by (10) and (14) to (18) the maximum principle [8] is applied.

The optimal control laws are derived from the Hamiltonian of the problem:

$$u_1(t) = \begin{cases} u_{1m} & \text{for} \quad x_1(t)p_2(t) > 0, \\ u_{1M} & \text{for} \quad x_1(t)p_2(t) < 0, \end{cases} \tag{19}$$

$$u_2(t) = -u_{20}\,\mathrm{sgn}\,p_2(t). \tag{20}$$

The costate variable $p_2(t)$ satisfies the adjoint system

$$\dot{p}_1 = u_1 p_2, \quad \dot{p}_2 = -p_1. \tag{21}$$

The optimal control laws (19) and (20) are of bang-bang type. For $p_2(t)=0$ a switching of both controls occurs while for $x_1(t)=0$ only u_1 switches. These control laws mean that an instantaneous change of the tip masses between their extreme positions is necessary and also that the positions of the two panels must be changed instantaneously by opposite rectangular turns. In Figure 1 all these possible configurations are drawn.

It may be remarked that due to the control law (20) the roll and yaw motions of the satellite are not influenced by aerodynamic torques. Therefore, it can be assumed that with respect to roll and yaw the satellite is in the position of relative equilibrium.

Now, by means of the optimal control laws the optimal solution can be determined. With $u_1=$ const and $u_2=$ const the integration of (14) and (21) leads to

$$u_1 \left[x_1 + \frac{u_2}{u_1} \right]^2 + x_2^2 = \text{const}, \tag{22}$$

$$p_1^2 + u_1 p_2^2 = \text{const}. \tag{23}$$

Defining the ratios

$$\alpha_m = \frac{u_{20}}{u_{1m}}, \quad \alpha_M = \frac{u_{20}}{u_{1M}} \tag{24}$$

and taking $u_1 > 0$ into account the candidates for the optimal trajectories (22) in the x_1, x_2 plane consist of ellipses with the centers $Q_1(\alpha_m, 0)$, $Q_2(\alpha_M, 0)$, $Q_3(-\alpha_M, 0)$, $Q_4(-\alpha_m, 0)$.

Considering the family of optimal trajectories (22) and the optimal control laws (19), (20) it follows that for the last step either the controls $u_1=u_{1m}$ and $u_2=-u_{20}$ or the controls $u_1=u_{1m}$ and $u_2=+u_{20}$ are possible. The first combination corresponds to the target trajectory for $x_1>0$ and the second combination to the target trajectory for $x_1<0$. Knowing that the optimal trajectories in the state plane are symmetric with respect to the origin the optimal synthesis can be given.

5. Optimal Synthesis

Considering the costate trajectories (23) we see that the target trajectories are two half ellipses $E1$ and $E2$ with the centers Q_1 and Q_4 as shown in Figure 2. Then, originating from points on these two half ellipses the switching curve $p_2(t)=0$ and thus the optimal synthesis can be obtained by backwards integration. The switching curve $p_2(t)=0$ consists of increasing arcs ending on the x_1 axis at points with the distances σ_ν ($\nu=0, 1, 2, \ldots$) from the origin of the state plane. Only the two arcs $E1$ and $E2$

are ellipses. A detailed investigation shows that the distances σ_ν satisfy the recurrence formula

$$\sigma_{\nu+1} = \alpha_m \left\{ 1 + \sqrt{ 1 + \frac{(\sigma_\nu + \alpha_M)^2}{\alpha_m \alpha_M} - \frac{\alpha_M}{\alpha_m} } \right\}, \quad \nu = 0, 1, 2, \ldots \quad (25)$$

$$\sigma_0 = 0$$

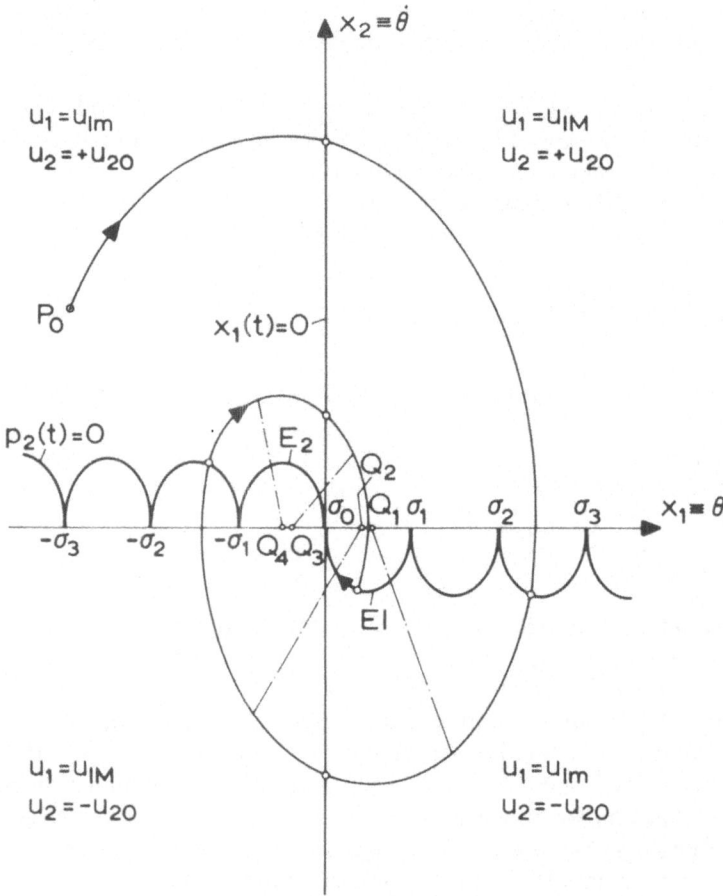

Fig. 2.   Optimal synthesis.

The two switching curves $x_1 = 0$ and $p_2 = 0$ separate four regions of different combinations for the optimal controls as shown in Figure 2.

Finally, in Figure 2 for one given initial point $P_0 = P_0(x_{10}, x_{20})$ the optimal trajectory is drawn and the centers of corresponding pieces of ellipses are indicated.

6. Conclusions

Relationship (25) and thus the optimal synthesis leads to the following result: The number of switchings of the controls and, consequently, the duration for the alignment of the satellite depend on the ratios α_m and α_M only. This is obvious since, for example, for increasing values of α_m and α_M the arcs of the switching curve $p_2 = 0$ also increase. It means that the number of switchings of the controls is reduced and the duration for the alignment becomes essentially smaller. Physically, this corresponds to increasing influence of the aerodynamic torque. For different ratios α_m and α_M optimal syntheses and typical optimal trajectories are shown in Figure 3. In all three

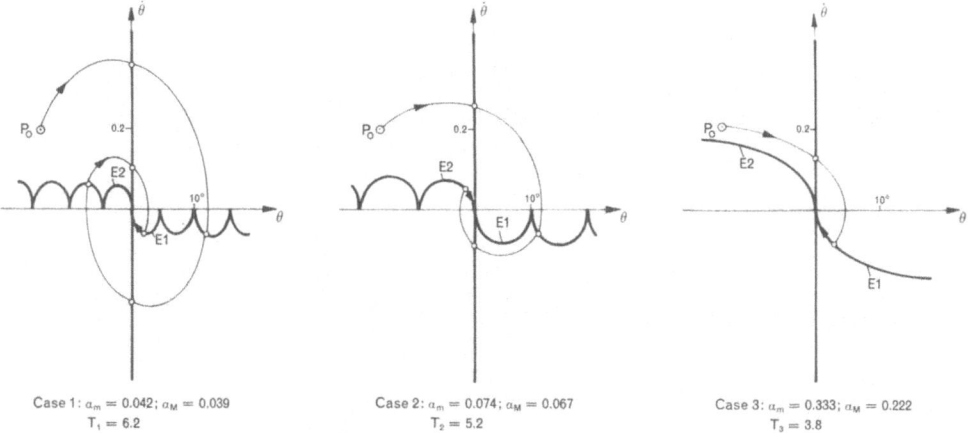

Case 1: $\alpha_m = 0.042$; $\alpha_M = 0.039$
$T_1 = 6.2$

Case 2: $\alpha_m = 0.074$; $\alpha_M = 0.067$
$T_2 = 5.2$

Case 3: $\alpha_m = 0.333$; $\alpha_M = 0.222$
$T_3 = 3.8$

Fig. 3.   Optimal synthesis and typical trajectories for different ratios

$$\alpha_m = \frac{u_{20}}{u_{1m}} \quad \text{and} \quad \alpha_M = \frac{u_{20}}{u_{1M}}.$$

cases the same initial point P_0 is chosen to demonstrate the reduction of the durations T_i ($i = 1, 2, 3$) for the alignment of the satellite.

The optimal synthesis of case 3 in Figure 3 is of mathematical interest inasmuch as it shows the tendency that for negligible gravitational torque – for example very large panels – the optimal solution of the stated problem is given by the time-optimal solution for the double integrator plant (the switching line $x_1 = 0$ vanishes and $E1$ and $E2$ become parabolas).

Naturally, for a technical realization the instantaneous changes of the satellite configuration as required by the bang-bang control laws are impossible. However, these control laws are in close agreement with reality, since the motion of the satellite about its center of mass is slow compared to the time needed for changing the position of the tip masses and of the panels.

Additionally, from a practical point of view the proposed attitude control has the great advantage that no mass is lost.

References

[1] Beletskii, V. V., *Motion of an Artificial Satellite about Its Center of Mass* (transl. from Russian), Israel Program for Scientific Translations, Jerusalem, 1966.

[2] Lunderstaedt, R. and Mesch, F., in Aseltine, J. A. (ed.), *Automatic Control in Space*, Proceedings of the 3rd International Conference on Automatic Control in Space, a Publication of the IFAC, Distributed by Instrument Society of America, Pittsburgh, Penns., (1970), 206.

[3] Frik, M. A., *AIAA J.* **8** (1970), 1780.

[4] Hiller, M. H. and Sagirow, P. A., in Aseltine, J. A. (ed.), *Automatic Control in Space*, Proceedings of the 3rd International Conference on Automatic Control in Space, a Publication of the IFAC, Distributed by Instrument Society of America, Pittsburgh, Penns., (1970), 572.

[5] Shidlovskiy, V. P., *Introduction to the Dynamics of Rarefied Gases*, Elsevier Publishing Comp., New York, 1967.

[6] Mohler, R. R. and Rink, R. E., 'Multivariable Bilinear System Control', Preprints of IFAC Symposium on Multivariable Control Systems, Düsseldorf, 1968.

[7] Hofer, E. P., 'Zur Theorie und Anwendung optimaler bilinearer Systeme', Ph.D. Thesis, University of Stuttgart, 1970.

[8] Pontrjagin, L. S., Boltjanskij, V. G., Gamkrelidze, R. V., and Misčenko, E. F., *Mathematische Theorie optimaler Prozesse*, Oldenbourg, 2nd improved edition, 1967.

B. FLUID MECHANICS ASPECTS OF SPACE FLIGHT

HEAT TRANSFER AND PRESSURE DISTRIBUTION
ON SHARP AND FINITE BLUNTNESS BICONIC AND
HEMISPHERICAL GEOMETRIES AT VARIOUS ANGLES
OF ATTACK IN A MACH 15–20 FLOW*

B. E. RICHARDS

von Karman Institute for Fluid Dynamics, Rhode-Saint-Genese, Belgium

V. DICRISTINA

AVCO Corporation, Wilmington, Mass., U.S.A.

and

M. L. MINGES

U.S. Air Force Materials Laboratory, Dayton, Ohio, U.S.A.

Abstract. Heat transfer and pressure measurements and schlieren flow visualization photographs have been made on five vehicle nose shapes tested in the V.K.I. Longshot free piston wind tunnel at flow Mach numbers of 15 and 20. The models tested were two hemisphere configurations with a smooth and a chemically etched rough surface, and three 50°–8° half-angle biconic configurations with sharp- and blunt-nosed smooth surfaced versions and a sharp-nosed version with a heavily machined roughness. Comparisons have been made of the measurements of heat transfer rate with currently used engineering predictions of heat transfer rate. Measured pressure distributions and shock shapes agreed well with proven inviscid theories. The boundary layer measurements on the smooth models at $M = 20$ conformed to laminar behavior. Model surface roughness appeared to promote turbulent flow. The shock-wave on the 50° half-angle cone at 10° incidence remained unexpectedly attached to the nose even though subsonic flow was formed on the windward side.

1. Introduction

In the design of thermal protection components for re-entry vehicles, it is critical to have dependable heat transfer coefficient correlations available over the range of conditions encountered. Before tackling the problem of understanding the processes in for instance a highly blown boundary layer with ablative thermal protection, it is necessary to correlate basic pressure and heat transfer results under non-blown hypersonic flow conditions. However, there is a lack of dependable pressure distribution and heat transfer measurements on entry vehicle frontal sections under hypersonic freestream conditions with which to test analyses. Data is required particularly within the most critical regions of sub-orbital or orbital re-entry which are defined by maximum vehicle deceleration and maximum surface heating. These regions occur at aerodynamic conditions of high Reynolds number between Mach numbers of 15 and 20.

A test series, in which were taken pressure distribution, heat transfer rate distribution and Schlieren photograph flow visualization measurements on several axisym-

* This research has been sponsored in part by the Air Force Materials Laboratory through the European Office of Aerospace Research, OAR, United States Air Force, under Contract No. F61052-70-C-0031.

metric bodies, was conducted in the nitrogen test flow of Longshot. The models included: (1) two metallic hemisphere bodies with nose radius $R_N = 3.5$ in., one model having a smooth surface, the other having a uniform surface roughness (0.001 in.) typical of an ablated graphite material; (2) two sharp-nosed metallic 50°–8° biconic model with $R_N = 0.75$ in. and a smooth surface.

The experimental data recorded in this study is compared with current engineering theories suggested in a review of ablation phenomenology by Minges [1]. Laminar heat transfer rates are compared with the reference enthalpy method of Eckert [2] and the local similarity solution of Lees [3] using the stagnation point correlation of Fay and Riddell [4]. The heat transfer measurements are also compared against two turbulent theories, the Sommer and Short reference enthalpy method [5] and the semi-empirical method of Spalding and Chi [6]. All these theories have proven useful in many compressible flow applications and the interest here is to examine their validity under more severe conditions. For instance the boundary layer on a 50° half-angle cone tested in the $M = 20$ flow of Longshot develops in an $M = 1.5$ flow-field with a stream temperature of the order of 2300 K and a wall temperature of 300 K. These conditions are outside the range in which these empirical theories were originally correlated or verified.

2. Model Instrumentation and Calibration

The five models supplied by AVCO Corporation were fitted with heat flux gauges and pressure taps. Ten heat gauges were mounted axially along the model surface beginning at or near the geometric stagnation point. Seven pressure taps on each model were similarly spaced along the surface but at 180° around the model from the row of heat transfer gauges. The heat sensor was a 0.004 in. copper disc on the back face of which a chromel-alumel thermocouple junction approximately 0.0015 in. diameter had been welded using 0.001 in. wires. The discs had diameters of both 0.125 in. and 0.110 in. and were bonded to an insulating holder. All gauges were contoured to the local body configuration. Output signals from the calorimeters were recorded on Tektronix oscilloscopes after pre-amplification. Typical heat flux assemblies were calibrated in the AEDC (U.S.A.) radiant heat flux calibration facility. The purpose was to obtain for a given set of standard material properties, the average gauge assembly effective thickness to be used in the reduction analysis of the heat flux data. Of six gauges tested, this average thickness was found to be 5% below the actual disc thickness. The maximum deviation, however, varied from $+4.8\%$ to -12.6% in actual thickness.

The transducers used for sensing pressures from the models and from a Pitot probe situated near the models were Hidyne types W and HR variable reluctance diaphragm differential pressure gauges. The reference side of the transducer was kept at about 1 μ of mercury during a test. The lengths of the pipes to the pressure taps on the models were kept to a minimum such that the internal volume of the complete pressure system was small enough to give a pneumatic response which matched the electronic response time of less than a millisecond. Each complete measuring system was calibrated before each test against a mercury or a Betz-type water manometer with the reference side of the gauge at atmospheric pressure. The calibration pressure was applied to the transducers such that the diaphragm deflected in the same direction as in the test. The electronic amplification of each system was adjusted in each test such that the predicted pressure would deflect the oscillograph trace by 20 to 30 mm. The calibration plots were generally so linear that errors incurred in using a constant calibration factor were small.

The time dependent heat transfer rates are normally determined from the temperature-time trace by measuring the signal slope at the selected time and multiplying it by the instrumentation calibration factor. Because of the computational inaccuracies of numerical differentiation, a more satisfactory method was devised in which an equation for the expected qualitative heat transfer variation (estimated from tunnel calibrations) was integrated to generate a set of transient temperature readings which were compared to the experimental output of the heat gauge. Excellent re-construction of the thermocouple output could be achieved after several selections of the peak heat transfer rate had been made. The method is described in full in [8]. The accuracy of the pressure and heat transfer measurements are within $\pm 5\%$ and $\pm 10\%$ respectively. The accuracy of the heat transfer measurements is conditioned primarily by the difficulties in welding the thermocouples to the calorimeter discs, and was estimated from calibrations of sample gauges as described above.

3. Test Facility

The V.K.I. Longshot test facility was used for this program. Longshot differs from a conventional gun tunnel in that a heavy piston is used to compress the nitrogen test gas to very high pressures and temperatures [7]. The test gas is then trapped in a reservoir at peak conditions by the closing of a system of check valves. The flow conditions decay monotonically during 10 to 20 ms running times as the nitrogen trapped in the reservoir flows through the 6° half-angle conical nozzle into the pre-evacuated open jet test chamber. The maximum supply conditions used in these tests are approximately 4000 atm. at 2400 K. These provide unit Reynolds numbers of 6×10^6 and 3×10^6 per ft at nominal Mach numbers of 15 and 20. Tests at $M = 20$ with a lower Reynolds number of 2×10^6 per ft were also employed in this series. The two Mach numbers were obtained at the 14 in. diameter nozzle exit plane by using throat inserts with different diameters.

4. Measurement Results

Typical pressure and heat transfer distributions measured on each of the five models are presented in Figures 1 to 8. A more complete record of the test results is given in [8]. The results are compared to appropriate theories more fully explained in the following section. Such comparisons have been made to assess the quality of the measurements and to illustrate the effects of conicity on them. Each model configuration is examined in turn.

A. HEMISPHERE MODELS

Figure 1 shows the pressure distribution for all the five test runs on the smooth and rough hemispheres, non-dimensionalized by an estimated stagnation point pressure. The results were compared with the theory of Belotserkovskii [9] with and without a correction for flow conicity. From examination of the theoretical curves, the correction for conicity can be seen to be important away from the nose of the model.

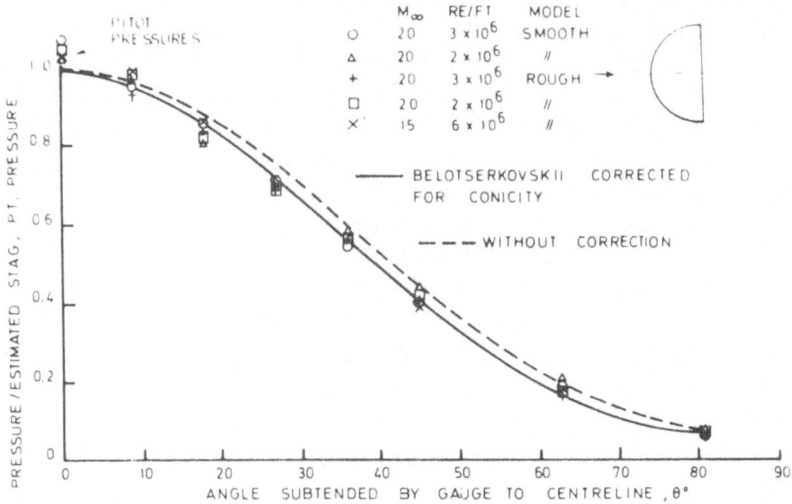

Fig. 1. Pressure distribution on hemisphere models.

The experimental spread of the non-dimensionalized data lay within a region of 3% of the stagnation point pressure and was in excellent agreement with the conicity-corrected theory. Typical heat transfer rates measured on the smooth hemisphere are shown in Figure 2 compared with the laminar similarity theory of Lees [3] corrected both for conicity and for Belotserkovskii instead of a Newtonian pressure distribution. Within the first 30 deg from the stagnation point, the measured values were higher than the theoretical value. This discrepancy, as yet unexplained, is contrary to tests on the blunt biconic model and in other reported tests [7]. At angles larger than 30°, the measured values were smooth and agreed well with laminar theory, as expected.

B. SMOOTH SHARP-NOSED MODEL

Figure 3 shows pressure measurements for the model at zero incidence, non-dimensionalized with respect to the freestream dynamic pressure in the plane of the nose and compared to Newtonian and tangent cone theories. These theories were corrected for source flow effects by including both changes of flow angularity at locations away from the nozzle center-line and changes in streamwise flow parameters, as was done in theories used for hemispheres. Numerical extrapolations of the tables of Jones [10]

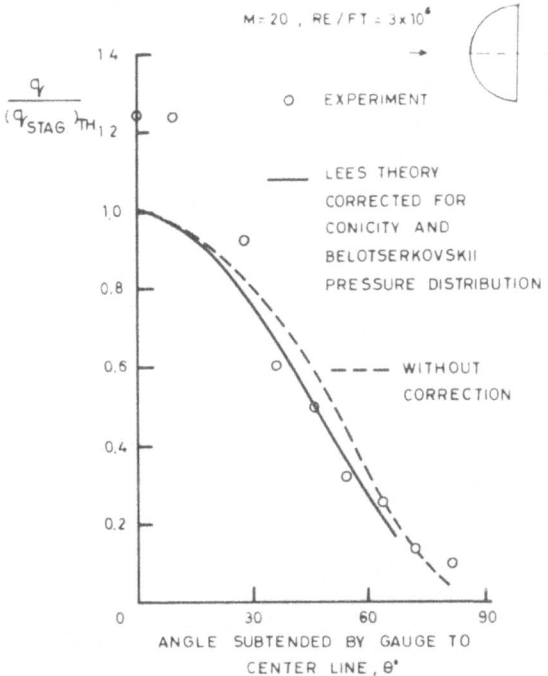

Fig. 2.   Heat transfer on smooth hemisphere.

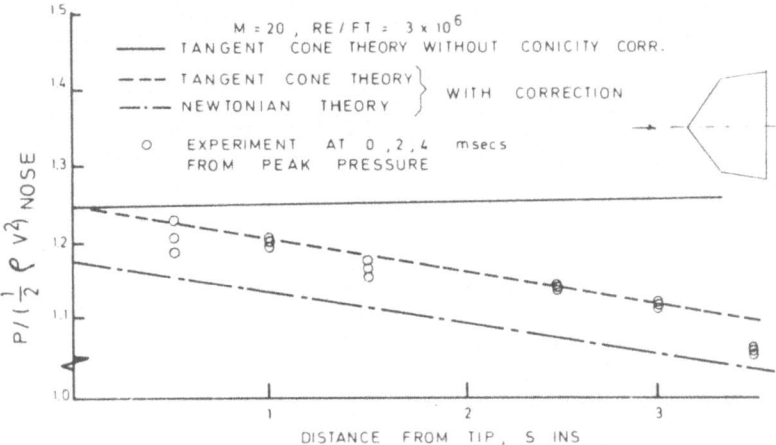

Fig. 3.   Pressure distribution on smooth pointed biconic model at zero incidence.

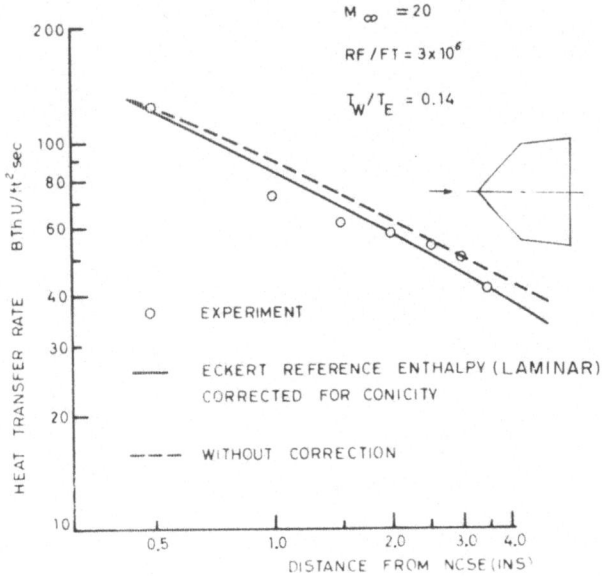

Fig. 4.   Heat transfer on smooth sharp-nosed biconic model at zero incidence.

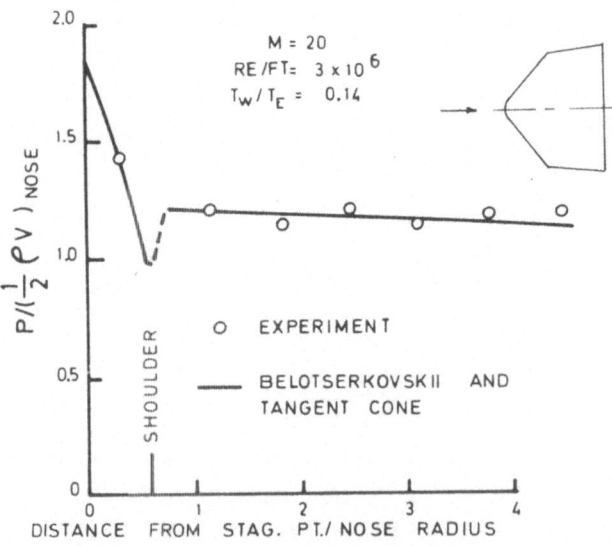

Fig. 5.   Pressure distribution on blunt-nosed model at zero incidence.

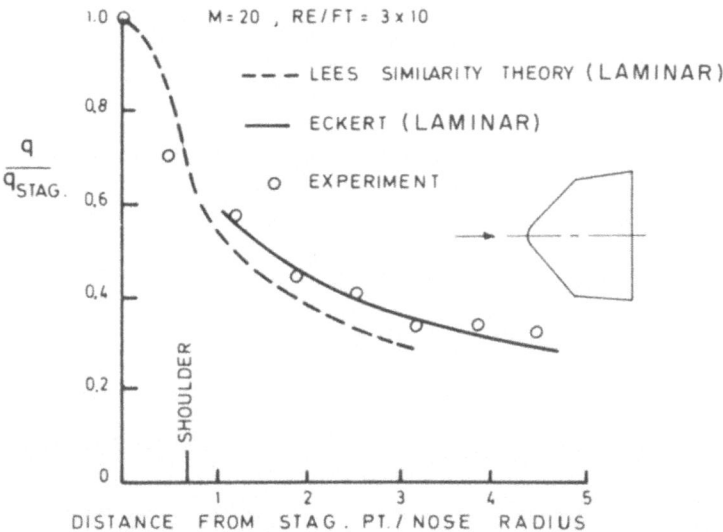

Fig. 6.   Heat transfer distribution on blunt-nosed model at zero incidence.

from cone half-angles of 40° were used for calculating the tangent cone theory. The conicity correction was as high as 10% at the end of the 50° cone surface. The measurements were smooth and agreed to within a few percent with the corrected tangent cone theory which itself predicted pressures roughly 5% above Newtonian theory. The heat transfer measurements for the same run is shown in Figure 4. Laminar Eckert reference enthalpy theory [2] was chosen to provide a qualitative comparison to the measurements. The measurements, if smoothed, gave the correct trend with theory; however, 'humps' existed which occurred consistently at the same position on the model for all the tests on this model. These 'humps' were not expected to be caused by freestream flow irregularities in view of the regularity of the pressure data, and were thought to be caused mainly by the neglect of individual heat gauge calibration.

C. SMOOTH BLUNT-NOSED BICONIC MODEL

A typical pressure distribution is shown in Figure 5 to agree well with Belotserkovskii theory (for the spherical nose surface) and tangent cone theory (for the cone) both being corrected for conicity. The flow appeared to recover from the over-expansion on the spherical surface before reaching the first gauge on the conic surface. The data scatter was small. Heat transfer rates shown in Figure 6 were reasonably smooth and consistent with Eckert theory. Stagnation point heat transfer rates agreed to within 5% of the theory of Fay and Riddell [4].

D. MACRO-ROUGHNESS BICONIC MODEL

The pressure measurements on this model are compared with smooth-model data in Figure 7. The large data scatter was probably due to viscous-inviscid interaction of

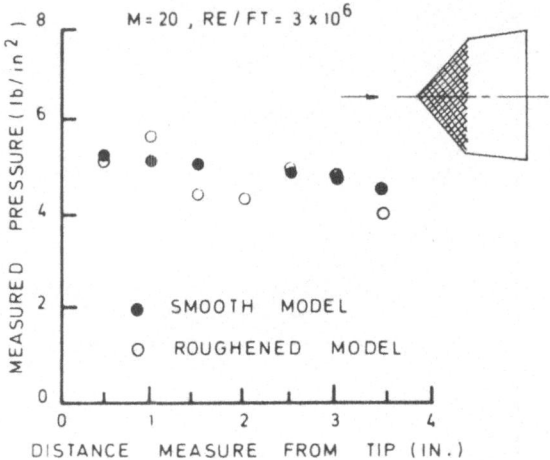

Fig. 7.   Pressure distribution on rough surfaced biconic model.

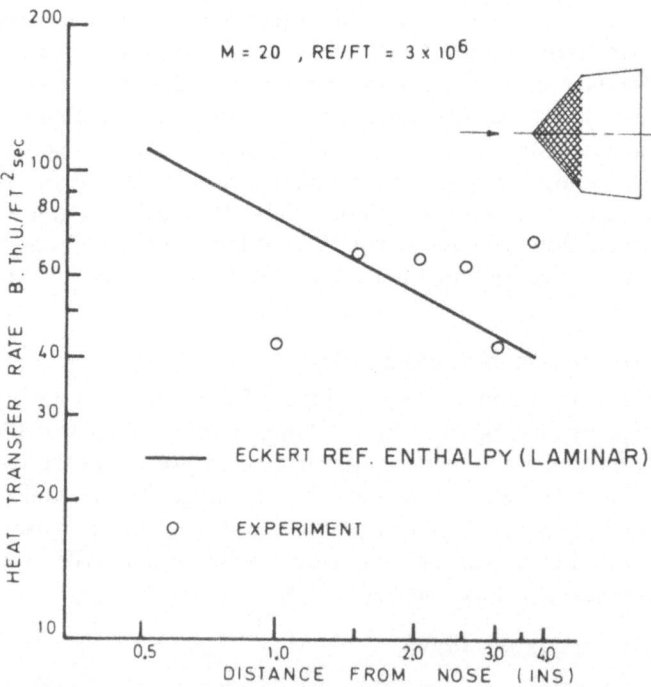

Fig. 8.   Heat transfer distribution on rough surfaced biconic model at zero incidence.

the large roughness elements with the flow since the pressure tapes were located in arbitrary positions with respect to the roughness elements. This latter result illustrated that extreme care must be taken in selection and location of instrumentation in order to gain illuminating data from rough-surfaced models. The heat transfer data shown in Figure 8 exhibits similar scatter. Although in this case the instrumentation was more carefully positioned.

From these assessments of data, it is concluded that the pressure measuring technique used produces results which show little data scatter and are consistent with well-proven prediction methods. It is likely that the larger data scatter observed with the heat transfer measurements can be reduced by individual calibration of the gauges. Conicity corrections of up to 10% of measured quantities were found necessary; however the simplified correction procedures employed appear to adequately describe the flow behaviour. It can be interpreted from this conclusion that conicity does not measurably affect the flow features under examination.

5. Discussion of Results

A. PRESSURE DISTRIBUTION AND SHOCK SHAPES

The measurements that are the most accurate and that would be expected to agree with well-proven and insensitive predictions are those of pressure and shock shape. A simplification in application of theory lay in the boundary layer thickness being so

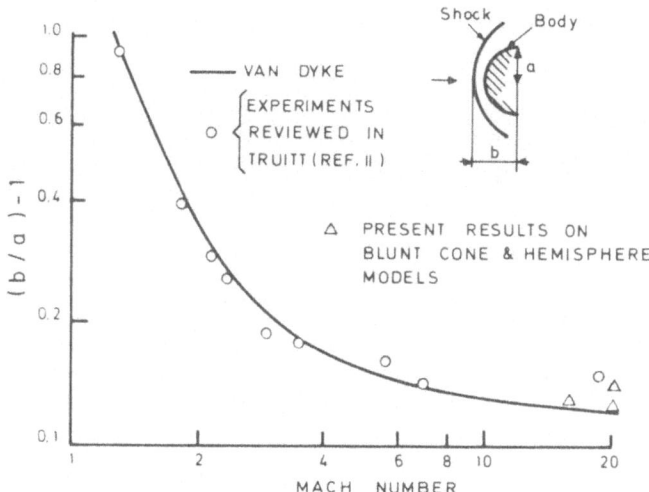

Fig. 9. Shock stand-off distances.

insignificant under these test conditions that the flow was considered inviscid. Success in such comparisons provided the ground work for the more interesting but less well understood viscous phenomena of convective heat transfer. It was shown in the last section and in further tests on cone models at $\pm 10°$ incidence [8] that pressure measurements agreed satisfactorily well with theory. Also in [8], it was shown that the shock angles to all the sharp and blunt cone surfaces (in the latter case, far down-stream from the blunt nose) at angles of attack 0° and ±10° examined agreed with cone tables of [10] to within ½ deg. The measured shock stand-off distance for each test on the blunt models (biconic and hemisphere) is shown to compare well with the theory of Van Dyke and other experiments [11] in Figure 9.

An interesting flow feature brought out from the flow visualization over the 50° half-angle surfaces of the sharp biconic model at a positive angle of attack to the stream (i.e., surface at 60° to the flow) was that the shock remained attached to the nose, despite the fact that a 60° half-angle cone in a Mach 20 flow would have a

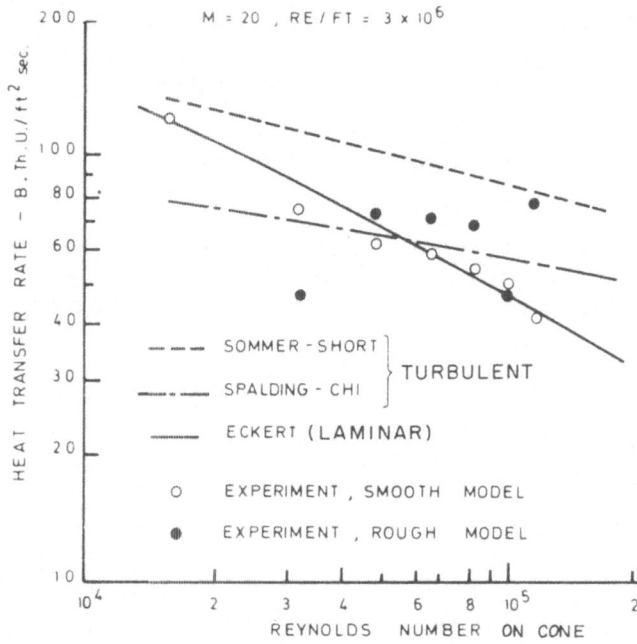

Fig. 10.   Heat transfer on sharp-nosed biconic model – comparison with theory.

detached shock. The flow behind the attached shock on the windward side was sub-sonic, while flow on the 'leeward side' was supersonic. This implies that there oc-curred a sonic line (or surface) similar to that around a cylinder (at right angles to the flow) or sphere as the flow expanded around the cone. At even higher angles of attack, the shock would undoubtedly detach from the nose forming a third flow phase. As yet, no measurements (and it is believed no calculations) have been made to locate the boundaries of these flow phases.

B. HEAT TRANSFER ON SMOOTH AND ROUGH BICONIC MODELS

In Figure 10 three heat transfer theories are compared with experimental results on the smooth sharp-nosed biconic model at zero incidence in a Mach 20 flow. These

are the laminar Eckert theory, and the turbulent Sommer and Short reference enthalpy and Spalding and Chi semi-empirical theories. It was assumed for the turbulent
theories that the virtual origin of the turbulent boundary layer occurred at the nose
of the body. The measurements were generally in agreement with the laminar theory,
although many of the data points also agreed well with the Spalding-Chi theory.
The proximity of the laminar and turbulent predictions was a reflection of the fact
that the Reynolds number on the surface is so low* that turbulent flow was unlikely
to be naturally achieved. The same conclusion was made in [8] for all the tests on
both the pointed and blunt smooth surfaced models at $M=20$ at all incidences
examined. Data from the rough model was plotted on the same figure. Ignoring the
two very low data points (since these could have been caused by a separated flow)

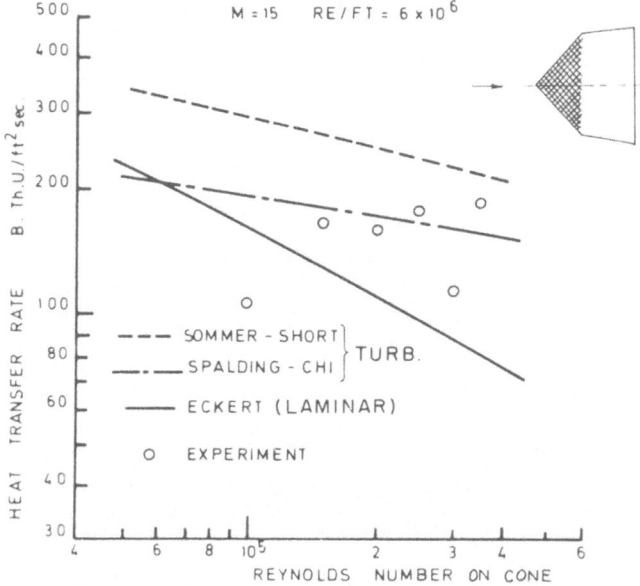

Fig. 11.   Heat transfer on rough-surfaced biconic model at $M=15$.

the measurements lay above the smooth model results. The spread of the two turbulent
heat transfer predictions encompassed portions of the data as well as portions of the
laminar curve thus making it difficult to assess whether the boundary layer was
laminar or turbulent. One interpretation is that the boundary layer on the rough
model was turbulent, however it is not unknown that high values of heat transfer
can be caused by roughness elements while the flow remains laminar. Richards [12]
has put forward the explanation that roughness elements can transfer, by mixing processes, high energy flow from the outer part of the laminar boundary to the layers
immediately adjacent to the surface thus causing higher heat transfer rates but without causing turbulent flow. A test carried out in the higher Reynolds number flow

* Because the high incidence of the surface to the free stream produced high static temperature
(~ 2000 K).

of Longshot at $M=15$ however showed more convincing evidence that the macro-roughness elements caused turbulent flow (Figure 11). The heat transfer results were more easily discernible to be in agreement with the turbulent Spalding-Chi theory. It is intended to carry out additional tests at an even higher Reynolds number at $M=15$ on the smooth model when it is expected that measurements of naturally occurring turbulent boundary layers will help to confirm this conclusion.

C. HEAT TRANSFER ON SMOOTH AND ROUGH HEMISPHERE MODELS

Measurements of heat transfer on the chemically etched rough hemisphere body at $M=15$ are illustrated in Figure 12. These rough body results again showed consider-able scatter (as in similar tests at $M=20$). Higher non-dimensional heat transfer

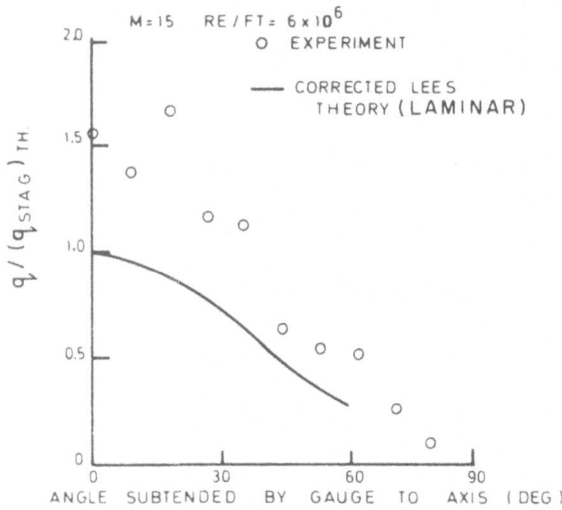

Fig. 12.   Heat transfer distribution on roughened hemisphere at $M=15$.

rates than on the smooth model have been measured which may be attributed to turbulent boundary layer behavior. In this case the roughness elements were smaller than the boundary layer thickness, and the scatter is ascribed to the mixing process in the boundary layer caused by the elements.

6. Conclusions

The combination of the Longshot facility and its instrumentation has generated good quality pressure, heat transfer and shock shape data on models resembling entry vehicle frontal sections at the high Reynolds number, high Mach number conditions necessary to simulate the critical region of atmospheric re-entry. Source flow effects due to the use of a 6° half-angle conical nozzle were not negligible; however, for these model configurations, they appeared not to be large enough to obscure the main features of the flow under examination. Simple corrections to the theories adequately compensated for the observed discrepancies.

Pressures and shock shapes were in good agreement with well proven inviscid theories. This was in accord with the information that the boundary layer thickness on these surfaces at high incidences to the flow was negligibly small. The shock shapes on biconic models at incidence indicated that the flow on the windward side of the model was subsonic even though the shock remained attached at the nose. This suggests a flow phase unknown to the authors which deserves further study.

The heat transfer measurements on the smooth bodies at $M = 20$ were representative of those expected of a laminar boundary layer. Roughness could have caused turbulent flow at the $M = 20$ cases; however, the turbulent predictions of heat transfer were too close to laminar theory to differentiate the behavior of the data in this flow region. The turbulent theories differed as much between themselves as between the laminar theory. At $M = 15$, when the Reynolds number was higher, the heat transfer results on the roughened model suggested that a turbulent boundary layer was present.

Acknowledgements

The authors wish to thank Mr Jos Slechten of V.K.I. and Dott. Salvatore Culotta of the University of Palermo for their inputs into this programme. The earlier work of Professor Kurt Enkenhus (now of NOL, Maryland, U.S.A.) on the Longshot contributed greatly to the success of the testing capabilities. Mr Jean Huge and Mr Fernand Vandenbroeck of V.K.I. ensured the smooth running of the Longshot.

References

[1] Minges, M. L., *High Temperature – High Pressures* 1 (1969), 607.
[2] Eckert, E. R. G., 'Survey on Heat Transfer at High Speeds', University of Minnesota, ARL 189, Dec. 1961.
[3] Lees, L., *Jet Propulsion* 26 (1956), 259.
[4] Fay, J. A. and Riddell, F. R., *J. Aerospace Sci.* 25 (1958), 73.
[5] Sommer, S. C. and Short, B. J., 'Free Flight Measurements of Turbulent Boundary Layer Skin Friction in the Presence of Severe Aerodynamic Heating at Mach Numbers from 2.8 to 7.0', NACA TN 3391, 1955.
[6] Spalding, D. B. and Chi, S. W., *J. Fluid Mech.* 18 (1964), 117.
[7] Richards, B. E. and Enkenhus, K. R., *AIAA J.* 8 (1970), 1020.
[8] Richards, B. E., Culotta, S., and Slechten, J., 'Heat Transfer and Pressure Distributions on Re-Entry Nose Shapes in the VKI Longshot Hypersonic Tunnel', A.F.M.L. Report No. 71–200, June 1971.
[9] Hayes, W. D. and Probstein, R. F., *Hypersonic Flow Theory*, 2nd ed., Vol. 1, Academic Press, 1966, p. 423.
[10] Jones, D. J., 'Tables of Inviscid Supersonic Flow About Circular Cones at Incidence $\gamma = 1.4$', AGARDograph 137, November 1969.
[11] Truitt, R. W., *Hypersonic Aerodynamics*, The Ronald Press Company, 1959, p. 269.
[12] Richards, B. E., *Aeronaut. Quart.* 18 (1967), 237.

HYPERSONIC LOW TEMPERATURE ABLATION
AN EXPERIMENTAL STUDY OF
CROSS-HATCHED SURFACE PATTERNS

HANS W. STOCK* and JEAN J. GINOUX**

von Karman Institute for Fluid Dynamics, Rhode-Saint-Genèse, Belgium

Abstract. Cross-hatching has been studied experimentally at a free stream Mach number of 5.3, using two different low temperature ablative materials, camphor and wax, which respectively sublime and liquefy.

The surface pattern parameters (i.e., the cant angle ϕ and the streamwise spacing λ) have been correlated with flow field properties. The effect of the exposure time under ablation conditions has been studied intensively. The present results are compared with other available data.

A brief review of previous experimental work on cross-hatching is included.

List of Symbols

M	Mach number
p	Pressure
Re	Reynolds number
T	Temperature
t	Time
α	Angle of attack
ε	Expansion angle
λ	Streamwise spacing of the cross-hatched pattern
π	Streamline inclination of the inviscid flow
π^x	Pattern inclination angle
θ	Total cone angle
θ^x	Total flare angle
ϕ	Cant angle of the cross-hatched pattern
ϕ	Cone azimuthal angle, measured with respect to the most windward meridian

Subscripts

e	Conditions at the outer edge of the boundary layer
r	Recovery condition
ST	Stagnation condition
w	Wall condition
∞	Upstream infinity condition

1. Introduction

The heating rates during Earth atmosphere re-entry or in rocket motors are such that effective thermal protection techniques are required. Values of 100–10 000 Btu ft^{-2} s^{-1} (0.113–11.3 kW cm^{-2}) are encountered in atmosphere re-entry depending on trajectory and vehicle configuration [1].

---

* Research assistant.
** Professor, Brussels University and head of VKI department of supersonics and hypersonics.

L. G. Napolitano et al. (eds.), Astronautical Research 1971, 105–120. All Rights Reserved.

Ablation has been found to be satisfactory for thermal protection in high heating conditions of finite duration. The complex ablation phenomenon has been defined as 'a self-regulating heat and mass transfer process in which incident thermal energy is expended by sacrificial loss of material' [2].

Although the technological interest in ablation phenomena is relatively new, the ablative aspect of meteorites and tektites has been studied for decades. In particular, surface patterns created by differential mass transfer rates during the ablation period of earth atmosphere penetration were investigated by Chapman and Larson [3]. They showed that ablation patterns on tektites which exhibited ring-wave flow ridges were similar to patterns observed in laboratory experiments on hypervelocity ablation. The regmaglypt pattern on some wind tunnel models resembled the ablation surface structure on certain meteorites [4].

The investigation of surface ablation patterns detected on recovered re-entry bodies of controlled initial shape and wind tunnel models is more recent. Three different kinds of patterns were distinguished, streamwise grooves [4, 5, 6, 10] turbulent wedges [7, 9, 10] and cross-hatching.

This paper will be concerned only with the third type of surface pattern. Cross-hatching consists of nearly straight grooves of regular spacing, running obliquely to the external flow direction in both senses and crossing, thus producing a highly ordered diamond shaped pattern.

The important cross-hatching pattern parameters are:

(1) The cant angle ϕ; the half angle between a left and right running groove;

(2) The streamwise spacing or pattern wavelength λ; the streamwise length of a cell of the diamond shaped pattern.

The considerable interest in the study of surface patterns, and especially in cross-hatching, comes partly from the fact that such ablation surface structures produce instabilities in the rolling moment of slender re-entry bodies. The vehicles start to oscillate or to spin up. Consequently, a part of the research activity is concentrated on the technological problem of avoiding cross-hatching. Apart from this, the basic aspect of determining the physical mechanism which creates this phenomenon has been studied experimentally and theoretically.

2. Review of Experimental Work on Cross-Hatching

Canning et al. [12] gave perhaps the first insight into the cross-hatched pattern phenomenon. They used 30° half angle cones made of Delrin and Lexan in a ballistic range. Plexiglas models of a geometry which allowed a study of the effect of surface pressure gradients have been tested in a hypersonic facility by Canning et al. [7]. Larson and Mateer [4] did a systematic experimental study of the development of cross-hatching using cones of various angles, thus producing locally subsonic or supersonic Mach numbers. The ablation materials tested were Lexan and Lucite. Williams [6] measured the rolling moment coefficient on ablation camphor and korotherm cones. McDevitt [9] determined the oscillation and spinning frequency during ablation on ammonium chloride, camphor and korotherm cones in a hypersonic wind tunnel.

2.1. CROSS-HATCHING HAS BEEN OBSERVED FOR:

(1) Different flight conditions
    a. Re-entry   ⎫ High enthalpy⎫ Time integrated
       vehicles [4, 5].   ⎰ environment. ⎰ ablation process,
    b. Ballistic range   ⎫   flow parameters
       models [8, 12].   ⎰ Low enthalpy⎰ are time dependent.
    c. Wind-tunnel   ⎰ environment.
       models [4–7, 9, 10, 13–17]⎭
(2) All ablation modes
    a. Melting [4–8, 13, 15].
    b. Melting and vaporizing [12].
    c. Subliming [5, 6, 9, 10, 13–17].
    d. Charforming [5].
(3) Different types of materials
    a. Acrylics [4, 5, 7, 8, 12].
    b. Phenolics [5].
    c. Teflon [5, 14, 16, 17].
    d. Camphor [6, 9, 10, 13, 15].
    e. Wood [5].
    f. Wax [13, 15].
    Etc.
(4) Different types of model configuration
    a. Two-dimensional models [4, 5, 13].
    b. Axisymmetric models [4–10, 12, 13, 15–17].
    c. Inside circular tubes [14].

2.2. REQUIREMENTS FOR THE APPEARANCE OF CROSS-HATCHING

(1) Requirements for the gas flow
    a. Supersonic boundary layer [4].
    b. Transitional or turbulent boundary layer [4, 12, 13].
    c. The static pressure and/or the heat transfer must exceed critical values [6].
(2) Requirements for the body material
    a. For melters the Reynolds number of the liquid film $U_l h / \nu$ (U_l being the liquid
       velocity at the liquid-gas interface, h the height and ν the viscosity of the liquid
       film) must be below a certain value [17].
    b. Under ablation conditions the material must be sufficiently viscous (apparently
       for sublimers) [17].
(3) Requirements for the initial interface between the body and the boundary layer.
    By grooving the initial surface perpendicular to the main flow (sinusoidal perturba-
    tions) the pattern can be suppressed [17].

2.3. CROSS-HATCHING FORMATION IN WIND-TUNNEL TESTS

(1) The pattern is spatially fixed on the model surface for charforming ablators [5].
(2) The pattern moves slowly downstream on liquefying ablators [13, 15].

(3) The spacing of the pattern λ stays relatively constant in the streamwise direction on a given model, for either wedges or cones [5].

(4) The streamwise spacing λ and the cant angle ϕ are independent of the run time [5, 13].

(5) Cross-hatching appears nearly simultaneously over much of the model surface [5].

(6) Longitudinal grooves frequently develop upstream of cross-hatching [4–10, 12].

A review of theoretical work has been made in [13].

3. Experimental Technique

3.1. TEST FACILITY

The present tests were carried out in the hypersonic blow down facility at the von Karman Institute at a Mach number of 5.3. The test section has the dimensions 14 cm × 14 cm. The tunnel stagnation conditions are:

$$T_{\mathrm{ST}} = 390 - 620^\circ\mathrm{K}$$

$$p_{\mathrm{ST}} = 12 - 33 \ \mathrm{kg_f/cm^{-2}}$$

giving unit free-stream Reynolds numbers of

$$0.85 - 5.1 \times 10^7/\mathrm{m^{-1}}$$

3.2. MODELS

Cones of 10 to 62° total vertex angle, and 10° cones with 12 to 40° total angle flares were tested at zero angle of attack. For one model configuration, the influence of the angle of attack was examined. The dimensions of the models and the pointed steel noses are given in [13]. These pointed steel noses were used to avoid apex deformation by ablation. Some cones consisting entirely of ablation material were tested to study nose blunting effects. A few runs were made using flat plate models at various angles of attack.

3.3. ABLATION MATERIALS

Two ablation materials were tested, natural wax without seedings, which liquefied under test conditions without vaporizing, and camphor, a purely subliming material. Camphor models were manufactured by sintering the powdered material in a vacuum under high compression (500 $\mathrm{kg_f \ cm^{-2}}$) using the technique of Charwat [18].

4. Test Results and Discussion

4.1. FLOW FIELD PARAMETERS

In Figure 1 the flow field properties on cones for an upstream Mach number of 5.3 are shown as a function of the total cone angle θ. The local Mach number at the outer edge of the boundary layer M_e and the local static pressure p_e were calculated by the tangent cone method ignoring viscous interaction and surface deformation caused by ablation. The conical properties are also used for cone-flare configurations. The effects of overcompression and overexpansion of the inviscid Mach number and static pressure distribution are small at this high Mach number. Because of the presence of the boundary layer, the actual Mach number and static pressure distributions are even less different from the conical ones.

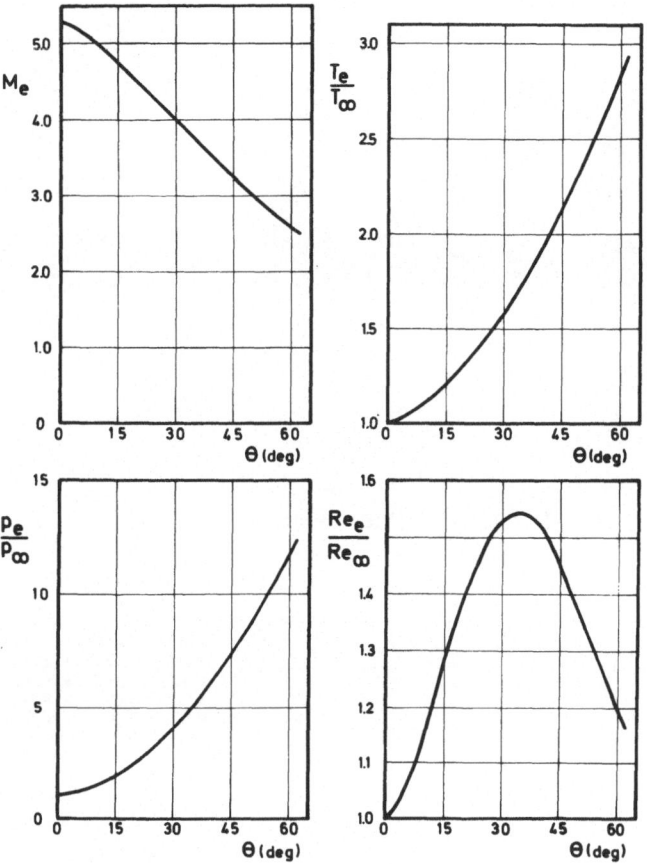

Fig. 1.   Flow field properties on cones for $M_\infty = 5.3$ (θ total cone angle).

4.2. CROSS-HATCHING PATTERN PARAMETERS

Typical test results are given in Figure 2 for axisymmetric and two-dimensional models. The cant angle ϕ and the streamwise spacing λ have been measured on photographs taken after the runs.

4.2.1. *Cant angle* ϕ

The cant angle ϕ is shown versus the local Mach number M_e in Figure 3 and compared with the Mach angle (solid curve). Available wind tunnel data and free flight data

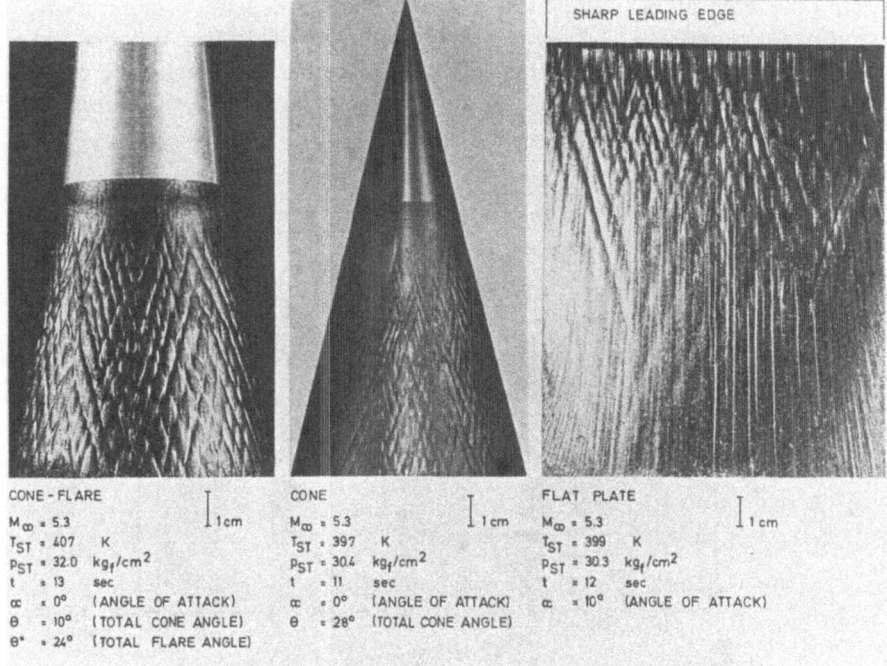

Fig. 2.   Typical test results on wax models.

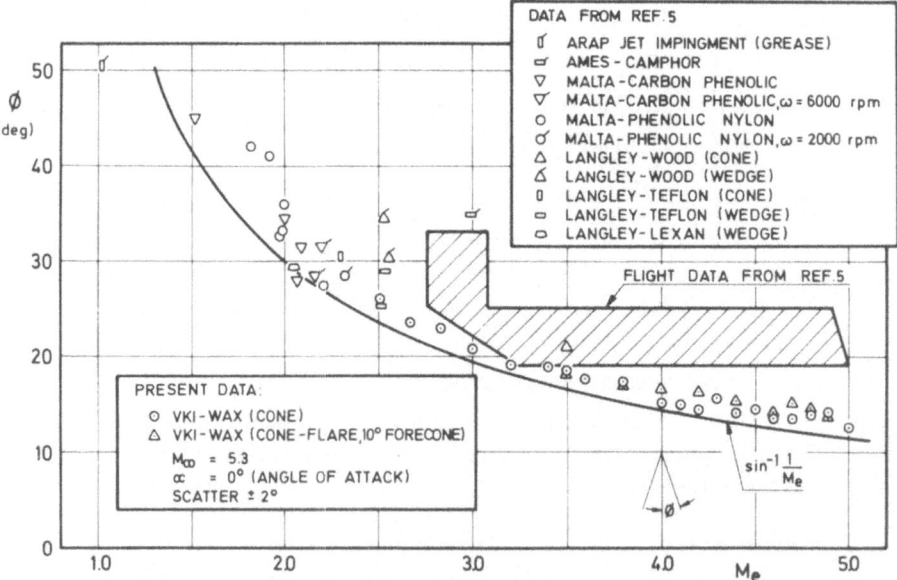

Fig. 3.   Influence of the local Mach number M_e on the cant angle ϕ.

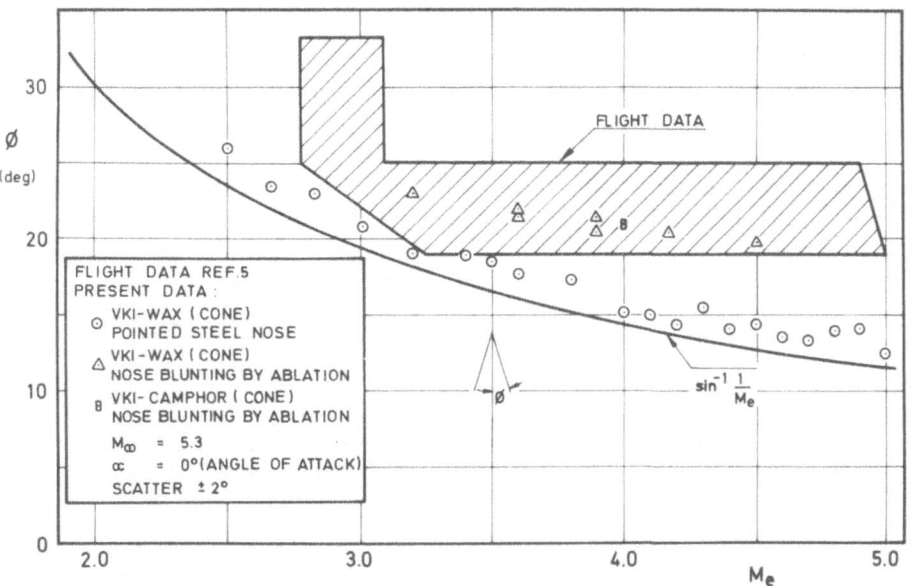

Fig. 4.   Influence of the Mach number M_e on the cant angle $\phi - M_e$ calculated for unblunted cones.

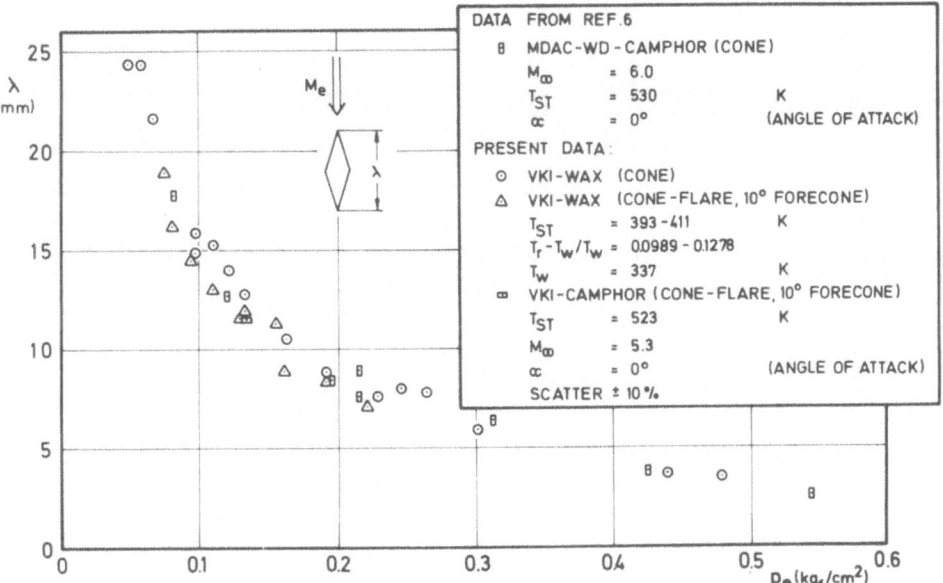

Fig. 5.   Influence of the local static pressure p_e on the streamwise spacing λ.

from [5] are shown for comparison. As seen, the present results follow the Mach angle trend in the Mach number range 2.5–5.0 contrary to the flight data for which 'freezing' was observed above a Mach number of 3. An explanation for the difference was suggested in [15], that nose blunting occurred in free flight, whilst wind tunnel tests were made with models having pointed noses.

Figure 4 shows the cant angle ϕ on self-blunting cones, consisting entirely of ablation material, versus the Mach number M_e, where M_e is calculated for sharp nose cones. Data for cones with pointed steel noses and flight data [5] are shown for comparison. The blunt nose data agree with free flight results and thus prove that the apparent

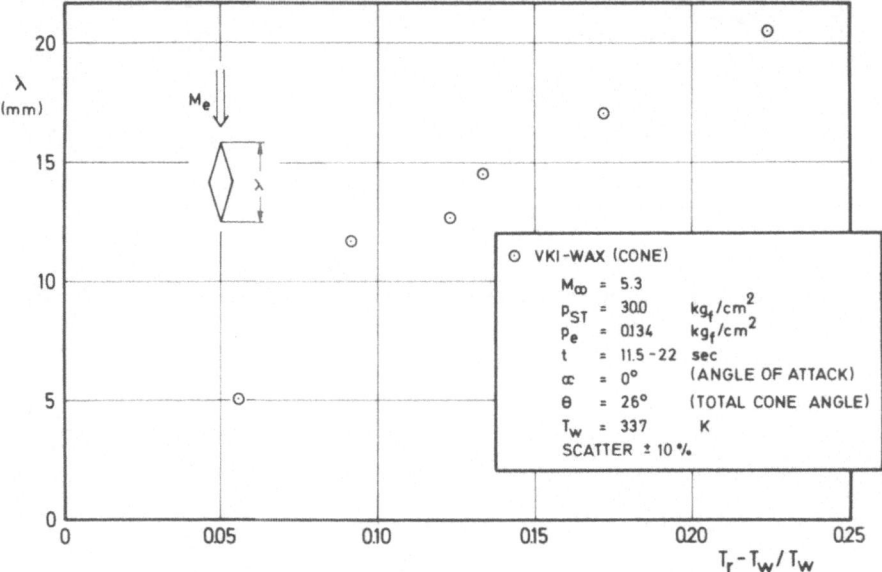

Fig. 6.   Influence of the driving temperature ratio $T_r - T_w/T_w$ on the streamwise spacing λ.

'freezing' of ϕ for Mach numbers M_e larger than 3 is due to nose blunting, i.e., improper use of M_e instead of the actual local Mach number on the blunt nose cones.

The cant angle ϕ seems to depend uniquely on the local Mach number. The unit free stream Reynolds number, Re_∞, and the Reynolds number based on conditions at the outer edge of the boundary layer, Re_e, the static pressure p_e, the driving temperature $T_r - T_w$ and the run time did not seem to have any influence. The recovery temperature T_r was calculated by assuming a turbulent recovery factor of 0.895. $T_w = 337$ K is the temperature at which the ablation material wax liquefies, independent of p_e.

4.2.2. *Streamwise spacing* λ

The effect of the local surface pressure p_e on the streamwise spacing λ is shown in Figure 5 for a nearly constant temperature ratio $T_r - T_w/T_w = 0.0989 \div 0.1278$ for wax models. The results agree quite well with those of Williams [6], extending the range to lower static pressures and greater values of the streamwise spacing. The surface pressure p_e was varied both by changing the cone or flare angle and thereby the local Mach number, and by altering for some tests the tunnel stagnation pressure.

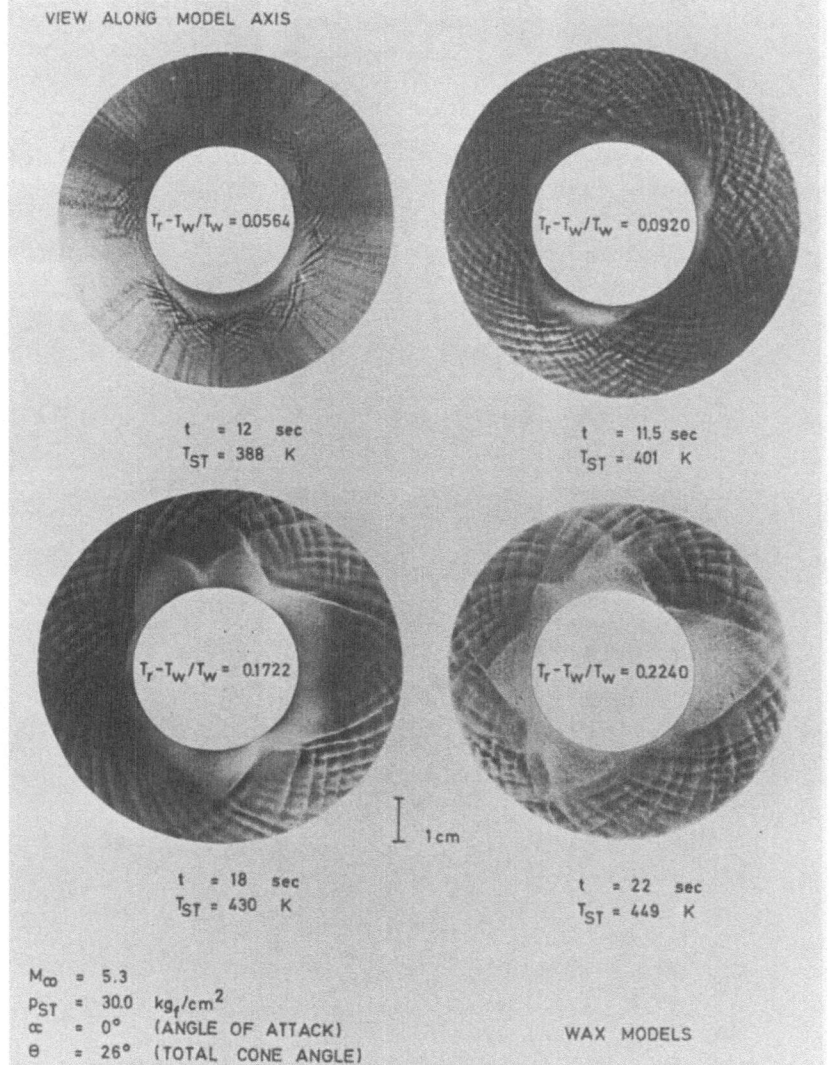

Fig. 7. Influence of the driving temperature ratio $T_r - T_w/T_w$ on the streamwise spacing λ.

The effect of the driving temperature ratio $T_r - T_w/T_w$ on the streamwise spacing λ measured on wax cones is shown in Figures 6 and 7 for constant values of M_e and p_e.

The local Mach number M_e, the Reynolds numbers, Re_∞ and Re_e, and the run time did not appear to have any influence on the spacing λ, when the static pressure p_e and

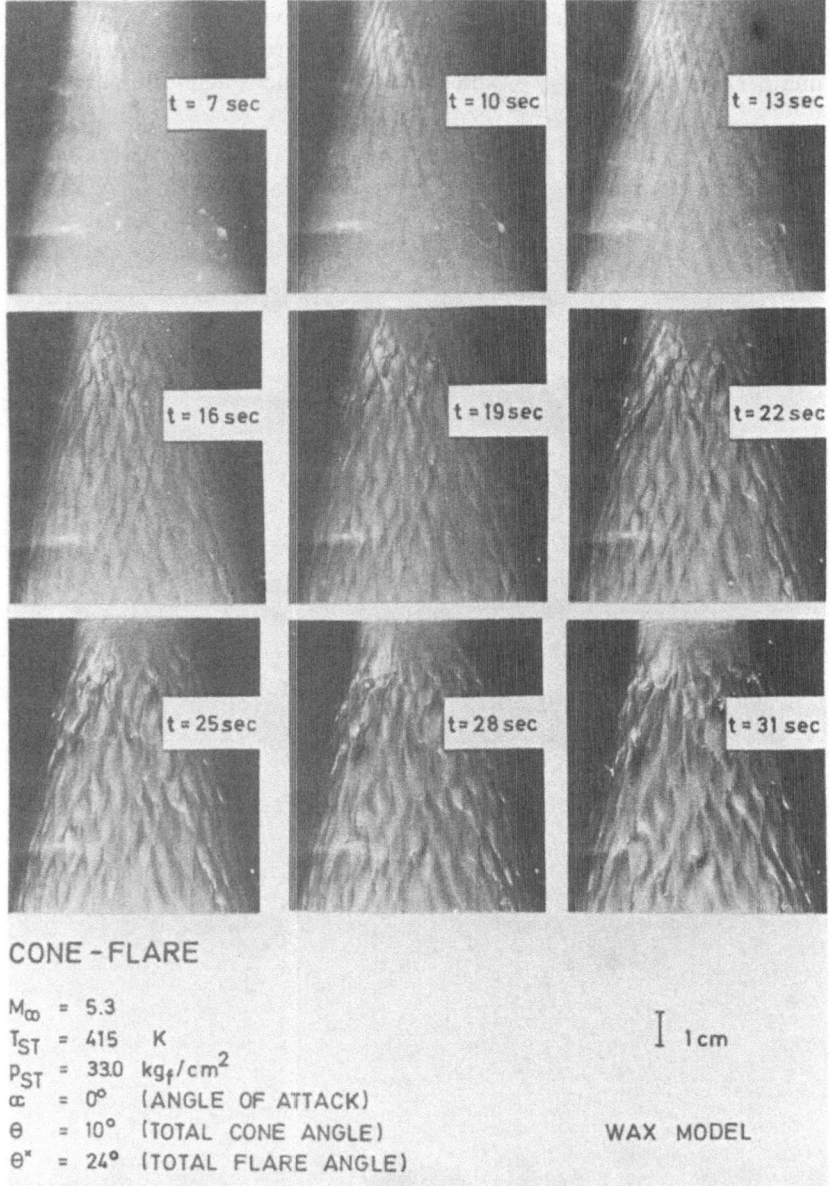

CONE - FLARE

M_∞ = 5.3
T_{ST} = 415 K
P_{ST} = 33.0 kg_f/cm^2
α = 0° (ANGLE OF ATTACK)
θ = 10° (TOTAL CONE ANGLE)
θ^x = 24° (TOTAL FLARE ANGLE)

I 1cm

WAX MODEL

Fig. 8.   Influence of the run time on the cross-hatching development.

the driving temperature ratio were held constant. No influence of nose blunting on λ could be seen, contrary to the effect on the cant angle ϕ. This may be due to the fact that the static pressure approaches its conical value after a distance of a few nose radii, whereas the local Mach number reaches its conical value only for large distances from the nose. The body size does not seem to be a scaling factor for λ. Williams [6] used models which were three times larger than those used at VKI, and for the camphor test no difference in λ could be seen, Figure 5.

4.3. RUN TIME

Figure 8 shows photos of the cross-hatching development reproduced from a film taken during the test. Several distinct time intervals can be defined corresponding to different stages in the pattern development on wax models:

1st time interval: $t_0 \rightarrow t_1$

    From tunnel start until the model surface reaches the liquefaction temperature and starts to ablate.

2nd time interval: $t_1 \rightarrow t_2$

    From the onset of ablation until the moment when cross-hatching starts to appear.

3rd time interval: $t_2 \rightarrow t_3$

    From the first appearance of cross-hatching until the surface pattern is fully developed, showing the maximum height difference between the bottom of the grooves and the enclosed hills.

4th time interval: $t_3 \rightarrow t_4$

    After being fully developed, the pattern starts to disintegrate showing a regmaglypt pattern resembling those on meteorites shown in [4].

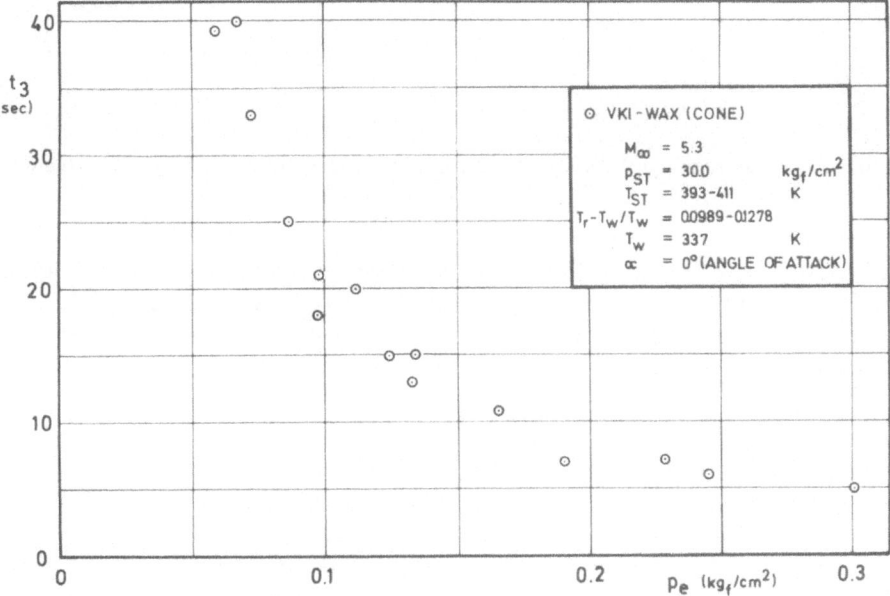

Fig. 9.   Run time required for a developed cross-hatched pattern as a junction of the local static pressure p_e.

In the test shown in Figure 8, cross-hatching started to appear at $t_2 = 7$ and was fully developed at $t_3 = 13$ to 16 s. The long run time photos show the pattern disintegration process. As may be seen during the time period t_2 until t_3, the cant angle ϕ and the streamwise spacing λ were independent of time.

It was found that the run time t_3 on wax models was dependent on the local static pressure p_e and the driving temperature ratio $T_r - T_w/T_w$. Figure 9 shows the run time t_3 as a function of the static pressure p_e, for a nearly constant temperature ratio $T_r - T_w/T_w = 0.0989 = 0.1278$. Figure 10 shows t_3 for constant static pressure p_e as a function of $T_r - T_w/T_w$.

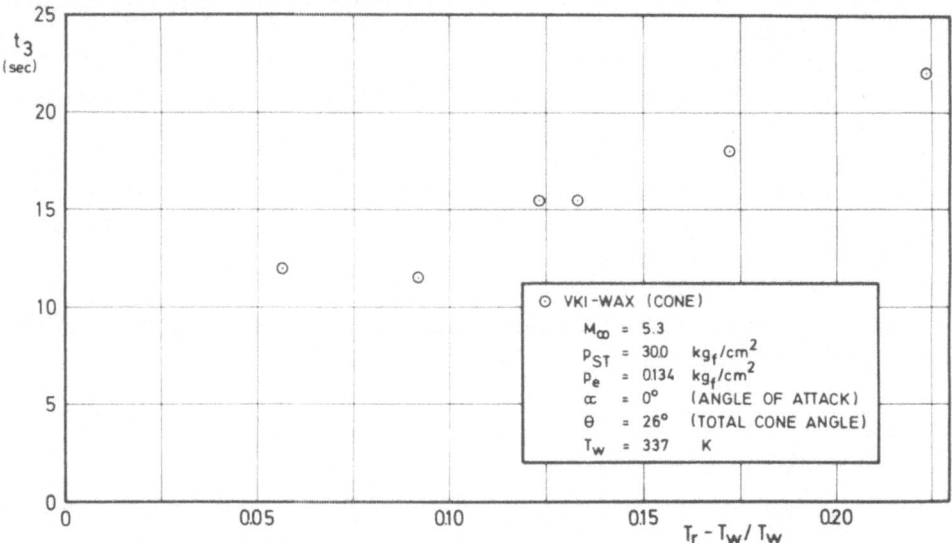

Fig. 10. Run time required for a developed cross-hatched pattern as a function of the driving temperature ratio $T_r - T_w/T_w$.

For a few tests with wax models (both axisymmetric and two-dimensional) a movie was taken during the complete testing period. It showed a slow 'creeping' motion of the whole cross-hatched pattern in the streamwise direction of the order of one streamwise spacing in 10 s.

4.4. EXPANSION CORNERS

The influence of an expansion wave on the development of cross-hatching by testing double cone models as sketched in Figure 11 was examined and the results were compared with that on a single cone.

The runs were stopped after 15 s, when a developed pattern appeared on the force-cone. As may be seen, the pattern on the after-cones does not appear for increasing expansion angles ε. This is a striking result of the influence of run time. Indeed, the static pressures get progressively lower on the after-cones, because of the expansion corners of increasing ε, resulting in longer required run times for the pattern development, according to Figure 9.

This is further demonstrated in Figure 12 which shows the results obtained by testing the same double cone model ($\varepsilon = 4°$) for 15 s and for 25 s, i.e., the run times t_3 required based on the fore-cone and after-cone static pressures respectively, according to Figure 9. On the short run time model cross-hatching appeared only on the fore-cone, the after-body being pattern free. In the longer run time case, the after-part is covered with cross-hatching. However, this pattern is less clear than on a single cone of 18°, which is shown for comparison. The reason is that the boundary layer flow is highly perturbed by the surface irregularities on the fore-cone when the pattern starts to develop on the after-body.

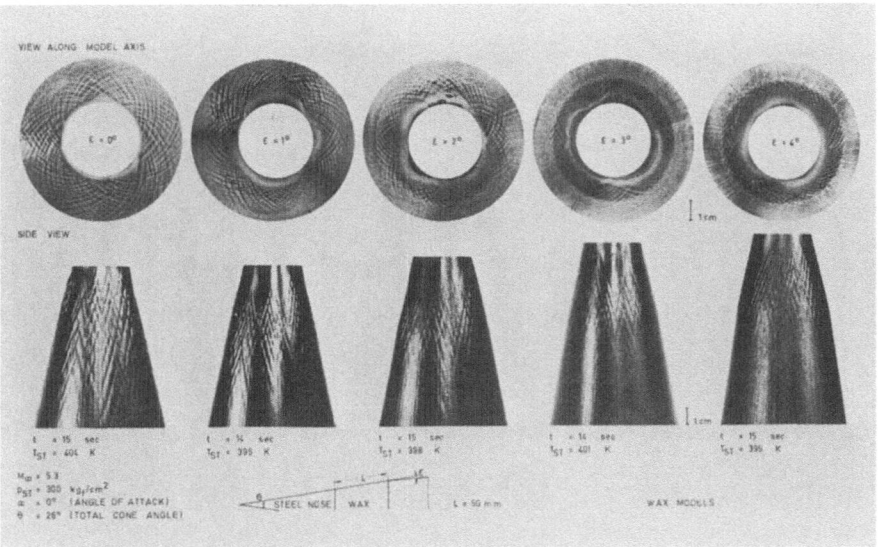

Fig. 11. Influence of flow expansion on the cross-hatching formation.

4.5. ANGLE OF ATTACK

Figure 13 shows the test results on cones at angle of attack from 0°–6° at 2° intervals. The tunnel was stopped when a marked pattern appeared on the windward side of the models, where the static pressure increases with increasing angle of attack, thus leading to a decreasing run time t_3, according to Figure 9.

On the model sides, the pattern became progressively inclined relative to the side-meridian, as the angle of attack was increased. (The pattern inclination or orientation is characterized by a line which bisects the angle between left and right running grooves.) This effect is caused by the inclination of the streamlines of the inviscid flow around cones at angle of attack.

The inclination angle of the pattern was measured on the model sides and is plotted versus the angle of attack in Figure 14. The streamline inclination calculated by a characteristics method for slender bodies [10] is shown for comparison. As seen, the pattern orientation follows closely the streamline direction of the inviscid flow.

A similar observation of pattern orientation following the streamline direction of the inviscid flow is qualitatively described in [5]. On ablating cones spinning at 2000

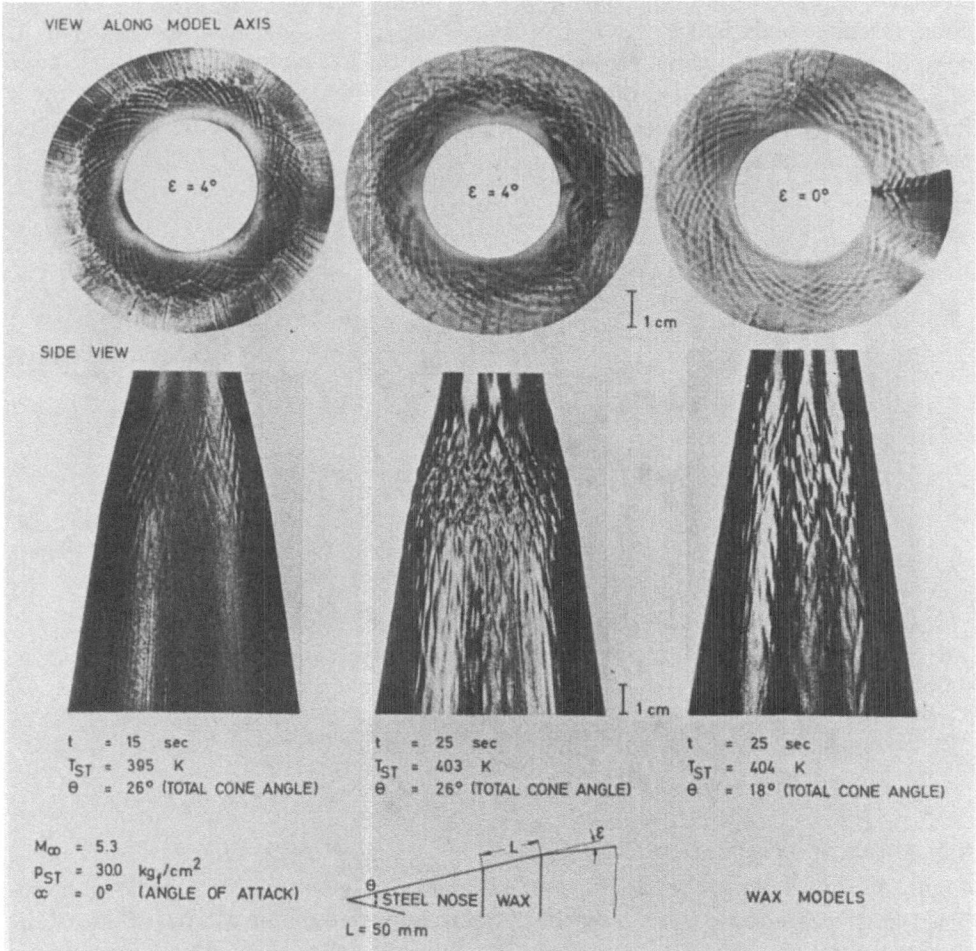

Fig. 12.   Influence of run-time on the cross-hatching formation (flow expansion models).

and 6000 rpm, the cross-hatched pattern was shifted in a direction consistent with the local cross flow.

In [13] test results are described on models showing initial surface perturbations like slots, holes and wavy walls. Tests results are given trying to verify or support hypotheses for the origin of cross-hatching.

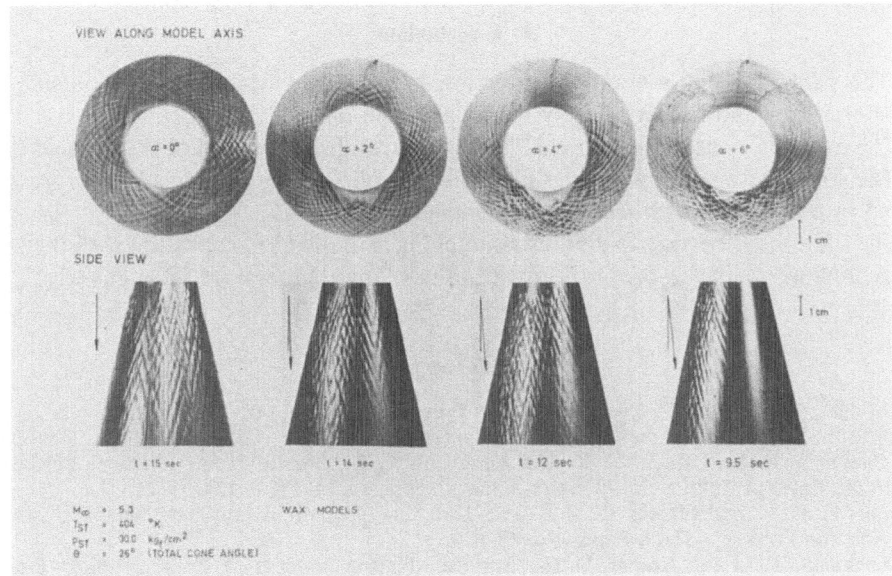

Fig. 13.   Influence of the angle of attack on the cross-hatching formation.

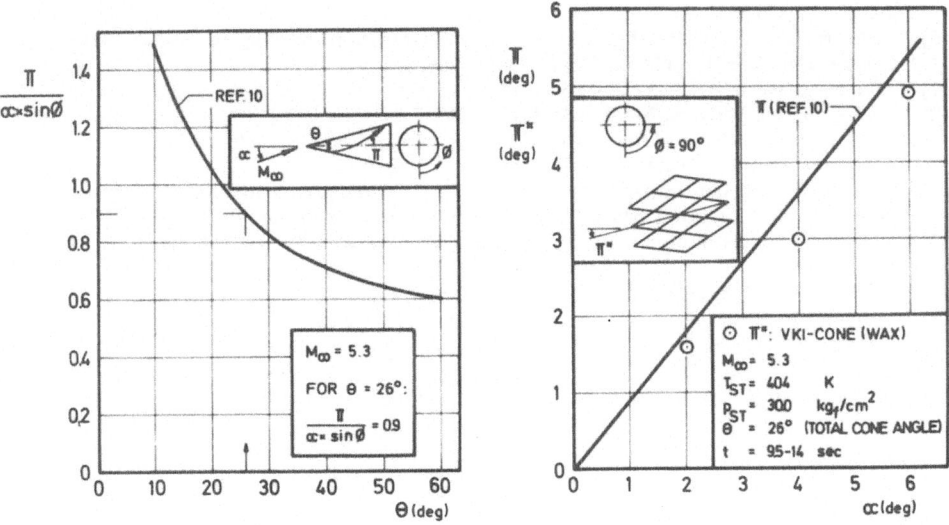

Fig. 14.   Comparison of the pattern inclination angle π^x with the streamline inclination angle π^x of the inviscid flow for cones at angle of attack on the model side ($\phi = 90°$).

5. Conclusions

(1) The cant angle ϕ is a unique function of the local Mach number outside the boundary layer and follows closely the local Mach angle.

(2) The pattern orientation follows closely the streamline direction of the inviscid flow.

(3) The streamwise spacing λ varies in inverse proportion with the local static pressure p_e and in proportion with the driving temperature ratio $T_r - T_w/T_w$.

(4) The run time necessary to give cross-hatching is in inverse proportion with the local static pressure p_e and in proportion with the driving temperature ratio $T_r - T_w/T_w$.

References

[1] Minges, M. L., *High Temperatures – High Pressures* **1** (1969), 607.

[2] Wilson, K. H. and Koubek, F., NOL TR 68–126, 169.

[3] Chapman, D. R., Larson, H. K. and Anderson, L. E., 'Aerodynamic Evidence Pertaining to the Entry of Tektites into Earth's Atmosphere', NASA TR R-134, 1962.

[4] Larson, H. K. and Mateer, G. G., 'Cross-Hatching – A Coupling of Gas Dynamics with the Ablation Process', AIAA Paper No. 68–670.

[5] Laganelli, A. L. and Nestler, D. E., 'Surface Ablation Patterns: A Phenomenology Study', AIAA Paper No. 68–671.

[6] Williams, E. P., 'Experimental Studies of Ablation Surface Patterns and Resulting Roll Torques', AIAA Paper No. 69–180.

[7] Canning, T. N., Tauber, M. E. and Wilkins, M. E., 'Orderly Three-Dimensional Processes in Turbulent Boundary Layers on Ablating Bodies', AGARD CP No. 30 and Suppl., *Hypersonic Boundary Layers and Flow Fields*, May 1968.

[8] Canning, T. N., Wilkins, M. E. and Tauber, M. E., 'Ablation Patterns on Cones having Laminar and Turbulent Flows', *AIAA J.* **6** (1968), 174.

[9] McDevitt, J. B., 'An Exploratory Study of the Roll Behaviour of Ablating Cones', *J. Spacecraft* **8** (1971), 161.

[10] McDevitt, J. B. and Mellenthin, J. A., 'Upwash Patterns on Ablating and non-Ablating Cones at Hypersonic Speeds', NASA TN D 5346, 1969.

[11] Wilkins, M. E., *AIAA J.* **3** (1965), 1963.

[12] Canning, T. N., Wilkins, M. E. and Tauber, M. E., 'Boundary Layer Phenomena Observed on the Ablated Surfaces of Cones Recovered After Flights at Speeds up to 7 km s^{-1}', AGARD CP No. 19, Vol. 2; *Fluid Physics of Hypersonic Wakes*, May 1967.

[13] Stock, H. W. and Ginoux, J. J., 'Hypersonic Low Temperature Ablation – An Experimental Study of Cross-Hatched Surface Patterns', von Karman Institute, TN 64, July 1971.

[14] Winkler, E. M., Humphrey, R. L., Madden, M. T. and Koenig, J. A., *AIAA J.* **8** (1970), 1895–1896.

[15] Stock, H. W. and Ginoux, J. J., *AIAA J.* **9** (1971), 971.

[16] Laganelli, A. L. and Zempel, R. E., *AIAA J.* **8** (1970), 1709.

[17] Nachtsheim, P. R. and Larson, H. K., AIAA Paper 70–769.

[18] Charwat, A. F., 'Exploratory Studies on the Sublimation of Slender Camphor and Naphtalene Models in a Supersonic Wind Tunnel', RM-5506-ARPA, July 1968, Rand Corporation.

THE SCATTERING OF GAS MOLECULES
FROM METAL SURFACES

LEON TRILLING

Dept. of Aeronautics and Astronautics, Massachusetts Institute of Technology,
Cambridge, Mass., U.S.A.

Abstract. The physical process of energy and momentum exchange between inert gas molecules and clean metal surfaces in the range between 0.05 eV and several eV is described and computed accommodation coefficient and scattering pattern shifts are compared to measurements. Several features of time of flight measurement techniques are described.

The study of rarefied gas flows which has been greatly stimulated by space exploration activities, has brought back an interest in the physics of gas–solid interfaces. Gas surface interactions at low densities determine the specification of boundary conditions for solutions of the Boltzmann equation [1–5] and the structure of the Knudsen sublayer in transition flow. They play a crucial role in the momentum transfer which causes the drag of space vehicles. In the energy range of one to ten electron volts, they also condition the solid surface and modify its physical properties.

This paper describes some recent findings in the mechanics of gas–surface interaction, and some experimental techniques developed for their laboratory study. It stresses the exchange of energy and momentum at the surface and incidentally suggests some of the molecular interactions which occur at orbital or entry velocities.

The discussion is limited to 'clean' surfaces, or surfaces which are free of a layer of adsorbed impurities. Such surfaces do not readily occur in the conventional atmospheric environment; they must be prepared by cleaning and baking in vacuum at pressures of order 10^{-8} Torr and even then, except in the case of cleaved ionic crystals, the experimenter has great difficulty in providing a precise description of the surface with which he works. On the other hand, clean surfaces are easier to theorize about than 'engineering' surfaces and one may hope that they are present in the hard vacuum space environment.

The goal of the inquiry is to answer the question: given a flux of molecules which approach a surface from a direction (θ_i, φ_i) at a velocity u_i, what is the probability of their leaving the surface in a direction (θ_r, φ_r) at a velocity u_r? The answer may be sought either by impinging a calibrated molecular beam on a target and measuring the resulting scattering pattern, or by computing the trajectories of an appropriate sample of molecules as they fly by a surface.

The essential physical parameters which characterize the interaction are a set of energies: the gas temperature or kinetic energy kT_g, the internal energy of the gas molecule; the interaction potential between the gas molecule and an atom or set of atoms on the surface D; the temperature of the solid kT_s; the binding energy of the solid E_s. In addition, the ratio μ of the mass of the gas molecule m to that of a solid atom M, the ratio of a characteristic interaction length l (range of interaction potential or amplitude of thermal oscillation) to the lattice spacing d and the ratio of the time of interaction $2l/u_i$ to the shortest period of the lattice oscillations $\bar{h}/k\theta_D$ play important roles; finally one may wish to specify some geometrical variables (direction of approach of molecules, etc. . . .).

For most of the phenomena of interest here, the energy transfer per collision is large enough so that the laws of classical physics accurately represent the situation; exceptions being the interactions which involve internal (rotational, vibrational or electronic) degrees of freedom.

The following sections describe some calculations of energy and momentum transfer and of scattering patterns and some experimental techniques and results which together summarize the state of the art.

1. Interaction Theory: Energy Exchange

When a gas molecule flies by a solid surface, it interacts with the set of bound atoms which constitutes the solid lattice; the interaction is defined by a combination of attractive long range forces and repulsive short range forces; in general, the attractive interaction involves a number of lattice atoms which undergo uncorrelated thermal oscillations. It is therefore conservative. The repulsive interaction essentially takes place with a single nearest atom. That atom is in motion both because of thermal oscillations and because of the interaction; it therefore exchanges energy with the gas molecule. The amount of energy exchanged is described by an energy accommodation coefficient defined as:

$$\alpha_e = \frac{E_{in} - E_{out}}{E_{in} - (E_{out})_s}, \tag{1}$$

where $E_{in} - E_{out}$ is the energy change experienced by a set of gas molecules and $E_{in} - (E_{out})_s$ is the energy change the same set of gas molecules would have experienced if it had left the surface in thermal equilibrium with it. The numerator is the net work done by the gas molecules on the surface while the denominator is proportional to the difference between gas and surface temperature.

The first task is to evaluate the work done by the gas on the solid. At gas temperatures so high that the kinetic energy of the gas molecules is large compared to the interaction well-depth, and the interaction is fast enough compared to the Debye period to be considered instantaneous, the interaction is well-simulated by the collision of two hard spheres, so that the accommodation coefficient is a function of molecular mass ratio μ alone [6, 7]:

$$\alpha_e = \frac{2.4\mu}{(1 + \mu)^2}, \tag{1a}$$

where the numerical factor results from averaging over all directions.

As the gas energy decreases, but remains much larger than the well-depth, the lattice spring constant comes into play and the effective mass of the solid set in motion increases (several atoms have time to participate in the interaction): α_e decreases.

On the other hand, a very slow gas molecule is trapped in the potential well and fully accommodated; for it:

$$\lim_{T_g \to 0} \alpha_e = 1.0 \tag{1b}$$

It follows that the curve of accommodation coefficient as a function of gas temperature must have a minimum which calculation and experiment place at a temperature slightly higher than the interaction potential well-depth (Figure 1).

A number of detailed calculations of the energy exchange have been carried out for a classical square lattice with nearest neighbor interaction [8, 9, 10, 11] or for an elastic semi-infinite continuum [12, 13]. While the details of the calculations vary,

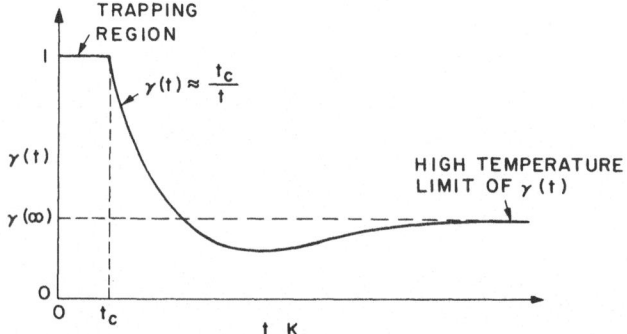

Fig. 1. Typical single molecule accommodation curve.

the general nature of the results is consistent both among models and with experimental results. To illustrate, such a calculation is summarized with the following assumptions.

(i) The solid surface initially oscillates as a result of its finite temperature.

(ii) The solid is a homogeneous constant property linear medium of specific elastic constants (Young's modulus or Debye temperature and Poisson ratio).

(iii) The solid surface is stress-free except for a small disk whose diameter is of the order of the lattice spacing and which is under the action of the interaction force.

(iv) All collisions are head-on normal collisions; since in this model, lattice orientation is not accounted for in detail, surface features are averaged out; in many cases, this is consistent with the level of information available from experiments. A Morse potential defines the interaction.

Under such assumptions, the trajectory of a typical gas molecule satisfies the equation

$$m\ddot{y}_g + F(R) = 0, \tag{2}$$

where $R(t)$ is the gas-surface separation at any time. Then

$$\frac{y_g(t)}{l} = \log \left| \frac{2\delta}{1 - \delta^2} \left(\cosh \frac{u_0 t}{l} - \delta \right) \right| \tag{3}$$

and the force history is:

$$\frac{lF_0(t)}{mU_0^2} = \frac{1 - \delta \cosh(u_0 t/l)}{(\cosh(u_0 t/l) - \delta)^2}$$

(4)

where l is the potential range, u_0 the initial gas molecule velocity and δ is defined in terms of the potential well-depth by

$$\delta = \left(\frac{2D}{2D + mu_0^2}\right)^{1/2}$$

(5)

The displacements X_i of the elastic solid then satisfy the equation:

$$\frac{\partial^2 X_i}{\partial t^2} - C_1^2 \frac{\partial^2 x_i}{\partial x_j \partial x_j} - C_2^2 \frac{\partial^2 x_j}{\partial x_i \partial x_j} = 0,$$

(6)

where C_1, C_2 are the propagation speeds in the solid, and are related to the Young's modulus and the Poisson ratio in the conventional manner.

The boundary conditions on the surface stress are given by:

$$\sigma_{xy}(x, 0, z, t) = \sigma_{yz}(x, 0, z, t) = 0$$

(7a)

$$\sigma_{yy}(x, 0, z, t) = \frac{F_0(t)}{\pi r_0^2} \qquad (0 < r < r_0)$$

(7b)

$$= 0 \qquad (r_0 < r),$$

where $F_0(t)$ is given by (4) and r_0 defines a surface disk of diameter equal to the lattice spacing. This problem is solved to give $Y_s(t)$ the surface displacement history. The work done on the surface is the integral

$$\frac{W}{mu_0^2} = \int_{-\infty}^{\infty} \frac{F_0(t)}{mu_0^2} \frac{dY_s}{dt} \, dt.$$

(8)

If in addition, the surface has a thermal motion, a similar integral, averaged over all phases gives the work done by the thermal motion of the surface on the gas molecule. The result of such a calculation is that the net work done by the gas on the surface is of the form

$$\frac{W}{mu_0^2} = \frac{6\Psi(\sigma)}{\sqrt{\mu}} \left(\frac{D}{k\theta_D}\right)^{3/2} \left(\frac{2h^2}{Mk\theta_D l^2}\right)^{3/2} (u - v) f(u)$$

(9)

u, v are gas and surface velocities respectively made non-dimensional by division through $(2D/m)^{1/2}$, and $f(u)$ is a function determined from a solution of the differential equations. Here, $\Psi(\sigma)$ is a function of the Poisson ratio, $\Psi(0.25) = 0.6$; μ is the mass ratio; D is the well-depth, k the Boltzmann constant and θ_D the Debye temperature of the solid.

By integration over suitable distribution functions $p(u, v)$ one obtains the energy and momentum accommodation coefficients in terms of gas and surface temperature and molecular properties. The distribution functions depend on whether the interaction time is long or short compared to the Debye period of the solid [14]. For slow interactions, one expects to find

$$p(u, v) = uP(u)Q(v) = uP(u)\, \frac{e^{-\beta v^2}}{\sqrt{\pi \beta}} \qquad (10)$$

whereas for fast interactions, one has:

$$p(u, v) = (u + v)P(u)Q(v). \qquad (11)$$

The actual distribution functions are complicated, but by requiring zero energy exchange when gas and surface are at the same temperature, one obtains the result:

$$\alpha_e \cong \frac{2\tilde{\alpha}_e(T_g)}{1 + (T_s/T_g)^{1/2}} \qquad (T_s, T_g \ll D/k) \qquad (12a)$$

$$\alpha_e \cong \tilde{\alpha}_e(T_g) \qquad (T_s, T_g \gg D/k) \qquad (12b)$$

in agreement with measurements [17] where $\bar{\alpha}_e$ is the accommodation coefficient at equilibrium. Computed and measured [15, 16] values of $\bar{\alpha}_e(T)$ for various combinations of materials are shown on Tables I, II. In these computations, solid properties were taken from standard tables, potential well-depths and ranges were selected to fit the data for each gas at one temperature of one class of metals (alkali or non-alkali)

TABLE I

Comparison of measured and computed values of accommodation of inert gases on non-alkaline metals

System	m	M	D/k (K)	l (Å)	θ (K)	α calc	α meas.
Hw–W	4	184	55	0.60	315	0.0165	0.0164
Ne–W	20	184	200	0.60	315	0.0431	0.043
A–W	40	184	1000	0.60	315	0.25	0.26
Kr–W	84	184	1600	0.69	315	0.40	0.42
Xe–W	131	184	2560	0.69	315	0.65	0.66
He–Mo	4	96	50	0.60	380	0.023	0.022
Ne–Mo	20	96	200	0.60	380	0.045	0.045
A–Mo	40	96	1000	0.60	380	0.27	0.27
Kr–Mo	84	96	1600	0.69	380	0.43	0.43
Xe–Mo	131	96	2560	0.69	380	0.69	0.69
He–Al	4	27	50	0.60	400	0.066	0.073
He–Ni	4	59	50	0.60	320	0.068	0.073
He–Pt	4	195	50	0.60	230	0.040	0.038
Ne–Fe	20	57	200	0.60	425	0.056	0.053
He–Be	4	9	50	0.60	1000	0.141	0.145
Ne–Be	20	9	200	0.60	1000	0.316	0.315

All data and calculations presented on this table are for $T_g = 300$ K.

TABLE II

Accommodation coefficients of inert gases on alkaline metals

(a) Lithium	83 K		195 K		273 K	
	α comp.	α meas. [15]	α comp.	α meas. [15]	α comp.	α meas. [15]
He–Li	0.031	0.028	0.048	0.047	0.062	0.060
Ne–Li	0.070	0.051	0.107	0.069	0.128	0.082
A–Li	0.282	0.310	0.206	0.223	0.195	0.222
Kr–Li	0.495	0.480	0.321	0.320	0.280	0.248
Xe–Li	0.845	0.775	0.522	0.493	0.448	0.410

(b) Sodium	78 K		90 K		195 K		298 K	
	α comp.	α meas.	α comp.	α meas.	α comp.	α meas.	α comp.	α meas.
He–Na	0.044	0.035	0.049	0.038	0.069	0.065	0.089	0.09
Ne–Na	0.11	0.06	0.13	0.065	0.15	0.12	0.22	0.20
A–Na	0.51	0.43	0.46	0.39	0.36	0.43	0.45	0.46

(c) Potassium	78 K		90 K		195 K		273 K	
	α comp.	α meas.	α comp.	α meas.	α comp.	α meas.	α comp.	α meas.
He–K	0.039	0.041	0.043	0.043	0.062	0.062	0.074	0.077
Ne–K	0.147	0.067	0.156	0.071	0.174	0.120	0.187	0.178
A–K	0.448	0.480	0.413	0.435	0.390	0.386	0.441	0.440
Kr–K	0.704	0.715	0.639	0.623	0.532	0.512	0.545	0.575
Xe–K	0.98	0.88	0.92	0.77	0.81	0.64	0.79	0.68

Note: The well-depths used in the calculations are: 16 K for He on Na, K, Li; 40 K for Ne on Na, K, Li; 200 K for A on Na, K; 320 K for Kr on Na, K; 512 K for Xe on Na, K; 400 K for A on Li; 640 K for Kr on Li; 1024 K for Xe on Li.

and then used to predict values at other sets of conditions as indicated. It appears that the model satisfies an internal consistency test.

2. Interaction Theory – Scattering Patterns

A more sensitive test of the model outlined above is whether it predicts scattering patterns. When a molecular beam strikes a metal surface, it is re-emitted in a lobular pattern; the orientation of the lobe peak is a function of angle of incidence of the beam, and of the beam and surface temperature; the width of the lobe is determined by the velocity distribution of the incident beam and the macroscopic and molecular roughness of the surface [18–28].

At thermal energy, very little tangential momentum is exchanged between beam and surface, and the lobe peak shift from the specular direction is largely the result of energy exchange between the normal component of beam velocity and the normal component of the thermal motion of the surface. Calculations based on this idealization were performed by Logan, Stickney and Keck who used a 'soft' cube model in which a single spring-mass system models the solid [29–30]. Similarly, the energy exchange model described above can be applied [31]. Simple geometrical considerations yield the lobe shift from the specular:

$$\tan \Delta\theta = \frac{(W - 2\alpha_t)\sin\theta_i}{(1 - W)\cos\theta_i + \sqrt{\cos^2\theta_i - W}}$$

$$\times \left[1 - \frac{\alpha_t(W - \cos 2\theta_i)}{\cos^2\theta_i - W + (1 - W)\cos\theta_i\sqrt{\cos^2\theta_i - W}}\right], \quad (13)$$

where θ_i is the angle of incidence of the beam, measured from the normal to the surface, W is the nondimensional energy exchange and α_t is a (small) tangential momentum accommodation coefficient. One may then integrate Equation (13) and predict

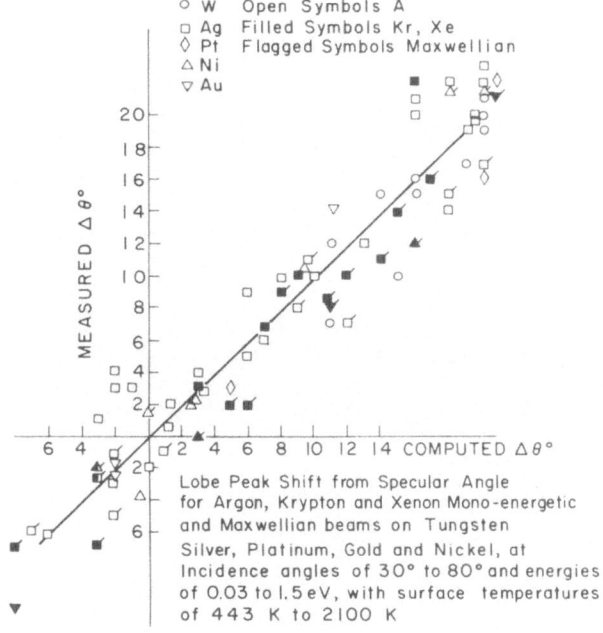

Fig. 2.   Shift in Lobe Peak theory and experiment.

the resulting angular shift $\Delta\theta$. Extensive comparisons of such predictions with measurements over a wide range of material combinations, incident beam directions and velocity distributions, and gas and surface temperatures, are shown on Figure 2.

Numerical calculations [32, 33, 34] and recent experiments [35, 36, 37] suggest that when the gas energy increases to several electron volts and becomes comparable to the binding energy of the solid, the surface no longer 'looks flat' to the impinging

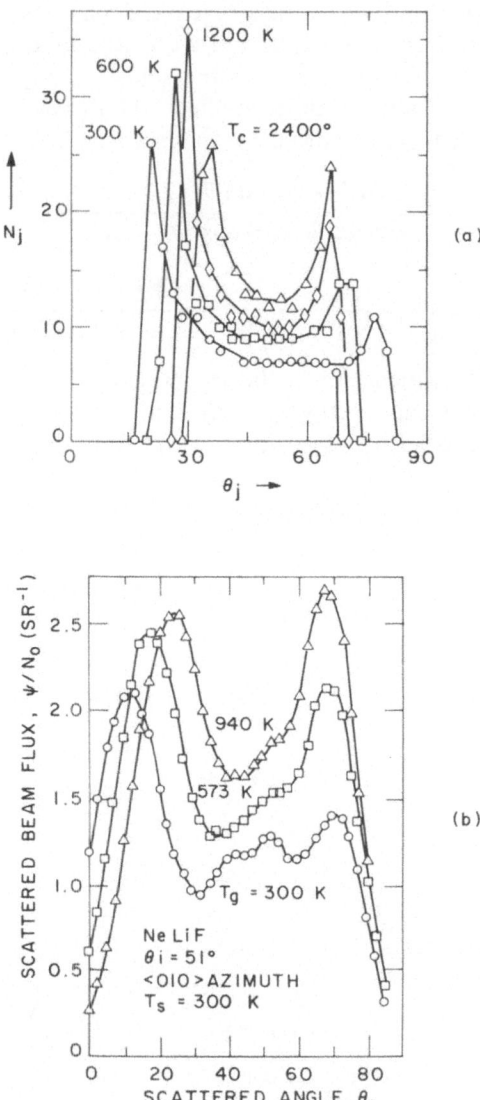

Fig. 3. Comparison of McClure's 2-D results with experimental data for neon on the ⟨010⟩ azimuth of LiF, McClure (1969).

gas molecule, the fly-by is no longer slow enough for the average tangential force to be negligible as a result of the periodicity of surface force-centers. A more complex 'structural' scattering then takes place; it is characterized by appreciable tangential momentum transfer and a multiple-lobe (usually two-lobe) re-emission pattern (Figure 3).

A number of models have been proposed to account analytically for bi-lobular re-emission patterns, and they all emphasize the role of geometrical focussing which results from a wavy surface. The crudest representation of this effect is a sinusoidal solid specular wall [38]; more appropriately, one may compute the interaction of a stream of gas molecules with an Einstein solid [33–34, 39, 40]; a 'rainbow' effect then results from the dependence of re-emission angle on impact point location within a surface lattice cell. Another alternative is to add to the normal interaction force discussed above a tangential force which varies as a harmonic function of target point location on the surface [41] or to replace the surface by a fixed force field with a tangential sinusoidal component [42]. In all these calculations, the intensity of the rainbow effect is related to the ratios of gas kinetic energy to solid binding energy and interaction potential well-depth. But the calculations are carried out for a simple square surface lattice on an ideal crystallographic plane, where actual solid surfaces are more complex. The agreement between calculation and measurement is qualitative. Further computations and systematic detailed measurements over a range of energies are needed to assess the calculations.

This discussion has stressed the classical features of the interaction. Much important work on quantum interactions has been done by Feuer and her associates [43] and most especially by Goodman [44–45] who has succeeded in fitting a quantitative description of molecular beam diffraction within the lattice theory. Lattice scattering theory may also be extended to surface charge exchange [46] or reaction e.g., of an oxygen beam on a graphite or tungsten surface [47–48] and the field of adsorption, condensation and evaporation [49–52]; in all of those important problems, the model of molecules interacting dynamically with an elastic lattice provides new detailed information and suggestive ideas.

3. Experimental Devices and Procedures: The Beam Source

In most scattering experiments, a calibrated beam of neutral molecules is directed at a target and the angular distribution of the flux of mass momentum and other properties of the re-emitted molecules is measured. The experiments are carried out in vacuum (generally of order 10^{-7} to 10^{-8} Torr). The key components of the apparatus, in addition to the vacuum system, are: a beam source, a movable target mount, a modulator (or chopper) which identifies a sample of test molecules, and a movable detector which collects and counts the test molecules. The key measurement is a time of flight over a measured course. Some technique for characterizing the properties of the target surface while the experiment is in process is desirable but not readily available. Attempts have been made to work with specified faces of cleaved (LiF) or carefully grown single crystals; for example, Smith and Saltsburg have epitaxially grown their silver or gold target by vapor deposition while the experiment was in progress but did not measure surface properties [53]. On the other hand, Merrill and Smith [54] have performed scattering experiments while a LEED apparatus was monitoring the surface, but their data is hard to analyze because LEED records

are not sufficiently sensitive to surface impurities and are difficult to assess unam-
biguously.

Attention in this survey is focussed on the beam, the chopper and the detector.
Since the angular distribution of molecules scattered from a surface is to be measured,
one requires the ability to identify beams whose intensity is a small fraction (1% or
less) of the original beam; the tagged molecules which are counted are a fraction of
all the molecules in the beam; signal to noise requirements, and the pressure attain-
able in the vacuum chamber then dictate the required order of magnitude of the
incident beam intensity. It is generally in the range of 10^{16} to 10^{18} particles per square
centimeter per second.

At low energies, neutral beams of this intensity may be obtained by efflux from a
Knudsen cell. To increase the energy of the beam, one can heat the source up to
temperatures of 2500 K or 0.25 eV; if a substantially mono-energetic beam is needed,
the supply gas may be expanded through a nozzle and the core of the expanded jet
may be selected by a skimmer collimator system; Kantrowitz and Grey proposed such
a scheme in 1951 [55] and it has received wide use in Germany [56–57], France
[58–59], Italy [60], the U.S.S.R. [61], the U.S. [62–64], and Canada [65, 23]. While
nozzles do give nearly mono-energetic (high Mach number) beams, they are tricky
to design; in particular, the optimum size, shape and location of the collimator-
skimmer is a critical function of supply pressure [59]; off-design operation may create
a bow shock ahead of the skimmer; and condensation or polymerization may occur
at the skimmer lip [66, 67]. An alternative scheme was suggested and tested by
Moran [68]. He found that if the ratio of supply pressure to vacuum chamber pressure is sufficiently high ($\sim 10^8$), a uniform beam of Mach number 8 and of required
intensity is obtained by passing the supply gas through a small tube or even through
an orifice. The performance of the beam is not sensitive to the geometry of the orifice
or of the collimation device.

Whether one uses a Grey–Kantrowitz nozzle beam or a simple orifice beam, the
energy available is limited by the permissible gas stagnation temperature to a value
substantially below 1 eV. To attain a higher energy, a light gas beam may be seeded
with a heavier component which is entrained so that its energy is higher by a factor
m_1/m_2. Thus the argon in a helium–argon beam may reach ten times the average
beam energy [69, 70, 71]. Because of its smaller amplitude of thermal wandering,
the heavy gas concentrates near the jet axis. The performance of an arc-heated argon–
helium beam is described by Young and Knuth [72] who obtained argon energies
up to 21 eV with an initial mole fraction of 0.02. The price to be paid in a mixed beam
is the need of a mass-spectrometer in the detection apparatus, to insure that one collects only the desired species. A simple mobile mass-spectrometer specially designed
for use with a mixed beam was described by Wang [73]. An alternative approach for
the production of high intensity neutral beams in the 1 to 10 eV energy range is the
neutralization of ion beams. Ion beams of high energy are easily produced and
channelled, but charge exchange by collision in a neutralization chamber becomes
less efficient as the beam energy decreases and does not appear very promising for
the production of neutral beams of the required intensity below 20 eV.

Another method, useful if an alkali atomic beam is appropriate, consists in bombarding a potassium target with a 6 keV argon ion beam and collimating the stream of sputtered potassium atoms which are in the range of 0.5 to 35 eV [74].

4. Experimental Procedure: The Chopper-Detector System

An essential component of any time of flight measurement is a device to identify the molecules whose flight is timed over a path of given length. That identification is done, for example, by chopping the beam mechanically: a rotating wheel is located in a plane normal to the beam so that its rim interrupts and scatters the beam except during the very short time intervals when the beam goes through a slot cut out at the rim of the wheel. The molecules which are all allowed through at the same time from the observed sample and the history of their time of arrival at a detector allows the determination of the molecular velocity distribution. First, the incident beam in a scattering experiment is sampled by this technique. Then, the scattered field is sampled by scanning over re-emission directions both in the principal plane (defined by the incident beam and the normal to the surface) and out of plane. The scanning is most easily accomplished (especially out of the principal plane) by rotating the target about the fixed incident beam.

While chopping with a slotted wheel (or with two properly phased slotted wheels for velocity selection) is a classical technique [75–77], the analysis of data which it provides is complicated by the fact that the detector does not read the rate of arrival density but rather its convolution with a triangular 'shutter function' which defines the fraction of the beam allowed through the shutter as the shutter sweeps by [78–80]; that convolution integral must be inverted, and when the shutter opening time is not very short compared to the measured flight time, the inversion of the experimentally determined curve leads to accuracy problems of some seriousness. In general, the convolution is not inverted directly, but zero, first, and second velocity moments of the flux are computed [81]. If the motor which drives the chopper wheel rotates at 20 000 rpm, the wheel radius is 2 cm and the slot has a width of 0.30 cm, the opening time is 75 μs. If the beam is helium at room temperature, the peak intensity of a Maxwellian distribution travels 15 cm in 280 μs. These numbers indicate that, especially at energies higher than thermal, a compromise must be struck between higher wheel speed (which present problems and operational problems), narrower slots (which reduces signal intensity) and a longer flight path (which implies a larger vacuum system and larger pumping requirements).

The mechanically chopped beam must then be detected at the end of its prescribed travel path. One often used method of detection [81] consists in intersecting the beam with a stream of electrons which ionizes a fraction of the beam molecules; these ions are pulled out by means of a set of magnetic lenses and counted on arrival at an electron multiplier system.

An alternative detector particularly appropriate for high energy mechanically chopped beams of moderate intensity relies on the sputtering of adsorbed alkali metal ions from a surface. If all adsorbate atoms which receive energies in excess of the ion

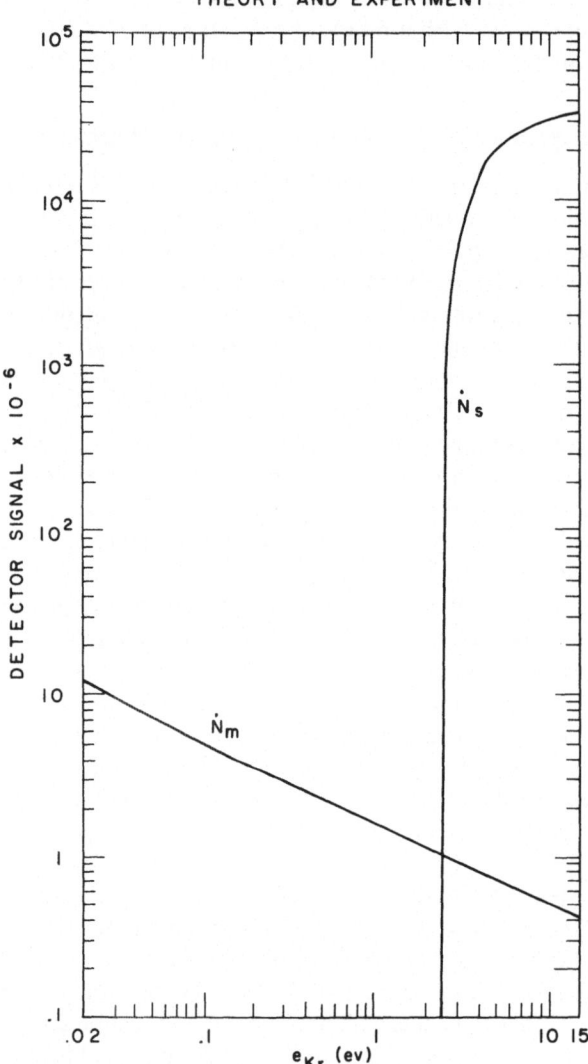

Fig. 4. Theoretical detector signal intensity vs incident beam energy for a krypton beam vs. N_s is signal from mechanically chopped beam with sputtering detector N_m is signal from electronically chopped beam with electron multiplier detector.

desorption energy are sputtered, and if the beam adsorbate interaction at the high energies involved is as in hard sphere collision, then

$$\frac{\dot{N}_s}{\dot{N}_i} = P \frac{\theta}{\cos \theta_i} \left[1 - \frac{(1 + \mu)^2}{4\mu} \frac{\varphi_p}{e_i} \right], \tag{14}$$

where $\dot{N}_s$ is the flux of sputtered ions, $\dot{N}_i$ is the desired incident flux, θ is the adsorbate coverage fraction, θ_i is the angle of incidence of the beam, φ_p is the ion desorption energy (of order 2 eV for the heavier alkalis) and e_i is the incident beam energy. A comparison of the signal intensity to be expected from the two detectors is shown on Figure 4.

An alternative to mechanical chopping which obviates some of the shutter function difficulties has been proposed and demonstrated by French and his associates [82] and has also been used in the MIT laboratory [83] and at the University of Michigan [84]. It involves intersecting the test beam with an electron beam to excite a fraction of the beam molecules to a higher metastable state. The metastables are identified at the end of their flight path by the fact that on striking an electron multiplier plate, they are de-excited and cause the emission of an electron from the multiplier plate. The exciting pulse is square and may be made quite short (or order 20 μs) so that the data reduction process is more reliable for high speed beams. However, the electron impacts impart a finite lateral velocity to the metastable test molecules and a measurable drift is therefore present. On the one hand, the amount of drift provides an independent check on time of flight distribution measurement; on the other hand, particularly the angular distribution of scattered molecules is sought, the test beam must be carefully collimated before it meets the exciting electron flux.

References

[1] Cercignani, C., *Mathematical Methods in Kinetic Theory*, Plenum Press, N.Y., 1969, pp. 96–98.
[2] Cercignani, C., RGD 7, Pisa, 1970*.
[3] Kuscher, I., RGD 7, Pisa, 1970.
[4] Shen, S. F., RGD 7, Pisa, 1970.
[5] Klavins, A., 'Some Aspects of the Surface Boundary Condition in Kinetic Theory', Univ of Illinois, Tech. Report AAE-70-3.
[6] Baule, B., *Ann. Phys.* **44** (1914), 175.
[7] Goodman, F. O., RGD 5, Oxford, 1968, p. 35*.
[8] Cabrera, N. B., *Disc. Farad. Soc.* **28** (1959), 46.
[9] Zwanzig, R. W., *J. Chem. Phys.* **32** (1960), 1173.
[10] Goodman, F. O., *J. Phys. Chem. Solids* **23** (1962), 1269.
[11] Goodman, F. O., *Surface Sci.* **11** (1968), 283.
[12] Karamcheti, K. and Scott, L. B., in M. Rogers and H. Saltsburg (eds.), *Fundamentals of Gas Surface Interactions*, Academic Press, 1967, p. 536.
[13] Trilling, L., *Surface Sci.* **21** (1970), 337–365.
[14] Trilling, L., *Surface Sci.* **27** (1971), 112.
[15] Thomas, L. B., RGD 5, Oxford, 1966, p. 163.
[16] Kouptsidis, J. and Menzel, D., *Ber. Bunsen Gesellsch. Phys. Chem.* **71** (1967), 720.
[17] Kouptsidis, J. and Menzel, D., *Ber. Bunsen Gesellsch. Phys. Chem.* **74** (1970), 512.
[18] Datz, S., Moore, G. E., and Taylor, E. G., RGD 3, Paris, 1962, p. 347*.
[19] Smith, J. N. and Fite, W. L., RGD 3, Paris, 1962, p. 430.
[20] Hinchen, J. J. and Foley, W. M., RGD 4, Toronto, 1964, p. 505*.
[21a] Saltsburg, H. and Smith, J. N., Gen. Atomics Rept. GA 6740, Dec. 1965.

* RGD 3, 4, 5, 6, 7 – Proceedings of International Symposium on Rarefied Gas Dynamics, Academic Press, N.Y.

[21b] Smith, J. N. and Saltsburg, H., *Fundamentals of Gas Surface Interactions*, Academic Press, 1967, p. 370.

[21c] Smith, J. N. and Saltsburg, H., RGD 6, Cambridge, Mass., 1968, p. 1141*.

[22]   Bishara, M. N., University of Virginia Report AEEP-4038-109-69U, March 1969.

[23a] French, J. B. and O'Keefe, D. R., RGD 4, Toronto, 1964, p. 299.

[23b] O'Keefe, D. R. and French, J. B., RGD 6, Cambridge, Mass., 1968, p. 1278.

[24]   Knuth, E. L. and Kuluva, N. M., in I. Glassman (ed.), *Rec. Adv. Aerothermochemistry*, AGARD, 1967, p. 277.

[25]   Jakus, K., Un. Calif. Report As 68-7, 1968.

[26a] Moran, J. P., Ph.D. Thesis, MIT, Feb. 1968.

[26b] Moran, J. P., Wachman, H. Y., and Trilling, L., *Phys. Fluids* 12 (1969), 987.

[27]   Yamamoto, S., Ph.D. Thesis, MIT, Aug. 1969.

[28]   Romney, M. J., Ph.D. Thesis, Princeton Un., Jan. 1969.

[29]   Logan, R. M. and Stickney, R. E., *J. Chem. Phys.*, 44 (1966), 195.

[30]   Keck, J. C. and Logan, R. M., *J. Chem. Phys.* 49 (1968), 860.

[31]   Trilling, L., RGD 7, Pisa, 1970.

[32]   Oman, R. A., RGD 5, Oxford, 1966.

[33]   Oman, R. A., *J. Chem. Physics* 48 (1968), 3919.

[34]   Lorenzen, J. and Raff, L. M., *J. Chem. Phys.* 49 (1968), 1165.

[35]   Smith, J. N., O'Keefe, D. R., and Palmer, R. L., Gulf Gen. Atomic Rep. Ga-9410, 1969.

[36]   Fisher, S. S., Bishara, M. M., Kuhlthau, A. R., and Scott, J. E., RGD 6, Cambridge, Mass., 1968, 1247.

[37]   Callinan, J. P. and Knuth, E. L., RGD 6, Cambridge, Mass., 1968, p. 1247.

[38]   Healy, T. J., in M. Rogers and H. Saltsburg (eds.), *Fundamentals of Gas Surface Interactions*, Academic Press, 1967, p. 435.

[39]   McClure, J. D. and Wu, Y., RGD 6, Cambridge, Mass., 1968, p. 1191.

[40]   McClure, J. D., Boeing SRL Report D-1-82-0885, Aug. 1969.

[41a] Falcovitz, J., Ph.D. Thesis, MIT, June 1970.

[41b] Falcovitz, J., Trilling, L., Wachman, H. Y., and Keck, J. C., RGD 7, Pisa, 1970.

[42]   Chrisman, C. C. and Sentman, L. H., University of Illinois Tech. Report AAE-71-3, 1971.

[43]   Feuer, P. and Osburn, C., RGD 6, Cambridge, Mass., 1968, p. 1095.

[44]   Cabrera, N., Celli, V., Goodman, F. O., and Manson, R., *Surface Sci.* 19 (1970), 67.

[45]   Goodman, F. O., *Surface Sci.* 19 (1970), 93.

[46]   Trilling, L., Falcovitz, J., Moran, J. P., and Wachman, H. Y., Entropie No. 30, 1969, p. 65.

[47]   Batty, J. C. and Stickney, R. E., RGD 6, Cambridge, Mass., 1968, p. 1165.

[48]   Shih, W., Ph.D. Thesis, MIT, 1971.

[49]   Armand, G., RGD 6, Cambridge, Mass., 1968, p. 1055.

[50]   McCarroll, B. and Ehrlich, J., *J. Chem. Phys.* 38 (1963), 523.

[51]   Levenson, L., Ph.D. Thesis, University of Paris, 1968.

[52]   Pagni, P., Ph.D. Thesis, MIT, 1970.

[53]   Saltsburg, H. and Smith, J. N., Gen. Atomics Rept. GA-6740, 1965.

[54]   Smith, D. L. and Merill, R. P., RGD 6, Cambridge, Mass., 1968, p. 1159.

[55]   Kantrowitz, A. and Grey, J., *Rev. Sci. Instr.* 22 (1951), 328.

[56]   Becker, E. W. and Henkes, W., *Z. Physik* 146 (1956), 320.

[57]   Bier, K. and Hagena, O., RGD 4, Toronto, 1964, p. 260.

[58]   Campargue, R., RGD 4, Toronto, 1964, p. 279.

[59]   Campargue, R., RGD 6, Cambridge, Mass., 1968, p. 1003.

[60]   Gazzola, C., Scoles, G., and Torello, F., RGD 6, Cambridge, Mass., 1968, p. 977.

[61]   Rebrov, A. K. and Sharafutdinov, R. G., RGD 6, Cambridge, Mass., 1968, p. 965.

[62]   Anderson, J. B. and Fenn, J. B., *Phys. Fluids* 8 (1965), 780.

[63]   Bossel, U., Hurlbut, F. C., and Sherman, F. S., RGD 6, Cambridge, Mass., 1968, p. 945.

[64]   Zapata, R. N., Ballard, R., and Cabrera, N., RGD 6, Cambridge, Mass., 1968, p. 997.

[65]   Govers, T. R., LeRoy, R. L., and Deckers, J. M., RGD 6, Cambridge, Mass., 1968, p. 985.

[66]   Becker, E. W., Bier, K., and Henkes, W., *Z. Physik.* 146 (1956), 333.

[67]   Beck, E. W., Klingelhöfer, R., and Mayer, H., RGD 6, Cambridge, Mass., 1968, p. 1349.

[68] Moran, J. P., *AIAA Jl.* **8**, No. 3 (1970), 539.
[69] Becker, E. W., Beyrich, W., Bier, K., Burghoff, H., and Zigan, F., *Z. Naturforsch.* **12a** (1957), 609.
[70] Waterman, P. C. and Stern, S. A., *J. Chem. Phys.* **31** (1959), 405.
[71] Reis, V. H. and Fenn, J. B., *J. Chem. Phys.* **39** (1963), 3240.
[72] Young, W. S. and Knuth, E. L., *Entropie* No. 30, 1969, p. 25.
[73] Wang, J. C., SM Thesis, MIT, 1968.
[74] Politiek, J., Rol, P. K., Los, J., Onderlinden, D., and Schiffer, J. J. M., RGD 5, Oxford, 1966, p. 1443.
[75] Stern, O., *Z. Physik.* **2** (1920), 49.
[76] Costa, J. L., *et al.*, *Phys. Rev.* **30** (1927), 349.
[77] Eldridge, J. A., *Phys. Rev.* **30** (1927), 931.
[78] Scott, P. B., Bauer, P. H., Wachman, H. Y., and Trilling, L., RGD 5, Oxford, 1966, p. 1353.
[79] Alcalay, J. A. and Knuth, E. L., *Rev. Sci. Instr.* **40** (1969), 438.
[80] Amend, W. E. and Hurlbut, F. C., RGD 6, Cambridge, Mass., 1968, p. 1205.
[81] Scott, P. B., MIT Thesis, 1965; also MIT–FDRG Report 65-1.
[82] French, J. B. and Locke, J. W., RGD 5, Oxford, 1966, p. 1461.
[83] Wang, J. C., Ph.D. Thesis, MIT, 1971.
[84] Donnelly, D. P., Pearl, J. C., and Zorn, J. C., RGD 6, Cambridge, Mass., 1968, p. 1319.

LIFTING BODY CONFIGURATIONS
FOR SUSTAINED HYPERSONIC FLIGHT

G. G. CHERNYI and A. L. GONOR

Institute of Mechanics, Moscow State University, Moscow, U.S.S.R.

Abstract. This paper deals with the calculation of hypersonic conic flow about a configuration of the 'wing-body' type.

Solution of a boundary problem by 'two-approximations' method is shown for the case of flow about a yawed triangular plate. Then flow about a wing of finite thickness is considered. It is proved that in the latter case the flow structure is qualitatively different from the case of flow about a triangular plate.

The last part of the paper deals with hypersonic flow about a conic body having an inverted V-shaped cross section.

In case of large V-angles, when a system of curvilinear Mach type shocks is formed, the aforementioned 'two approximations' permits the construction of a closed solution. For arbitrary geometrical wing parameters a scheme of numerical calculation with artificial viscosity is used. The possibility of deriving a solution with a strong bow shock is discussed. Aerodynamic wing characteristics are calculated under the assumption that maximum lift-to-drag ratio is attained at V-angle less than 180°.

Results of experimental investigation of four models of V-shaped wings at Mach number 6 are reported. Forces are measured on an aerodynamic balance and the maximum value of lift-to-drag ratio is obtained.

The possibility of using aerodynamic lift during the re-entry of a vehicle into an atmosphere and also for sustained hypersonic flight is more and more drawing the attention of investigators. Studies are under way both in the direction of investigation of hypersonic flow about a wing with a geometry close to the traditional configuration of supersonic wings, and in the direction of search for new designs of vehicles in which the body simultaneously plays the role of wing producing lift in a nonsymmetrical flow. The so-called 'two-approximations' method has proved to be effective in the study of flow about conic wings. The essence of the method is clearly seen in the case of hypersonic flow about a triangular flat plate. In the coordinate system r, θ, φ, shown in Figure 1, the equations of conic flows for variables ψ, φ have the following form:

$$\frac{w}{\cos\theta}\frac{\partial u}{\partial\varphi} - v^2 - w^2 = 0$$

$$\frac{w}{\cos\theta}\frac{\partial v}{\partial\varphi} + uv + w^2 \ \mathrm{tg}\ \theta = -\frac{1}{\rho\theta_\psi}\frac{\partial p}{\partial\psi}$$

$$\frac{\kappa}{\kappa-1}\frac{p}{\rho} + \frac{u^2+v^2+w^2}{2} = C, \quad \frac{\partial}{\partial\varphi}\left[\frac{p}{\rho^\kappa}\right] = 0 \tag{1}$$

$$\frac{\partial}{\partial\varphi}\ln(\rho w\theta_\psi) + 2\frac{u}{w}\cos\theta = 0, \quad w\theta_\varphi = v\cos\theta.$$

L. G. Napolitano et al. (eds.), Astronautical Research 1971, 137–146. All Rights Reserved.

In these equations all variables are dimensionless and referred to a velocity of undisturbed flow U, density ρ^0 and dynamic pressure $\rho^0 U^2$ correspondingly. Variable ψ satisfies the relation $v\psi_\theta + \omega\psi_\omega/\cos\theta = 0$ and represents the stream surface in a conic flow.

A solution will be sought under the assumption that the region of disturbed flow consists of a highly pressurized thin layer. Then according to traditional estimates of the shock layer theory (see [1]) it will be convenient to introduce the following transformations

$$\theta = \varepsilon\bar{\theta}, \qquad v = \varepsilon\bar{v}, \qquad \rho = \varepsilon^{-1}\bar{\rho}. \tag{2}$$

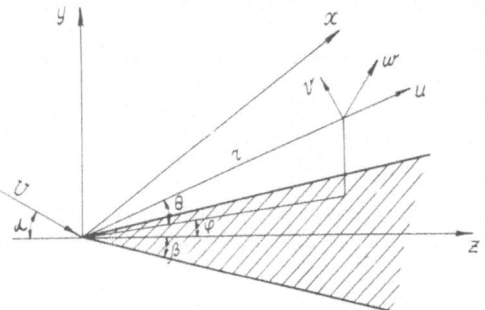

Fig. 1.   Coordinate system.

In the variables (the line is dropped) one has the following system:

$$\frac{w}{\cos\varepsilon\,\theta}\frac{\partial u}{\partial\varphi} - \varepsilon^2 v^2 - w^2 = 0, \qquad \frac{\kappa p}{(\kappa+1)\rho} + \frac{u^2 + \varepsilon^2 v^2 + w^2}{2} = C$$

$$\left(\frac{w}{\cos\varepsilon\theta}\frac{\partial v}{\partial\varphi} + uv + w^2\varepsilon^{-1}\,\mathrm{tg}\,\varepsilon\theta\right)\varepsilon = -\frac{1}{\rho\theta_\psi}\frac{\partial p}{\partial\psi}$$

$$\frac{p^{1/\kappa}}{\rho} = \delta(\psi), \qquad \kappa = \frac{1+\varepsilon}{1-\varepsilon} \tag{3}$$

$$\frac{\partial}{\partial\varphi}[\ln(\rho w\theta_\varphi)] + 2\frac{u}{w}\cos\varepsilon\theta = 0, \qquad w\theta_\varphi = v\cos\varepsilon\theta.$$

The second Equation (3) allows us to represent pressure as a sum of two parts

$$p = p_1(\varphi, \varepsilon) + \varepsilon p_2(\psi, \varphi, \varepsilon). \tag{4}$$

Now let us simplify expressions (3) and omit members of the order of ε^2 and higher in them and in the boundary conditions. Then the functions we are seeking for will satisfy the following equations:

$$\frac{\partial u}{\partial \varphi} - w = 0, \qquad w \frac{\partial v}{\partial \varphi} + uv + w^2 \theta = - \frac{1}{\rho \theta_\psi} \frac{\partial p_2}{\partial \psi}$$

$$u^2 + w^2 = \Delta^2(\psi), \qquad \frac{\partial}{\partial \varphi} [\ln (\rho w \theta_\psi)] + 2 \frac{u}{w} = 0, \qquad w \theta_\varphi = v.$$

(5)

In this approximation the boundary conditions for the shock wave $\theta = \theta^*(\varphi)$ will have the following form:

$$u^* = \cos \alpha (\cos \varphi - \varepsilon \theta^* \operatorname{tg} \alpha)$$

$$v^* = - [\theta_\varphi^* \cos \alpha \sin \varphi + (1 + m_0) \sin \alpha$$
$$+ \varepsilon \cos \alpha (\theta^* \cos \varphi - \theta_\varphi^* \sin \varphi) + \varepsilon \sin \alpha \theta_\varphi^{*2}]$$

$$m_0 = 2/(\kappa - 1) M_\infty^2 \sin^2 \alpha$$

$$w^* = -\cos \alpha (\sin \varphi + \varepsilon \theta_\varphi^* \operatorname{tg} \alpha)$$

(6)

$$p^* = \sin^2 \alpha + \varepsilon p_1^*, \qquad \rho^* = (1 + m_0)^{-1}$$

$$p_1^* = \sin 2\alpha (\theta^* \cos \varphi - \theta_\varphi^* \sin \varphi) - \sin^2 \alpha - M_\infty^{-2}.$$

Upon the surface of the wing defined by equation $\theta = 0$ one gets from the condition of attached flow

$$v = 0.$$

(7)

While deriving the third member of Equations (5) we used an expression for pressure (6) and the fourth member of Equations (3) which can be written as follows:

$$\frac{1}{\rho} = \frac{1}{\rho'} \left[1 + \frac{\varepsilon}{\sin^2 \alpha} (p_1^{*'} - p_1^* - p_2) \right].$$

(8)

The primed variables correspond to the point of intersection of the shock wave with a streamline ($\varphi = \varphi'$ in Figure 2). Pressure $p_1(\varphi)$ can be calculated directly by its

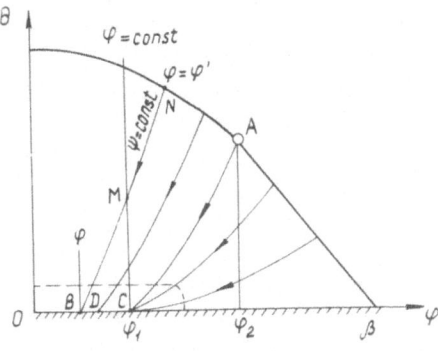

Fig. 2.   Crossflow pattern.

value behind the shock wave (6) by putting $p_1 = p^*$ then p_2 is zero upon the shock wave.

System (5) allows us to find velocity components u and w

$$u = \Delta(\varphi') \cos [\varphi + \alpha(\varphi')], \qquad w = -\Delta(\varphi') \sin [\varphi + \alpha(\varphi')]$$

$$\Delta(\varphi') = \cos^2 \alpha - \varepsilon \sin 2\alpha(\theta^{*\prime} \cos \varphi' - \theta_\phi^{*\prime} \sin \varphi') \tag{9}$$

$$\alpha(\varphi') = \varepsilon \operatorname{tg} \alpha(\theta_\phi^{*\prime} \cos \varphi' + \theta^{*\prime} \sin \varphi').$$

The other unknown functions are defined in [2]. Omitting details we will only state that all of them were received in the form of integrals, depending upon configuration of the shock wave surface which is defined by the following integro-differential equation:

$$z + \delta[2L_2(z \cos \varphi - z' \sin \varphi) - L_1]$$

$$= 2\varepsilon\delta\{(z + z'')[2 \sin \varphi L_3 - \cos \varphi L_4] - (z' + z''') \sin \varphi L_4\}$$

$$z = \varepsilon \operatorname{ctg} \alpha \theta^*(\varphi), \qquad \delta = \varepsilon \cos \alpha (1 + m_0)$$

$$L_1 = \int_\beta^\varphi \frac{w}{w'^2}\left\{1 + \frac{3 + 2m_0}{1 + m_0} [z(\varphi') \cos \varphi' - z'(\varphi') \sin \varphi']\right\}d\varphi'$$

$$L_2 = \int_\beta^\varphi \frac{w}{w'^2}d\varphi' \tag{10}$$

$$L_3 = \int_\beta^\varphi \frac{w}{w'^2}d\varphi' \int_{\varphi'}^\varphi \left(\frac{w^3}{w'^2}\right)_{\varphi'=\xi} d\xi \int_\beta^\xi \left(\frac{u}{w'^2}\right)_{\varphi'=\eta} d\eta$$

$$L_4 = \int_\beta^\varphi \frac{w}{w'^2}d\varphi' \int_{\varphi'}^\varphi \left(\frac{w^3}{w'^2}\right)_{\varphi'=\xi} d\xi \int_\beta^T \left(\frac{w}{w'^2}\right)_{\varphi'=\eta} d\eta.$$

The coefficients L_i are complex functions of the form of shock wave z but in a number of cases, for instance in the case of a plane shock wave, the integrals are easily estimated and all L_i are represented by final expressions. This is used later while constructing the solution. The third order derivative present in the equation allows for smooth matching of a flat wave with a curvilinear one. This is according to Equations (10) accompanied by break of curvature smoothness at point A (Figure 2). Let us now find the solution for Equations (10). The equation of the shock wave surface in the region $0 \leqslant \varphi \leqslant \varphi_2$ will be sought for in the form of the following series:

$$z(\varphi) = \sum_{n=0}^\infty \frac{z^{(n)}(\varphi_2)}{n!} (\varphi - \varphi_2)^n. \tag{11}$$

Values of functions $z(\varphi_2)$ and $z'(\varphi_2)$ are defined by the condition of smooth matching with homogeneous flow. At point φ_2 coefficient L_i is continuous and known; therefore connection between $z''(\varphi_2)$ and $z^{(4)}(\varphi_2)$ can be easily established. By successive differentiation of Equations (10) one can find a similar connection between $z^{(4)}(\varphi_2)$,

$z^{(5)}(\varphi_2)$ etc. As a result series defined by Equation (11) will be dependent only upon an arbitrary constant which can easily be found from the condition in plane of symmetry $z'(0) = 0$. In the constructed solution regions of homogeneous, potential and rotationary flows are matched continuously (see [2]).

Another important result is the topology of stream surfaces.

Study of streamlines crossing a curvilinear shock shows that the former can cross wing surface OC becoming singular in this region (Figure 2). Here, as in the case of the theory of hypersonic conic flows, an entropy layer exists which, being taken into account, changes velocity distribution in the region limited in Figure 2 by a dashed line. Results of the corresponding calculation of transverse velocity taking into account entropy layer are shown in Figure 3 by a solid line.

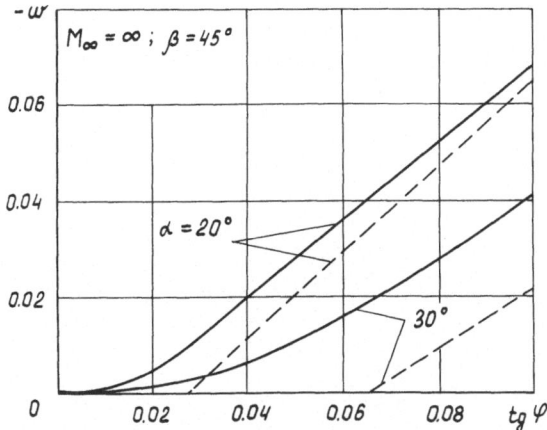

Fig. 3.   Distribution of the velocity in enthropy layer.

Let us consider some data for pressure distribution upon the wing surface. Figure 4 shows the pressure coefficient upon the wing surface calculated in the first approximation and accounting for P_2 (dashed line) and comparison is made with corresponding values calculated in [3] by the finite differences method and Newton's theory (dashed solid line). In the same figure the results of the comparison with the experiment at Mach number $M = 5$ and the shock wave configuration are shown. One can conclude from the distribution of curves that pressure changes slowly in the vicinity of the axis and the pressure gradient normal to the plate is negative.

It is of interest to generalize this method for the case of the finite thickness wing. Let us consider a flow about a triangular wing with a rhombus cross section (Figure 5). Investigating the expression for the transverse velocity w, one can conclude that there exists a plane (other than the symmetry plane) where the transverse velocity vanishes (Figure 6). Thus the finite thickness of wing 'a' results in the displacement of merging line from the plane of symmetry, and the flow has a qualitatively different structure as compared with the flow about a thin triangular wing. This results, in particular, in an ambiguous definition of entropy upon the wing surface and in the necessity of defining

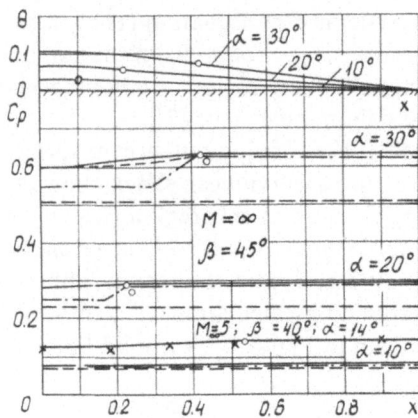

Fig. 4.   Pressure distribution on the wing surface.

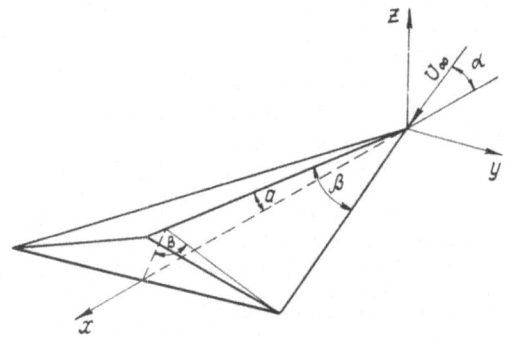

Fig. 5.   Wing with rhombus cross section.

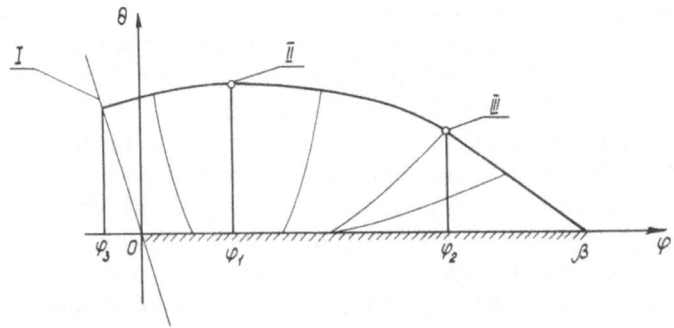

Fig. 6. Crossflow pattern.

the solution in each region separately. If one does not consider this factor in one's calculations then erroneous results will be obtained [4]. Spanwise pressure distribution curves are shown in Figure 7. A new fact consists in displacement of minimum pressure from the symmetry plane towards the leading edge. Curves of the distance

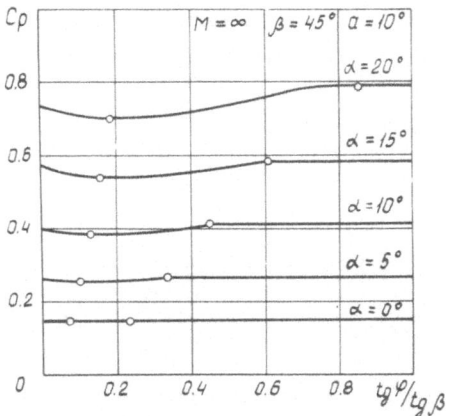

Fig. 7.   Pressure distribution on the wing surface.

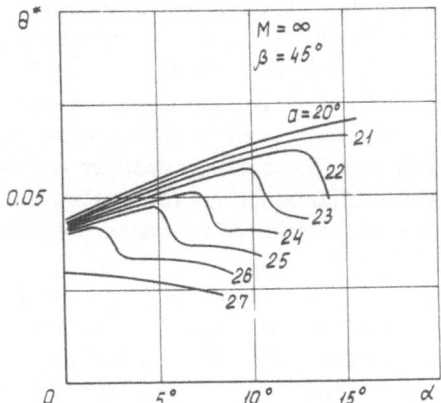

Fig. 8.   Curves of the shock wave detachment.

of the shock wave detachment are shown in Figure 8 for different values of wing thickness. The curves show that beginning with a certain angle of attack α, the shock wave detachment in the symmetry plane starts to diminish. Such behaviour of the shock wave is evidently associated with the growth in intensity of the spanwise flow of gas from the symmetry plane when the angle of attack becomes larger. The solution given in [5] permitted us to study the behaviour of the lift-to-drag ratio of a wing with a rhombus cross section under the assumption that the friction coefficient upon the wing surface is C_τ = const. It was found that with a fixed length and cross section S (or volume) the lift-to-drag ratio is maximal for a wing whose rhombus cross section degenerates into a triangle. The base of the triangle forms the down-wind plane parallel to the undisturbed flow velocity. The dependence of maximum lift-to-drag ratio upon the cross section area at $C_\tau = 2 \times 10^3$ is shown in Figure 9. The dashed line shows corresponding values based on Newton's formula. These data show that Newton's theory definitely gives lower values of lift-to-drag ratio.

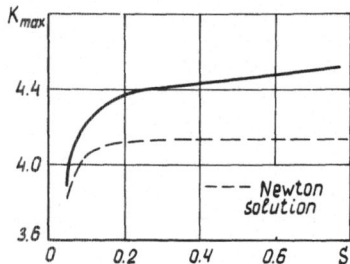

Fig. 9.   Dependence of maximum lift-to-drag ratio upon cross section area.

Another generalization of the theory developed above was a method of calculation of the flow about a wing with a V-shaped cross section. The use of such wings due to the creation of an interfering system of shocks seems to be in prospect for hypersonic flow. Theoretical results received in this field were either based on a rough scheme of Newtonian collision or referred to a narrow range of flow about wings of special classes. Here results are obtained for hypersonic flow about an arbitrary V-shaped wing with large V-angles, when according to experimental studies the Mach configuration of curvilinear shocks is formed. The boundary problem is reduced to the solution of a system of two integro-differential equations for configurations of bow and internal shock waves. An approximate solution of the system is obtained which allows us to effectively find characteristics of flow [6].

The results of the calculation of the pressure behind the bow shock for $M = \infty$, $\beta = 45°$, $\gamma = 15°$ and angles of attack $\alpha = 10, 15, 20°$ are shown in Figure 11 (curves 1–3). The average pressure distribution behind the shock wave is shown by a dashed line; it has been calculated by the method of establishment [7]. In the upper part of the figure the general shock pattern is presented for the cross section with the same values of parameters as in the case of curve 3. It is of interest to find the dependence of lift-to-drag ratio upon the V-angle of the wing. The calculation was done under the assumption that the volume coefficient, $\tau = V/S^{3/2}$ (V = wing volume, S = plane area) and the wing area S_1 or lift coefficient C_y were fixed. A characteristic dependence of the lift-to-drag ratio on angle γ is shown in Figure 12. One can see that the maximum lift-to-drag ratio is obtained at V-angles less than 180°.

We conclude the paper with the results of the measurements of forces acting on V-shaped wings. The experiments were conducted on a 3-component strain balance. Four models were studied at Mach number $M = 4$ with apex angle $\beta = 60°$ and angle

$2\gamma = 0$, 10°, 20°, 30° [8]. The variation of lift-to-drag ratio with the angle of attack is shown in Figure 13. Looking at the curves one comes to the conclusion that the maximum lift-to-drag ratio is obtained at V-angles less than 180°. The measurements have also shown that the value of the lift coefficient changes insignificantly when the

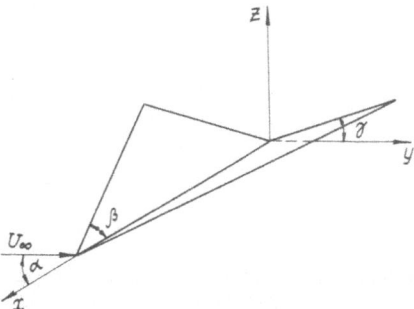

Fig. 10.   Caret wing.

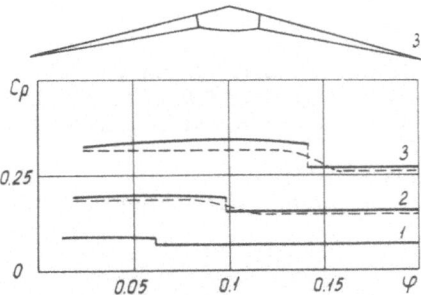

Fig. 11.   Shock pattern and pressure distribution.

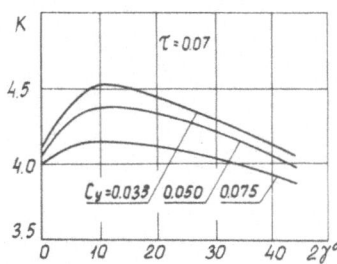

Fig. 12.   Dependence of lift-to-drag ratio upon V-angle of wing.

angle of V of V-shaped wings diminishes. At the same time the drag coefficient becomes smaller. Thus the growth of the lift-to-drag ratio is a result of the same effect which is responsible for the lower drag of bodies with star-like cross section.

Fig. 13.   Dependence of lift-to-drag ratio upon angle of attack.

References

[1] Gonor, A. L., *Izv. S.S.S.R. Mech.* (1959), No. 1.
[2] Gonor, A. L., *P.M.M.* (1970), **34**, No. 3.
[3] Babaev, D. A., *J.V.M.M.F.* (1962), **2**, No. 6.
[4] Voscresenskiy, G. P., *Izv. S.S.S.R. M.J.G.* (1968), No. 4.
[5] Gonor, A. L. and Ostapenko, N. A., *Izv. S.S.S.R. M.J.G.* (1970), No. 3.
[6] Gonor, A. L. and Ostapenko, N. A., Institute of Mechanics, Moscow University, Rep. No. 1164, 1970.
[7] Lapygin, V. I., *Izv. S.S.S.R.* (1971), No. 3.
[8] Gonor, A. L., Shwets, A. I. and Kaskov, M. N., Institute of Mechanics, Moscow University, No. 1, 1970.

C. BIOASTRONAUTICS

SUMMARY OF A 90-DAY MANNED TEST OF A REGENERATIVE LIFE SUPPORT SYSTEM

ALBIN O. PEARSON

NASA Langley Research Center, Hampton, Va., U.S.A.

and

JOHN K. JACKSON

McDonnell Douglas Astronautics Company

Abstract. A 90-day manned test of a regenerative life support system was completed on September 11, 1970. The test was performed with a crew of four men and featured closed chamber operation with no resupply. All food, makeup water, spare parts, and tools were stored onboard at the start of the test. This paper presents the test objectives, describes the life support subsystems, outlines some of the operating procedures, and reviews some of the more significant accomplishments of the test. Conclusions and some recommendations for additional future efforts are presented.

1. Introduction

The National Aeronautics and Space Administration (NASA) is vitally involved in developing the life support technology required to support future long-duration manned space flight. As part of this life support program, extensive research and development is being pursued on individual life support subsystems [1]. Periodically, however, it becomes a necessary part of the program to assemble all of the various elements into a total test situation which can provide an insight into how individual subsystems interact with each other and with man, and how all elements operate as a unified, integrated system. Such tests simulate the closed ecology of an actual spacecraft with the men and machines interacting to provide the life support functions. These simulations have become increasingly sophisticated over the years, the degree of closure with relation to passing of materials in and out of the simulator has improved, and more advanced and effective subsystems have become available. These tests have expanded over the years from experiments lasting only a few days [2] to a recent test of 60 days [3 and 4], and now to a 90-day manned test.

It is emphasized that the 90-day manned test experiment was a research and development effort oriented primarily toward the development of life support hardware; however, physiological, psychological, and microbiological studies of man were conducted to assure the health and safety of the crew and as secondary efforts not to interfere with the hardware development tests. The subsystems used in the test incorporated advanced concepts applicable for use in space flights of the near future, but they were not fully flight qualified. These subsystems were maintained and repaired as necessary by the crew members, using only tools and spare parts stored onboard. The test was performed by McDonnell Douglas Astronautics Company under NASA Contract NAS1-8997 managed by the Langley Research Center,

L. G. Napolitano et al. (eds.), Astronautical Research 1971, 149–162. All Rights Reserved.

and was completed on September 11, 1970. This paper presents an overview of the 90-day manned test including a review of the more pertinent results and most significant accomplishments.

2. Test Objectives

The following objectives were established to provide the necessary goals and guidelines for test planning and performance:

(1) To demonstrate a capability to operate a multiman life support system in a continuous regenerative mode for a 90-day period without resupply. The system must provide a habitable atmosphere, food and water for nutritional support, and personal accommodations consistent with man's needs in the areas of personal hygiene, waste management, comfort, and health. The system will include regenerative oxygen and water loops. It will be a goal to minimize the amount of stored, expendable materials required for the test.

(2) To evaluate a number of advanced life support subsystems, using the proven subsystems from earlier tests as backup, and obtain operating experience and performance data for continuous testing with realistic conditions of manned loads and subsystem interaction.

(3) To obtain total life support system and subsystem performance characteristics which include a material balance, a thermal balance, and power requirements.

(4) To operate with no materials passed into or out of the test chamber for the maximum duration possible. If resupply were required, it would continue to be an objective to hold the passing in and out of materials to a minimum and, whenever feasible, materials to be passed into the chamber would be sterilized.

(5) To demonstrate man's capability to perform in-flight maintenance as a means of increasing system reliability to demonstrate the capability for in-flight monitoring of the necessary human, environmental, and systems parameters.

(6) To obtain data which will assist in determining the precise role of man in performing in-flight experiments, assist in determining the practical benefits of manned activity in space, and assist in validating mathematical models of space missions.

(7) To obtain data on physiological and psychological effects of long-duration exposure of the crew to confinement in the cabin atmosphere, on long-term group dynamics, and on crew work–rest cycles.

3. Equipment and Procedures

The 90-day manned test was performed with a crew of four carefully selected and well-trained men in a space station simulator (SSS) (Figure 1) which was a double-walled horizontal cylinder, 3.66 m (12 ft) in diameter and 12.2 m (40 ft) in length. The 116 m^3 (4100 ft^3) steel chamber was operated at the design environmental conditions shown in Table I. In order to maintain a realistic chamber environment with respect to control of outside contamination, the annular space between the simulator

Fig. 1.   Space station simulator for 90-day manned test.

inner and outer walls was evacuated and retained at a pressure approximately 1.24 kN m^{-2} (5 in. of water) below that of the cabin, thus assuring that the normal cabin leakage was outboard. The outer chamber wall was coated with 10.2 cm (4 in.) of thermal insulation to minimize thermal, and to some extent, acoustic transmissions.

An airlock with a volume of 4.5 m^3 (160 ft^3) was provided at one end of the chamber for crew entrance and egress. A pass-through port containing an autoclave was used for weekly passout of various samples for analysis. It was sterilized prior to each

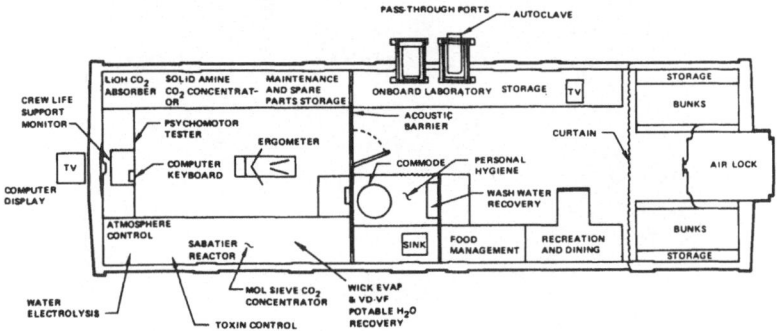

Fig. 2.   Space station simulator interior arrangement for 90-day manned test.

use to prevent microbial contamination of the test chamber from external sources. The airlock and pass-through port were normally maintained at the annulus pressure of 1.24 kN m^{-2} (5 in. of water) below cabin pressure.

TABLE I
Space station simulator environmental design criteria

Total pressure	68.9 ± 2 kN m^{-2} (517 ± 15 mm Hg)
Oxygen partial pressure (nitrogen diluent)	20.7 ± 0.67 kN m^{-2} (155 ± 5 mm Hg)
CO$_2$ partial pressure	0.506 kN m^{-2} (3.8 mm Hg)
Cabin temperature	294° ± 2.8 K (70° ± 5 °F)
Relative humidity	40 to 70%

Figure 2 shows the interior arrangement of the simulator. The main cabin area was separated into two compartments by means of an acoustic barrier. Results from prior tests, such as the 60-day test of [3] and [4], have shown the need for realistic noise control and lead to establishing test design noise levels of noise criteria adjusted (NCA) 60 [5] in the main equipment area and NCA 50 in the crew living area.

Fig. 3.   Space station simulator interior equipment area, viewed from the entrance to the crew living area.

Fig. 4. Space station simulator crew living area, viewed from the entrance to the equipment area.

The equipment area (Figure 3) contained the command center located at one end and composed of the crew life support monitor, a psychomotor test console, and a computer-linked keyboard. All of the mechanical equipment of the environmental control system and the majority of the life support equipment and instrumentation were also located in this equipment area.

The crew living area (Figure 4) included space for food storage and preparation, a folding table and chairs for use during eating and recreation activities, an enclosed area for waste management and personal hygiene functions, and an onboard laboratory for basic microbiological and biochemical analysis of water and blood. Four bunks were arranged in double-deck fashion, two on each side of the airlock, and were isolated from the main living area by fire retardant Armalon draperies.

The environmental control/life support system (EC/LSS) evaluated during the 90-day test is shown schematically in Figure 5. The system included advanced subsystems being evaluated for the first time in a manned test, with redundancy provided by subsystems well proven by prior manned testing.

In order to obtain maximum assurance of accomplishing the test objectives, it was decided early in the program planning to incorporate the greatest possible realism in all phases of test operations. Crew activities were carefully timelined and, as near as possible in a test situation, were representative of tasks likely to be performed on

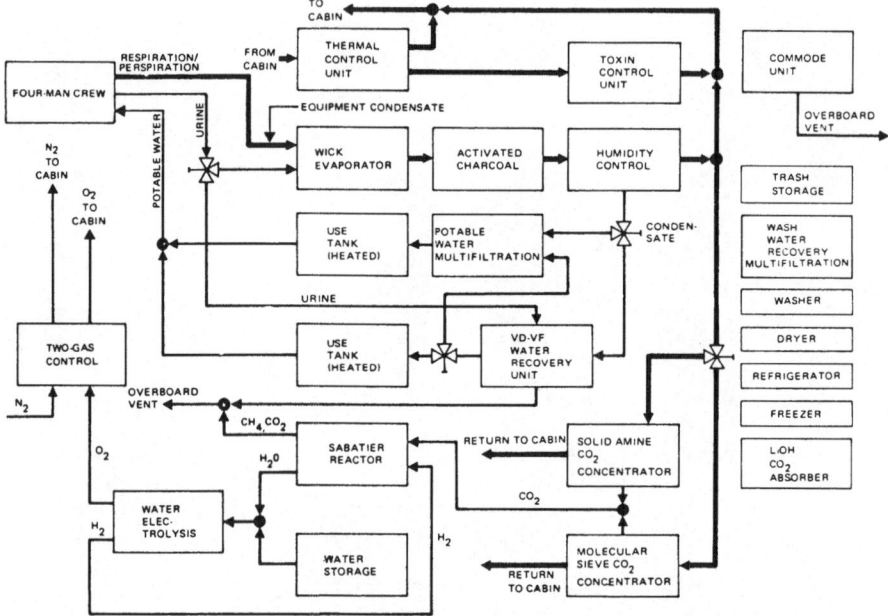

Fig. 5.   Block diagram of the life support system for the 90-day manned test.

actual flight missions. This realism was felt to be an important factor which might have an effect on such basic parameters as food consumption, task performance, physiological changes, motivation, and morale. An important aspect of effectively establishing mission realism was the incorporation of a significant number of experiments and studies other than those associated with routine test, maintenance, and repair of life support equipment.

The acquisition and handling of data was a combined task involving both the onboard crew and the outside test monitors. Sufficient instrumentation was provided onboard for the crew to operate the life support system and manage their own expendables. An onboard computer link was used to maintain up-to-date inventories of water, food, and spare parts. The life support system engineering performance data were recorded on magnetic tape for periodic processing on a computer. Man–system data were collected relative to mission task analysis, acceptance of habitability features, and overall physiological and psychological performance.

4. Test Results

4.1. LIFE SUPPORT SYSTEM

The system provided all of the crew's needs for the entire 90 days of continuous operation, with no requirement for pass-in operations through the autoclave pass-through port to resupply expendables, spare parts, or tools. The results obtained during the performance of the test are presented in detail in [6]. Some of the more significant accomplishments of this test, however, are presented briefly in this paper and show the extent to which some of the stated pretest objectives were met.

Throughout the test, the simulator environment was carefully monitored for trace contaminant buildup. In all, only 13 contaminants were detected of the nearly 100 compounds for which chromatograph calibrations existed. Table II shows the types

TABLE II
Atmospheric trace contaminants

	Measurement threshold	Quantity		Test contingency level
		Median[a]	Maximum	
	(ppm)	(ppm)	(ppm)	(ppm)
Metabolic products of crewmen				
Methane	10.0	150.0	290.0	—
Carbon monoxide	2.0	17.0	26.0	100.0
Aldehydes	0.05	0.32	0.47	15.0
Ethyl alcohol	0.2	0.5	1.5	300.0
Acetone	0.05	0.5	2.39	—
Residues of pretest cleaning or solvents				
Freon 113	0.20	4.0	11.50	150.0
Toluene	0.05	0.10	0.15	30.0
Methylethyl ketone	0.05	0.10	0.27	—
Dichloroethane	0.05	0.10	0.25	—
Decomposition products				
Ammonia	0.05	1.5	4.0	75.0
Oxides of nitrogen	0.05	Trace	0.15	1.5
2-ethyl butanol	0.05	0.2	0.45	20.0
2-ethyl hexanol	0.05	0.2	0.65	—

[a] Median of non-zero data from daily samples.

and concentrations of each of these compounds. The levels of the contaminants were well below the pretest established limits, which showed that the cabin atmosphere was relatively clean. Freon TF (113) was present in the environment, although well within the physiological limit, and was apparently concentrated by the CO_2 concentrator during removal of carbon dioxide and resulted in poisoning of the Sabatier reactor catalyst bed. In general, there were only occasional notifications from any of the crew members regarding detectable or obnoxious odors which were of a transient nature.

All water used during the entire 90 days was processed from onboard sources. The potable water was reclaimed from urine and humidity condensate and a total of 1070 kg (2357 lb) was certified for crew consumption. Of this total, approximately 58% was processed by the radioisotope powered Vacuum Distillation/Vapor Filtration (VD/VF) advanced subsystem and the balance by the wick evaporator unit. Table III presents details of water amounts collected for reclamation and also a listing of water amounts dispensed. It may be seen that humidity condensate accounts for about 66% of the total collected for processing with the balance made up predominantly of urine plus urine flush water. It is pointed out, however, that the humidity condensate was greater than might normally be expected due to an additional

TABLE III
Potable water production

	Totals		Average rate	
	kg	lb	kg/day	lb/day
Humidity condensate				
Respiration and perspiration	477.8	1053.3	5.31	11.70
Wash water evaporation	254.2	560.3	2.82	6.23
Subtotal	731.9	1613.6	8.13	17.93
Cabin humidity caused by solid amine operation[a]	487.7	1075.1	7.01[a]	15.45[a]
Total humidity	1219.6	2688.7	13.55[b]	29.87[b]
Urine and urinal flush	561.1	1236.9	6.23	13.74
Recycled water samples	19.7	43.4		
misc. sources	39.6	87.2		
Total, all sources	1840.0	4056.2	20.44	45.07
Distribution prior to certification				
Makeup to wash water	232.7	512.9	2.59	5.70
Makeup to solid amine unit[a]	520.7	1148.0	7.49[a]	16.52[a]
Samples passed overboard	10.0	22.0		
Miscellaneous	2.0	4.4		
Urine accumulator net increase	9.1	20.0		
Holding tanks net decrease	−3.5	−7.8		
Total distribution	770.9	1699.5	8.56[b]	18.88[b]
Recovered, certified potable water	1069.0	2356.7	11.88	26.19
Losses after certification	141.6	312.1	1.57	3.47
Consumed by crew	927.4	2044.6	10.31	22.72

[a] Solid amine unit based on 69.50 days of operation.
[b] Based on overall 90-day totals.

amount of moisture discharged into the cabin by the solid amine CO_2 removal subsystem. The VD/VF unit was used for a total of 63 days and produced 614 kg (1353 lb) of water for a recovery rate of 94.3%. A total of 318 kg (701 lb) of this water was further processed by multi-filtering to improve the taste or to remove microbial contamination which occurred during the latter portion of the test in the unit condenser. The major problems of the VD/VF unit were caused by urine overflow on three separate occasions. On each occurrence, the system required flushing and then sterilizing by means of an onboard steam sterilizer. This procedure eventually could not maintain condensate sterility, necessitating reprocessing of the product water in the multifiltration unit. No problems were encountered with the $^{238}PuO_2$ radioisotope heat sources. Five of these units produced a total thermal output of 350 W and during 76 manipulations of the capsules resulted in crew radiation exposures well below design safety limits.

The potable water storage and distribution system operated satisfactorily for the entire test except for the cold-water dispensers which became bacterially contaminated and were not used after the third test day. The stringent requirement of less than 10 micro-organisms per milliliter of water could not be met at these dispensers, hence the drinking water was drawn from a hot dispenser and chilled in the refrigerator prior to consumption.

It became apparent that improved techniques for determining potability of water are required, particularly in the area of chemical analysis. To accomplish the analysis presently recommended requires considerable instrumentation and a relatively large amount of water for samples.

The wash water multifiltration recovery unit processed a total of 5077 kg (11 182 lb) of water as shown in Table IV. The bulk of this water, 4744 kg (10 448 lb), was used for washing and included 44 loads of clothing. On test day 35, the system was drained and the water changed because of crew complaint and rejection of the water. The water again became objectionable and, on day 40, a carbon column was changed with a distinct improvement in the water quality. Examination of the felt retainer pad in the carbon column revealed an odorous, confluent microbial growth believed to be responsible for the poor water quality. The carbon and resin columns were changed on day 52, and by the end of the test were in need of another change.

TABLE IV
Wash water system summary water balance
for four men, 90 days

	kg	lb
Water produced:		
Multifiltration unit	5077	11 182
Water used:		
Including evaporation		
loss – 254 kg (560 lb)		
Washing	4744	10 448
Reprocess	227	500
Urinal flush	94	207
Phase change	40	88
Miscellaneous	6	13
Inventory change	– 34	– 74
	5077	11 182

CO_2 was removed from the simulator cabin by a solid amine unit during the majority of the first 80 days of test. The backup molecular sieve CO_2 concentrator unit was used during test days 14 through 25, 33, 34 and 81 through 90. The variation of CO_2 partial pressure with test day is given in Figure 6. It may be seen that the average CO_2 partial pressure for the 90-day period was about 0.67 kN m^{-2} (5 mm Hg) with the maximum peak in the order of 1.51 kN m^{-2} (11.3 mm Hg). The major problems encountered with the amine unit were maintaining correct percentage of water in the beds during absorb and desorb cycles and keeping the inlet air at the correct temperature for the bed conditions. A number of mechanical problems were encountered such as a sticking valve that required manual cycling, a solenoid valve sticking open, and clogging of the water drain in the exhaust condenser. When all adjustments were correct, the unit demonstrated the capability to maintain the CO_2 level at a partial pressure below the pretest desired level of 0.506 kN m^{-2} (3.8 mm Hg). The molecular sieve CO_2 concentrator unit performed satisfactorily in its backup role with only minor problems encountered. There was no requirement to operate the LiOH unit.

The recovery of oxygen was performed by electrolysis of water which was produced by concentrating exhaled carbon dioxide from the simulator environment and

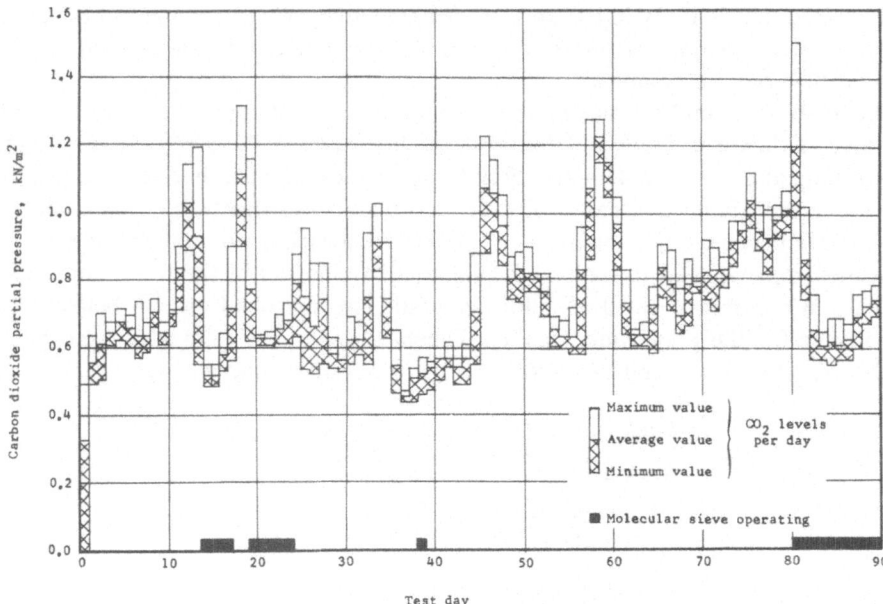

Fig. 6. Variation of carbon dioxide partial pressure with time.

combining it with hydrogen in a Sabatier reactor. Two experimental water electrolysis units were tested. One of these, a vapor feed unit, was located inside the simulator. The second unit, a circulating electrolyte type, was physically located outside the chamber but was installed in a closed-loop fashion so that it was supplied with water directly from inside the chamber and fed its gases back to the simulator in a return loop. A third unit was a commercial electrolyzer located outside the simulator and used as a backup to eliminate the need for a large number of backup gas storage bottles. A summary of the gases produced by each of these units is given in Table V. Approximately 65.7% of the oxygen required for leakage and metabolic consumption was provided by the circulating electrolyte unit and about 2.6% by the vapor feed unit. The balance was provided by the commercial electrolyzer except for a small amount of oxygen used from the backup gaseous storage to meet transient needs.

The onboard vapor feed unit operated for a total time of only 98 h. Although a number of attempts were made to repair and restart this unit after its shutdown on

TABLE V

Summary of oxygen and hydrogen supplied to the space station simulator

Source	Oxygen		Hydrogen	
	kg	lb	kg	lb
Vapor feed electrolysis	10.17	22.40	1.27	2.80
Circulating electrolyte	256.97	566.02	25.04	55.15
Electrolyzer	117.27	258.30	10.46	23.00
Gaseous storage	6.95	15.30	0.00	0.00
Total	391.36	862.02	36.77	80.95

test day 3, they were all unsuccessful and the unit was secured and permanently shut down on test day 20.

The circulating electrolyte unit located outside the chamber operated for 70.1 days during which it supplied gases to the simulator for 62.2 days, and in a standby mode vented the generated gases to ambient for 7.9 days. A number of problems were encountered with this unit, but its accessibility permitted repair by skilled personnel and hence its operation was continued.

TABLE VI
Maintenance and repair summary

Activity	Items	Hours
Waste management		
Commode	4	6.2
Urine phase separator	8	14.3
Water management		
VD–VF	21	23.9
Wick evaporator	4	0.3
Humidity control	33	16.2
Potable multifilter	4	5.6
Wash water recovery	19	4.8
Atmosphere purification		
Solid amine concentrator	47	33.4
Molecular sieve concentrator	8	4.4
Toxin control	0	0.0
Thermal control	0	0.0
Atmosphere supply and pressurization		
Sabatier reactor	12 (8)[a]	12.6 (6.0)[a]
Electrolysis (vapor feed)	46 (1)	25.5 (1.0)
Electrolysis (circulating electrolyte)	0 (16)	0.0 (83.2)
Two-gas control	0	0.0
Mass spectrometer	5	4.1
Baseline two-gas control	0	0.0
Baseline two-gas sensors	1	0.5
Less subtotal	212 (25)	151.8 (90.2)
Miscellaneous items		
Defrost refrigerator	9	4.5
Cleaned cabin floor	9	9.0
Cleaned dewpoint mirrors	4	1.5
Corrected leak in trash container	2	0.3
Repaired psychomotor pedal	1	1.0
Deactivated smoke alarm head	2	0.7
Repaired radiation monitor	1	0.3
Repaired particle counter	1	0.5
Replaced videcon tube	1	0.5
Repaired TV coaxial cable	5	2.5
Installed intercom call light	1	1.0
Attempted repair of visual sensitivity tester	4	6.7
Scheduled maintenance		23.0
Miscellaneous and scheduled subtotals	40	51.5
Totals	252 (25)	203.3 (90.2)

[a] Numbers in parenthesis represent activity by test monitors on equipment located outside of the space station simulator.

The crew performed 212 onboard maintenance and repair tasks on the life support system (Table VI) and 40 tasks on various support equipment. The total time involved in those activities amounted to 203.3 h for an average of 2.3 h/day. The crew clearly demonstrated the capability to perform onboard maintenance as a means of increasing system reliability.

4.2. EVALUATION OF HABITABILITY PROVISIONS

Although there is no doubt as to the importance of the life support system in sustaining man in the hostile environment of space, the significance of various habitability features is not completely understood with respect to acceptance for long-term space missions. A number of these habitability features were evaluated during the 90-day test including personal hygiene and acoustic level. The evaluations were based on crew reactions and comments during the test and their comments in the post-test debriefings.

Personal hygiene was accomplished by wet washcloth bathing with the use of 'Basic H' cleanser if desired. Negative comments were directed at the degradation in wash water quality previously mentioned and, by some crew members, at the apparent difficulty in removing the cleanser from the skin and a resultant feeling of oiliness. One crewman discontinued use of this cleanser. All remarked that shower bathing would have been desirable but not mandatory.

Measured ambient noise levels corresponded to approximately NCA 70 in the equipment area, NCA 55 in the living quarters, and NCA 45 in the bunk area which is equivalent to an overall sound pressure level of 80, 74, and 69 dB, respectively. In general, the crew members expressed some irritation with noise levels in the equipment area. Other complaints included sleep disturbance by intermittent noises from equipment such as pumps, and from conversation of on-duty crew members.

4.3. BIOMEDICAL AND MICROBIOLOGICAL STUDIES

Although the 90-day test was primarily an evaluation of life support equipment, biomedical and microbiological studies were conducted toward surveillance of crew health. Included in this effort were pre- and post-test physical examinations, daily measurements of body weight, pulse, blood pressure, and oral temperature, weekly analysis of urine samples, and biweekly measurement of blood chemistry and hematological indices. In addition, the microbial flora of the crewmen, the atmosphere, and various surfaces in the SSS were determined and the water recovery system and product water were monitored to assure conformity with established potability criteria. In summarizing the results of these studies, no significant adverse changes in crew health status were revealed. Distinct microbial contamination of the reclaimed potable water was detected on several occasions, but the contaminated waters were reprocessed to meet essential sterility criteria prior to crew consumption. The results of the biomedical, psychological, and microbiological data tend to indicate that the test environment was generally neutral and not unduly stressful to the crew. Interpersonal transfer of some micro-organisms took place during the test, but no clinical illness resulted.

5. Conclusions and Recommendations

An operational 90-day manned test of a life support system was successfully completed on September 11, 1970. This test was performed with a crew of four carefully selected and well-trained men in a space station simulator (SSS) which had a two-gas atmosphere maintained at a total pressure of 68.9 kN m^{-2} (517 mm Hg) and composed of oxygen at a partial pressure of 21 kN m^{-2} (155 mm Hg) with nitrogen as the diluent. The test was planned to provide data in a closed ecology, similar to that of a space station. All crew equipment and expendables were stored onboard at the start of the test to eliminate the need for pass-in operations. During the course of the test, passout operations were conducted on a once-per-week basis, primarily to provide samples to verify the health of the crew and to obtain basic medical data. On the basis of the results from this 90-day test, the following conclusions and recommendations can be made:

(1) The life support system used in this test provided all the crew's needs for the 90-day period. Although shown as practical, the recovery of oxygen from atmospheric CO_2 requires additional development and improvement of water electrolysis units.

(2) A clean, physiologically acceptable environment was maintained by proper selection of materials used inside the simulator chamber and by proper operation of the trace contaminant removal system.

(3) Recovery of water from urine and atmospheric humidity was accomplished and met the potability standards of this test. Methods for onboard monitoring, particularly under flight conditions, are not apparent and reevaluation of the criteria and instrumentation required for flight are recommended.

(4) Radioisotope sources for thermal energy were safe and practical as handled during this test.

(5) Sufficient wash water to accommodate the crew's needs for personal hygiene and for washing of clothing and bedding was reclaimed. Microbial growth in the wash water multifiltration system was a recurring event.

(6) A solid amine carbon dioxide concentrator was used in a manned test for the first time and demonstrated that it was effective in maintaining an acceptable carbon dioxide level. Further work, however, is required to improve the performance and reliability prior to additional testing.

(7) All maintenance and repair tasks were performed by the crew using tools, spare parts, and equipment stored onboard at the start of the test. The crewmen clearly demonstrated man's capability to perform both routine and unscheduled repair and, in many instances, reflected a high degree of ingenuity in performing these tasks.

(8) Washcloth bathing was acceptable provided the wash water quality did not degrade and if a suitable cleanser were available. In general, however, the crew would have appreciated a shower facility.

(9) Measured noise levels of NCA 70 in the equipment area created some irritation to the crew members. The lower levels of NCA 55 and NCA 45 in the crew living quarters and bunk area, respectively, appeared satisfactory, although intermittent

noises from equipment and on-duty crew member conversations were a source of complaint.

(10) The results of the biomedical, psychological, and microbiological data tend to indicate that the test environment was generally neutral and not unduly stressful to the crewmen. No significant adverse changes in health were revealed.

References

[1] Ingelfinger, A. L. and Secord, T. C., 'Life Support for Large Space Stations', AIAA Paper No. 68-1032, presented to AIAA Fifth Annual Meeting and Technical Display at Philadelphia, Pennsylvania, October 1968.

[2] Pecoraro, J. N., Pearson, A. O., Drake, G. L., and Burnett, J. R., Contribution of a Developmental Integrated Life Support System to Aerospace Technology. AIAA Paper No. 67-924, presented to AIAA Fourth Annual Meeting at Anaheim, California, October 1967.

[3] Bonura, M. S. et al., '60 Day Manned Test of a Regenerative Life Support System With Oxygen and Water Recovery. Part I – Engineering Test Results', McDonnell Douglas Astronautics Company – Western Division. NASA CR-98500, December 1968.

[4] Taliaferro, E. H. et al., '60 Day Manned Test of a Regenerative Life Support System with Oxygen and Water Recovery. Part II – Aerospace Medicine and Man Machine Test Results', McDonnell Douglas Astronautics Company – Western Division. NASA CR-98501, December 1968.

[5] Beranek, L. J., Noise Reduction, McGraw-Hill, New York, 1960.

[6] Jackson, J. K. et al., 'Test Report, Test Results, Operational 90-Day Manned Test of a Regenerative Life Support System', McDonnell Douglas Astronautics Company. NASA CR-111881, May 1971.

ADVANCED METHODS OF RECOVERY FOR
SPACE LIFE SUPPORT SYSTEMS

S. V. CHIZHOV, B. A. ADAMOVICH, YU. E. SINYAK,
V. B. GAIDADYMOV, Z. P. PAK, M. I. SHYKINA,
I. N. FETIN and V. V. KRASNOSHCHEKOV

U.S.S.R. Academy of Sciences, ul. Vavilova 32, Moscow B-312, U.S.S.R.

To reclaim water from water containing wastes various physical and physico-chemical methods could be employed. The selection of the optimum method of water recovery is a very complicated task. It takes long-term medical and engineering experiments, such as the one-year Soviet experiment or the 23-, 60- and 90-day NASA sponsored tests in the USA to obtain truly reliable evaluation results for different water recovery assemblies.

Water sources in spacecraft and space stations include water-containing wastes, such as atmospheric humidity condensate, urine, wash and personal hygiene water, higher plants transpiration on water, chlorella reactor condensate, higher and lower plants nutrient solution wastes, as well as certain technological solutions, such as electro-chemical generator (fuel cell) condensate, hydrogen peroxide decomposition condensate and carbon dioxide catalytic hydration condensate. The quantitative ratio and chemical composition of some of them are shown on Tables I and II. The data indicate the diversity of sources and complexity of chemical composition of moisture-containing wastes.

TABLE I

Quantitative ratio of the moisture content wastes in manned spaceships

No.	Water sources for regeneration	Quantity, litres	Note
1.	Atmospheric water condensate	1–2	
2.	Urine	1–1.5	
3.	Feces	0.1–0.3	
4.	Sanitary hygienic water	1.8–10	up to 25 l in extended spaceflights
5.	Kitchen water	0–3	
6.	Transpiration higher plant moisture	85	
7.	Chlorella reactor condensate	10–15	
8.	Fuel cell moisture release	11	with 1 kW power
9.	Hydrogen peroxide decomposition water	0.9	when receiving 0.8 kg O_2

It is characteristic for moisture-containing wastes that they include a wide range of organic substances which are (or tend to be) highly contaminated by micro-organisms. The composition of trace contaminants, as well as a number of additional require-ments specific to regenerative life support systems in space, make the reclamation of

L. G. Napolitano et al. (eds.), Astronautical Research 1971, 163–169. All Rights Reserved.

TABLE II

Main properties of the formulation of some moisture-content wastes

No.	Physico-chemical indices	Units	Water sources						
			Electro-chemical generator condensate	Atmospheric water condensate	H_2O_2 condensate	Urine	Sanitary hygienic water	Kitchen water	Green-house transpiration water
1.	Total quantity of contaminants	%	0.1	0.1	0.1	5	3	5	0.1
2.	Permanganate oxidability	(mg/eq O_2)/l	7–100	4–450	1	20 000	450–600	1200	12
3.	Urea	mg l^{-1}	0	0	0	10 000	100	—	0
4.	Sodium chloride	"	100–2500	0	0	—	300	150	0
5.	KOH	"	20	8–160	0	400–1200	60	—	0
6.	NH₃	"	10	0–480	0	0	0	0	53
7.	Alcohol	"	0.1–52	15–30	0	1300	—	—	0
8.	Acids	"	0	0–2	0	0	180–800	600–2000	—
9.	Fat	"	0	0–0.8	0	60–200	traces	—	0
10.	Calcium	(mg/eq)/l	0	0	0	0	0	0	0.5–14
11.	Hardness	mg l^{-1}	0	0	0	0	0	0	0.2–0.9
12.	Ag	mg l^{-1}	0		2.2	0	0	0	0
13.	Sn				10				0
14.	Microbial contamination	microbe ml^{-1}	0	10^5	0		1.2×10^5		—

moisture-containing wastes an extremely complicated business. These requirements include high water recovery efficiency and high quality of produced potable water, the capability to operate in zero-g environment, simplicity of maintenance, reliability and low equivalent weight.

These requirements are best met by the system based on sorption processes. However, since the sorbent bed capacity is limited this method of water regeneration could be successfully applied to slightly contaminated wastes only. It is also the main technique for the final purification of water produced by any other regeneration method.

In short-term missions and missions with regular resupply it could be used for wash and hygienic water recovery. The sorption technique is also applicable to mission durations exceeding one year with onboard sorbent supplies or possibilities for onboard sorbent regeneration.

TABLE III

Averaged data on the composition of the atmospheric water condensate in the year-long medico-technical experiment

No.	Physico-chemical indices	Units	Concentration of the contaminants before incorporation of the green-house	Concentration of the contaminants after incorporation of the green-house
1.	Permanganate oxidability	(mg/eq O_2)/l	35	5
2.	Bichromate oxidability	,,	160	48
3.	Ammonia	mg/l^{-1}	40.2	12
4.	Nitrate	,,	0.2	4.4
5.	Nitrite	,,	0.2	0.12
6.	Chloride	,,	54	7.6
7.	Hardness	(mg/eq)/l	0.3	0.88

The sorbents used in the sorption concept of water regeneration should meet a number of specific requirements. The ion-exchange resins and activated charcoal should possess high sorption capacity and adequate resistance against mechanical, radiation and thermal exposures. The best choice in this respect are strong acid and strong base ion-exchange resins (such as KU-2 and AV-17) and certain kinds of activated charcoal. It should be noted that activated charcoal layers should be the last sorbent beds along the water flow to provide the better sorption of non-ionized organic molecules and higher degree of water purification from ion-exchange resin destruction products. This technique of water reclamation was used to recover all above-mentioned types of moisture-containing wastes in laboratory conditions and the produced potable water conformed to the established water quality standards. It was also used to recover humidity condensate during the one-year medico-engineering experiment. The sorbent beds used were cation exchanger KU-2 × 8, anion exchanger AV-17 and AN-31 and activated charcoal AG-5. Table III shows the average data for the humidity condensate constituents during the one-year experiment. It should be noted that the incorporation of the green-house affected the condensate composition lowering the contamination level. Before the green-house incorporation it took about 10 ml of sorbent to recover 1 l of water while after the incorporation 4 or 5 ml of sorbent was enough for the same amount of water. The recovery of water from the humidity condensate at the same time contributes to the general purification of the spacecraft atmosphere from trace contaminants. During the one-year experiment the humidity condensate purification assembly removed 105.1 gm of organic matter, 78.2 gm of ammonia and its compounds, 3.6 gm of ketones, 3.4 gm of aldehydes. During the same time the atmospheric purification assembly removed 125 gm of organic matter, 0.7 gm of ammonia and its compounds, 1 gm of ethanol and 1.4 gm of acetone. In all the humidity condensate purification assembly processed throughout the experiment 1621 l of water and produced 1312 l of good quality potable water.

The most difficult problem is water recovery from highly contaminated wastes: urine, wash and personal hygiene water. In laboratory conditions the following methods of water recovery were used to produce the potable water: vacuum distillation, sublimation, electrodialysis, reverse osmosis, membrane diffusion and air evaporation in conjunction with catalytic oxidation of volatile trace contaminants. A comparative analysis for all these concepts has been carried out. The minimum equivalent weight of a water recovery assembly was selected as the main criterion. The calculation technique included the determination by approximate methods of energy loads on the main units of the assembly: evaporators, condensators, catalytic reactors, pumps, fans, etc. From obtained values of energy loads and rates the masses of respective units were estimated by the use of specific mass values. The specific mass calculations for the equipment necessary to remove the excessive heat to space were done in the same way. The masses of other units (vessels, post-treatment units, resupplies) were taken from experimental data. The results of the equivalent weight calculations for missions of different duration showed at the moment the most advanced concept of water recovery from water-containing wastes of high and moderate degree of contamination is the concept of air evaporation with catalytic oxidation. Its additional advantages include low power requirements (the use of the thermal energy isotope sources, such as plutonium 230 is possible), low post-sorbent requirements for recovered water post-treatment and high potability of the produced water. It decided the choice of an assembly to use for water reclamation from the urine during the one-year experiment. The assembly included an evaporator connected to the air loop, catalytic reactor, condenser, pumps, system of water post-treatment and conditioning and a quality sensor. During the whole experiment 1500 l of urine were processed to produce 1297 l of potable water. Table IV shows the chemical composition of the water recovered from urine and wash-personal hygiene waters.

For water recovery from the wash and personal hygiene water, apart from the sorption method that has limited application (in short-term missions and general distillation processes), the methods of coagulation and reverse osmosis could be used. The former was used for regeneration of wash and personal hygiene water during the one-year experiment. The 10% aluminium sulphate solution was utilized as the coagulating agent. Preliminary laboratory experiments demonstrated that the highest degree of purification from particulate and microbial contaminants is achieved at the coagulating agent expenditure rate of 350–500 mg per litre while using fatty detergents. Results of the one-year experiment confirmed this conclusion. Using this technique, 587 l of wash and personal hygiene water were processed, of which 575.5 l were returned for further use.

The specificity of the problem of supplying the astronauts with the high quality potable water obtained through regeneration processes is related to the fact that the obtained water is too similar to distilled water, therefore is biologically deficient and needs secondary treatment.

The process of water conditioning includes enriching it with the main mineral constituents, to obtain better palatability and physiological integrity, followed by sterilization. The desirability and even necessity of enriching the regenerated water

TABLE IV

Chemical composition of water regenerated from urine and sanitary wastes

No.	Physico-chemical indices	Units	Water regenerated from urine	Regenerated sanitary-hygienic water
1.	Active reaction 'pH'		7–8	5.5–6
2.	Transparency	cm	30	30
3.	Colourity	degree	20	20
4.	Smell at 20 °C	points	0	1
5.	Taste at 20 °C	,,	0	—
6.	Total hardness	(mg/eq)/l	3.1	0.75
7.	Chlorion content	mg/l^{-1}	90.8	15.0
8.	Ammonia nitrogen content	,,	0.3	2.5
9.	Permanganate oxidability	(mg/eq O_2)/l	0.72	3.7

chemical composition was confirmed by practical and experimental observations indicating to the adverse effect the melted snow and desalted water have on humans and animals because of its deficit of important macro- and micro-elements. Water supply system in space, as distinct from the ground-based, operates during relatively short periods of time, yet a number of specific peculiarities of space environment must be taken into consideration that can aggravate or provoke some changes in the body physiological functions in response to the prolonged consumption of the water deficient in minerals. Thus astronauts were found to display the dehydration or decalcination syndromes and other water-mineral balance changes.

While developing techniques for artificial mineralization of the recovered water the principal criterion, in our opinion, should be the duration of the mission. For 15- to 30-day missions it is enough to add some small amounts of mineral compounds to achieve better palatability of the regenerated water. They include cations responsible for water hardness, such as calcium or magnesium, and a number of anions, e.g., bicarbonates, chlorides and sulfates. For missions longer than 30 days it is desirable to add to the regenerated water a wider range of mineral salts, including fluorine and iodine. The most perspective techniques in this respect are those based on utilization of solid-state mineral compounds, such as mineral filters (artificial silicates, granulated dolomite, etc.) and mineral pills containing principal macro- and micro-elements. The introduction of controlled amounts of highly concentrated mineral solutions containing ions of calcium, magnesium, chlorine, sulfate, potassium, sodium, fluorine, iodine and bicarbonate could also be successfully used.

The microbiological purification of the recovered water is achieved by physico-chemical methods of sterilization. For water supply assemblies meant to operate during long-term missions the best technique is water sterilization and preservation with electrolytically produced silver ions. The experimental studies conducted in order to determine the magnitude and duration of the bactericidal effect of the electrolytic solution of silver during prolonged water storage showed that in presence of 0.1 mg l^{-1} of silver ions the bactericidal action of the water kills 10 000 millions of E. coli in one hour. The bactericidal effect persists for 3 yr if water is stored in glass vessels. When polymeric films are used for this purpose the rate of silver ion sorption by the container walls is increased which means an increase in preservative concentration from 0.1 to 0.2 mg l^{-1} for longer water storage. Along with the high bactericidal effect of silver its other significant advantage is neutrality towards palatability and physico-chemical indices of the potable water.

The intake, even during prolonged periods of time, of such insignificant amounts of silver ions with water does not result in their accumulation in or any adverse effect on animal or human body. It was confirmed by the toxicological experiments with administration to the animals of silver ions in amounts 200 to 500 times greater than those used for preservation.

The final stage of hygienic studies carried out to evaluate the biological and physiological acceptability of the recovered water and work out the recommendations for its use were extensive toxicological investigations with acute and chronic experiments on animals using modern biochemical and physiological techniques. Observations of hydrobiological objects (Daphnia rubra and the water tank fish gupi) placed in tested water, the isolated frog heart technique and the gauge method for tissue respiration study were used as express methods to reveal any possible adverse effect of the recovered water upon living organisms.

During the chronic experiment with warm-blooded animals living on the recovered water the hematologic indices, blood serum SH-group activity and liver antitoxic function were investigated. Different load tests were used and histochemical analyses of animal organs were carried out.

Integrated hygienic investigations of various recovered water samples permitted evaluation and laying down recommendations for practical use of a number of physico-chemical methods of water reclamation from water-containing human and technological wastes. They include concepts of catalytic oxidation, sorption, vacuum distillation, sublimation and electrodialysis. Water produced by these methods fulfil the hygienic requirements to the potable water. The technological feasibility of the main concepts of water reclamation, including the post-treatment conditioning, were repeatedly tested in animal chronic experiments and successfully used during manned experiments up to 1 yr long.

PART II

ENGINEERING AND MANAGEMENT ASPECTS OF SPACE TECHNOLOGY

A. SPACE TRANSPORTATION: EARTH TO ORBIT

SPACE TRANSPORTATION REPORT

DALE D. MYERS

National Aeronautics and Space Administration, Washington, D.C., U.S.A.

The need for economical space transportation has been recognized since the earliest days of the space age. Studies of reusable space transportation have been made in several countries for more than a decade. But now, with the technological progress of the past few years, we feel confident that this major forward step can be made in the 1970's and utilized operationally in the 1980's.

A fairly specific approach to such a system has been studied for more than two years. Basically, reusable space transportation does three things shown in Figure 1.

SPACE TRANSPORTATION SYSTEM BENEFITS

- A MAJOR REDUCTION IN THE DIRECT COSTS OF SPACE FLIGHT ACTIVITIES

- THE ABILITY TO MORE RAPIDLY EVOLVE MEANINGFUL AND CRITICAL SPACE USES

- THE STIMULATION OF ADDITIONAL USES WHICH ARE PRESENTLY UNECONOMICAL OR UNFORESEEN

- AN EXCELLENT PROGRAM FOR INTERNATIONAL COOPERATION BOTH IN THE R&D AND OPERATIONS PHASES

Fig. 1.

First, it offers the opportunity for a major reduction in the direct costs of space flight activities. Second, it provides the ability to evolve meaningful and critical space uses more rapidly. Finally, it will be a stimulus for additional uses of space that are now uneconomical or unforeseen. It also affords an excellent medium for international cooperation in both the research and development and operational phases and will provide a continued stimulus of world technology.

In considering reusable space transportation we must assume that the utilization of space is critical to the well being and future progress of man on our own planet. Space provides a global observation post, and a means of almost unlimited global

L. G. Napolitano et al. (eds.), *Astronautical Research 1971, 175–183. All Rights Reserved.*
Copyright © 1973 by D. Reidel Publishing Company, Dordrecht-Holland.

information flow. In space, we have access to a unique physical environment. And it is a primary arena of future exploration, discovery and expansion.

More is being learned about this potential every day. For example, at this very conference there are four space applications sessions chaired by Leonard Jaffe of the United States and K. Y. Kondratyev of the Soviet Union.

A simpler, more flexible and economical access to space is required if we are to exploit its utility to the fullest. The ways in which the space transportation system will influence payload design and program costs have been discussed [1] in other sessions of this congress [2]. It seems evident that with these possibilities the shuttle,

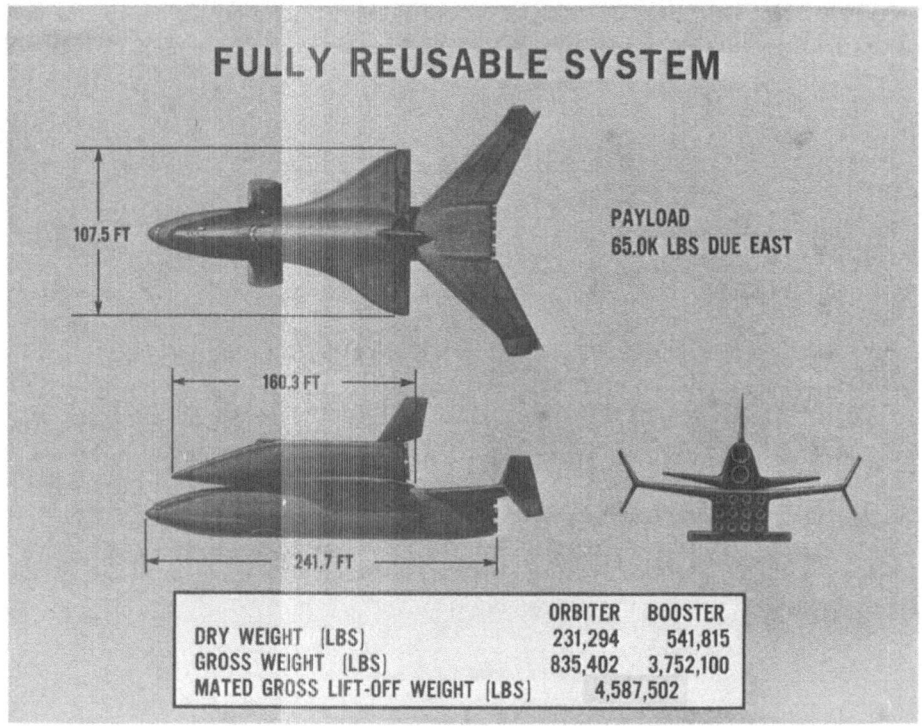

Fig. 2.

when it becomes truly operational, will result in a broad expansion of the use of space in a manner that no one can clearly foresee at this time. Despite the impressive list of activities projected for space in the future, it may well be that the most significant benefits from space are yet to be conceived and realized.

The principal elements of the space transportation system between Earth and Earth orbit consists of a space shuttle and a space tug. The shuttle will operate between the Earth and low Earth orbit. The tug will operate between low Earth orbit and higher orbits that include geostationary altitudes. NASA is well advanced in the work on the definition of the shuttle component of this system but our work on the space tug has proceeded only through the phase of preliminary studies. We are following with interest the work going on in Europe. The rest of this paper is devoted to the space shuttle.

The definition of the shuttle system that has evolved from studies over the past 15 months provides an interesting interplay between technology and economics.

It is important to keep in mind that a program baseline is expected to change as tradeoffs are studied. This is just what has occurred in the space shuttle definition. At the beginning, in June 1970, a shuttle was employing a fully reusable system, requiring minimum refurbishment. These requirements resulted in a very large orbital

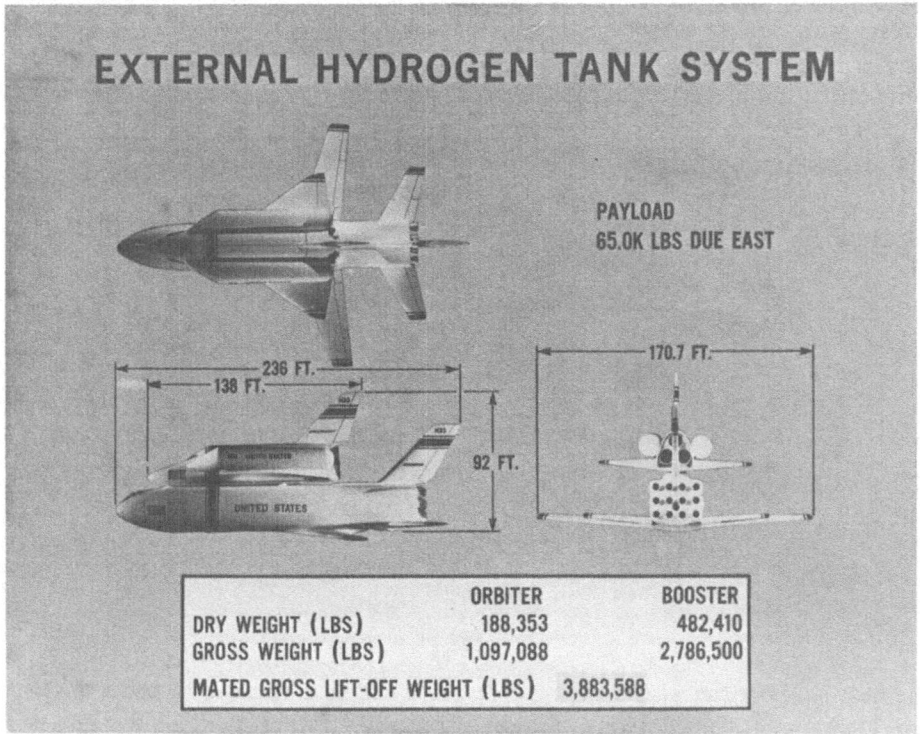

Fig. 3.

vehicle because of the necessity of enclosing all the hydrogen tankage within the body of the spacecraft (Figure 2). Other features of this baseline shuttle are a reusable thermal protection system, high-pressure main engines burning hydrogen and oxygen, auxiliary propulsion burning hydrogen and oxygen, advanced integrated avionics, crossrange capability of 1100 nautical miles, a payload bay 15 ft in diameter and 60 ft long, payload capability of 65 000 lb in a due-east orbit, and abort to orbit.

Such a system offers the lowest operational cost. But its size and technical sophistication projected very high development costs. The servicing and inspection of internal cryogenic hydrogen tanks also posed unique problems. The projected high development costs plus concerns about tank inspection procedures and safety then led to study of alternate approaches that would maintain to a high degree the low operational costs of the fully reusable shuttle. Our studies were, therefore, directed toward configurations in which the hydrogen tanks were outside the orbiter. These tanks would be carried into orbit then released. One such design would utilize external hydrogen tanks (Figure 3). This configuration resulted in considerably lower system weights and, consequently, lower development costs. But they would be accompanied by a modest increase in the operational costs because of expending the tanks.

The improved structural efficiency of the orbiter with the external tanks also resulted in a lower optimum staging velocity for the system which has made possible a booster with heat sink type of thermal protection. A heat sink thermal system absorbs a large increment of heat energy by distributing the energy throughout its entire structural mass, raising temperatures uniformly by a relatively small amount.

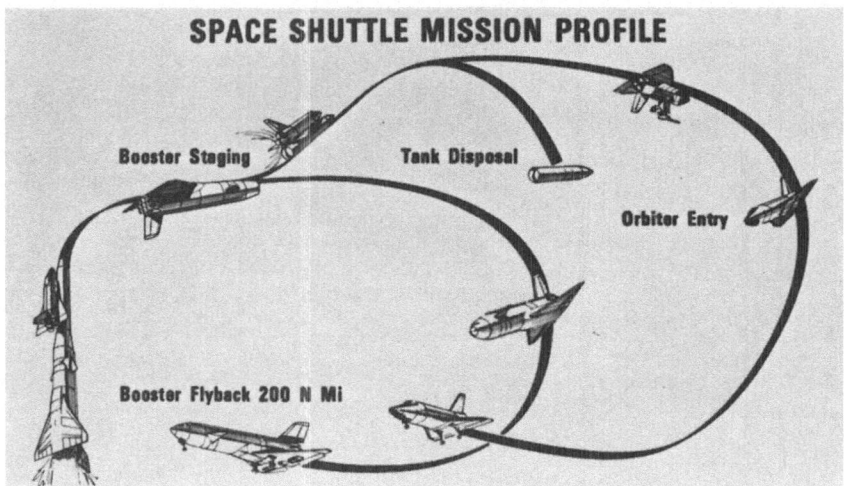

Fig. 4.

The heat sink booster also lowers the development and operational costs of the system. Operational considerations for the orbiter for certain abort modes also led to the use of three engines. Estimated dry weight savings of about 43 000 lb in the external hydrogen tank orbiter resulted from this system trade. The booster weight was also reduced substantially. The overall impact on the system design is about 700 000 lb reduction in gross lift-off weight of the system.

The lower staging velocity also influences the booster operations in a favorable way (Figure 4). In a typical mission profile associated with a drop tank orbiter, the booster can return to the launch site after staging at about 6000 ft s^{-1} only 200 mile downrange from the launch site. This contrasts with the 450 mile downrange distance for the fully reusable system. After staging, the orbiter, of course, proceeds into its planned Earth orbit. The release of the tanks in orbit can be programmed in such a way that they fall within a limited area in a remote region of the Indian Ocean when launched from anywhere in the continental United States, regardless of launch azimuth.

The orbiter in this illustration is shown with a single external tank in contrast with the twin tank arrangement illustrated in the earlier slide. It represents some current efforts on the possibilities of further design and operational advantages that might accompany removing oxygen tanks also. For example, the ability to separate the

Fig. 5.

orbiter from all propellant tanks at any selected point during ascent affords a number of optional abort modes beneficial to both crew safety and orbiter design.

By the conclusion of our Phase B studies at the end of October, we expect to have all the information necessary to select the best combination of technical and programmatic approaches for the Space Shuttle Program. At present, we have in hand detailed design studies of the external hydrogen tank orbiter coupled with the reusable heat sink booster, both employing advanced design hydrogen–oxygen rocket engines. We have also examined in depth the effects on the program of such important variables as: payload weight and compartment size; time phasing of the orbiter and booster development; and the suitability of various liquid and solid rocket expendable boosters, in event a phased development of the reusable booster is chosen. For the remainder of the study, we will concentrate on the external hydrogen and oxygen tank orbiter configuration, and the possibility of phasing systems on the orbiter including selected subsystems, engines and technology in a Mark I/Mark II fashion. In addition, the possibility of employing Saturn F-1 engines in the reusable booster is being evaluated. We are confident that a highly effective program can be defined in the near future

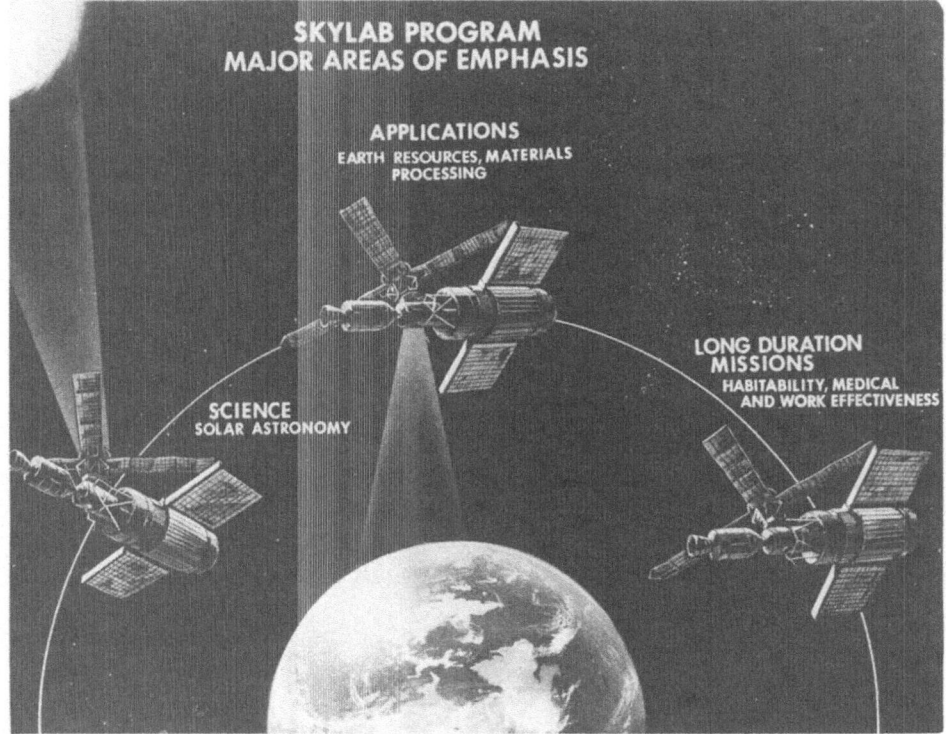

Fig. 6.

to permit the achievement of the first manned orbital flight of the shuttle system in 1978.

To give a more complete understanding of our program of the 1970's leading up to the shuttle era, ongoing approved manned space flight programs should be mentioned. The Apollo Program is continuing through two additional missions, the last of which is Apollo 17, scheduled for December 1972. The crew of Apollo 17 will include Geologist Jack Schmitt, the first U.S. astronaut who has concentrated his entire education and career in a scientific field. As this audience knows, the scientific community is tremendously excited by the data that is coming out of the Apollo 15 mission (Figure 5) and from the instruments left on the Moon on the Apollo 11, 12, and 14 missions. The results of the Apollo Program will keep scientists busy for years seeking and answering questions about Sun–Moon–Earth relationships that have written our own planet's history and will determine its future. The Apollo 15 mission commander, Dave Scott, and Dr Charles Berry are making reports on Apollo to this congress.

Following Apollo, we will conduct the flight program with Skylab, our first experimental space station (Figure 6). The Skylab crews, which will include scientist astronauts, will concentrate on science and applications investigations and study the ability of man to perform during long-duration space flights. The missions will also add to our store of space technology. Three separate crews will man this space workshop for periods up to four weeks on the first mission and up to eight weeks on two subsequent missions. It can be anticipated that very substantial progress in solar astronomy, earth resources surveys and space medicine will come from this highly significant program. Skylab reports to this congress include those being given by Leland F. Belew of the NASA Marshall Space Flight Center and John H. Disher of NASA Headquarters.

The results of the Skylab missions will go far toward putting us in a position to decide the proper mix of manned and unmanned missions in the shuttle era. As we see it today, manned and unmanned systems will seldom be in competition for the same job. Where automatic, repeated remote-control operations are feasible, practical and effective, they will be selected. Man will be used when there is some special need related to the program objectives or where his unique intellect, multiple senses or manipulative dexterity make a substantial contribution.

As reduced costs permit expanded use of human beings in space for such purposes, we believe they will find roles in large and complex multi-use vehicles, facilities and operations. Typical examples of tasks for men would include maintenance, repair, updating and reconfiguration of spacecraft and space equipment.

In addition, the system and operational complexity, long life and mission diversity of the shuttle will require that human pilots participate in its operation frequently to protect flight systems and payloads or prevent mission failures just as they do in the operation of aircraft.

Now I would like to touch on the possible international participation in the development of the space transportation system and associated endeavors. International cooperation is a major objective of the U.S. space program and has been repeatedly encouraged by President Nixon. Special efforts have been made for almost two years to provide information on our plans and studies to those countries with the most obvious potential for post-Apollo participation. Our purpose has been to help other countries involve themselves at a very early stage of the program and to be in a position to decide whether they wish to commit their own resources in working with us. Initial responses have been gratifying. Several million dollars have been invested by countries other than the United States in such studies of the transportation system. Five industrial concerns in Britain, France, and West Germany funded by their own governments have been working with NASA contractors in the space shuttle definition studies. The European Launcher Development Organization has sponsored studies that have begun an examination of whether Europe might develop the space tug as an integral part of the space transportation system. Industrial concerns in eight countries have participated. European groups have also been involved with basic technology programs and have participated in working sessions in such areas as aero-thermodynamics and configurations, structures and materials, propulsion, and integrated electronics. Work has also been undertaken in important allied areas such as the space station and experiment modules (Figure 7). This slide illustrates some of the ideas for sortie research application modules that are being considered for the shuttle.

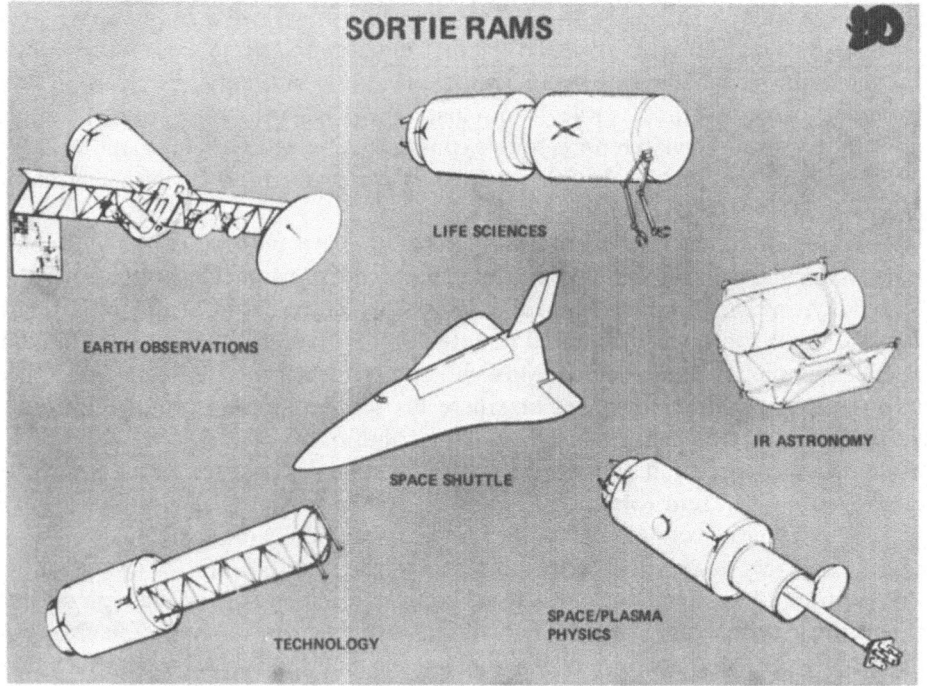

Fig. 7.

Nevertheless, it is recognized that it is difficult for any country to determine what its participation might be at this stage when the shuttle is not firmly defined. Thus, we are prepared and plan to enter into further substantive technical discussions on cooperation with Europe. It should be emphasized of course that we are also prepared to discuss cooperation with countries outside of Europe on the same basis.

In summary, NASA is completing the definition of a two-stage reusable space shuttle system. Our plans call for proceeding with the development of the shuttle in 1972 with the goal of first manned orbital flight in 1978. In view of budget considerations, we are studying various ways of sequencing the development, test, and operations of this system. The United States has restated its interest in international cooperation in the space transportation system program and we are currently exchanging information. We believe the space shuttle and its attendant tug offer an unprecedented opportunity to expand and exploit activities in space in such a manner as to improve the lot of man on the Earth.

References

[1] Hunter, Max, 'Space Shuttle Influences on Payload and Spacecraft Designs', in XXII IAF Congress Proceedings.
[2] Donlan, C., 'Impact of the Shuttle on the Reduction of Cost of Space Operations', in XXII IAF Congress Proceedings.

A NONLINEAR PROGRAMMING APPROACH TO SPACE SHUTTLE TRAJECTORY OPTIMIZATION

RICHARD G. BRUSCH

Convair Aerospace Division of General Dynamics, San Diego, Calif., U.S.A.

1. Introduction

The solution of constrained optimal control problems has received considerable attention in recent years, motivated primarily by the need for solutions to practical problems in the aeronautical sciences. In particular, trajectory design for the reusable space shuttle vehicle will require optimization of highly constrained trajectories. Structural integrity and human factors necessitate the consideration of state variable inequality constraints on the instantaneous normal and axial accelerations, on the instantaneous heat flux, and on the total heating during both launch and entry. Thrust, roll, and pitch controls must obey complex inequality constraints during periods of high dynamic pressure for reasons of stability. The practical considerations of range safety, tracking, and weather also constrain the flight path.

This paper considers the solution to such problems using the nonlinear programming method of Fiacco-McCormick [1]. Several authors [2–8] have successfully applied the technique to constrained optimal control problems of a limited scope. Reference [9] surveys many of the recent applications of mathematical programming to optimal control problems. The present study expands the theory to encompass a general mathematical model for trajectory optimization capable of directly handling six types of equality and inequality constraints. Results of the numerical solution of a highly constrained space shuttle trajectory optimization problem are presented.

2. Modeling Concepts

This section considers a systematic approach that facilitates description of many unrelated optimization problems to a single optimization program. The general mathematical model presented provides a flexible skeletal framework for describing a wide spectrum of complex optimal control problems in terms of problem-oriented functions. Before describing the mathematical model, it is important that the user have a complete understanding of several concepts fundamental to trajectory simulation and modeling. The concepts of trajectory sectioning, branched trajectories, state variable discontinuities, subarc elimination, and control function are discussed in detail in the following sections.

A. TRAJECTORY SECTIONING

Trajectory sectioning is a method of subdividing the time history of a trajectory simulation into parts relevant to the description of the simulation. The concept is an

L. G. Napolitano et al. (eds.), *Astronautical Research 1971*, 185–199. *All Rights Reserved.*
Copyright © 1973 by D. Reidel Publishing Company, Dordrecht-Holland.

extremely useful tool for systematically describing the following changes in the simulation model:

(1) Changes in the differential equations of motion. (The variables, whose time rate of change are defined by a set of first-order ordinary differential equations, are termed state variables).

(2) Changes in the control model (those functions and parameters that the user is free to manipulate).

(3) Changes in the various trajectory constraints.

(4) Changes in the objective function (a function that gives a measure of the performance to be maximized or minimized).

(5) Discontinuities in the state variables (instantaneous changes in the state variables; e.g., a weight discontinuity caused by jettisoning an expended stage).

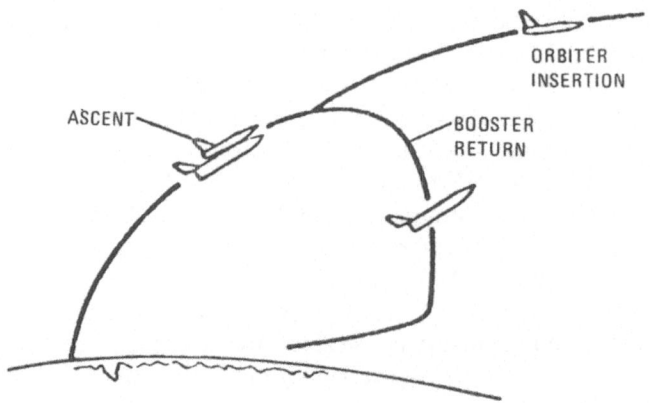

Fig. 1.   Sectioning and branched trajectories for space shuttle simulation.

A section is defined as any portion of the trajectory in which the mathematical model is of a given form and the state variables $x_i(t)$ are continuous functions of time. Section endpoints are chosen to coincide with points at which the differential equations of motion, the control model, or the trajectory constraints change form; or at which the state variables experience a discontinuity. Trajectory sectioning provides a skeletal framework that may be molded by users with widely varying problems to facilitate description of their particular problem to a general mathematical model.

B. BRANCHED TRAJECTORIES [10]

For many problems the simulation sections are time sequential; that is, it is often implied that simulation section 'I' is integrated immediately following the integration of section 'I-1'. This need not be the case – consider a typical flight of the reusable space shuttle shown in Figure 1. During the first portion of the simulation, the booster and orbiter vehicles are attached. After the booster has terminated thrusting, the two vehicles separate, the orbiter ignites and flies to orbital insertion, while the booster enters the atmosphere and cruises back to the launch site. It is necessary to simulate both branches of this trajectory since the optimization of each branch will have the effect of modifying the initial flight phase. Both branches of the trajectory cannot be identified by the same number, since the separated vehicles have different simulation models. Either the booster or the orbiter must have a section number which is not contiguous with the last section of coupled flight.

To simulate trajectories similar to that in Figure 1, a vector called LINK may be used. This vector specifies the relationship of each trajectory section to the other sections. LINK(I) = J specifies that Section I follows Section J. J = 0 specifies that Section I follows the initial conditions. For example, the multiply branched trajectory shown in Figure 2 corresponds to the LINK vector: (0, 1, 2, 3, 2, 5, 5). A normal unbranched trajectory is described by the LINK vector: (0, 1, 2, 3, 4, 5).

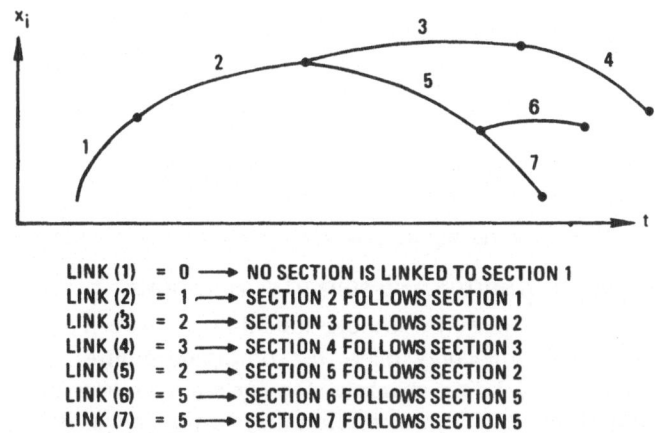

LINK (1) = 0 ⟶ NO SECTION IS LINKED TO SECTION 1
LINK (2) = 1 ⟶ SECTION 2 FOLLOWS SECTION 1
LINK (3) = 2 ⟶ SECTION 3 FOLLOWS SECTION 2
LINK (4) = 3 ⟶ SECTION 4 FOLLOWS SECTION 3
LINK (5) = 2 ⟶ SECTION 5 FOLLOWS SECTION 2
LINK (6) = 5 ⟶ SECTION 6 FOLLOWS SECTION 5
LINK (7) = 5 ⟶ SECTION 7 FOLLOWS SECTION 5

Fig. 2.   Definition of LINK vector for multiply branched trajectory.

The subarcs will be integrated in order, and it is assumed that the initial conditions for a given subarc are a function of the final state and time of the subarc physically preceding it. (Note that numerical order does not imply physical order.)

Due to the order of subarc integration and the assumed relation between initial and final conditions of linked sections, it is implied that subarc I cannot follow physically subarc J, where I < J. In terms of LINK, this implies LINK(I) < I. This restriction implicitly eliminates the possibility of circular linking.

A striking example of the importance of the capability to stimulate branched trajectories is computing an optimal ascent trajectory constrained by abort considerations. After each ascent trajectory simulation, the hypothetical abort trajectories are simulated starting from a state defined at the termination of any ascent simulation section, as shown in Figure 3. The abort trajectory is subjected to realistic constraints. To meet such constraints, the ascent trajectory must be modified; thus the hypothetical abort directly influences optimization of the ascent trajectory. The booster return trajectory can be treated similarly.

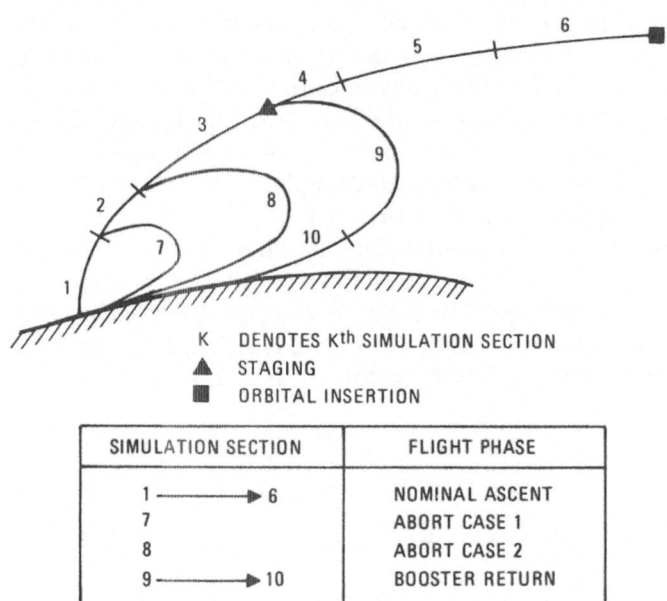

Fig. 3.   Typical ascent trajectory with booster abort/return constraints.

C. STATE VARIABLE DISCONTINUITIES AND SUBARC ELIMINATIONS

Jumps or discontinuities in the state variables are allowed between any two simulation sections. Esoteric applications may make use of discontinuities in both the state variables and the time [11]. For example, for certain segments of the trajectory, analytic integration of the equations of motion may be possible as is the case for two-body coasting arcs. These arcs are 'eliminated' by making a jump in state and time. A

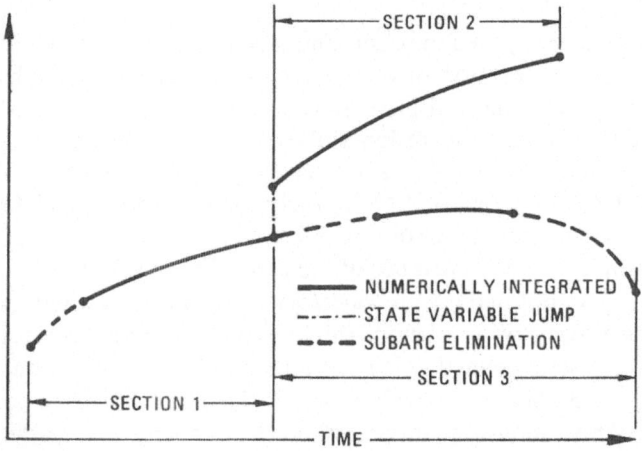

Fig. 4.   General branched trajectory with state variable discontinuities and analytic subarc elimination.

portion of the trajectory, which otherwise would require numerical integration, is eliminated by analytic functions, giving the state after the jump as a function of the state before the jump and elapsed time. This technique is referred to as 'subarc elimination'. Neither dynamic nor integral constraints are permitted on eliminated subarcs.

Figure 4 shows a trajectory with a state variable discontinuity at the start of Section 2, and subarc elimination at the start of Sections 1 and 3 and after the integration in Section 3.

D. CONTROL MODELING

A general mathematical model should be capable of incorporating two classes of independent variables, which are to be chosen to extremize some objective function. The first class includes independent variables that are functions of time (e.g., thrust magnitude or pitch angle of attack). They are termed dynamic control variables and designated $u_k(t)$. Second, it is desirable to include design variables, d_p, which are constant with respect to time (e.g., wing area, gross liftoff weight, or propellant mass fraction).

To formulate a trajectory optimization problem as a general problem in nonlinear programming, it is necessary to approximate the dynamic control variables by a finite set of parameters. This is because a continuous function of time is sought as the solution to the trajectory optimization problem, whereas the solution to the nonlinear programming problem is represented by a point in Euclidian N-space. The dissimilarity is resolved by approximating the continuous function of time by a function of n independent parameters. These n parameters then become a subset of the N independent variables in the nonlinear programming formulation.

There are two ways in which the n parameters can conveniently be chosen to describe the control. The n parameters can specify the control magnitude at specified points in time. An interpolation device, such as simple linear interpolation or spline interpolation [12] can be used to define the control at intermediate points in time. Alternatively, the parameters can be regarded as coefficients of some mathematical model, which is either a function of time (open-loop control) or a function of the current state (closed-loop control).

3. General Mathematical Model

A mathematical model encompassing these capabilities is detailed below. If the subscript j denotes the trajectory section, then the problem is to

$$\text{minimize}_{u_k(t),\, d_p} J = \Phi(t_j^0, t_j^f, x_{ij}^0, x_{ij}^f, d_p) + \sum_{j=1}^{l} \int_{t_j^0}^{t_j^f} L_j[x_{ij}(t), u_{kj}(t), d_p, t]\, \mathrm{d}t$$

(1)

subject to the differential equations

$$\dot{x}_{ij}(t) = f_{ij}[x_{ij}(t), u_{kj}(t), d_p, t]; \quad i = 1, 2, \ldots, n; j = 1, 2, \ldots, l$$

(2)

the boundary conditions

$$\pi_s(t_j^0, t_j^f, x_{ij}^0, x_{ij}^f) = 0; \quad s = 1, 2, \ldots, n_\pi \le n \tag{3}$$

and equations relating the initial state and time of each section to the terminal state and time of any previous section,

$$x_{ij}^0 = \Gamma_{ij}(x_{ir}^f, t_r^f); \quad j = 2, 3, \ldots, l \tag{4}$$

$$t_j^0 = \Omega_j(x_{ir}^f, t_r^f); \quad 1 \le r < j \tag{5}$$

where $x_{ij}(t)$ = the ith state variable in Section j; $i = 1, 2, \ldots, n$;
$u_{kj}(t)$ = the kth dynamic control variable in Section j; $k = 1, 2, \ldots, w \le n$;
d_p = the pth design parameter $(\partial d_p/\partial t = 0)$; $p = 1, 2, \ldots, q$;
t = the independent variable, hereafter called time.

and the superscripts '0' and 'f' denote evaluation at the initial and final values of time of the corresponding section. A superscribed dot denotes differentiation with respect to time.

The general mathematical model incorporates six types of problem-oriented equality and inequality constraints. Two fundamentally different types of constraints can be identified. Those that are functions of the time-varying system variables are termed 'dynamic' constraints. Those that are functions of the design parameters and the initial and final points of each trajectory section are termed 'parametric' constraints. For user convenience, two additional constraints depending upon integrals of the time-varying system variables are provided. These constraints are defined by the functions below.

Dynamic inequality:

$$\xi_{sj}[x_{ij}(t), u_{kj}(t), d_p, t] \ge 0; \quad s = 1, 2, \ldots, n_\xi. \tag{6}$$

Dynamic equality:

$$\eta_{sj}[x_{ij}(t), u_{kj}(t), d_p, t] = 0; \quad s = 1, 2, \ldots, n_\eta. \tag{7}$$

Parametric inequality:

$$\xi_s(t_j^0, t_j^f, x_{ij}^0, x_{ij}^f, d_p) \ge 0; \quad s = 1, 2, \ldots, n_\xi. \tag{8}$$

Parametric equality:

$$\Psi_s(t_j^0, t_j^f, x_{ij}^0, x_{ij}^f, d_p) = 0; \quad s = 1, 2, \ldots, n_\Psi. \tag{9}$$

Integral inequality:

$$\sum_{j=1}^{l} \int_{t_j^0}^{t_j^f} Q_{sj}[x_{ij}(t), u_{kj}(t), d_p, t] \, dt - D_s \ge 0; \quad s = 1, 2, \ldots, n_Q. \tag{10}$$

Integral equality:

$$\sum_{j=1}^{l} \int_{t_j^0}^{t_j^f} P_{sj}[x_{i_j}(t), u_{k_j}(t), d_p, t]\, dt - C_s = 0; \quad s = 1, 2, \ldots, n_P, \tag{11}$$

where ξ_{sj}, η_{sj}, Q_{sj}, and P_{sj} denote independent functions applicable during trajectory Section j. Equations (1) and (2) (subject to Constraints 3 to 11) comprise the general optimization model.

Dynamic inequality constraints, $\xi_{sj} \geq 0$, include the familiar state and control variable inequality constraints as a subset. The parametric equality constraints Ψ_s allow the specification of the boundary conditions on the differential equations, as well as constraints on the design variables. Also, note that Equations (3) to (5) are of the same form as Equation (9) and, consequently, may also be treated as parametric equality constraints.

The design parameters, d_p, may serve a variety of useful purposes apart from modeling obvious design parameters; e.g., booster-to-orbiter mass ratio. Undetermined initial conditions on state variables at the start of any section may be treated as design variables [13]. Thus, problems whose initial state is confined to some subset of state space by equality and inequality constraints (Equations (8) and (9)) are easily treated. The time duration of any section or the time discontinuity between any two sections may be treated as a design parameter. Certain applications permit the modeling of dynamic control variables as a function of several design parameters without significant loss in accuracy. For example, the coefficients of a straight line representing a control variable as a function of time may be treated as design variables. A surprisingly similar mathematical model has recently been published by Rosendaal [14], which also includes the capability of modeling general branched trajectories.

A. LIMITATIONS

The mathematical model is limited to the solution of problems that depend upon the integration of a set of first-order ordinary differential equations. All functions describing the performance measure, the equations of motion, and the constraints must be C_1 continuous over a finite set of time intervals. The range of the control function must be at least piecewise continuous with respect to time, and at any given time the range of the control functions must form a simply connected manifold of E_w. The range of the design parameters must also form a simply connected set. Problems at any given time with a design parameter or a control function having a range described by a disconnected or discrete set cannot be solved.

The technique used to describe the branching topology and intersectional discontinuities is limited to forward branching, i.e., problems in which the initial conditions for any section depend upon the final condition of two or more previous sections cannot be handled.

4. A Nonlinear Programming Computational Algorithm

The Sequential Unconstrained Minimization Technique (SUMT) of Fiacco and McCormick [1] seeks the solution to this constrained optimal control problem as the limit to a sequence of unconstrained optimization problems. At each stage, a penalty function, **P**, must be minimized. The penalty function is formed by adding to the performance index, a penalty term for each equality and inequality constraint – Equations (6) through (11). The penalty terms are always of such sign as to increase or 'penalize' the performance index. The magnitude of each penalty term is proportional to the degree to which the corresponding constraint is unsatisfied. Thus, in the limit, minimizing the penalty function minimizes the performance index and drives the constraints into satisfaction.

A. PENALTY FUNCTION

The form of the penalty function for the constrained optimal control problem described in the preceding section was developed based upon an extension of the SUMT method by Lasdon *et al.* [7] to inequality constrained optimal control problems. In this formulation, the differential equality constraints (Equation (2)) are not included in the penalty function; rather, they are satisfied directly through numerical integration. References [2] through [6] replace the differential equations with a set of finite-difference equations, which are incorporated into the penalty function as equality constraints.

Before exhibiting the function **P** for the trajectory optimization problem, it is convenient to transform the integral constraints, Equations [10] and [11], into parametric constraints on the final values of the following pseudo-state variables:

$$\dot{x}_{(n+i),j} = P_{ij}[x_{ij}(t), u_{kj}(t), d_p, t]; \quad x^0_{(n+i),1} = 0; i = 1, 2, \ldots, n_p \tag{12}$$

with corresponding parametric constraints

$$x^f_{(n+i),1} - C_i = 0 \tag{13}$$

and

$$\dot{x}_{(n+n_p+i),j} = Q_{ij}[x_{ij}(t), u_{ij}(t), d_p, t];$$
$$x^0_{(n+n_p+i),1} = 0; \quad i = 1, 2, \ldots, n_Q \tag{14}$$

with corresponding parametric constraints

$$x^f_{(n+n_p+i),1} - D_i \geq 0. \tag{15}$$

Now the **P** function for the general trajectory optimization problem may be formed by adding to the objective function, Equation (1), penalty terms for constraints, Equations (6)–(9), (13), (15).

$$\mathbf{P} = G(t^0_f, t_f, x^0_{ij}, x^f_{ij}, d_p) + \sum_{j=1}^{l} \int_{t^0_f}^{t} L^*_j[x_{ij}(t), u_{kj}(t), d_p, t]\,dt \tag{16}$$

where

$$G = \Phi(t_j^0, t_j^f, x_{ij}^0, x_{ij}^f, d_p) \qquad \text{parametric performance}$$

$$+ r^{-1/2} \sum_{s=1}^{n_\Psi} \Psi_s^2(t_j^0, t_j^f, x_{ij}^0, x_{ij}^f, d_p) \qquad \text{parametric equality}$$

$$+ r^{-1/2} \sum_{i=1}^{n_P} [x_{(n+i), l}^f - C_i]^2 \qquad \text{transformed integral equality} \qquad (17)$$

$$+ r \sum_{s=1}^{n_\zeta} 1/\zeta_s(t_j^0, t_j^f, x_{ij}^0, x_{ij}^f, d_p) \qquad \text{parametric inequality}$$

$$+ r \sum_{i=1}^{n_Q} 1/[x_{(n+n_p+i), l}^f - D_i] \qquad \text{transformed integral inequality}$$

and

$$L_j^* = L_j[x_{ij}(t), u_{kj}(t), d_p, t] \qquad \text{integral performance}$$

$$+ r^{-1/2} \sum_{s=1}^{n_\eta} \eta_{sj}^2[x_{ij}(t), u_{kj}(t), d_p, t] \qquad \text{dynamic equality} \qquad (18)$$

$$+ r \sum_{s=1}^{n_\xi} 1/\xi_{sj}[x_{ij}(t), u_{kj}(t), d_p, t] \qquad \text{dynamic inequality}$$

B. EVALUATING THE PENALTY FUNCTION

Several different computational procedures are available for evaluating the penalty function, depending upon the treatment of the endpoint/cornerpoint conditions (boundary conditions), Equation (3); and the 'jump conditions', Equations (4)–(5). One procedure is outlined below.

The boundary conditions, Equation (3), are treated as parametric equality constraints. Thus, these conditions are included in the penalty functions through the parametric equality terms of Equation (17). The so-called 'jump conditions' of Equations (4)–(5) could also be included as penalty terms in the case where the initial states and times of each section were considered to be unknown 'design variables' whose values were to be selected during the optimization to satisfy Equations (4) and (5); however, this generality has the disadvantage of the model user accidentally prescribing an inconsistent set of Equations (3)–(5); there are $3(n+1)$ potential constraints relating the initial point of one section to the final point of a previous section and only two $(n+1)$ endpoint variables involved. To render the algorithm foolproof, the following restrictions have been imposed:

(a) Boundary conditions may be placed on initial points of only the first section.

(b) At least one boundary condition must be specified on the final points of all sections.

(c) Jump conditions must specify the values of all initial states and times of each section, except the first, as a function of the final state and time of a previous section.

The need to include the jump conditions (Equations (4) and (5)) as parametric equality constraints is obviated by the computational procedure described below. The procedure also eliminates any need to include the state equations (Equation (2)) in the penalty function since they are automatically satisfied. The computational procedure is:

(1) If the initial point (x_{i1}, t_1) is fixed, the values are known and are inputs; if the initial point is free, but constrained to some initial manifold or subspace by Equations (8) and (9), the free initial variables are treated as unknown 'design variables' and initial estimates are input. These variables are adjusted to satisfy Equations (8) and (9) and to obtain the most extreme objective function. In either case, initial values or estimates are known for the initial states and times.

(2) State differential Equation (2) are integrated forward, using a standard fourth-order Runga-Kutta-Merson variable stepsize integrator to the final time and state t_1^f, $x_{t_1}^f$. Any terminal boundary condition may terminate the section; there must be at least one per restriction (b). All initial conditions for Section 2 will be specified because of restriction (c).

(3) The final states and times are stored in a table and the initial conditions for Section 2 (x_{i2}^0, t_2^0) are computed using Equations (4) and (5). These values are also stored in a table.

(4) Steps 2 and 3 are repeated until all sections have been integrated.

As can be seen, Equations (2), (4), and (5) have been automatically satisfied and there is no need to include them in the penalty function. All information needed to evaluate the penalty function is contained in either the table of initial and final states for each section or in the final pseudo-state vector. The pseudo-state vector provides automatic integration of the functions of the dynamic and integral constraints needed in evaluating the integral terms of the penalty function.

C. SUMT ALGORITHM

The numerical algorithm requires minimization of the penalty function (Equation (16)) with respect to the independent variables $u_k(t)$ and d_p. However, recall that the dynamic control variables $u_k(t)$ have been modeled by finite sets of parameters. These parameters, together with the parameters d_p constitute a vector of independent variables y.

The algorithm begins at a point $y^{(0)}$ in the Euclidian N-space at which all inequalities are satisfied (a feasible point). A systematic method for obtaining an initial feasible solution is given by Fiacco-McCormick [15]. The constant r_1 is selected and the point $y^{(1)}$ is found so that $P(y^{(1)}r_1)$ is a minimum. A new value, $r_2 < r_1$ is then selected and a point $y^{(2)}$ is found so that $P(y^{(2)}r_2)$ is a minimum, etc. The limit

$$\lim_{r_k \to 0} \left[\min_y P(y^{(k)}, r_k) \right] \tag{19}$$

approaches the solution to the nonlinear programming problem. Each iteration requires minimization of the function **P** for a given value of r. This problem is termed the **P**-problem and its solution is fundamental to the method of Fiacco-McCormick. The computational algorithm implemented minimizes the function **P**, using a Fletcher-Powell [16] function minimization algorithm in conjunction with the simultaneous golden-section-and-cubic-fit, one-dimensional search of Johnson and Meyers [17].

It can be seen from the form of the penalty term that a minimization technique will avoid points that cause the inequality constraints to go to zero and become negative since the corresponding penalty term would increase without bound. Clearly, the initial point $y^{(0)}$ must be feasible. It is also clear that the minimization of $\mathbf{P}(y, r_k)$ will force the equality constraints to go to zero, otherwise these terms would increase without bound as $r^{1/2}$ goes to zero.

A sequence of minimization is performed with $r_1 > r_2 \ldots > r_k \ldots > r_f$, rather than just one minimization of $\mathbf{P}(y, r_f)$, because the latter minimization problem is very difficult to solve from a numerical standpoint. Starting with a large r_k results in a relatively easy minimization problem. The solution of the **P**-problem at each stage then provides a good initial estimate for the solution of the **P**-problem at the following stage.

5. Space Shuttle Re-Entry Example

Recently, Kamm, Johnson, and Sullivan [18–21] have shown that the space shuttle payload inserted into a 51 by 100-nautical mile polar orbit could easily be increased by 20% by flying an optimized lifting ascent in the sensible atmosphere, rather than the usual 'gravity turn'. Rosendaal [14] has shown similar results for the complete problem, including the booster entry and flyback branch. References [22–24] present other branched trajectory results applicable to space tug operations. In summary, there is a growing body of evidence underlining the performance rewards to be gained by optimizing space shuttle trajectories.

To simulate and optimize space shuttle ascent and entry trajectories, a general trajectory optimization program (GTOP) was developed [18]. This general-purpose computer program combines a flexible, high-speed, trajectory simulation module with a highly reliable nonlinear programming optimization driver. GTOP represents an implementation of all features of the general mathematical model presented previously. Details of the program and the modeling used to simulate the space shuttle vehicle are available from the author (Mail zone 986–30, P.O. Box 1128, San Diego, Calif. 92112).

A nominal space shuttle trajectory consists of three branches as shown in Figure 1: (1) ascent to booster staging, (2) orbiter flight to orbital injection, and (3) booster entry and return to the launch site.

Consider the problem of space shuttle booster entry and flyback. After the booster has separated from the orbiter, the following conditions are typical:

Altitude = 244,784 ft   Longitude = 239.343 deg
Velocity (relative) = 10,824 fps   Wing area = 8451 sq. ft

Gamma (relative) = 5.654 deg   Weight = 771,492 lb
Heading azimuth (relative) = 182.495 deg   Wing loading (W/S) = 91.29 fps
Latitude = 32.788 deg   Staging time = 216.36 sec

The problem is to minimize the distance of the space shuttle booster from the launch
site after the booster has completed re-entry (altitude = 20,000 ft). Minimizing this
flyback distance minimizes the jet fuel required to execute the powered return to the
launch site. Clearly, the less flyback fuel required, the greater the payload injected into
orbit.

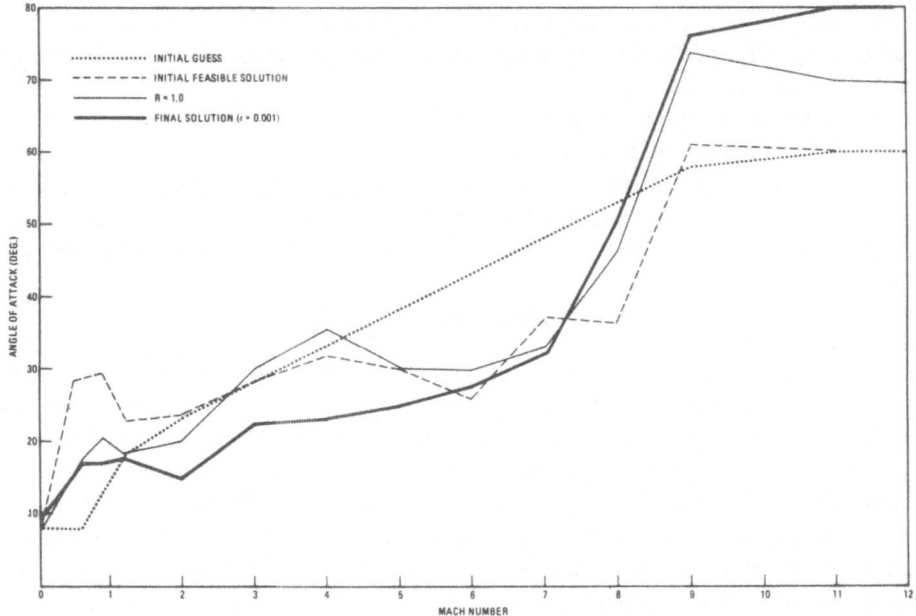

Fig. 5.   Angle of attack versus Mach number.

The entry trajectory for the B-9U booster configuration was constrained by five
state variable inequality constraints:

Total acceleration load factor $\leq 4g$
Dynamic pressure $(\frac{1}{2}\rho V^2) \leq 500$ lb/ft^2
Instantaneous heating parameter $(\frac{1}{2}\rho V^3) \leq 3.2 \times 10^6$ lb/s ft
Angle of attack ≥ 0 deg
Angle of attack ≤ 90 deg

The last two constraints were due to the unavailability of aerodynamic data outside
of this region.

The initial angle of attack and bank angle were each modeled by 12 parameters as a
function of Mach number, a monotonic function of time for the entry. This independent variable yields nearly uniform sensitivities of the performance index (flyback

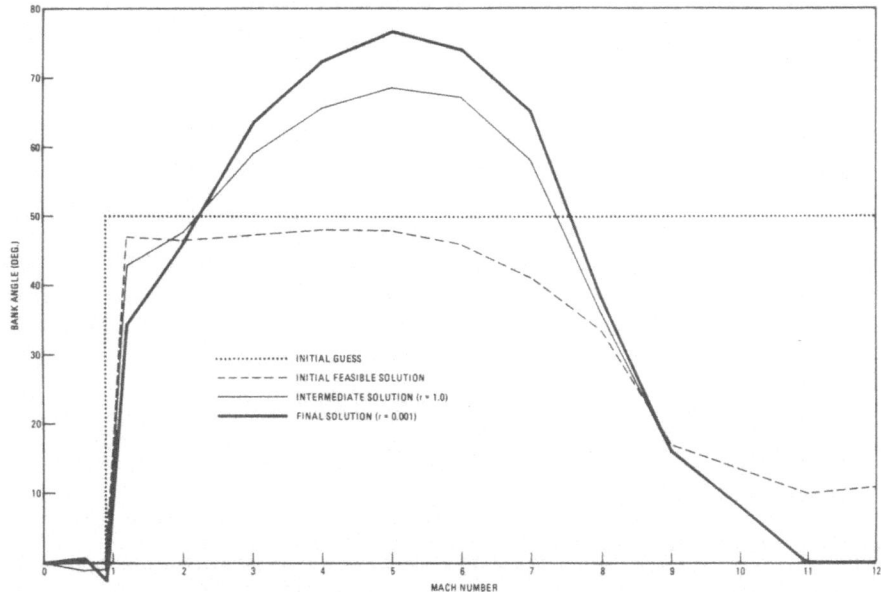

Fig. 6.   Bank angle versus Mach number.

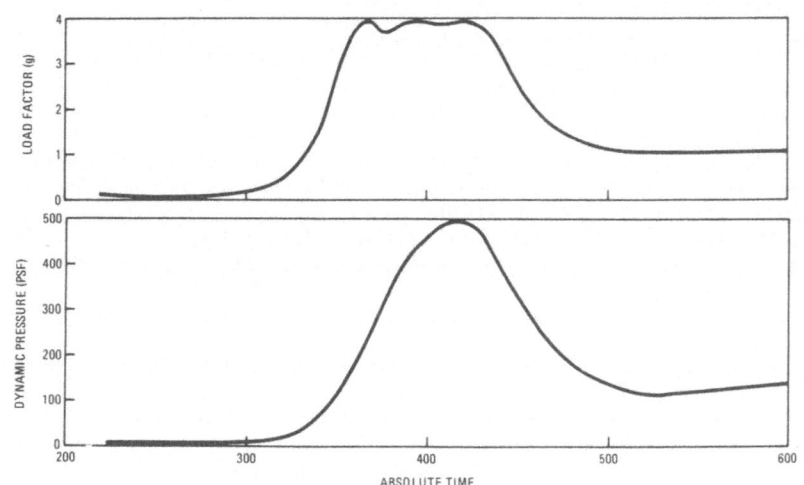

Fig. 7.   Acceleration load factor and dynamic pressure constraints.

distance) to variations in the control modeling parameters. Time is unsatisfactory in this respect as an independent variable. The controls were tabulated at the following Mach numbers: 0, 0.6, 0.9, 1.2, 2, 3, 4, 5, 6, 7, 8, 9, 11, and 50.

The initial guess on the optimal control histories was intentionally poor (Figures 5 and 6) to demonstrate the insensitivity of the algorithm to initial guesses. The trajectory corresponding to the initial control estimate resulted in a flyback distance of

408.7 n.mi and experienced a maximum acceleration load factor of 5.0 g. Two one-dimensional searches were required to obtain a solution satisfying all constraints (called 'initial feasible solution' in Figures 5 and 6). Since no attempt is made to maximize performance when generating the initial feasible solution the flyback distance for the initial feasible solution was 465.6 n.mi. However, after one unconstrained function minimization ($r = 1.0$) the flyback distance was reduced to 395.3 n.mi.

After three more unconstrained function minimizations (with $r = 0.1$, 0.01, and 0.001, respectively) the final control history shown in Figures 5 and 6 was obtained with a corresponding flyback range of 369.8 n.mi. The relevant trajectory parameters are shown in Figure 7. Note that both the acceleration load factor and the dynamic pressure state variable constraints are simultaneously active. The ripples in the load factor constraint are due to the discretization of the control.

References

[1] Fiacco, A. V., and McCormick, G. P., *Nonlinear Programming: Sequential Unconstrained Minimization Techniques*, John Wiley & Sons, Inc., New York, 1968.
[2] Cannon, M. D., Cullum, C. D., and Polak, E., *Theory of Optimal Control and Mathematical Programming*, McGraw-Hill Book Company, New York, N.Y., 1970.
[3] Tabak, D., and Kuo, B. C., *Optimal Control by Mathematical Programming*, Prentice-Hall, Englewood Cliffs, N.J., 1971.
[4] Balakrishnan, A. V., *J. Opt. Theory Appl.* **4** (1969).
[5] Jones, A. P., and McCormick, G. P., *SIAM Journal of Control* **8**, (1970), 218.
[6] Pierson, B. L., *Astron. Acta* **14** (1969), 157.
[7] Lasdon, L. S., Warren, A. D., and Rice, R. K., *IEEE Transactions on Automatic Control*, Vol. AC-12, No. 4, August 1967.
[8] Brusch, R. G., and Schappelle, R. H., 'Solution of Highly Constrained Optimal Control Problems Using Nonlinear Programming', *AIAA Guidance, Control and Flight Mechanics Conference*, Santa Barbara, August, 1970.
[9] Tabak, D., *IEEE Transactions on Automatic Control*, AC-15, Dec. 1970.
[10] Mason, J. D., *Some Optimal Branched Trajectories*, NASA CR-1331, May 1968.
[11] Vincent, T. L., and Mason, J. D., *J. Opt. Theory Appl.* **3** (1969), 263.
[12] Ahlberg, J. H., Nilson, E. N., and Walsh, J. L., *The Theory of Splines and Their Applications*, Academic Press, New York, 1967.
[13] Kelly, H. J., 'Methods of Gradients', *Optimization Techniques*, (ed. by G. Leitmann) Academic Press, New York, 1962.
[14] Rozendaal, H. L., 'A General Branched Trajectory Optimization Algorithm With Applications to Space Shuttle Vehicle Mission Design', *AAS/AIAA Astrodynamics Specialists Conference*, AAS No. 71–326, Ft. Lauderdale, Fla., August 17–19, 1971.
[15] Fiacco, A. V., and McCormick, G. P., *Management Sci.* **10** (1964), 360.
[16] Fletcher, R., and Powell, M. J. D., *Computer Journal* (1963).
[17] Guilfoyl, G., Johnson, I., and Wheatly, P., 'One-Dimensional Search Combining Golden Section and Cubic Techniques', *NASA* CR-65994, Jan. 1967.
[18] Tramonti, L. G., Brusch, R. G., Schappelle, R. H., *Hypersonic Vehicle Trajectory Optimization*, Convair Aerospace Report GDC-ERR-1604, 1970.
[19] Johnson, I. L. and Kamm, J. L., 'Parameter Optimization and the Space Shuttle', *Joint Automatic Controls Conference*, St. Louis, Mo., August, 1971.
[20] Johnson, I. L. and Kamm, J. L., 'Near Optimal Shuttle Trajectories Using Accelerated Gradient Methods', *AAS/AIAA Astrodynamics Specialists Conference*, AAS/AIAA Paper No. 328, Ft. Lauderdale, Fla., August 17–19, 1971.

[21] Kamm, J. L. and Johnson, I. L., 'Optimal Shuttle Trajectory-Vehicle Design Using Parameter Optimization', *AAS/AIAA Astrodynamics Specialists Conference*, AAS/AIAA Paper No. 329, Ft. Lauderdale, Fla., August 17–19, 1971.
[22] Sullivan, H. C. and Uzzell, B. R., 'Optimal Boost Trajectories for the Shuttle Vehicle', *AAS/AIAA Astrodynamics Specialists Conference*, AAS No. 71–324, Ft. Lauderdale, Fla., August 17–19, 1971.
[23] Armstrong, E. S., Childs, A. G., and Markos, A. T., 'Applications of a Mathematical Programming Technique to Finite Thrust Rocket Trajectory Optimization', *AAS/AIAA Astrodynamics Specialists Conference*, AAS No. 71–308, Ft. Lauderdale, Fla., August 17–19, 1971.
[23] Thurneck, W. J., Jr., 'Branched Trajectory Optimization by the Method of Steepest Descent', *AAS/AIAA Astrodynamics Specialists Conference*, AAS No. 71–309, Ft. Lauderdale, Fla., August 17–19, 1971.
[25] Mason, J. D., 'Space Tug Performance Optimization', *AAS/AIAA Astrodynamics Specialists Conference*, AAS No. 71–330, Ft. Lauderdale, Fla., August 17–19, 1971.

B. SPACE TRANSPORTATION: ORBIT TO ORBIT

THE PROBLEM OF DOCKING WITH A PASSIVE ORBITING OBJECT WHICH POSSESSES ANGULAR MOMENTUM*

MARSHALL H. KAPLAN**

The Pennsylvania State University, University Park, Penn., U.S.A.

Abstract. The problem of docking with a passive orbiting object which is either spinning or tumbling has recently developed interest for several reasons. Retrieval and repair of spent satellites will be technically feasible and economically desirable with the advent of the space shuttle system. Manned space stations will be circling the globe within a few years; thus, indicating that elimination of at least part of the space debris may be imperative to insure safety from collisions. Many of the discarded items are spinning and have a wide variety of shapes and sizes. Furthermore, if a collision or other mishap should occur, a manned vehicle may become uncontrolled and enter a tumbling mode of attitude motion. In order to rescue the crew or repair the spacecraft, this tumbling must be controlled and eliminated. Many studies have been carried out to formulate rescue schemes and requirements, develop uncooperative rendezvous techniques, and investigate satellite manipulation via remote operators. However, several specific problems related to docking with objects possessing angular momentum have not previously been considered and require new technology for solution. Techniques for retrieving uncooperative spinning objects are presented. A conceptual spacecraft for remotely despinning and retrieving debris of moderate size is discussed in depth. In addition, the more complex problems of detumbling a large vehicle are described. Specific aspects of rescue from such a vehicle are presented, although a general solution has not yet been formulated. The major objective is to identify the particularly important problems of docking with a spinning or tumbling object and present the current status of some promising solutions.

1. Introduction

During the past 14 yr man has placed in excess of 5200 objects into terrestrial, lunar, and solar orbits [1]. Of these, more than 2300 are currently in orbits about the Earth. There are only a few hundred operational satellites among all the terrestrial captives, with the remainder consisting of launch vehicle upper stages, payload fairings, and other components associated with spacecraft insertions. Although the average number density of objects is negligible, most are confined to just a few popular orbits, e.g., synchronous and low, inclined orbits. In the near-future manned space stations will be orbiting the Earth as well as an increasing number of unmanned satellites and debris.

Many orbiting objects possess angular momentum in the form of steady spin or tumbling because of stabilization mode, attitude jet failure, collision, etc. Tumbling motion will always dissipate into steady spin due to energy absorption resulting from structural flexibility, propellant sloshing, and nutation dampers. Due to the increasing probability of collision, retrieval or elimination of many discarded items is becoming an urgent concern with regard to success of future orbital missions and restoration of repairable satellites. A significant problem arising in missions dealing

* This research is supported by NASA Grant NGR 39-009-162.
** Associate Professor of Aerospace Engineering.

L. G. Napolitano et al. (eds.), *Astronautical Research 1971*, 203–217. *All Rights Reserved.*
Copyright © 1973 by D. Reidel Publishing Company, Dordrecht-Holland.

with passive objects such as satellites and disabled space bases is docking with a target which possesses angular momentum. In the case of a discarded satellite, steady spin can be expected if angular momentum is present, because a minimum energy state is assumed to have been reached. However, a disabled station requires immediate attention and cannot be left to dissipate tumbling motion into steady spin. Unfortunately, tumble is the general result of a major uncorrected disturbance resulting from control system failure, collision, or explosion.

A considerable amount of information is available in the literature on handling passive, non-spinning objects [2]. Limited work has been done with capturing targets of moderate size using a modified Apollo command and service module [3]. A preliminary analysis to determine the causes of tumbling and resulting acceleration components has recently been carried out [4]. Thus, properties of bodies possessing angular momentum are well known, but promising capture and control techniques for spinning objects and tumbling space bases have not previously been studied in depth. The work presented here deals primarily with docking problems associated with these two situations. Of particular concern are techniques and devices for eliminating angular momentum, capturing spinning objects of moderate size, and rescuing the crew of a tumbling space vehicle. Dynamics and control aspects of a device for retrieving spinning objects are presented in detail. This remotely controlled vehicle may be used in conjunction with the space shuttle system [5]. Finally, attitude dynamics and control of tumbling motion are also discussed in reference to the problem of rescue from a disabled space base and other manned vehicles.

2. Classification of Angular Momentum States

For the purpose of defining docking problems associated with objects having angular momentum, attitude motion can be classified into three categories: simple spin, nutation, and general tumbling. Simple spin is defined as the case in which the angular velocity and momentum vectors are colinear, and motion is assumed to be about the principal body axis only. Nutation is defined as a perturbed state of steady spin in which the transverse components of angular velocity are small. Thus, the deviation of motion from simple spin is only slight, allowing considerable simplification of the equations of motion. Finally, the general case of attitude of motion is referred to as 'tumbling', since transverse components of angular velocity may be large and there is no preferred axis of rotation.

General torque-free tumbling motion of rigid bodies has been well established and may be described analytically or geometrically. For an unsymmetrical body the equations of motion are non-linear and cannot be solved without difficulty. A geometrical interpretation has been established by Poinsot [6]. In fact, the 'Poinsot ellipsoid' represents the locus of all possible values of angular velocity of the body which satisfy the constant kinetic energy condition. This imaginary ellipsoid is fixed to the body and moves with it. Attitude motion can then be described as the Poinsot ellipsoid rolling on an inertially fixed plane with its center at a fixed distance from this plane. If the body is symmetric, the geometric interpretation of motion is much simpler. A 'body cone' whose apex is at the center of mass and is fixed to the body rolls on an inertially fixed 'space cone' whose axis is the angular momentum vector. The common cone element is colinear with the angular velocity vector.

Classification of angular momentum states is associated with the types of missions in which docking with passive objects is required. Simple spin is usually associated with retrieval of old satellites and other orbiting items of interest. Nutation may arise in capturing a spinning object if the target is accidently perturbed by the retrieval package. Tumbling may arise in connection with space rescue operations involving large manned space vehicles which have suffered a loss of attitude control. Problems of docking during retrieval and rescue are discussed in detail below.

3. Operational Aspects of Orbital Retrieval

Of primary concern here is the retrieval of unmanned payloads which can be stored in the shuttle cargo bay, e.g., weather and scientific data collecting satellites. Such satellites and existing orbiting objects do not necessarily incorporate docking provisions, and possibly spin in such a passive state that docking equipment cannot be used. It is assumed that objects to be retrieved are spinning about their major principal axes of inertia. This is the state attained after a period of time in which kinetic energy has been dissipated by liquid sloshing, elastic deformation of structural members, or nutation dampers. For objects of the size considered here, the time required to reach a stable state is, at most, of the order of weeks, after significant attitude disturbances have stopped. However, large objects which are tumbling may require several months to dissipate passively enough energy to reach a stable spin state. Techniques used for spinning bodies would not be directly applicable to rescue from a tumbling vehicle, primarily because tumbling motion does not have an inertially oriented spin axis associated with it, which is assumed in normal retrieval operations.

Since the most economical means of reaching an object in a low orbit for retrieval, repair, or elimination will be by using the space shuttle, many aspects of the operational sequence assume a man will be controlling the events and maneuvers. Automatic control systems are assumed whenever practical, but the nature of such missions requires the ability of a man to make decisions and perform functions which are prohibitive in an automated mode because of the inherent uncertainties involved. Missions to high altitude orbits would require very similar maneuvers. However, shuttle performance is expected to limit it to low altitude orbital missions and an orbit-to-orbit vehicle may also be required for high altitude retrieval.

The initial task in a retrieval sequence is considered to be locating and identifying the object upon arrival in its predetermined vicinity. The size and shape of the search area will greatly influence the method of search and identification, and time to search. Furthermore, a successful rendezvous requires extreme accuracy in orbit determination. A non-cooperative radar rendezvous system is considered to be available using current technology. It is understood that the accuracy of orbit determination by ground tracking will permit the shuttle to be guided very close to the target object. In fact, the search region within which the target is predicted to be located is assumed to be a cone with a 4 mile (6.4 km) diameter, 5 mile (8 km) long (1σ), up to a possible 12 mile (19.3 km) diameter, 15 mile (24 km) long (3σ), with the shuttle at the apex [5]. A non-cooperative radar rendezvous should have a system range of at least 35 mile (56 km) with a range accuracy of about 1% and angular accuracy of 2 mrad (3σ). Once in the search cone the system can automatically scan the region and should detect an object within minutes.

One possible method of identification upon initial acquisition of an object employs a television camera guided by the radar tracking system. A zoom lens can be used to receive an image with limited resolution. If the object is spinning a 'frozen' scene television display can be produced. This is equivalent to using an electronic strobe, except the picture is 'flashed' on a storage tube. The image can be held or reinforced at the spin rate of the object. However, the degree of resolution is somewhat un-certain with such a system. To determine the dynamic state and physical condition of a spinning body with assured accuracy and image quality, an optical or electronic strobe is very effective once the shuttle is within a few hundred feet of the target [5].

After locating and identifying the object, orbital maneuvers are executed to ap-proach and acquire a stand-off position relative to the target. If the object is not spinning or has some minimal spin rate and is of acceptable size, a direct docking may be attempted. Otherwise, the shuttle must maintain a stand-off or parking posi-tion while the retrieval package is deployed to eliminate angular momentum and capture the target. Objects which are too large for retrieval are not considered re-usable. In many cases it is desirable to simply eliminate such large pieces of 'space junk' from orbit. To accomplish this a remotely fired retro-pack could be attached to an object after despinning. Of course, this device would reenter with the object and would not be reusable. Furthermore, large objects, such as empty upper stages and payload fairings, in low orbits decay rapidly due to pronounced drag effects.

Standard rendezvous maneuvers are anticipated during approach of the shuttle to the stand-off position. Upon arriving at this position an autopilot will be enlisted to maintain spatial relationship with the target while the attitude control system maintains required orientation. Figure 1 illustrates one possible situation in which the shuttle is positioned along the orbital path of the object. Assuming a despin maneuver is required, the retrieval package must make an orbital transfer from the shuttle to the position of the target such that the axis of this remotely controlled space-craft is in line with that of the object spin axis. Fine adjustments are to be made after one of the shuttle crewmen checks alignment via a remote bifocal television system on the package. This terminal situation is represented in Figure 2, illustrating the retrieval of OSO 1.

Position of the shuttle and transfer trajectory of the retrieval package during cap-ture operations are very important factors to mission success. An artificial reference frame must be provided for the crew in order to remotely perform capture maneuvers and satisfy viewing constraints for continuous communications and observations. In terms of propellant requirements of the shuttle to maintain such a position, the ideal location would be along the orbital path of the target, as shown in Figure 1. Propellant expenditure would be for correcting perturbations only. Excessive fuel

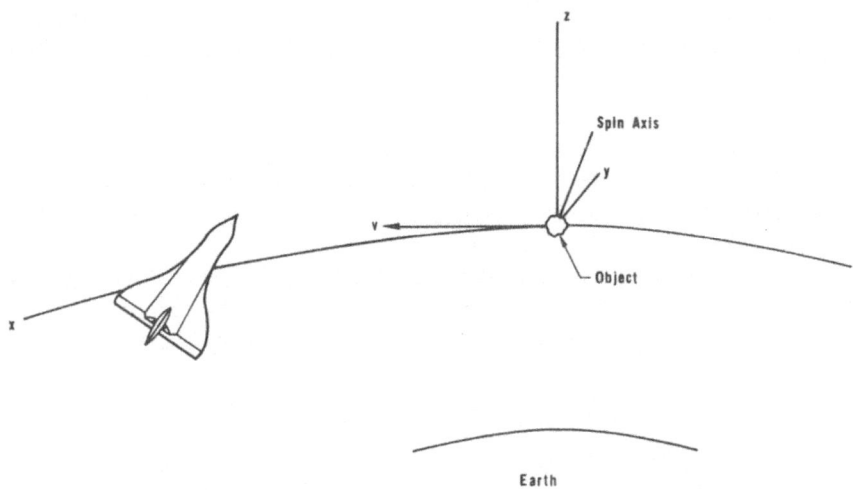

Fig. 1.   Shuttle/object relative position nomenclature.

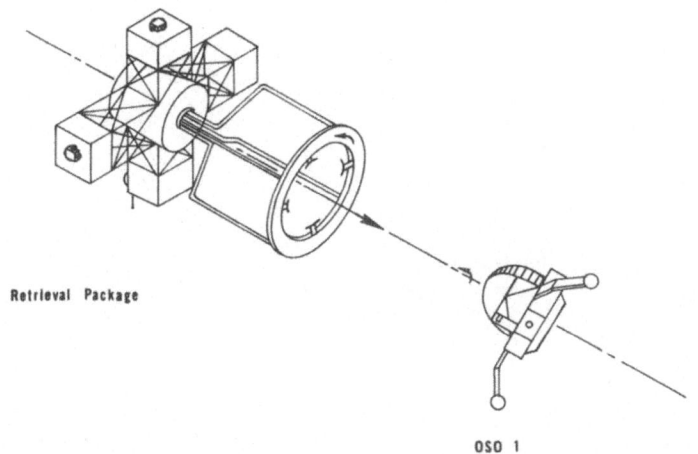

Fig. 2.   Retrieval package aligned with spin axis of OSO 1.

would be required to maintain an out-of-plane position; thus, it is assumed that the shuttle will be coplanar with the target during retrieval maneuvers. Final selection of a stand-off position for a given situation will depend on spin axis orientation of the target with respect to the orbital plane.

The transfer trajectory of the retrieval package to the target will, in turn, depend on shuttle position and object spin orientation. In general, this spin axis will not lie in the orbital plane of motion. Therefore, an out-of-plane transfer must be assumed for proper terminal alignment. A minimum of three thrust impulses is required for a transfer time of less than half an orbital period. If the shuttle is maintained along the orbital path of the target, then initial and final transfer conditions on orbital energy and altitude are identical, but a change in orbital position relative to the object has taken place. A typical transfer sequence is illustrated in Figure 3. At point *A* the package leaves the orbital plane on a path corresponding to an orbit with a different period. A midcourse thrust at *B* is necessary to bring the package back to the original plane in less than half a period. The final impulse at *C* will stop the package, and terminal capture procedures may then commence.

Actual attachment to and despin of the object is carried out by a ring which is aligned with the object spin axis. This ring is spun up to the same angular speed as the target, while the main body of the despin package remains 3-axis oriented in an

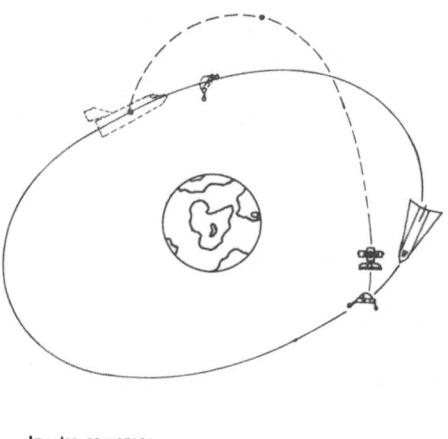

Impulse sequence:
A — Initial impulse geometry
B — Midcourse correction
C — Terminal impulse geometry

Fig. 3.   Typical three impulse transfer geometry for retrieval package.

inertial frame as shown in Figure 2. After the ring is synchronized through the use of an axially mounted television camera, the entire spacecraft is translated along its axis until the ring-plane reaches a position near the center of mass of the object. At that point attachment arms are extended from the ring toward its center. Once the target is secured by these arms, despin is executed while momentum is dumped via attitude jets on the main body of the package. A docking device on this inertially oriented body allows the shuttle to rendezvous and recapture the package with object in hand, and stow it in the cargo bay for return or repair.

Deployment of payloads which are to be spin stabilized can be performed in a similar manner. The retrieval package would extract this payload from the cargo bay, maneuver to a safe position, spin it up, and release it. The shuttle could stand by to check system operation before continuing with the mission.

4. Retrieval Package

The problem of despinning an uncooperative object has been cited as one of primary interest in this study. The decision to propose a separate device for this operation is the result of safety and performance considerations with respect to the shuttle. With a despin package maneuvering about the object at a 'safe' distance away from a manned vehicle, the risk of bodily harm from a mishap is minimized. Crewmen on the shuttle can maneuver a small spacecraft with great ease, especially when aligning the package with the target spin axis.

The despin package is conceived to consist of two major components; tender and despin ring. The tender provides all functions required for orbital transfer and alignment with the target, while the despin ring performs the actual capture of the object.

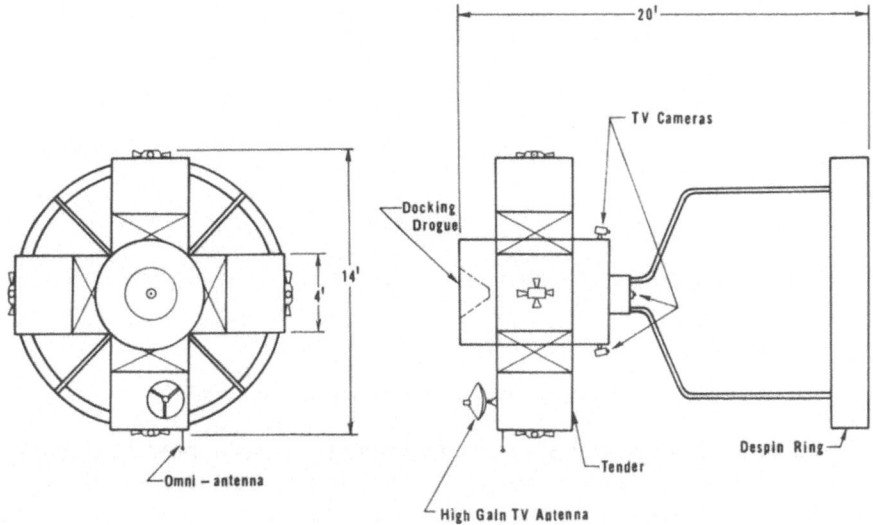

Fig. 4. Details of retrieval package configuration.

A configuration based on constraints and mission objectives were formulated. The complete spacecraft is illustrated in Figure 2, with tender and despin ring shown in greater detail in Figure 4. Overall size of the despin package is limited by the shuttle cargo bay dimensions, proposed to be 15 ft (4.6 m) in diameter and 60 ft (18.3 m) long. This restricts the ring size and, in turn, limits the size of objects which can be considered for retrieval. The tender is configured to allow maximum applied torque from the reaction control system while permitting large values of inertia about the ring axis. These innovations will minimize propellant requirements and effects of disturbances associated with the despin sequence. Thus, the major dimensions of this main body were selected as $14 \times 14 \times 8$ ft ($4.3 \times 4.3 \times 2.4$ m). Compartments in the four arms contain the power system, monopropellant reaction control equipment, and command and telemetry system. The central cylinder houses the ring spin motor, twin-gyro controllers for the attitude control system, a reserve propellant tank, and shuttle docking drogue. The despin ring as conceived here has an inner diameter of 13 ft (3.9 m) and an outer diameter of 14 ft (4.3 m). The four docking arms are assumed to extend inward another foot (0.3 m) when in the retracted position. These arms are independently operated to permit capture of arbitrarily shaped objects and can each extend to 6 ft (1.8 m) in length. Therefore, objects as small as 1 ft (0.3 m) and as large as 11 ft (3.3 m) in diameter can be handled by the same despin ring. Structural support of this ring is provided by four structs illustrated in Figures 2 and 4.

Attitude and orbit control systems, in conjunction with remote commands from the shuttle, provide maintenance of position and orientation during transfer and docking. Twin-gyro controllers were chosen for momentum exchange and mono-propellant reaction jets for momentum dumping. Since the tender is inertially oriented in the attitude maintenance mode there is no first-order cross-coupling. A momentum

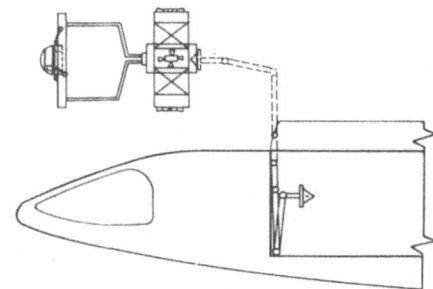

Fig. 5.   Shuttle/retrieval package docking arrangement.

wheel could be used during ring spin-up and despin to conserve propellant. However, this would introduce more complexity into the system. Furthermore, the reaction jet tanks are easily refillable before each retrieval sequence. This reaction control system also provides thrust impulses for transfer to the target and aligns the package with the target spin axis under commands from the crew. Once the package is aligned, capture should be accomplished in a short period of time. Therefore, relative position drift is very slight during this time interval, and an automated position control system is not required.

The command and telemetry system incorporates television cameras and two antennas, an extendable omnidirectional type for the command and telemetry link, and a high gain directional dish for television transmission. The high gain dish imposes some constraint on the shuttle stand-off position in order to satisfy the viewing requirements for this antenna. One camera is mounted along the ring axis at the center of the support struts, and spins with the ring. This offers a very convenient means of synchronizing ring and target spin rates and provides a check on final alignment. Two other television cameras mounted diametrically opposed on the central cylinder are used for relative position maintenance through bifocal viewing of the target. These cameras will be focused at a given distance from the target which will permit rough alignment and stand-off parking simultaneously. When the angle between the spin axis and the center line of the despin ring is zero, the axes will be aligned. Slight spin desynchronization will cause an apparent rotation of the target about its actual spin axis with frequency equal to the difference between object and ring spin rates. This can be used to make fine adjustments for final synchronization. The despin sequence may then proceed.

The shuttle/package docking apparatus, shown in Figure 5, consists of a folding arm mechanism with probe, docking latches, and tender drogue. The folding arm extends a docking probe to a position which is easily observable by the pilot. The shuttle then maneuvers to the retrieval package, and upon completion of docking. it is deactivated and retracted into the cargo bay.

5. Relative Motion of a Passive Object During Retrieval

During terminal maneuvering and capture operations the retrieval package may perturb the steady spin of the target. The resultant motion must be anticipated in order to determine attitude control requirements of the package. Disturbance torques may result from the despin ring striking an appendage or one capture arm making contact before the others. Such a situation can be considered through the use of the Euler moment equations. As an example, consider the case of a symmetric or near symmetric target [6],

$$A\dot{\omega}_1 = \tau_1$$
$$B\dot{\omega}_2 + \omega_1\omega_3(A - B) = \tau_2$$
$$B\dot{\omega}_3 - \omega_1\omega_2(A - B) = \tau_3$$

where A, B are the principal moments of inertia; ω_1, ω_2, ω_3 are angular velocity components about the body principal axes; and τ_1, τ_2, τ_3 are the disturbance torque components. Solving for transverse accelerations yields

$$\dot{\omega}_2 = \frac{\tau_2}{B} - \omega_1\omega_3 \left(\frac{A - B}{B}\right)$$

$$\dot{\omega}_3 = \frac{\tau_3}{B} + \omega_1\omega_2 \left(\frac{A - B}{B}\right)$$

It is apparent that a disturbance about the axis of symmetry (1-axis) will change the spin rate without affecting spin axis orientation. Nevertheless, such a disturbance torque is highly unlikely since the ring would be synchronized with the target about this axis. There may, however, be disturbances about the transverse principal axes, which would result in simple harmonic oscillation of both ω_2 and ω_3 corresponding to steady precession. The frequency of oscillation would be $(A - B)\omega_1/2\pi B$. Amplitude of motion is proportional to the ratio of disturbance impulse to transverse moment of inertia, i.e., $\int \tau \, dt/B$. Thus, an observer on the ring spinning at a rate of

ω_1 would see only ω_2 and ω_3 components of angular velocity, i.e., target would appear to wobble without spin.

Disturbances resulting from docking maneuvers are considered to be small and their effects can be handled by a combination of capture arm mechanisms and the retrieval package attitude control system. Furthermore, such disturbing torques can be minimized by observing three procedural suggestions: (1) make axis alignment as accurate as possible, (2) close with target at a very low rate, and (3) capture target about its center of mass, if possible.

6. Attitude Control of the Retrieval Package

In order to determine the feasibility of using a remotely controlled retrieval package for docking with spinning objects, an attitude control system should be synthesized and responses associated with the maneuvers obtained. Twin-gyro controllers were chosen as the momentum exchange devices appropriate for this spacecraft [7]. Selection was based on their inherent advantages, including first-order decoupling and large operational range of gimbal angles. The uncoupled equations of motion for a twin-gyro controlled spacecraft assuming small gimbal angles, δ_x, δ_y, δ_z are [8]

$$I_x \dot{p} = -2C_z \Omega_z \delta_z + L_x$$

$$I_y \dot{q} = -2C_x \Omega_x \delta_x + L_y$$

$$I_z \dot{r} = -2C_y \Omega_y \delta_y + L_z,$$

where I_x, I_y, I_z are the vehicle principal moments of inertia; p, q, r are the corresponding angular velocity components; C_x, C_y, C_z and Ω_x, Ω_y, Ω_z are the axial moments of inertia and angular velocities, respectively, of the gyros associated with the three torque units; and L_x, L_y, L_z are the disturbance torque components. Thus, the x axis is controlled by the z gyrotorquer, etc. Use of identical gyros on all axes permits further simplification,

$$C = C_x = C_y = C_z$$

$$\Omega = \Omega_x = \Omega_y = \Omega_z.$$

Control equations now differ only in vehicle moments of inertia and disturbance torque components. Thus, the three control systems for the x, y, and z axes are identical except for the system gains which vary with moments of inertia.

Consider the roll (or ring) axis controller, taken as the z axis here. The rate of change of the gimbal angle δ_y is given by the characteristic equation of a twin-gyro device [8],

$$2(I_a + A_g)\ddot{\delta}_y + K_t \dot{\delta}_y + K_c K_g \delta_y = -K_c K_B \frac{r_E}{s} + 2C\Omega r,$$

where r_E is the roll rate error, I_a is the sum of transverse gimbal and gearing moments of inertia; A_g is the transverse rotor moment of inertia, K_t, K_c are torque-speed and torque-voltage constants, respectively, and K_g, K_B are input gains of gimbal angle and roll rate error, respectively. Combining this with the roll equation of motion yields the basic control system block diagram given in [8]. The uncompensated system transfer functions are:

(a) Command:

$$\frac{r(s)}{r_c(s)} = \frac{K_0\omega_c^2}{s^3 + 2\xi_c\omega_c s^2 + \omega_c^2 s + K_0\omega_c^2}$$

(b) Disturbance:

$$\frac{r(s)}{L_z(s)} = \frac{s^2 + 2\xi_c\omega_c s + \omega_c^2}{I_z s(s^2 + 2\xi_c\omega_c s + \omega_c^2) + K_0 I_z \omega_c^2}$$

(c) Error:

$$\frac{r_E(s)}{r_c(s)} = \frac{s^3 + 2\xi_c\omega_c s^2 + \omega_c^2 s}{s^3 + 2\xi_c\omega_c s^2 + \omega_c^2 s + K_0\omega_c^2}$$

where

$$K_0 = \frac{2C\Omega K_B K_c}{4C^2\Omega^2 + I_z K_c K_g}$$

$$\omega_c^2 = \frac{4C^2\Omega^2 + I_z K_c K_g}{2I_z(I_a + A_g)}$$

$$\xi_c = \frac{K_f}{4\omega_c(I_a + A_g)}.$$

With the aid of the final value theorem, steady state response relations are immediately available. All responses to an impulsive input are zero. For a unit step input

$$\left.\frac{r}{r_c}\right|_{ss} = 1.0$$

$$\left.\frac{r}{L_z}\right|_{ss} = \frac{1}{I_z K_0}$$

$$\left.\frac{r_E}{r_c}\right|_{ss} = 0.$$

Thus, it is desirable to have system gain K_0 as large as possible. The transient response and system stability can be obtained by root locus methods once the values of constants are determined. Some form of compensation may be required to obtain a desirable combination of damping, response time, and steady state values. Furthermore, the spinning ring has a stiffening effect on the control system and must be included in a detailed system synthesis as a momentum device.

7. Despin Ring Control System

Synthesis of a spin control system to maintain synchronization of the despin ring and the target vehicle is considered here. The ring is physically constrained such that motion is possible only about the major principal axis of the package, so the single equation of motion is

$$J_z \dot{\rho} = f_z + L_z,$$

where J_z is the ring moment of inertia about the roll axis, ρ is the angular velocity or spin rate of the ring, and f_z, L_z are control and disturbance torques about the z axis, respectively. The control torque is provided by the spin motor and can be expressed as

$$f_z = K_c V_m,$$

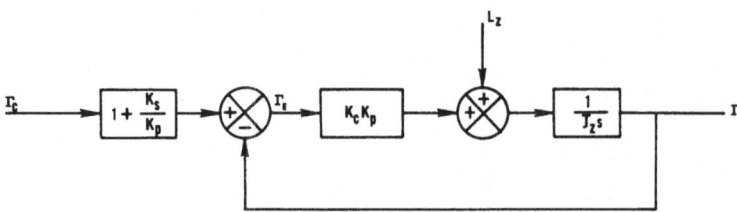

Fig. 6.   Despin ring control system.

where K_c is a torque-voltage constant, and V_m is the input signal to the motor. Input voltage is proportional to the command spin rate and the spin rate error, i.e.,

$$V_m = K_s \rho_c + K_p \rho_E,$$

where $\rho_E = \rho_c - \rho$, and ρ_c is the commanded spin rate.

The equation of motion in Laplace form then becomes

$$s J_z \Gamma(s) = K_c K_s \Gamma_c(s) + K_p K_c \Gamma_E(s) + L_z(s).$$

This leads to the control system block diagram shown in Figure 6. Closed-loop transfer functions can be obtained easily;

(a) Command;

$$\frac{\Gamma}{\Gamma_c} = \frac{K_c K_p}{J_z s + K_c K_p}$$

(b) Disturbance:

$$\frac{\Gamma}{L_z} = \frac{1.0}{J_z s + K_c K_p}$$

(c) Error:

$$\frac{\Gamma_E}{\Gamma_c} = \frac{J_z s}{J_z s + K_c K_p}$$

Steady state responses to unit impulse and step inputs are obtained directly. Response to an impulse is zero in all three cases. A unit step causes the following steady state situation:

$$\left.\frac{\Gamma}{\Gamma_c}\right|_{ss} = 1.0$$

$$\left.\frac{\Gamma}{L_z}\right|_{ss} = \frac{1}{K_c K_p}$$

$$\left.\frac{\Gamma_E}{I_c}\right|_{ss} = 0$$

This indicates that it is desirable to have K_c and K_p as large as possible in order to minimize the disturbance effects.

A root-locus analysis shows that this system is stable for all positive values of $K_c K_p / J_z$. Since this is a first-order system, it has an exponential transient response. Compensation may be required to achieve desired system characteristics.

8. The Tumbling Space Base

In the operation of manned orbiting vehicles there is always the small, but finite, probability that an accident will occur which renders the vehicle disabled. Therefore, an emergency must be anticipated in which evacuation of personnel is required. In general, such space missions may be divided into three phases [9]: rescue alert and rendezvous with the disabled vehicle, rescue operations proper, and return of the rescue vehicle. The second phase is of primary concern here, since a major part of the operation may very well involve detumbling of the uncontrolled spacecraft before escape of the occupants or repair of the control system. Tumbling is the general result of a significant attitude perturbation to an uncontrolled vehicle. This situation is coupled with continuous angular motion of all three principal body axes, i.e., no inertially oriented axis. Of course, a tumbling body would reach a stable spin state after a sufficient amount of energy has been dissipated. However, large bodies such as space stations have relatively low dissipation rates and may require many weeks or months to passively stabilize.

Astronauts trapped in a tumbling spacecraft could not easily escape and may not even be able to move about inside due to the changing nature of accelerations. Therefore, rescue from such a situation represents a very difficult problem and one which could arise as the result of an explosion or collision. Either event would probable cause tumbling and possible loss of attitude control. Preliminary estimates indicate that residual spin rates of the order of 4 rpm may result [10]. It is apparent that spacecraft attitude motion would have to be arrested before rescue operations could be carried out. However, elimination of angular motion of a large tumbling body is complicated, because it must be done from a non-tumbling frame. A 'detumbling package' could not move in such a way as to eliminate relative motion, because both the target and detumbling system cannot simultaneously possess the same center of mass location, principal moments of inertia, and angular momentum components.

Although techniques for controlling tumbling motion of a large vehicle have not yet been fully developed, a few promising ideas are worthy of identification. Techniques for reducing tumbling motion must involve either the activation of a momentum transfer device on the disabled vehicle or a tumbling-arrester system on the rescue craft [10]. Examples of momentum transfer devices include a yo-yo or rocket system appropriately located in anticipation of an emergency. Tumbling arrest from a rescue vehicle involves application of torque, even though hard docking does not seem feasible prior to stabilizing the distressed craft. The proper application of a fluid jet eminating from the rescue ship is one promising technique for applying detumbling torque. If spin or tumble can be reduced to a low value, hard docking or grappling may be attempted such that no further damage results.

9. Conclusions

A thorough understanding of the motion of orbiting bodies with angular momentum is essential before attacking the problems of docking with such passive objects. Solutions of these problems are becoming increasingly urgent as progress is made in establishing space bases and developing reusable boosters. Many of these problems have been identified and some solutions offered. Two major conclusions can be drawn:

(a) Spinning objects of moderate size can be retrieved from at least low orbits in a safe manner through the use of a remotely commanded vehicle operated in conjunction with the space shuttle.

(b) A situation in which a large manned space base is in an uncontrolled tumbling mode may result from a mishap or control failure. Rescue from such a vehicle is a very complex operational problem that must be solved to provide complete space rescue capability.

Of course, further efforts will be required to completely establish retrieval and rescue procedures, hardware requirements, and mission profiles. Results could significantly influence future space shuttle designs, operations, and applications.

References

[1] 'Satellite Situation Report', NASA Goddard Space Flight Center, Vol. 11, No. 5, May 31, 1971.
[2] Interian, A. and Kugath, D., *Astronaut. Aeronaut.* 7 (1969), 24.
[3] Tobey, W. H., French, R. T., and Adams, D. M., 'Experimental Material Handling Device, Final Report', Martin Marietta Corp., Denver Division, Report No. MCR-69-414, September 1969.
[4] 'Preliminary Analysis of Escape from a Tumbling Space Station', General Electric Co., Apollo Systems Division, Report No. SS-TR-060-4, June 1970.
[5] Kaplan, M. H., Yarber, W. H., Creehan, E. J., and Thoms, E. C., 'Dynamics and Control for Orbital Retrieval Operations Using the Space Shuttle', presented at the NASA Shuttle Technology Conference, Phoenix, Arizona, March 1971. Also in *Space Shuttle Technology Conference*, Vol. 1: *Operations, Maintenance, and Safety*, NASA Kennedy Space Center TR-1113, May 1971, p. 175.
[6] Thomson, W. T., *Introduction to Space Dynamics*, Wiley and Sons, 1963, p. 113.
[7] Yarber, W. H., 'Conceptual Design, Dynamics, and Control of a Device for Despinning Orbiting Objects', Astronautics Research Report No. 71-2, Department of Aerospace Engineering, The Pennsylvania State University, May 1971.
[8] Greensite, A. L., *Analysis and Design of Space Vehicle Flight Control Systems*, Spartan Books, 1970, pp. 385 and 413.
[9] Wild, J. W. and Schaefer, H., 'Space Rescue Operations', presented at the 3rd International Symposium on Space Rescue at the XXI International Astronautical Congress, Constance, Germany, October 1970.
[10] *Space Rescue Operations*, Vol. II: *Technical Discussion*, The Aerospace Corp., Report No. ATR-71 (7212-05)-1, May 1971, p. 22.

USE OF ORBIT-TO-ORBIT SHUTTLES FOR HYPERBOLIC RENDEZVOUS WITH RETURNING PLANETARY SPACECRAFT

DAVID R. BROOKS and EDWIN F. HARRISON

National Aeronautics and Space Administration,
Langley Research Center, Hampton, Va. 23365, U.S.A.

Abstract. An Earth-return mode for interplanetary spacecraft via hyperbolic rendezvous is described. In this mode an orbit-to-orbit shuttle leaves a circular Earth orbit, performs a rendezvous with a returning interplanetary spacecraft approaching Earth on a hyperbolic trajectory, docks, and performs desired transfers, and deboosts back into a circular Earth orbit while the spacecraft continues on its hyperbolic path. The initial mass in Earth orbit required for a rendezvous system utilizing chemical or nuclear propulsion compares favorably with the initial masses in Earth orbit chargeable to transport of a retrobraking Earth-return system to a target planet and back. The maneuver also has considerable merit as a backup and rescue system for any planetary mission even when not suitable as a primary recovery mode.

1. Introduction

Within the framework of the space transportation system proposed by NASA, missions requiring maneuvering in space will probably be performed by reusable orbit-to-orbit shuttle (OOS) vehicles designed specifically to spend their entire operational lives in a space environment. Chemical versions of such vehicles are being considered for lunar orbit-to-surface support missions, placement, servicing and retrieval of satellites in Earth orbit, space station support, and the injection of some unmanned planetary probes [1, 2]. Larger nuclear-powered versions of the shuttles are being planned primarily for use in translunar operations [2]. Conceptually, at least, the mission, operational, and hardware requirements of the reusable OOS are understood; justification for hardware development lies ultimately in demonstrating more and more areas in which the capabilities of such vehicles can be utilized in a unique or cost-effective way.

The area of particular concern for this study is the Earth capture or return phase of interplanetary missions, which may involve velocities almost twice Earth's escape velocity. In this paper, it will be shown that chemical and nuclear orbit-to-orbit shuttle vehicles have a definite role to play as either primary or backup Earth return systems for round-trip planetary missions, and that such a role is compatible with the capabilities of currently envisioned vehicles or their immediate successors. Typical planetary return missions have Earth approach hyperbolic excess velocities ranging from 4 to 18 km s^{-2} or corresponding atmospheric reentry velocities of 12 to 21 km s^{-1} [3]. Because Apollo heat-shield technology cannot cope with such reentry requirements without major redesign, it is clear that Earth return is a problem area which needs alternatives to be compared with development of new aerodynamic reentry vehicles.

Among many possible operational modes for Earth return of interplanetary space-

L. G. Napolitano et al. (eds.), Astronautical Research 1971, 219–236. All Rights Reserved.
Copyright © 1973 by D. Reidel Publishing Company, Dordrecht-Holland.

craft, the most demanding way in which an OOS might be used would be to provide the propulsive capability necessary for returning all or part of a spacecraft from a hyperbolic trajectory to a low circular Earth orbit. Such a maneuver will be referred to as 'hyperbolic rendezvous' and will be used to define upper performance limits for OOS vehicles serving as Earth return systems. Hyperbolic rendezvous, like other propulsive schemes for Earth return, is not a new concept [4], but its applicability has always been limited by the relatively primitive state of propulsion and space operational capabilities. It can be made feasible by the advent of a new era in space transportation, involving an Earth-to-orbit shuttle, highly developed chemical propulsion, nuclear propulsion, and vastly increased flexibility for Earth orbital maneuvering [2]. For the purposes of this study, such a space transportation system is assumed to be operational and available for support of hyperbolic rendezvous and other Earth return mission modes. Clearly, it is not possible to justify the development of an entire space transportation system, or even a component of it like the OOS, for the sole purpose of supporting round-trip planetary missions, but the operations discussed here can be combined with other studies to demonstrate the wide variety of activities which can be supported or made possible by orbit-to-orbit shuttle vehicles and to assure their maximum utilization once they are operational.

Nomenclature

a	semimajor axis, km
e	eccentricity, dimensionless
g_0	Earth sea-level gravitational acceleration, 0.0098 km s^{-2}
I_{sp}	specific impulse s
MEO	initial OOS mass in Earth orbit, kg
R	point at which rendezvous along a hyperbola is assumed to take place
r	radial distance or radius, km
t	time along a hyperbolic trajectory, h
V_∞	hyperbolic excess velocity, km s^{-1}
γ	flight-path angle on a conic trajectory, deg
ΔV	impulsive velocity increment required to perform either the outbound or inbound leg of the hyperbolic rendezvous maneuver, km s^{-1}
ΔV_1	impulsive velocity increment required to leave a 500-km circular Earth orbit on a transfer ellipse which intersects a hyperbolic trajectory, km s^{-1}
ΔV_2	impulsive velocity increment required to rendezvous with a hyperbolic trajectory from a transfer ellipse, km s^{-1}
$\Delta\gamma$	the difference between elliptic and hyperbolic flight-path angles at a given point, deg
Θ	angle between the elliptic and hyperbolic axes, deg
θ	true anomaly, deg
μ	Earth's gravitational constant, 3.9858×10^5 km^3 s^{-2}

SUBSCRIPTS

A	apoapsis
E	elliptic
EN	Earth entry – when applied to velocity, the velocity of a spacecraft at an altitude of 122 km
H	hyperbolic
P	periapsis
R	rendezvous point
$\oplus$	Earth
$\male$	Mars

2. The Hyperbolic Rendezvous Maneuver

A. MISSION DESCRIPTION

The assumed path for a round-trip hyperbolic rendezvous maneuver with a returning spacecraft is illustrated in Figure 1. The solid arrows indicate the trajectory for an OOS, while the spacecraft path along a hyperbola is indicated by the open arrows. The OOS leaves a circular Earth orbit at A after determining the proper transfer trajectory. The point B is selected as a result of compromise among competing demands imposed by a desire to minimize time on the transfer ellipse, while at the same time minimizing propulsion requirements and allowing sufficient time from periapsis

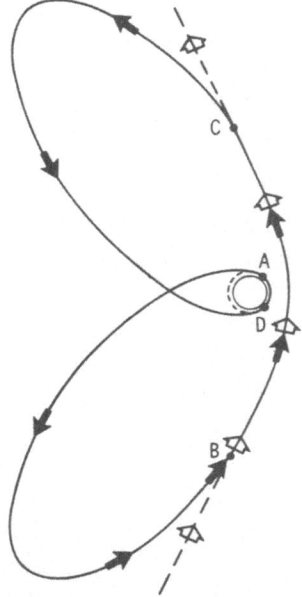

Fig. 1. Coplanar hyperbolic rendezvous maneuver. The four impulses are: A – leave circular Earth orbit; B – match hyperbolic trajectory, C – deboost to return trajectory, D – reenter circular Earth orbit.

time minimizing propulsion requirements and allowing sufficient time from periapsis of the hyperbola. At B, the OOS performs the final rendezvous and docking maneuver. Transfers of crew, modules, or samples are performed on the hyperbolic trajectory between B and C. At C, the remaining portion of the interplanetary spacecraft continues on the hyperbola, while the OOS deboosts into the return ellipse with its cargo and recircularizes at D. This mode provides a relatively long transit time on the transfer ellipse (possibly several days) so that the parameters of the transfer ellipse may easily be adjusted. Plane changes may be cheaply accomplished at the apoapses of the ellipses. Also, time phasing may be easily achieved, for example, on the return leg of the maneuver to bring the OOS into position for circularizing near a space station or some other site chosen for debriefing and quarantine, in the case of a manned mission, or simply isolation of the cargo in the case of an unmanned sample return.

The operational mode described by Figure 1 is feasible as a primary Earth return system for a round-trip planetary mission, although it could be argued that safety and reliability considerations would make retrieval from a highly elliptical orbit more desirable. However, in an emergency situation where, for example, a manned planetary mission had lost the propulsive or maneuvering ability to retrobrake or aerobrake into an Earth orbit, the hyperbolic rendezvous maneuver could be indispensable, and it deserves examination on the basis of its rescue capability alone. It will be shown in this paper that use of a hyperbolic rendezvous maneuver for Earth return can reduce the total weight which must be injected into Earth orbit for some manned planetary missions, compared to an onboard propulsive retrobraking system. Thus, the hyperbolic rendezvous maneuver can be competitive, at least on a weight basis, with other seriously considered Earth return options.

B. EQUATIONS

The general equations for hyperbolic rendezvous are derived in [5] using basic conic equations. From these, limits are established on the allowable transfer ellipses so that it is possible to compute the velocity changes necessary to perform the rendezvous

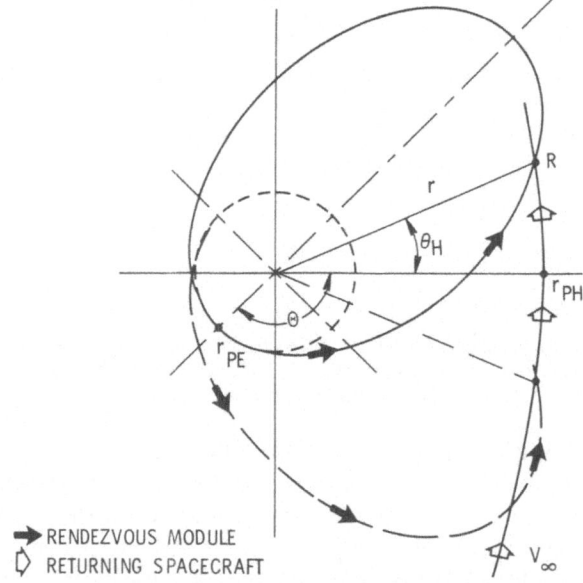

Fig. 2.   Geometry for derivation of hyperbolic rendezvous equations.

maneuver. The geometry for the problem is defined in Figure 2. For deriving the equations, it is assumed that the rendezvous vehicle (solid arrows) will leave a circular Earth orbit to meet the returning spacecraft (open arrows) on the incoming hyperbolic return trajectory at point R. The circular Earth orbit and the hyperbolic return trajectory are assumed to be coplanar, with the rendezvous vehicle always available at the right place and time in Earth orbit, that is, the problems of phasing are ignored.

The geometry of the circular Earth orbit is completely specified by its radius, which is also the periapsis of the transfer ellipse, r_{PE}. The hyperbolic return trajectory is specified in terms of the hyperbolic periapsis r_{PH} and the hyperbolic excess velocity V_∞.

It can easily be shown that the ellipses for transfer between Earth orbit and R may have the following range of eccentricities:

$$\frac{r_R - r_{PE}}{r_R + r_{PE}} \leq e_E < 1, \tag{1}$$

where r_R is the radial distance to the rendezvous point.

For each set of initial conditions, that is, values of r_{PE}, r_{PH} and V_∞, it is necessary to compute the velocity changes needed to perform the rendezvous for the range of transfer ellipses specified by Equation (1). The rendezvous maneuver is assumed to consist of two impulsive velocity changes: ΔV_1 removes the rendezvous vehicle from circular orbit and places it on the transfer ellipse, and ΔV_2 matches the velocity vector of the rendezvous vehicle with the return vehicle. ΔV_1 is the elliptical periapsis velocity minus the circular velocity and ΔV_2 is found by the law of the cosines:

$$\Delta V_1 = \sqrt{\mu \left(\frac{2}{r_{PE}} - \frac{1}{a_E}\right)} - \sqrt{\mu \frac{1}{r_{PE}}} \tag{2}$$

$$\Delta V_2 = \sqrt{V_{RE}^2 + V_{RH}^2 - 2V_{RE}V_{RH}\cos\Delta\gamma_R}, \tag{3}$$

where

$$\Delta\gamma_R = \gamma_{RH} - \gamma_{RE}$$

The sum $\Delta V_1 + \Delta V_2$ is affected by two choices involving time, both of which are important to the operational concept of the hyperbolic rendezvous mission. In Figure 3, time on the transfer ellipse (the phase of the mission between points A and B in Figure 1) is shown as a function of total one-way ΔV ($\Delta V = \Delta V_1 + \Delta V_2$). The assumed hyperbola has a hyperbolic excess velocity V_∞ of 6.1 km s^{-1} and a periapsis radius r_{PH} of 12 756 km ($r_{PH}/r_\oplus = 2$). The OOS is assumed to start from a 500-km circular orbit ($r_{PE} = 6878$ km) and the rendezvous point on the hyperbola occurs 2 h prior to hyperbolic periapsis passage. The minimum velocity increment for this case occurs as e_E approaches 1, that is, as the time spent on the ellipse approaches infinity, and restricting the size of the transfer ellipse leads to increases in ΔV. For example, with the conditions given in Figure 3, ΔV for an ellipse having $e_E = 0.9$ is about 9% greater (7.2 vs 6.6 km s^{-1}) than the minimum, and requires 46 h 12 min on the transfer ellipse.

The sum $\Delta V_1 + \Delta V_2$ is also affected by variations in the time of the rendezvous relative to hyperbolic periapsis passage and by the radius of the hyperbolic periapsis. The minimum value of ΔV (which may imply an infinitely large transfer ellipse) is

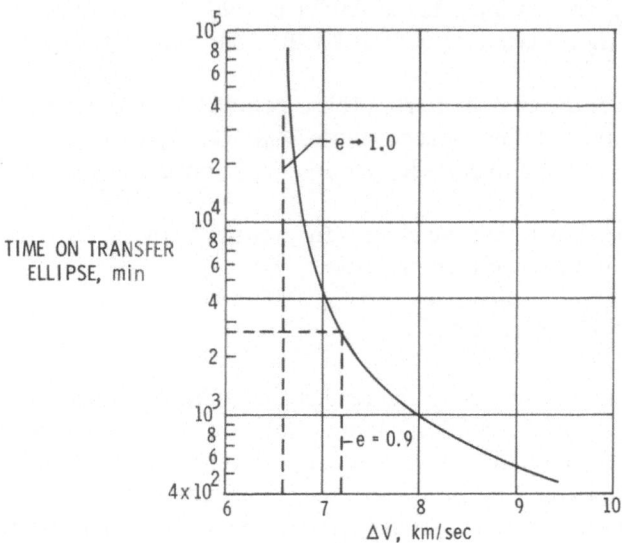

Fig. 3.   Time spent on transfer ellipse as a function of total one-way velocity increment.
$V_\infty = 6.1$ km s^{-1}; $t_{PH} = 2$ h; $r_{PH}/r_\oplus = 2$.

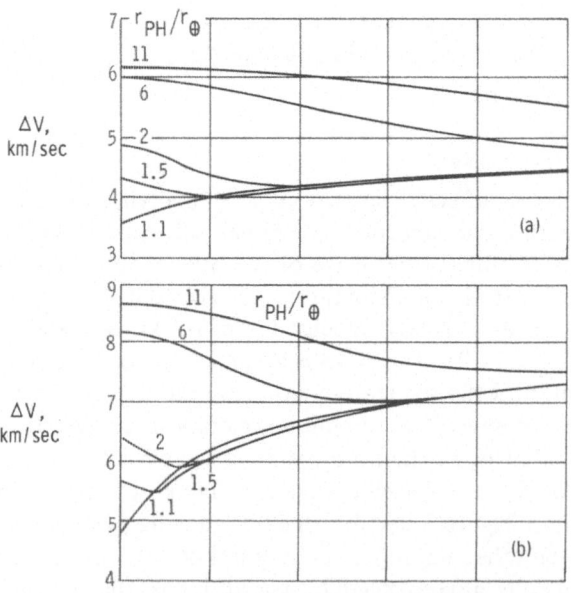

Figs. 4a–b.   Minimum one-way velocity increment as a function of time from hyperbolic periapsis for five values of the ratio of hyperbolic periapsis radius r_{PH} to Earth radius $r_\oplus$.

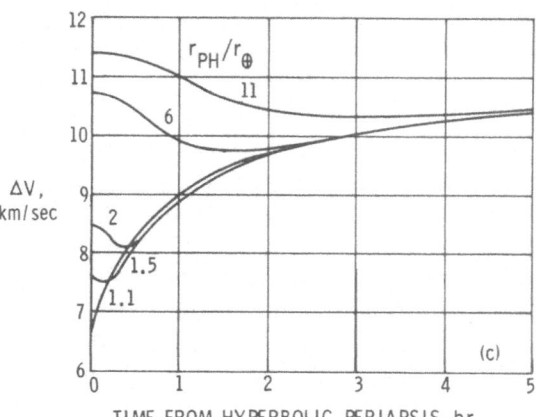

Fig. 4c.

shown as a function of time from hyperbolic periapsis for the hyperbolic excess velocities of 3.05, 6.10 and 9.15 km s^{-1} in Figures 4a, 4b and 4c, respectively. The hyperbolic periapsis is shown parametrically, in terms of the dimensionless ratio $r_{PH}/r_{\oplus}$. For example, for $V_{\infty} = 6.1$ km s^{-1}, Figure 4b shows that the minimum ΔV occurs at about 40 min from hyperbolic periapsis passage when $r_{PH}/r_{\oplus} = 2$. Figure 4 demonstrates that, especially for large values of time from periapsis, ΔV is not a strong function of the hyperbolic periapsis distance. This fact could be important in the case of a rescue operation involving a disabled spacecraft which had lost its guidance or propulsive capability for maneuvering into a particular hyperbolic trajectory – the lack of sensitivity to closest approach distance enables maneuvering and tracking to be done by the OOS and ground stations so that the returning spacecraft could be passive during the operation.

C. VEHICLE AND TRAJECTORY CONSTRAINTS AND ASSUMPTIONS

For the purpose of examining hyperbolic rendezvous as applied to planetary missions, several assumptions must be made concerning, first, elements of the hardware involved, and second, some details of the proposed trajectories to be followed by both the returning vehicle and the OOS. The rendezvous spacecraft is assumed to consist of four parts: a rendezvous module, a fixed inert stage mass, tanks, and propellant. Propulsion is either chemical or nuclear.

Rendezvous Module. – The rendezvous module houses the crew and contains all necessary systems for life support, guidance, communications, and so forth. Its mass is assumed to be between 2700 and 4500 kg and is independent of the type or amount of propulsion used or required for a particular mission. On the return trip, it has been assumed that an additional 900 kg of cargo is onboard, corresponding to a six-man crew and samples, for example.

Chemical Propulsion System. – A typical proposed OOS configuration* for rela-

* Unpublished data from Aerospace Corporation.

tively low energy missions has the following characteristics: $I_{sp} = 460$ s, MEO = 31 300 kg, propellant = 27 700 kg, tank and tank-related structure (micrometeoroid shielding and thermal insulation) = 1500 kg, inert stage weight = 2100 kg. For such a system, the tank fraction is 5.4% and this value has been assumed for this study. The inert stage weight is taken to be constant even though a much different configuration may be required for very high energy missions. For a small rendezvous spacecraft, RL-10 (or similar) engines with a thrust of about 67 000 N per engine are envisioned, and their mass is assumed to be included in the 2100 kg. For a larger stage, a J-2 class engine, with a thrust of about 900 000 N is envisioned, but the inert mass is still fixed at 2100 kg. This simplification is justified by the uncertainty allowed for the rendezvous module and further by the realization that so much propellant is required for a typical hyperbolic rendezvous mission that the tanks may be more massive than the inert stage weight.

Nuclear Propulsion System. – A solid core hydrogen-fueled reactor is assumed to provide an I_{sp} of 850 s with a thrust of 336 000 N. The total inert stage mass, including the engine and shielding (but not tanks) is 13 600 kg. The assumed parameters correspond to a NERVA configuration [6]. Tanks are assumed to weigh 15% of the propellant – 13% for structure and 2% for contingency [7].

TABLE I

Mission times for the rendezvous maneuver between the LM and CSM following lunar lift-off of Apollos 11 and 12

Maneuver	Apollo 11		Apollo 12	
	Mission time, h:min	Cumulative elapsed time, h:min	Mission time, h:min	Cumulative elapsed time, h:min
Terminal phase initiation	127:04	—	144:36	—
Terminal phase finalization	127:46	0:42	145:06	0:30
Docking completed	128:03	0:59	145:36	1:00
LM/CSM separation	130:10	3:06	148:00	3:24
Final separation maneuver completed	130:30	3:26	148:05	3:29

Mission Times. – There are two timing problems which must be settled before applying hyperbolic rendezvous to a hypothetical mission. First, the total mission time should be limited in some way, and second, the time on the hyperbola must be sufficiently long to allow whatever maneuvers are required. For the first problem, it is convenient to limit the transfer ellipses to 50 h. (For a periapsis of 6878 km, this corresponds almost exactly to an eccentricity of 0.9.) Such a limit restricts round-trip times to a few days without substantially increasing the ΔV requirements (recall Figure 3).

The problem of allowing sufficient time on the hyperbola for maneuvers and transfer procedures has been approached by referring to experience gained on Apollos 11 and 12 during the docking of the LM and CSM following lunar lift-off. Table I (ref. [8] and unpublished data from Manned Spacecraft Center) shows a time history of Apollos 11 and 12 CSM/LM docking procedures starting from initiation of rendezvous radar tracking and ending with the CSM/LM separation maneuver. These times are considered to be representative of the times required in the case under study. For Apollo, about $3\frac{1}{2}$ h were required for rendezvous, docking, and transfer of two men, their equipment, and samples. For representative proposed planetary missions, up to six men are involved; therefore it is assumed that a total time somewhat longer than for Apollo, about 4 h, should be allowed for rendezvous, docking, and transfer of crew and/or cargo. For the mission mode of Figure 1, this corresponds to a rendezvous at 2 h before periapsis. Since the allowed time can only be considered as a rough guess at best, it is fortunate that the velocity requirements are only weakly sensitive to an increase in the time-in fact, the velocity increments for longer times from periapsis actually decrease for some values of r_{PH} (see Figure 4).

Spacecraft Trajectory. – Since it has been previously shown that, in general, ΔV is relatively insensitive to r_{PH}, it is sufficient to select a representative value of r_{PH} to use in ΔV computations for a particular case. For this purpose $r_{PH}/r_{\oplus} = 2$ will be the nominal value. It is also assumed that for the nominal trajectories the spacecraft trajectory is coplanar with the rendezvous vehicle. Certainly this should be part of the mission plan for use of hyperbolic rendezvous as a primary return mode. However, in an emergency situation a plane change might be required. Of course, for an infinitely large ellipse, the velocity increment for the plane change approaches zero. For the 50-h transfer ellipse previously taken as nominal, a plane change of as much as 90° can be made with velocity increment of only a few hundred meters per second [5].

D. TRAJECTORY DATA FOR HYPERBOLIC RENDEZVOUS MANEUVERS WITH INTERPLANETARY SPACECRAFT

Trajectory data for a hyperbolic rendezvous maneuver have been computed for several values of V_{∞} up to 12.19 km s^{-1} (40 000 ft s^{-1}). These data are shown in Table II, which includes – for each value of V_{∞} – eccentricity of the spacecraft hyperbola e_H, the magnitude of difference between elliptic and hyperbolic flight-path angles at rendezvous $\Delta\gamma_R$, rendezvous radius r_R, velocities on the ellipse and hyperbola at rendezvous V_{ER} and V_{HR}, the velocity increment required for the second part of the rendezvous maneuver ΔV_2, the one-way velocity increment ΔV, and time spent on the transfer ellipse t_R. Since ΔV represents the one-way velocity increment, the total velocity increment required to perform the mission is $2 \times \Delta V$. Spacecraft trajectory assumptions have been utilized, as discussed above, and the starting orbit is circular, with an altitude of 500 km ($r_{PE} = 6878$ km) as previously assumed. For a 50-h ellipse ($e_E = 0.9$) starting from this altitude, $V_1 = 2.88$ km s^{-1}.

A sketch of a representative trajectory based on the data of Table II is shown in Figure 5, for $V_{\infty} = 6.1$ km s^{-1}. The distances involved are normalized in terms of the Earth's radius. The ellipse has a periapsis of 6878 km and an apoapsis of about 131 000 km. The rendezvous radius is 55 690 km.

TABLE II

Trajectory data for the hyperbolic rendezvous maneuver

$$\begin{bmatrix} r_{\mathrm{PH}}/r_{\oplus} = 2 \\ t_{\mathrm{PH}} - t_{\mathrm{R}} = 2\,\mathrm{h} \\ e_{\mathrm{E}} = 0.9\ (50\text{-h period}) \\ \Delta V_1 = 2.88\ \mathrm{km\ s^{-1}} \end{bmatrix}$$

V_∞, km s^{-1}	e_{E}	r_{R}, km	$\lvert \Delta \gamma_{\mathrm{R}} \rvert$, deg	V_{ER}, km s^{-1}	V_{HR}, km s^{-1}	ΔV_2, km s^{-1}	ΔV total, km s^{-1}	t_{R}, h:min
1.52	1.074	38 150	4.82	3.89	4.82	1.00	3.88	47:51
3.05	1.297	42 130	0.65	3.62	5.31	1.69	4.57	47:31
6.10	2.188	55 690	7.79	2.92	7.17	4.30	7.18	46:12
9.14	3.673	73 430	13.44	2.25	9.72	7.55	10.43	45:01
12.19	5.752	93 080	18.66	1.66	12.54	10.97	13.85	40:54

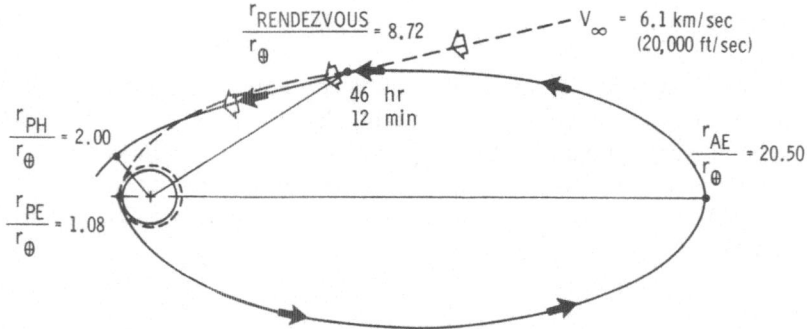

Fig. 5. Representative transfer ellipse for hyperbolic rendezvous maneuver based on data of Table II. $V_\infty = 6.1$ km s^{-1}.

E. OOS MASSES FOR THE HYPERBOLIC RENDEZVOUS MANEUVER

Using the constraints and assumptions discussed in the previous section, the initial mass in Earth orbit (MEO) required for an OOS to perform the hyperbolic rendezvous maneuver has been computed as a function of hyperbolic excess velocity for two operational modes: (a) propellant tanks retained for reuse, and (b) propellant tanks expended after each of the first three engine burns. The masses have been computed assuming impulsive burns, using the well-known ideal rocket equation, and the results are summarized in Figure 6, which shows MEO plotted as a function of V_∞ with expendable and reusable tanks, utilizing both chemical and nuclear propulsion. The upper and lower boundaries on each curve correspond to 4500 and 2700 kg rendezvous modules, respectively, indicating uncertainty about the requirements of such a module. The nuclear OOS, with its large inert weight, becomes superior to the lighter but much lower performance chemical OOS at a V_∞ of about 4 km s^{-1}.

3. Comparison of Earth Return Options

A. PLANETARY MISSIONS WITH CONVENTIONAL EARTH RETURN SYSTEMS

To be placed in proper perspective, the OOS mass requirements presented in Figure 6 need to be compared with other Earth return options for some representative manned planetary missions. For this purpose, a set of manned Mars missions have been examined. These utilize Venus swingbys on the outbound or inbound part of the

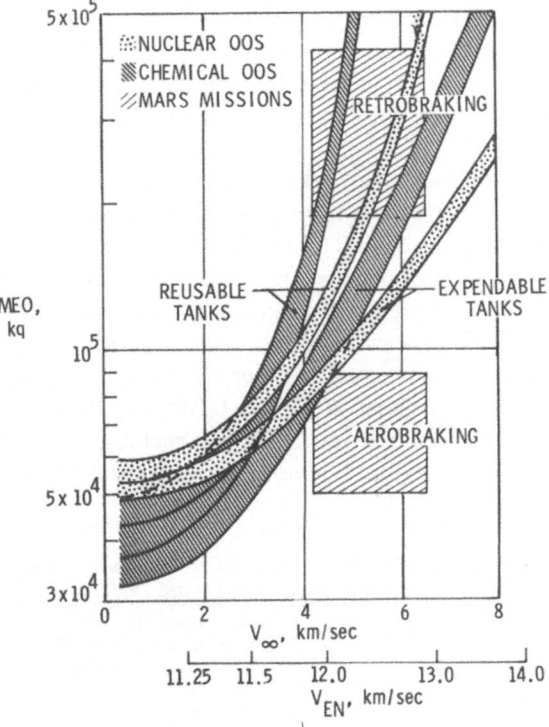

Fig. 6. Initial mass in Earth orbit as a function of hyperbolic excess velocity and Earth-entry velocity compared with manned Mars missions utilizing aerobraking or retrobraking for Earth return.

trip and have round-trip times of 550–570 days. Details of trajectory and operational aspects of this group of missions are given in Appendix A, the purpose of which is to establish Earth return system weights for comparison with the hyperbolic rendezvous maneuver and some intermediate OOS-based strategies.

The hatched areas of Figure 6 indicate the results of the analysis of Appendix A. For the Mars missions, the ordinate of Figure 6 refers to the additional mass in Earth orbit required to provide for the transport of aerobraking (bottom area) or retrobraking (top area) Earth return capability for a six-man crew to Mars and back. 'Additional mass' is with respect to a spacecraft with no onboard provision for Earth return, that is, one which will be returned by the hyperbolic rendezvous maneuver. Thus, the comparison to be made is between additional mass for onboard return systems and the MEO of an OOS capable of performing the hyperbolic rendezvous maneuver. As noted in Appendix A, the aerobraking mode includes an 8200-kg reentry capsule and retrobraking assumed a 2700-kg crew module which is circularized at an altitude of 500 km.

Clearly, the aerobraked Earth return represents the smaller MEO of the two conventional systems. The MEO chargeable to the Earth return systems consists only of the aerobraking capsule itself plus propellant to transport that weight to Mars and back. The weight of the aerobraking module is itself assumed independent of hyperbolic excess velocity, but the fact that it must be propelled throughout the mission along with the rest of the spacecraft components forces the additional MEO requirement for this system to vary over the range shown for the eight missions considered.

The MEO chargeable to the retrobraked system includes the mass of the crew module, propellant for the retrobraking maneuvers, and propellant for transporting the entire retrobraking system to Mars and back.

Figure 6 illustrates that hyperbolic rendezvous with either reusable or expendable tanks can be competitive on a weight basis with other primary Earth return modes for some planetary missions. The values of V_∞ for the Venus swingby Mars missions are representative of those encountered in other planetary missions, as shown in [3], and the masses in Earth orbit given in Figure 6 for retrobraking and aerobraking Earth return systems are representative of those expected for other planets as well. Therefore, the applicability of Figure 6 is not restricted to the Mars missions shown.

TABLE III

Initial masses in Earth orbit (thousands of kilograms) required for various Earth return modes at the termination of 1988 and 1990 manned Mars missions

Earth return mode	1988 ($V_\infty = 4.406$ km s^{-1})		
	S/C	OOS	Total MEO
Full retrobrake	697	—	697
24-h ellipse	622	20[a]	642
Minimum capture	613	26[a]	639
Aerobrake	547	—	547
No Earth return system and hyperbolic rendezvous	497	121[b] (82)[c]	618[b] (579)[c]
	1990 ($V_\infty = 6.371$ km s^{-1})		
Full retrobrake	1206	—	1206
24-h ellipse	1051	20[a]	1071
Minimum capture	1032	26[a]	1058
Aerobrake	885	—	885
No Earth return system and hyperbolic rendezvous	797	370[b] (144)[c]	1167[b] (941)[c]

[a] Chemical OOS, reusable tanks.
[b] Nuclear OOS, reusable tanks.
[c] Nuclear OOS, expendable tanks.

B. ADDITIONAL OOS-BASED EARTH RETURN OPTIONS

It was noted in section 1 that safety considerations, for example, might render the hyperbolic rendezvous maneuver unacceptable as a primary Earth return system (although its usefulness in an emergency situation makes such a capability desirable). There are other modes of operation, based on availability of a high energy orbit-to-orbit vehicle, which could combine an OOS with propulsion capability aboard an interplanetary spacecraft. To illustrate the trade-offs that are available in such a situation, Table III summarizes some masses in Earth orbit for Earth return options ranging from full retrobraking into a 500-km circular orbit to zero retrobraking. Data for the best (1988) and worst (1990) of the eight Mars missions from the point of view of MEO are used to bracket the requirements. For example, the fully retrobraked 1988 mission requires an MEO of 697 000 kg, while the hyperbolic rendezvous mode requires an MEO of 497 000 kg for the Mars spacecraft plus an OOS of 120 000 kg or 82 000 kg depending on whether reusable or expendable operation is assumed. Retrobraking to a 24-h or minimum capture ellipse with a periapsis of 500 km requires intermediate values of MEO for the spacecraft, as shown in Table III, plus a modestly sized OOS which is assumed to be a fully reusable chemical vehicle. For the 1988 mission, hyperbolic rendezvous even with a fully reusable nuclear shuttle requires less total MEO than any of the other propulsive modes shown. If the OOS is operated in an expendable mode, it happens that for this particular case the nuclear and chemical shuttles are of nearly equal mass, 82 000 kg, as can be seen by referring to Figure 6.

The mass breakdown for a fully reusable nuclear OOS for the 1988 mission is shown in Table IV. The OOS might consist of a NERVA engine, permanently attached to a propellant module, six additional propellant modules clustered around an assembly frame, astrionics packages, and a rendezvous module fitted with a docking adapter for the interplanetary spacecraft. The modularized approach allows piece-by-piece launch to Earth orbit by the Earth-to-orbit shuttle. With a total of seven propellant modules, such a vehicle could perform a hyperbolic rendezvous maneuver in a fully reusable mode up to a V_∞ of about 5 km s^{-1}. Jettisoning some of the clustered tanks as they are emptied would, of course, improve the potential performance.

The 1990 mission is more difficult to retrieve from a hyperbolic trajectory because of its higher hyperbolic excess velocity (6.371 vs 4.406 km s^{-1} for 1988), and requires an expendable operational mode to compare favorably with retrieval from an elliptical orbit. For the 1990 mission, chemical OOS vehicles are probably not practical for hyperbolic rendezvous because of the large propellant requirements but, of course, they may still be used to retrieve the spacecraft from elliptical orbits.

The nuclear OOS sizes indicated by Figure 6 and Table III are similar to those being studied for other uses, primarily support of advanced lunar exploration and exploitation. One such vehicle proposed by Lockheed Missiles and Space Company [9] would have, for the mission discussed in this paper, an initial mass in Earth orbit of 164 000 kg, including 122 400 kg of liquid hydrogen (LH$_2$) for main propulsion and a 2700-kg rendezvous module (RM). Another configuration, developed by McDonnell

TABLE IV

Mass breakdown of a reusable nuclear orbit-to-orbit shuttle sized for the hyperbolic rendezvous maneuver required at the termination of the 1988 manned Mars mission

System	Mass
Propellant 1	35 400
2	24 700
3	17 800
4	12 800
Tanks (15%)	13 600
Nerva engine	13 630
Rendezvous module	2 700
MEO	120 630

[a] 1 – Leave 500 km circular earth orbit, 2895 km s^{-1}.
2 – Rendezvous, 2.865 km s^{-1}.
3 – Deboost, 2.865 km s^{-1}.
4 – Recircularize at 500 km, 2.895 km s^{-1}.

Douglas Astronautics Corporation [10], would have an MEO of 172 945 kg, including 135 100 kg of LH$_2$ and a 2700-kg RM. Both vehicles are modularized for launch into Earth orbit by the space shuttle and subsequent orbital assembly, and both utilize a NERVA engine operating at a specific impulse of 825 s. These vehicles, as presently sized, could perform the hyperbolic rendezvous maneuver for hyperbolic excess velocities as high as 4.64 and 5.31 km s^{-1} respectively. (Operation of the NERVA engine at a specific impulse of 850 s^{-1}, as was assumed for the nuclear OOS of this study, would improve the performance to 4.85 and 5.52 km s^{-1}, respectively, without making any structural changes.) For purposes of comparison, the MEO of a nuclear system based on the assumptions in this paper are about 140 000 kg for $V_\infty = 4.85$ km s^{-1} and 200 000 kg for $V_\infty = 5.52$ km s^{-1}. Clearly, there is no conceptual hindrance to enlarging modular vehicles to the extent required for use in support of interplanetary missions as suggested in this paper.

4. Conclusions

Hyperbolic rendezvous with returning interplanetary spacecraft has been shown to be feasible for a primary or rescue Earth return mode if the availability of orbit-to-orbit shuttles as large as several hundred thousand kilograms (initial mass in Earth orbit) is assumed. Both chemical and nuclear systems have possible applications, although the chemical systems are restricted to missions involving small hyperbolic excess velocities. The hyperbolic rendezvous maneuver has been compared with conventional Earth return systems for a set of manned Mars missions, and the initial masses in Earth orbit required for hyperbolic rendezvous have been shown to compare favorably with the requirements for transporting an Earth retrobraking capability to Mars and back. The Earth return requirements of Mars missions are considered to be representative of other planetary missions as well.

The usefulness of hyperbolic rendezvous in a rescue operation is enhanced by the demonstrated lack of sensitivity of the ΔV requirements to location of the hyperbolic periapsis. This means that a disabled spacecraft can be tracked and recovered successfully with a minimum of maneuvering on its part. Even out-of-plane maneuvers can be accommodated when time is allotted for sufficiently large transfer ellipses.

Appendix A. Mass Requirements for Manned Mars Missions

It is the purpose of this appendix to document the procedures used for generating the manned Mars mission data used in Figure 6. Three cases are required: retrobraking, and no Earth return capability on the spacecraft. The last case is, of course, the hyperbolic rendezvous case.

The manned Mars missions chosen for comparing conventional Earth return methods, that is, aerobraking or propulsive braking to Earth orbit, with hyperbolic rendezvous are inbound (I) or outbound (O) Venus swingbys in the period 1982–1999. Eight opportunities are examined. The dates for arrivals and departures are given in Table V, along with the hyperbolic excess velocities associated with each maneuver [1]. The largest and smallest values of V_∞ at Earth return form the right- and left-hand boundaries of the hatched areas in Figure 6. Note that the Mars stay times are always 30 days, and the total trip time is not more than 580 days. Clearly, other equally valid criteria could have been used for selecting mission schedules.

There are many operational details and component masses which need to be specified to completely define a mission. For the purposes of this study, the following mission sequence is followed for a retrobraked Earth return:

(1) The initial mass in Earth orbit (MEO) just prior to departure consists of a 50.0×10^3 kg Mars excursion module (MEM), a 37.2×10^3 kg Earth retromodule (RM), four 13.6×10^3 kg nuclear engines (one for each burn), propellant for each of the four burns, tanks and propellant reserve for each burn which, taken together, weigh 20% of the respective propellant load, and meteoroid shields (an additional 20% of propellant) for each burn except for the first.

(2) Launch from 185 km circular Earth orbit and jettison engine.

(3) Jettison meteoroid shield just prior to Mars orbit insertion.

(4) Deboost into very low circular Mars orbit and jettison engine.

(5) Jettison meteoroid shield just prior to Mars orbit departure.

(6) Leave Mars orbit and jettison engine.

(7) Jettison meteoroid shield for retrostage just prior to Earth capture.

(8) Deboost into 500 km circular Earth orbit.

Using the ideal rocket equation,

$$\exp\left(\Delta V / I_{sp} g_0\right) = \frac{\text{Initial weight}}{\text{Final weight}}$$

TABLE V

Arrival and departure dates and hyperbolic excess velocities for selected manned Mars missions used in determining representative initial masses in Earth orbit

Mission opportunity	Julian date, 244xxxx or 245xxxx				Hyperbolic excess velocities (V_∞, km s^{-1})			
	Lv. $\oplus$	Ar. $\male$	Lv. $\male$	Ar. $\oplus$	Lv. $\oplus$	Ar. $\male$	Lv. $\male$	Ar. $\oplus$
1982(I)[a]	4970	5190	5220	5550	3.364	4.257	6.549	5.210
84(I)	5680	5960	5990	6230	3.840	3.572	8.336	6.549
86(O)[b]	6150	6530	6560	6710	4.168	5.120	4.495	3.930
88(I)	7350	7550	7580	7910	3.424	2.650	7.055	4.406
90(O)	7830	8160	8190	8390	4.168	5.478	8.395	6.371
93(O)	8510	8820	8850	9070	4.733	6.192	3.721	4.198
95(I)	9660	9890	9920	0220	3.751	4.227	5.954	5.061
99(O)	0840	1180	1210	1390	4.733	5.686	5.359	4.466

[a](I) – Venus swingby on inbound part of mission.
[b](O) – Venus swingby on outbound part of mission.

the initial mass in Earth orbit (MEO) can be obtained by working backwards from the final burn. The makeup of each stage just prior to a particular burn is shown schematically in Figure 7.

When aerobraking is assumed, an 8200-kg aerobraking capsule replaces the RM, tanks, propellant, and meteoroid shield for that maneuver. Steps 7 and 8 are replaced

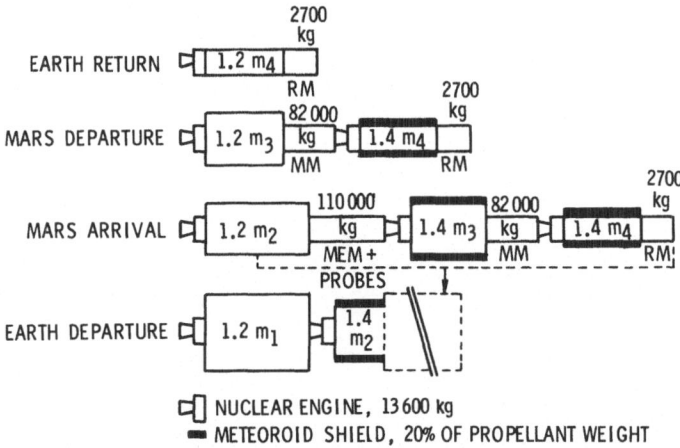

Fig. 7. Schematic representation of weights associated with each stage of a manned Mars mission just prior to the major rocket burns associated with the mission. MM = mission module; MEM = Mars excursion module; RM – retromodule.

by the aerobraking maneuver. The MEO required when hyperbolic rendezvous is used includes no Earth return capability, as mentioned previously. For retrobraking to an elliptical Earth orbit, the mission sequence proceeds as indicated above, but step 8 is replaced by a deboost into a 24-h or nearly parabolic orbit with a periapsis altitude of 500 km.

References

[1] Staylor, W. F., Brooks, D. R., and Pritchard, E. B., *J. Spacecraft Rockets* **7** (1970), 814.
[2] Mueller, G. E., *Astronaut. Aeronaut.* **8** (1970), 30.
[3] Eggers, A. J., Jr. and Swenson, B. L., *Astronaut. Aeronaut.* **5** (1967), 24.
[4] Luidens, R. W., Burley, R. R., Eisenberg, J. D., Kappraff, J. M., Miller, B. A., Shovlin, M. D., and Willis, E. A., Jr., 'Manned Mars Landing Mission by Means of High-Thrust Rockets', NASA TN D-3181, January 1966.
[5] Brooks, D. R. and Harrison, E. F., 'Use of Orbit-to-Orbit Shuttles for Hyperbolic Rendezvous with Returning Planetary Spacecraft', NASA TN D-6342, September 1971.
[6] Corrington, L. C., *Spacecraft Rockets* **6** (1969), 465.
[7] Dugan, D. W., 'The Role of Staged Space Propulsion Systems in Interplanetary Missions', NASA TN D-5593, December 1969.
[8] Mission Evaluation Team, 'Apollo 11 Mission Report', MSC-00171, NASA Manned Spacecraft Center, Houston, Texas. November 1969. Also available as NASA TM X-62633, November 1969.

[9] Anon., 'Nuclear Shuttle Systems Definition Study, Phase III', Lockheed Missiles and Space Company, Report No. LMSC-A984555, Vol. 1, May 1, 1971, NASA Contract NAS8-24715.

[10] Johnson, K. P. and Holl, R. J., *J. Spacecraft Rockets* **8** (1971), 600.

[11] Anon., *Space Flight Handbooks*, Vol. III. *Planetary Flight Handbook*, Part 6 – Mars Stopover Missions Using Venus Swingbys. Supplement B: Tabular Trajectory Data for 30 Day Stopover Time. NASA SP-35 (Supplement), Part 6, (No date).

C. PROPULSION

PLASMA THRUSTERS FOR SECONDARY PROPULSION

A. V. LA ROCCA*

General Electric Company, RESD – Philadelphia, Penns., U.S.A.

Abstract. Recent development in thrusters, propellants, trigger and feed systems have advanced the pulsed plasma thrusters to the readiness status for long life space applications. Lifetimes of several thousand hours have been repeatedly demonstrated in the laboratory and one flight experiment. Current development activities are described together with the applications of multi-thruster systems, complete with power conditioning and protection logic, to three axis stabilized and spin stabilized geostationary satellites.

1. Introduction

Pulsed plasma thruster employing solid propellants have recently reached an advanced state of development and constitute one of the most promising classes of electrical propulsion systems. The systems here discussed are intended for such functions as attitude control, station keeping, spin maintenance and precession, and trajectory adjustment. These applications are characterized by total impulse requirements ranging from a few hundred to several thousand pound seconds and by thrust levels from one micropound to a few millipounds, and constitute the so-called secondary propulsion functions for spacecraft.

In the last few years particular attention has been devoted to this field because of the recognition that improvement in spacecraft performance, such as pointing accuracy, and in reliability, life, weight and volume could be achieved by the adaptation of advanced electrical propulsion schemes.

A systematic effort planned specifically to cover this field was undertaken at the General Electric Company, in Philadelphia with the SPET (Solid Propellant Electrical Thruster) Program [1–5].

Figures 1 to 4 illustrate some of the SPET thrusters. A single stage configuration employing waxlike perfluorocarbon propellant fed by the surface action exerted by means of graded porosity wicking assembly is shown in Figure 1. The vacuum impregnated wicking carries the propellant to a slit formed by Lucalox† ceramic plates and terminated at the two ends by tantalum electrodes (orthogonal to the plane of the figure). An electrical potential in excess of the breakdown voltage of the propellant ribbon filling the slit, is applied by switching on to the electrodes a capacitor by means of a Krytron‡ or triggered spark gap. The propellant is impulsively exploded by the ensuing discharge, which continues in the plasma produced by that explosion. A combination of electrothermal and electromagnetic body forces results in the acceleration of the plasma while it expands at a velocity greatly in excess of thermal

---

* Consulting Engineer, Advanced Propulsion, Associate Fellow American Institute of Aeronautics and Astronautics.
† General Electric Company trademark.
‡ Cold Cathode Thyratron, EG&G Inc. trademark.

L. G. Napolitano et al. (eds.), Astronautical Research 1971, 239–264. All Rights Reserved.
Copyright © 1973 by D. Reidel Publishing Company, Dordrecht-Holland.

velocity in the diverging non-conductive nozzle. Specific impulse higher than 1000 s and thrust of several micropounds have been measured on single stage thrusters, by direct thrust measurements on torsion balances and weight loss determination [3–5]. Life in excess of 2.5×10^7 pulses have been demonstrated with this type of thruster.

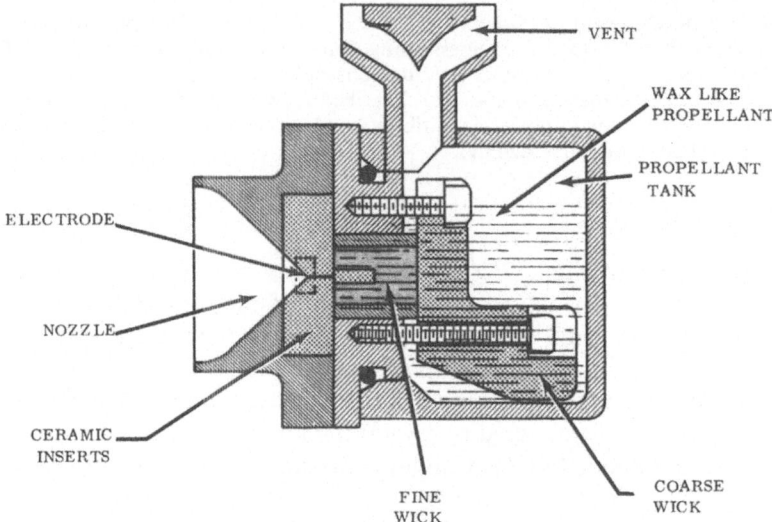

Fig. 1.   Single-stage thruster configuration, 108-C series.

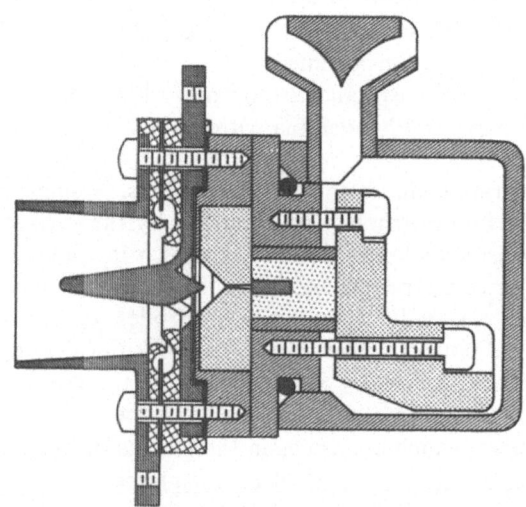

Fig. 2.   Two stage thruster configuration, 108-C series.

To increase both the thrust and specific impulse a second stage is added by cutting the non-conductive nozzle and replacing it with a coaxial arrangement consisting of a copper barrel and center electrode, separated from each other by an appropriate insulator assembly. This second stage is hard wired to another capacitor. The first

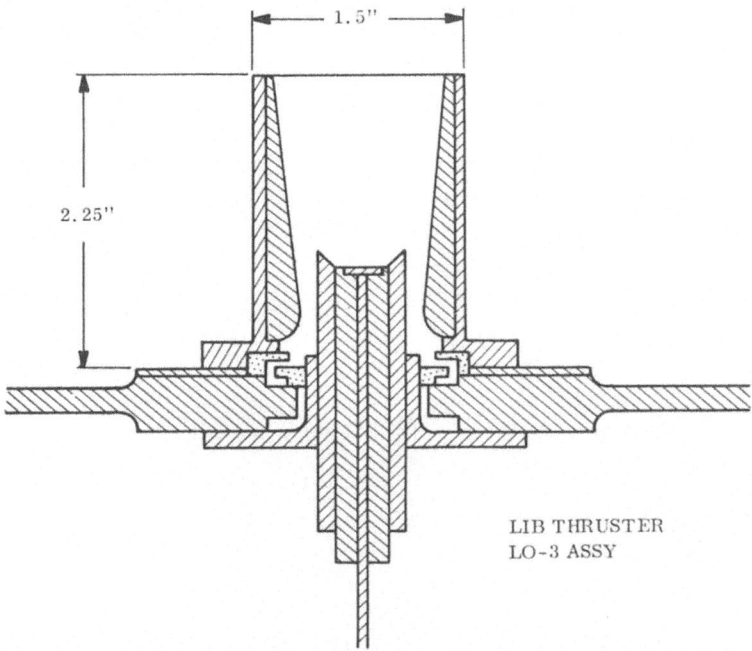

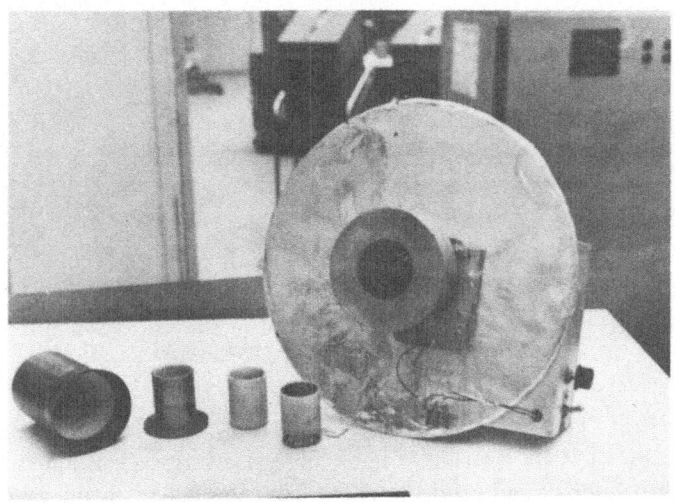

Fig. 3.   LIB thruster with high Q disc capacitor firing circuit, also showing alternate barrels and propellant inserts.

stage capacitance is reduced to deliver energy in the amounts solely necessary to generate the propellant plasma, which in turn sequentially triggers the second stage discharge of greater energy (typically by a factor of 50). In this manner the losses associated with the triggering devices are limited to a fraction of the small energy involved in the propellant change of phase and ionization. Specific impulses approaching 3000 s and thrust in the tens of micropounds have been measured with two stage thrusters [3–5]. Life of several million pulses have been demonstrated with these thrusters.

To further increase the thrust and improve the thrust to power ratio, an indicator of thruster and firing circuit efficiency at a given I_{sp} level, sleeves of solid propellant materials, which had been used in other phases of the program, were inserted in the

Fig. 4.   Alternate LIB thruster barrels and solid propellant sleeves. Shown in front row sleeves used with two stages SPET 108-C series.

second stage barrel. The success of these exploratory tests led to the design of the Large Impulse Bit (LIB-I) thruster, Figure 3, possessing larger dimensions, incorporating a new trigger design, and mated to a more efficient firing circuit. Impulse bits in excess of 60 μlb s were measured with a thrust to power ratio of 16.5 μlb W^{-1} at a specific impulse in the 350–500 s range, and thus with a thrust efficiency 12 to 18%. Figure 4 illustrates several LIB thruster barrels, and Teflon* and Delrin† plastic propellant inserts for the LIB and the two stage thrusters (smaller sleeves front row).

* DuPont DeNemours & Co. trademark for TFE fluorocarbon.
† DuPont DeNemours & Co. trademark for Acetal Resin.

During the development of the solid sleeve thrusters, attention was focused on to phenomena associated with the non-uniform ablation of the sleeve. The ablation would develop preferentially in one limited sector (90 to 110°) and progress with extensive burn through, which would expose part of the metal of the barrel. This would occur when the ring of propellant which typically would remain at the location of the breech shoulder, would finally be cut through, a situation reached after $\simeq 2$ million pulses for the LIB-I configuration. (See Figure 4, sleeve at extreme right of middle row.)

The non-uniform ablation could be explained by the fact that an azimuthally uniform electrical field was present within the barrel at the time of initiation of the discharge. The effective breakdown path would thus be easily affected by surface conditions which caused local non-uniformity. A small perturbation in surface geometry and/or condition would result in preferential striking at that location and this

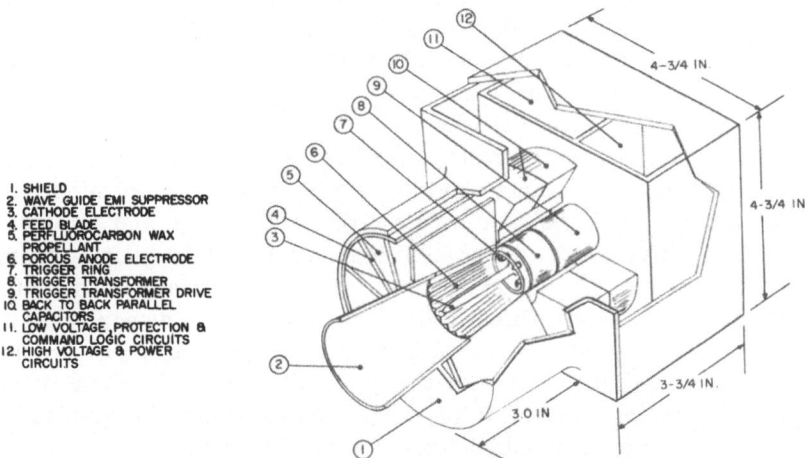

1. SHIELD
2. WAVE GUIDE EMI SUPPRESSOR
3. CATHODE ELECTRODE
4. FEED BLADE
5. PERFLUOROCARBON WAX PROPELLANT
6. POROUS ANODE ELECTRODE
7. TRIGGER RING
8. TRIGGER TRANSFORMER
9. TRIGGER TRANSFORMER DRIVE
10. BACK TO BACK PARALLEL CAPACITORS
11. LOW VOLTAGE, PROTECTION & COMMAND LOGIC CIRCUITS
12. HIGH VOLTAGE & POWER CIRCUITS

4-3/4 IN.
4-3/4 IN.
3-3/4 IN.
3.0 IN

Fig. 5. Single thruster subsystem (cutaway view).

situation would tend to grow worse with time, because the local alteration of geometry and surface conditions would favor successive initiation of the same location.

To obviate this situation and to permit uniform ablation, and thus full utilization of the propellant sleeve with constant thruster performance, a solution was formulated which consisted of the insertion of a number of radially directed metallic blades extending from the outer metal barrel to the propellant surface exposed to the discharge. In this manner the presence in the azimuthal direction of uniformly distributed peaks of the electrostatic gradients, associated with the sharp blade edges and the small blade to center electrode gaps, would make irrelevant any slight irregularity of the propellant surface and/or condition.

However the resulting configuration was amenable to the use with liquid or wax-like propellant. It was in fact very similar to the assembly of blades and porous wicking cone, which had been used in previous phases of the program to feed propellant from a tank to the barrel breech. Now this assembly could be removed from inside the tank, which had been in the more conventional location somewhat upstream of the thruster, and be placed outside surrounding the barrel to reconstitute the feed and tank assembly as shown in Figure 5. This configuration has been called the Inside-Out SPET.

2. Recent Developments

The Inside-Out SPET has been discussed in references [5] and [7]. The fuel, in the state of a viscous liquid or of a wax, is fed by the blades radially in toward the barrel surface constituted by a porous metallic liner. This material is coated with a film of propellant which is exposed and stabilized by surface action forces. The liner also performs the functions of equalizing the feed and distributing the ablative discharge. It also performs as an electrode (anode) while an axially located center electrode performs as the cathode. The shaping of the barrel causes azimuthal spreading of the electrical discharge by the presence of magnetic and fluid dynamic pressure gradients. The intersections of the equally-distributed radial blades with the conical liner constitute channels tapered toward the breech of the barrel, giving a feed component toward that region where a greater propellant generation is desirable. Shaping of the liner configuration (for example forming ridges to give reversed surface action for better film control), variation of its area exposed to discharge, and graduation of its porosity are variables which can be used together with pulse shaping to match propellant generation with the rate of energy release into the plasma. The initiation of the discharge occurs when the breakdown of the inner electrode gap is caused by either the injection of a small amount of energized gas particles or by an electrostatic pulse. Both schemes have been extensively tested for tens of millions of pulses, and have proved to be very effective in breaking inner electrode spacing of the order of one centimeter under a few hundred volts of applied potential.

The advantages offered by the Inside-Out design and by a novel capacitor configuration for pulsed plasma thrusters have been discussed elsewhere in the literature [5–8].

The performance advantages to be gained with a coaxial configuration in contrast with parallel plate, two-dimensional arrangements, are well known and derive from the self-contained magnetic field and the absence of fringing regions in the coaxial thruster. Other features include a more favorable initiation and propagation of the discharge, which can be enhanced by appropriate contouring of the electrodes to furnish favorable fluid dynamic and magnetic pressure gradients. These gradients constrain the discharge formation at the desired location at the breech and also improve the axial and azimuthal propagation to eliminate spoking and crowbarring of the discharge. As discussed in reference [9] the axisymmetric configuration is one of the few amenable to a complete, magnetogasdynamic analytical description that can be expended to include mass addition, energy addition, discontinuities in streamlines and wave structure, boundary layer shock interaction, multicomponent fluids, non-thermal equilibrium, etc.

More significant yet is the fact that the electromagnetic and electrothermal contributions to the acceleration process complement each other in one of the few (if not the only) configurations which are stable on the time scale of the pulsed devices. This situation is reflected in the results obtained with pulsed MPD thrusters which operate in a regime very close to SPET. Extensive diagnostic studies [10] indicate that a significant performance improvement can be obtained by proper combination of electromagnetic pinch and electrothermal heating.

In a coaxial device this combination exists in the central core region streaming around the cathode and exhausting from the barrel as a cathodic jet. The ions ejected radially inward from the anode are compressed in this region by a pinch effect resulting from one component of the self-induced electromagnetic field. Electrothermal effects, enhanced by magnetic containment, which greatly reduces the energy transfer to the cathode, give a contribution to the acceleration with a process akin to that of heat addition to a supersonic expanding stream. Thermalization of neutrals and ions is also favored by the higher pressure existing in the cathodic plume, where also most of the exhaust mass gets compressed, either by pinch or by the coalescing of the reacting (shocks with energy addition) waves [11].

The Inside-Out design permits the delivery of propellant without the problems associated with valving or the sputtering erosion of exploding film of limited exposure area and with confined discharges [12]. The propellant is delivered by establishing the discharge in an unconfined manner with a centrally located cathode. This electrode is external to the propellant film which possesses a large exposed area. As a result of the combination of low density and supersonic flow throughout (the propellant is injected in the fashion of an exploding film with energy greatly in excess of that required for change of phase [13], and thus at $M \gg 1$), a phenomenon akin to ablative protection eliminates the sputtering of the porous feeding substrate. The propellant delivery is self-regulating in a manner similar to that of electrical discharges seeking to entrain the critical or matched mass flow for stable operation [14].

The scheme can be controlled by design alterations of the contour of the feeding substrate, the interelectrode spacing and relative position of the electrodes, the porosity and extension of the exposed film-bearing substrate. All these factors also can be used to obtain the best matching between propellant delivery and rates of energy addition. For example, by varying the axial position of the triggering center electrode as was done on the LIB thruster of references [3] and [5], radial pinch effects can be applied to direct the ions in the cathodic region, where most of the acceleration seems to take place, in the fashion of a magnetically constrained supersonic electrothermal device. Another variable is that of reducing the length of the electrically conductive area of the outer electrode (anode) to enforce a standing wave at that location and thus permit a better thermalization of the exhaust stream on the lines discussed in references [10], [11] and [15].

These feed- and trigger-related parameters give a broad flexibility in design and the capability of matching propellant delivery far beyond the limitation imposed by a single pulse of an exploding propellant (either liquid or solid) or by an initial ablating discharge on a single exposed propellant surface located at the breach of the thruster [16, 17]. In those thrusters the trigger is located at a position which is downstream of the propellant-generating main discharge, and thus is vulnerable to coatings of propellant deposits. Because the characteristics of the trigger material are different from those of the deposits, malfunctions of the trigger can be encountered. The proposed design obviated this problem by proper location of the trigger and selection of compatible trigger material.

Figure 6 shows the engineering prototype of a 10 μlb Inside-Out thruster system on a vacuum torsion thrust balance. A close up picture of the same system is given in Figure 7. The internal housing accommodating the T.M. (Thruster Module)

Fig. 6.   SPET inside-out 10-μlb thruster system on vacuum torsion thrust balance.

consisting of main capacitor, trigger capacitor, spark gap and driver circuit is clearly visible together with the power conditioning, protection and trigger logic boards. Repackaging of these components would reduce the system volume to less than half of that shown in the photo and the weight to less than 2 lb. Expected performance of Inside-Out SPET thrusters is given in the curves of Figure 8, in which data points of several laboratory thrusters are also indicated. The Inside-Out 3 was incorporated in the systems illustrated in Figure 6.

The dependence of impulse bit on main discharge energy is illustrated in Figure 9. When impulse bit is proportional to energy, the slope of the curve is 45°. This proportionality at a constant specific impulse implies also constant thruster efficiency.

Over the range from 2 to 8 J this proportionality holds well for all the thrusters covered in the figure. Above 8 J the I-0.1 and I-0.2 thrusters begin to fall off. Below 8 J the impulse bit of I-0.3 and 4 are slightly less than for the smaller I-0.1 and 2 thrusters, however the curves continue to climb at the 45° or even greater slope for substantially higher energy level, indicating that peak performance had yet to be reached at the maximum energy of the tests. The specific impulses were in the range of 1000 to 1200 s with a tendency to increase slightly with increased energy discharge

Fig. 7. SPET system, engineering prototype, with top cover removed.

above the 5 J level. At discharges below one joule the specific impulse starts to increase sharply, while the impulse bit to power ratio decreases, indicating a threshold requirement for effective mass injection as well as a decrease in thruster efficiency (ratio of exhaust energy convertible in propulsive work to capacitor stored energy) a trend reflected also in the curves of Figure 8, which gives the corresponding efficiencies.

Another recent development is the ring ablation trigger. It has been known that the products of a low density arc tend to move away at high speed in a direction normal to the surface on which they have been generated. Even a very small and well-defined spark creates a plume moving away in a narrow angle about a direction normal to the surface.

15

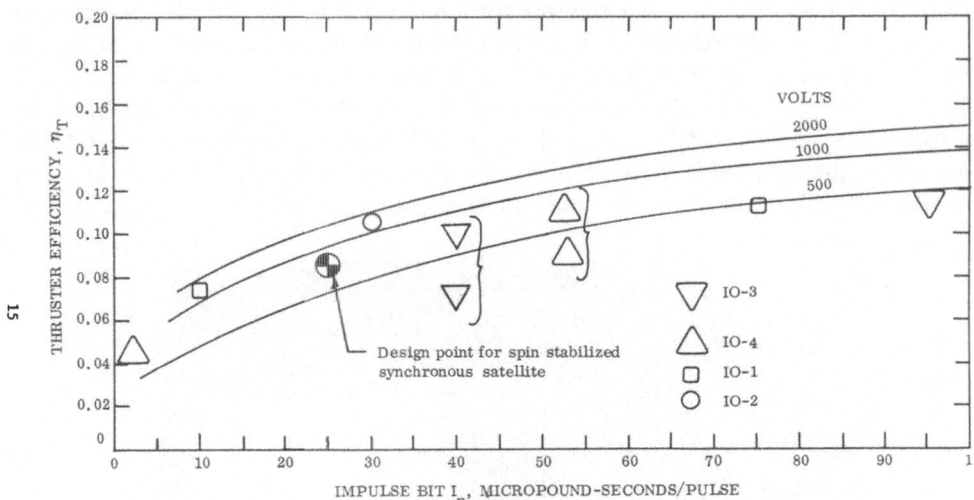

Fig. 8. Test data of laboratory thruster module on curves of projected performance of flight hardware.

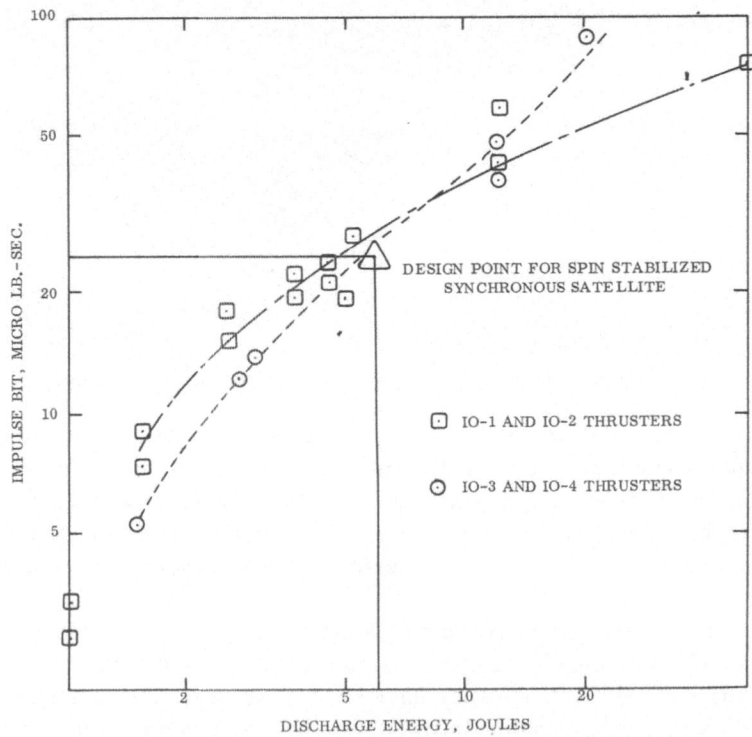

Fig. 9. Impulse bit vs. energy for several 'inside-out' SPET thrusters.

Conventional designs, including the LIB thruster discussed in the introduction, have not used this property in trying to transfer the discharge to the main gap. The motion of the generated particles has been directed axially away in the gap rather than radially out.

The modifications indicated in Figure 10 have been implemented in the trigger component illustrated in Figure 11. This scheme basically causes the triggering discharge to occur in the axial direction and the plasma plume due to bridge the gap radially out from the center to outer electrodes.

In this arrangement the cathodic spot of the triggering discharge also serves as the initiating cathodic spot of the main discharge. Thus it further enhances the initiation process, while at the same time the cathode spot is spread away in both the azimuthal and radial directions [18, 19], under the action of the self-induced electromagnetic

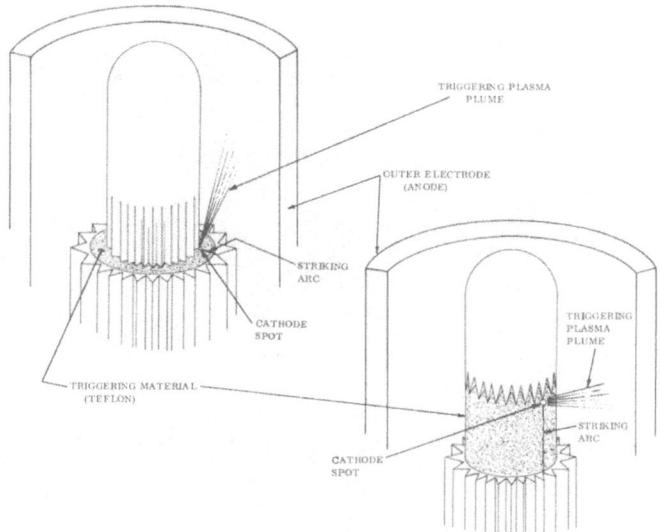

Fig. 10.   Plasma plume triggering.

gradient of the main discharge. The result is a significant improvement in the life of the trigger. The same phenomena affect the life of triggered vacuum spark gaps, which have survived more than 220 million pulses with no detectable performance change when designed to take advantage of cathode spot-EM field interaction. In a pulsed plasma thruster, proper operation of the trigger affects uniformity and propagation of the main discharge and therefore is of paramount importance for high thruster performance.

The mass injection trigger ablating a ring of Teflon has been used with a capacitance discharge and with an inductance coupled circuit. In the inductance coupled circuit a low voltage (200 V), low energy capacitor is switched by an SCR and dumped in the primary of a transformer. The turns ratio and the stray or additional lumped capacitance of the secondary loop are chosen to match the breakdown voltage and characteristics time (for breakdown) of the gap to be switched. The generation of a sufficient number of active particles also requires the capability of sustaining the current when the load impedance drops drastically after arc formation. The triggering can be facilitated by having it occur on the surface of a material possessing the desirable work function (and therefore Teflon) so as to enhance the initiation processes of both the triggering and main gaps. In a low density discharge the formation of a large number density of particle, sustained for sufficient times at the proper location, strongly influences the switching of the main gap.

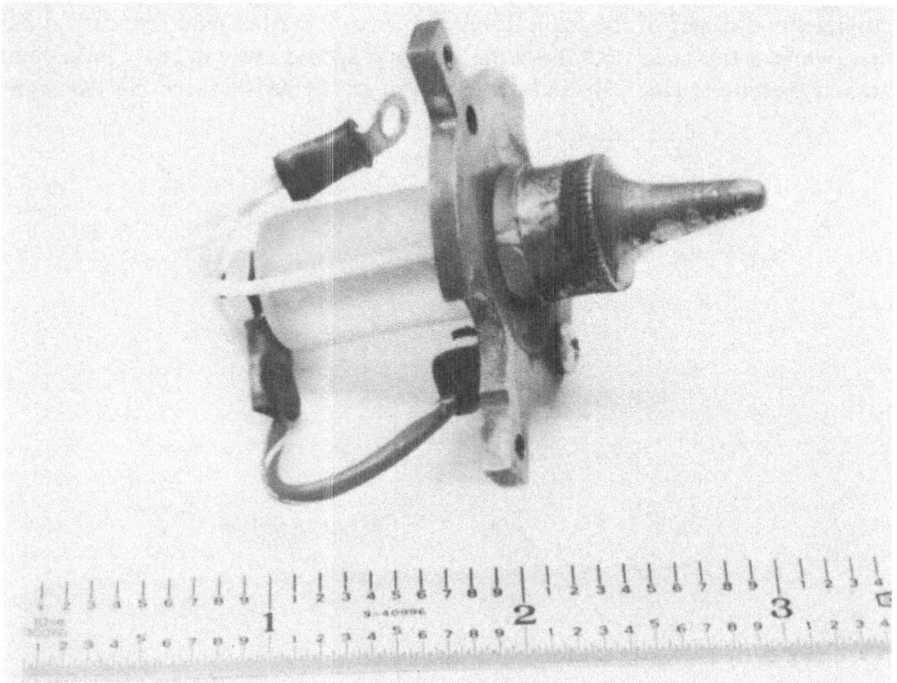

Fig. 11.   Ablation ring trigger after 10^8 pulses endurance test.

The inductance coupled ablation ring has proven as effective as the capacitance discharge. The selection of one of the two approaches depends on the availability of high voltage (capacitor circuit case) and a component trade off. The trigger energy has been in the 10 to 30 mJ range. This level of trigger energy has given reliable operation with thruster gap voltage as low as millions of pulses.

One trigger of the ablative ring type, shown in Figure 11, has been tested by itself in vacuum for about 10^8 pulses at 45 cps. At the end of the test it exhibited only a slight erosion which did not affect thruster operation. It should be noted that when the trigger is operated with the thruster, a mass transfer occurs between anode and trigger which, because of the similarity between the fuel and the material used in the trigger ring reconstitutes the trigger surface and thereby maintains the operation of the trigger unchanged. Thus operation of the trigger alone is a more severe test than when it is used to trigger the main discharge.

3. Typical Applications

A multithruster system complete with power conditioning, command protection logic, illustrates the application of the new design techniques to the station keeping and attitude control functions of a three axis stabilized, 2000 lb, Earth synchronous communication satellite. Figure 12 gives an artist's conception of the schematic of the

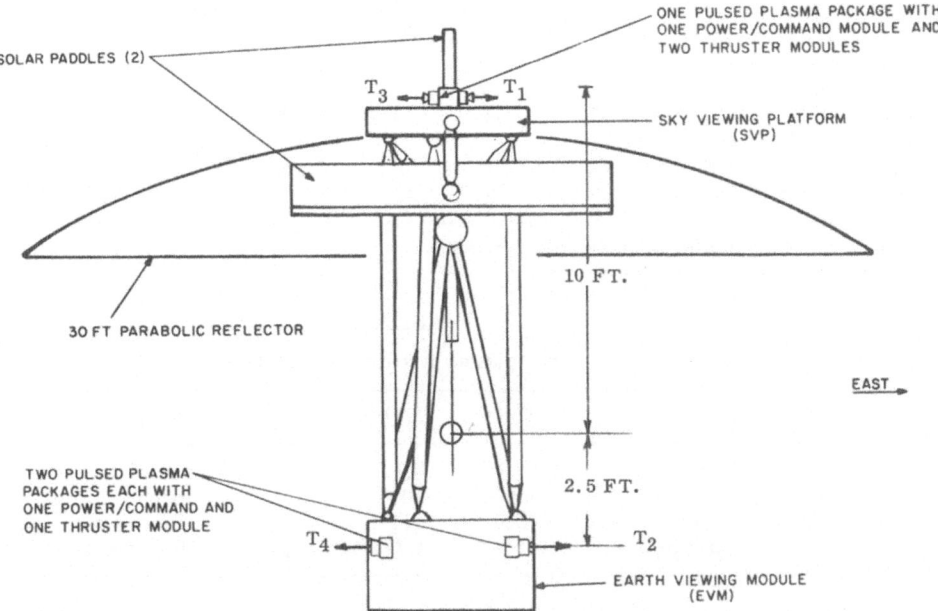

Fig. 12.   General arrangement of pulsed plasma system.

installation of the pulsed plasma system on that vehicle. Table I summarizes the functions performed and Table II gives the technical data of that system.

The application is intended as an experiment to demonstrate in the environmental condition of a geostationary satellite the operational capability of the pulsed plasma thrusters and their compatibility with the RFI and EMI interfaces of the communication function. As will be seen from the basic simplicity of the design the total impulse and therefore the duration of the control functions of the thruster system can be increased readily several times, requiring the addition of propellant weights, which are quite small because of the high specific impulse.

The system consists of three completely self-contained subsystems each comprising a set of identical basic power/command and thruster modules.

(1) A thruster module with integrated fuel supply and feed system of the Inside-Out design, mated to a high energy capacitor of special coaxial design which permits

optimum transfer of energy to the thruster. The hollow core of the capacitor accommodates the triggering elements, driving circuit and trigger transformer of the
thruster.

(2) A power/command module which can accept the thruster module on one or
more of its sides.

TABLE I

Pulsed plasma experiment summary

Functions	Duration	Resultant thrust	Peak power (watt)	Average power (watt)	Duty cycle	Commend (discretes)	Telemetry (analog channels)[a]
East–west station keeping	6 months	12	4	4	1.0	11[b]	3 to 10
Pitch axis attitude control	6 months	64	17	6.8	0.4	11	3 to 10
36° station change	74 days	150	38	38	1.0	11[b]	3 to 10

[a] Three for routine monitoring, ten for diagnostics.
[b] Six required while thrusting in one direction.

TABLE II

Pulsed plasma system technical data

Parameters	Total	EVM package	SVP package
Weight (lb)	14.8	4.2	6.4
Total impulse (lb s)	4000	1000	2000
Volume (in.³)	426	117	192
Envelope sizes (in.)		$6.75 \times 4.75 \times 4.75$	$10.75 \times 4.75 \times 4.75$

Two of the packages each containing one power/command and one thruster
module are installed in the east and west face of the Earth Viewing Module (EVM).
A third hardware package containing one power/command module and two thruster
modules are installed in the Sky Viewing Platform (SVP) with the thruster axis
pointing in the same directions as those of the EVM.

This multithruster system installation is a most attractive one, because it fully
utilizes the unique features of the pulsed plasma thruster: (1) minimum of interfaces,
(2) no thruster-on mode of failures, (3) no required thermal control, and (4) insensitivity to radiation and vacuum environments.

These features dispose of the usual constraints and permit installation of the
thruster packages at the vehicle locations most advantageous to impart, by vectorial
composition, forces, torque and couples for the control of the vehicle attitude and
station. Thruster induced disturbance torques are minimized, reducing the amount
of required controlling moment impulse. Furthermore, the variable thrust control-
lable simply by varying firing frequency, the small impulse bit characteristic of the

system can be exploited to reach near ideal performance for precise control of the vehicle. Migration and error in center of mass location can be compensated, as well as variable external disturbance torques to a degree of precision limited only by sensor and lead network characteristics. The reduction of the control moment impulse, the high specific impulse, the variable thrust and small impulse bit offer the capability of achieving long life, precise three axis control without the need of momentum wheels.

The proposed simple arrangement is recommended because it permits testing of the system performance in three of the most important functions for which the thrusters are designed: (1) East–west station keeping, (2) Attitude control about one axis, and (3) East–west station change. Attitude control about other axes could be easily accommodated by the addition of properly pointed thruster modules to the three power/command modules. However, for experiment simplicity, it is believed that the demonstration of one axis attitude control is adequate to assess the operational capability of the system and derive information pertinent to a three-axis control system.

Single axis (pitch) attitude control can be demonstrated conjunctly with east–west station keeping.

The east–west station keeping requires 0.47 fps velocity correction once every 32 days to maintain an accuracy of 0.1°. The average thrust to fulfil this requirement is 12 μlb. The proposed thruster system possesses an impulse bit of 24 μlb s for each thruster. Since a pair of thrusters will be operated conjunctly, the firing frequency for each thruster is a fraction of a cps. The proposed installation has four thrusters with their axis contained in the yaw-roll plane and aligned with the spacecraft roll axis so that, both on the EVM and on the SVP, one thruster can fire in the east and another in the west direction. This arrangement permits the application of a translatory force, such as that required for east–west station keeping or change, directly through the center of mass of the vehicle, thus avoiding the development of undesirable disturbance torques. This is obtained by firing a thruster pair facing in the same direction, at a level of thrust inversely proportional to their moment arms relative to the center of mass. Since the impulse bit is equal for all thrusters it follows that the firing frequencies have to be in the inverse ratio of the arms. Table III gives the pertinent operating parameters for the east–west station keeping and station change maneuver, as well as those of the pitch axis control function.

The same thrusters which fulfil the east–west station keeping at a very understressed level of operation, can be used at a fraction of their design power (which is employed only in the east–west station change) to perform several attitude control related functions. The simplest of these is momentum wheel unloading. To accomplish unloading requires exchanging the momentum stored in the pitch wheel with momentum imparted to the spacecraft by the thrusters. If a control torque of 400 μlb ft is assumed, the momentum wheel could be unloaded 2 ft^{-1} s in about one and a half hours. To develop the control torque described, each of the two thrusters operating as a couple must deliver 32 μlb. The station keeping the wheel unloading would aptly demonstrate the long term operating ability of the thrusters and the variable force output as a function of the analog frequency control (i.e., analog control of the thruster firing rates).

It is also possible to combine the station keeping and station change with the attitude control functions by implementing, in the operational computer program, simple routines which selectively choose appropriate sets of firing frequencies for a pair of thrusters. These routines are obtained by storing in the operational computer simple equations which enforce the simultaneous delivery of a constant resultant translatory force together with a control couple of a given value in the negative or positive directions. By implementing these schemes, the pitch axis control function could be

TABLE III

Requirements for the three control functions

Function	Six months station keeping[a]		Six months pitch control[c]	36° station change (74 days)[a]	
Thruster	T_1	T_2	T_2 and T_3	T_1	T_2
Thrust, μlb	2.4	9.6	32	30	120
Firing frequency, pulse s^{-1}	0.1	0.4	1.33	1.25	5
Total firings, millions	0.8	3.1	4.2	4	16
Total impulse, lb s	19	76	63[d]	96	384
Power[b], W					
Peak	1	3	8.5	8	30
Average	1	3	3.4	8	30
Total power, W					
Peak	4	4	17	38	38
Average	4	4	6.8	38	38
Total thrust, μlb	12	12	64	150	150

[a] Thruster called out are for east thrust, for west thrust calls are T_3 and T_4.
[b] Logic, protection and command functions included. Power conditioning efficiency 80%.
[c] Thruster called out are for + couple; for − couple calls are T_1 and T_4.
[d] Assumes single band limit cycle with 0.4 duty cycle at 5 times peak disturbance torque.

performed with no propellant expenditure during station change, and with a reduction of about 40% in propellant expenditure during station keeping. For more details about these techniques see reference [7], which discusses the application of a pulsed plasma system to the complete three axis attitude control, east–west and north–south station keeping functions of a similar satellite. The higher levels of thrust required for the north–south station keeping in that case are implemented with the addition of thruster barrels loaded with exothermic propellant and operating either in the low pressure detonator mode or in an augmented pulsed plasma mode.

Activation of the appropriate pulsed plasma thruster for station keeping can be accomplished simply by direct tie-in to the telemetry and command subsystem and power subsystem. With the transmission of the station keeping command, the command decoder would go to the thrust controller and logic unit which would select the thruster combination desired. A 'Pulsed Plasma System Enable' command would suffice to apply power to the logic unit. Once power is applied to the experiment, the specific function performed need not be restricted to the station keeping alone, but with the appropriate control logic, pitch attitude control can also be provided separately or simultaneously with station keeping as previously discussed. A block diagram system is shown in Figure 13. The east–west station change maneuver and associated routines can also be accomplished completely by ground command, provided the communication link is available. Insofar as adapting the pulsed plasma thruster system to an operational satellite, station changing of almost any magnitude could be implemented by storing the necessary timed events onboard the satellite, so that it becomes independent of a need for a ground link.

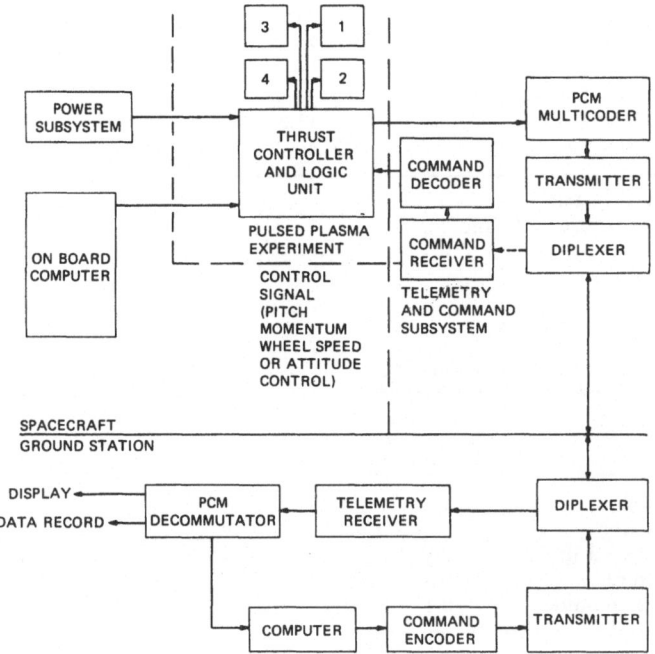

Fig. 13.   Pulsed plasma thruster system block diagram.

One of the EVM subsystems consisting of one thruster module and one power/command module is depicted in Figure 5. The thruster module specifications are given in Tables IV and V. The EVM and SVP subsystems, volumes and weights are given in Table VI. The requirements for the functions are summarized in Table III.

The application of a pulsed plasma system to a 550 lb, Earth synchronous, spin stabilized satellite is illustrated in the artist sketch of Figure 14. The functions to be performed are the 0.1° to 0.5° east–west station keeping and the spin axis precession control to 5″ for five years, plus an initial maneuver to remove an orientation error $\Delta\Theta = 0.01°$ in 15 min. In Figure 15 the impulse bits (in μlb s) are given vs the time (in seconds) required for spin axis precession control, having assumed a vehicle moment of inertia $I_z = 45$ slugs ft², and a thruster torquing arm $1 = 2.5$ ft. It is interest-

TABLE IV
Thruster module specification

Parameter	Value
Total impulse, lb s	1000
Impulse bit, μlb s	24
Specific impulse, s	1000
Firing frequency, cps	0.1 to 5
Thrust, μlb	2.4 to 120
Thrust efficiency, %	11
Thrust to power ratio, μlb W^{-1}	5
Design power, W	25
Power	
East west station keeping	1[a] to 4[b]
Pitch axis A/C	13.5
36° Station change	7[a] to 25[b]

[a] SVP thrusters T_1 and T_3.
[b] EVM thrusters T_2 and T_4.

TABLE V
Thruster module weights (lb)

Components	Weight (pounds)
Propellant	1.1
Capacitor	
4.8 J at 8 J lb^{-1}	0.6
Structure	0.4
Trigger module	0.1
Total	2.2

TABLE VI
Subsystem specifications

Modules	EVM packages		SVP packages	
	Weight (lb)	Volume (in.^{3})	Weight (lb)	Volume (in.^{3})
Thruster	2.2	44	4.4	88
Power/command	2.0	73	2.0	84
Total	4.2	177	6.4	192

ing to note that this relation is independent of spin rpm, because both gyrotorque and control torque are linear functions of spin rate.

The requirements for east–west station keeping at nominal longitudes $\lambda = 15°$ to $45°$ with a $\Delta\lambda = 0.1°$ to $0.5°$ results in an average $\Delta V = 7$ ft s^{-1} yr^{-1}. Assuming one firing for revolution, in the east–west direction, the required impulse bit vs. thruster duty cycle (D.C.) is given in Figure 16 where the effects of spin rpm are also indicated.

The total impulse and control torque for the fine precession axis correction are one order of magnitude smaller than those for east–west station keeping. The design is constrained by the initial $\Delta\Theta$ removal in the short time of 15 min. The resulting system is overdesigned for the remaining long duration functions. Thus its weight, volume and peak power are not the optimum. On the other hand it permits east–west

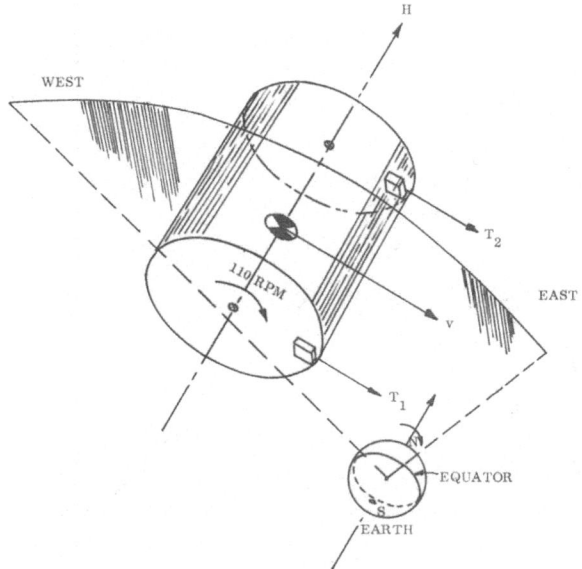

Fig. 14.   Spin axis precession control and east-west station keeping performed by pulsed plasma thruster system.

station keeping at a very low duty cycle, typically a few hours per week, which from an operational viewpoint, when the system has to be operated through a ground link, might show some advantage, by reducing the attention to be given by the ground crew to this function.

The design criteria for this application has to find a balance between the peak power handling capability and the weight and volume. The total impulse is rather small (700 lb s). A specific impulse in the 500 to 1000 s range, gives a satisfactory balance between propellant and power related components weight. By giving to each of the two subsystems full total impulse capability, complete redundancy can be obtained with the arrangement selected. In the case of failure of one subsystem, all the functions could be implemented by the other, by including in the command logic the capability of firing the subsystems in synchronism with each half vehicle revolution. This subsystem would operate still at a fraction of the design power, but for the case of the first corrective $\Delta\Theta$ maneuver, which would require the same nominal level. Figure 17 gives a sketch of the subsystem consisting of one power/command module and one thruster module (TM) integrated in a single package. The nominal design

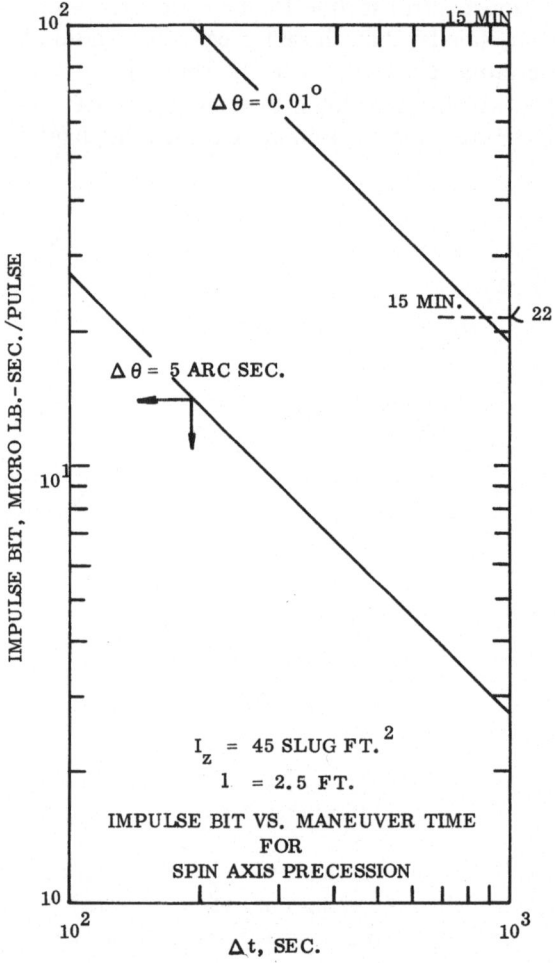

Fig. 15.   Impulse bit vs. time required for spin axis precession control.

point is selected at 25 μlb s impulse bit with a 350 lb total impulse. Because of the emphasis on reliability for a long life (5 yr) and also for consideration of future growth the design illustrated is that of 50 μlb s impulse bit, 700 lb s thruster, which in this application would be operated at half load condition and offer redundancy in total impulse. The power/command design rating is 20 W, with the capability of sustained operation at twice that power. The capacitor is grossly understressed at the nominal design point (4.8 J lb^{-1} and 8.9 W lb^{-1}).

The thruster module Figure 18 combines the thruster and main capacitor in a coaxial configuration which optimizes energy transfer, for improved performance and reduced dissipative heating. Four triggers and trigger driving circuits are accommodated for greater redundancy and reliability. The thruster module comes as a separate unit enclosed in its own aluminum housing and it is fitted in the front of another main housing, in which the complete pulsed plasma subsystem is contained as a single unit. With this approach greater RFI protection is derived from the double shielding of the more noisy components.

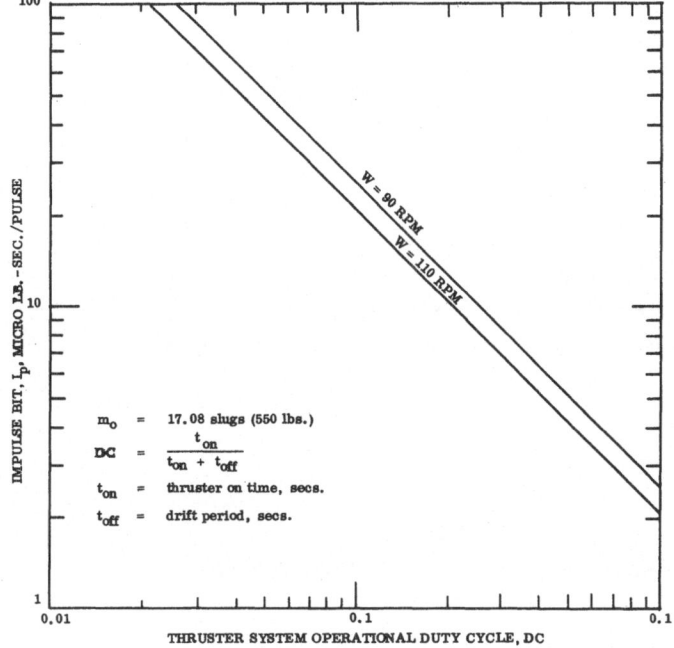

Fig. 16.   East-west station keeping thruster system duty cycle based upon Delta V of 7 ft s^{-1} per year.

The mechanical interface in between the TM and main housing consists of a mounting flange carrying a cowling complete with wave guide for EMI suppression extending in front of thruster barrel.

The cowling is potted together with the thruster barrel, propellant reservoir/feed assembly, but it is insulated from its outer surface. This provision allows the firing loop to float above ground, connected through an insulating resistor which bleeds electrostatic accumulated charges.

The thruster module has provisions for the attachment to its front mounting flange, of a bell-jar like device, which can be connected to a portable vacuum pump, to

TABLE VII

Component weights for Inside-Out flight prototype
subsystem

Component	Weight (gm)
Capacitor and can	635 (1.4 lb)
Propellant	210 (0.5 lb)
RFI cone	50
Electrodes	30
Coatings	50
Storage and feed[a]	70
Support mounting	175
Housing	75
Discharge initiating circuits (4)	180
Sub total	1475 (3.2 lb)
Control, logic and power conditioning	700 (1.55 lb)
TOTAL	2175 (4.79 lb)
Approximate total envelope volume (in.^{3})	230

[a] Sized for 700 lb s total impulse. If this capability is
desired, add 0.4 lb propellant.

TABLE VIII

Performance characteristics for Inside-Out flight
prototype subsystem

Parameter	Values
Impulse bit, μlb s	25
Firing frequency, Hz	1.0 to 1.84
Specific impulse, s	1000
Total impulse[a], lb s	700
Energy per pulse, J	6.75
Power input[b], W (at 1.84 Hz)	16
Power conditioning efficiency, %	80
η_T, %	8.2
Operating voltage	1300
Capacitance, μF	8

[a] Tank and feed assembly sized for 700 lb s.
[b] Includes telemetry and firing pulse discriminator logic
power 0.5 W.

permit test firing of the thruster with the subsystem under ambient conditions. The interface between the barrel assembly and the firing circuit assembly is vaccum tight and structurally capable of withstanding the pressure differential which appears solely on the interconnecting (thruster barrel-firing circuit) flange.

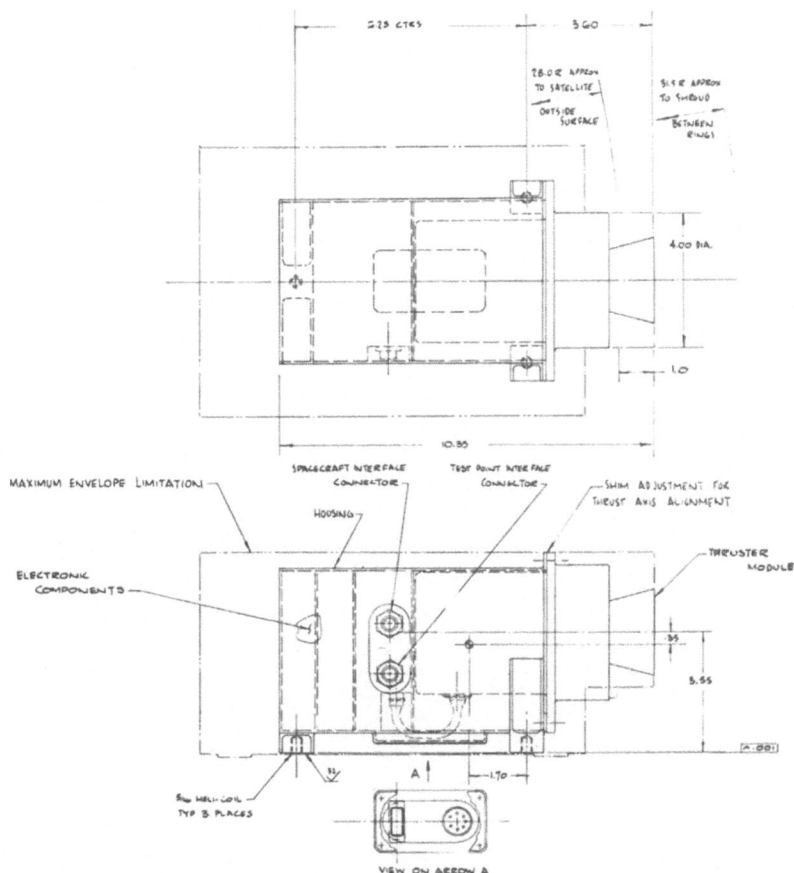

Fig. 17. Sketch of Inside-Out flight prototype subsystem.

The specifications for the subsystem are summarized in Tables VII and VIII. Weights, volumes and power of this preliminary design should be considered to be roughly double of those of a fully developed system.

4. Conclusions

Recent developments in pulsed plasma thrusters have been discussed and complete thruster systems, incorporating the new designs and techniques, have been illustrated in two applications: (1) an experimental 15 lb system for the east–west station keeping, one axis attitude control and a 36° station change of a 2000 lb geostationary three axis stabilized satellite; (2) an operational 9.6 lb system for the five year east–west station keeping, fine spin axis precession control and initial $\Delta\Theta$ removal for a 550 lb geostationary spin stabilized satellite.

In both examples the systems are grossly overdesigned, emphasizing redundancy, reliability, and long life which can be obtained at the limited cost of a few pounds. In the case of the three axis stabilized satellite, the pulsed plasma system intended as an experiment has the total impulse, thrust and power handling capability of a prime

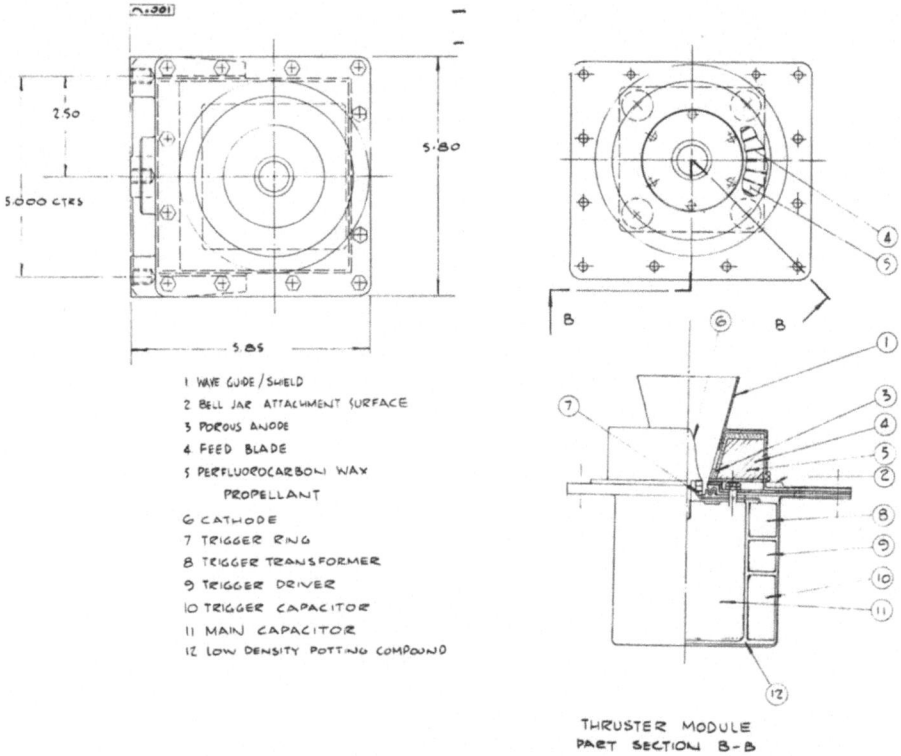

1 WAVE GUIDE / SHIELD
2 BELL JAR ATTACHMENT SURFACE
3 POROUS ANODE
4 FEED BLADE
5 PERFLUOROCARBON WAX
      PROPELLANT
6 CATHODE
7 TRIGGER RING
8 TRIGGER TRANSFORMER
9 TRIGGER DRIVER
10 TRIGGER CAPACITOR
11 MAIN CAPACITOR
12 LOW DENSITY POTTING COMPOUND

THRUSTER MODULE
PART SECTION B-B

Fig. 18.   Sketch of Inside-Out thruster module for flight prototype subsystem.

operational system. It is estimated [8] that the east–west station keeping and three axis attitude control for three years and two 36° station change maneuvers can be fulfilled by the system with the addition of two EVM packages and two thruster modules to the SVP package. The weight of the operational system would be under 30 lb. The system would have multiple internal and functional redundancy, would eliminate all pressurized and electromechanical components, including flywheels, while permitting precise pointing accuracy limited only by sensor and lead network capability.

The 9.6 lb pulsed plasma system for the spin stabilized satellite, is intended for operational use in a five year mission. Its weight, volume and power are roughly double of what at the present state of development would already be considered a conservative design. It is believed that the extra five pounds constitute a reasonable price to be paid for added redundancy and reliability, especially considering the other indirect saving in weights permitted by the simple interfaces.

Another aspect, which has not been discussed in this paper pertains to the very significant cost reductions which are permitted by these systems because of the elimination of pressurized and electromechanical long life components and associated dust contamination control and handling procedures. By taking into account the simplicity of the interfaces, the reduction in costs associated with integration, qualification and checkout procedures from first assembly to launch, it is estimated that the costs of these systems would be between one-third to one-fifth of that of the system conventionally used.

References

[1] La Rocca, A. V., 'Solid Propellant Electric Thrusters', AIAA Paper 65-619, Colorado Springs, Colo., 1965.
[2] La Rocca, A. V., 'Solid Propellant Electric Thrusters for Attitude Control and Drift Correction of Space Vehicles', AIAA Paper 66-229, San Diego, Calif., 1966.
[3] La Rocca, A. V., 'Solid Propellant Electric Thrusters for Space Applications', XVII International Astronautical Federation Congress, Madrid, Spain, Oct. 9–15, 1966.
[4] La Rocca, A. V. and Perkins, G. S., 'Development of Solid Propellant Electric Thruster Systems for Attitude Control and Station Keeping of Spacecraft', AIAA Paper 67-661, Colorado Springs, Colo., 1967.
[5] La Rocca, A. V. and Malherbe, P., 'A Development Program for a Solid Propellant Electric Thruster (SPET-A)', Final Report, JPL Contract 951591, Sept. 1966–Sept. 1967, Jet Propulsion Lab., Pasadena, Calif.
[6] La Rocca, A. V. and Perkins, G. S., 'Pulsed Plasma Microthruster Applications and Techniques', AIAA Paper 68-554, Cleveland, Ohio, 1968.
[7] La Rocca, A. V., J. Spacecraft Rockets 8 (1971), 709.
[8] La Rocca, A. V., 'Pulsed Plasma Thruster Systems for Attitude Control and Station Keeping of Spacecraft', First Western Space Congress, Santa Maria, Calif., October 27–29, 1970.
[9] Samaras, D. G., 'Theory and Application of Ion Flow Dynamic', Prentice-Hall, 1962.
[10] Jhan, R. G., Clark, K. E., Oberth, R. C., and Turchi, P. J., 'Acceleration Patterns in Quasi Steady MPD Arcs', Princeton University, Princeton, N.J., AIAA 8th Aerospace Science Meeting – New York, N.Y., AIAA Paper No. 70-165, January 1970.
[11] Malliaris, A. C. and Libby, B. R., 'Spectroscopic Study of Ion-Neutral Coupling in Plasma Acceleration', Avco Corp., Wilmington, Mass. – AIAA 8th Aerospace Science Meeting – New York, N.Y., AIAA Paper No. 70-166, January 1970.
[12] Grahm, J. R., Jr, 'Pulsed Vacuum Arc Thruster Study', Cornell Aero Lab. Inc., AFAPL TR-68-92-WPAFB-Dayton, Ohio, 1968.
[13] Chace, W. G. and Moore, H. K., Exploding Wires, Vol. 3, Plenum Press, New York, 1967.
[14] Hassan, H. A., AIAA J. 9 (1971), 578.
[15] Ducati, A., 'Repetitively Pulsed Quasi Steady Vacuum MPD Arc', Plasmadyne/Geotel Inc., Santa Ana, Calif., and R. G. Jhan, Princeton University, Princeton, N.J., AIAA 8th Aerospace Sciences Meeting, New York, N.Y., AIAA Paper No. 70-167, January 1970.
[16] Guman, W. J. and Nathanson, D. M., 'Pulsed Plasma Microthruster Propulsion System for Synchronous Orbit Satellite', AIAA Seventh Electric Propulsion Conference, Williamsburg, Virginia, AIAA Paper No. 69-298, March 1969.

[17] Vondra, R., Thomassen, K., and Solbes, A., 'Analysis of Solid Teflon Pulsed Plasma Thruster', Massachusetts Institute of Technology, Lincoln Laboratory, Lexington, Mass., *AIAA 8th Aerospace Sciences Meeting*, New York, N.Y., AIAA Paper No. 70-179, January 1970.
[18] Jhan, R. G., *Physics of Electrical Propulsion*, McGraw Hill, New York, 1968.
[19] Larson, A. V., 'Experiments on Current Rotations in an MPD Engine', General Dynamics, Convair, San Diego, California, *AIAA Electric Propulsion and Plasma Dynamics Conference*, Colorado Springs, Colo., AIAA Paper No. 67-689, Sept. 1967.

PRÉVISION DES DOMAINES DE FONCTIONNEMENT
INSTABLE D'UN PROPULSEUR À PROPERGOL SOLIDE*

MARCEL BARRÈRE et LIONEL NADAUD

Office National d'Études et de Recherches Aérospatiales (O.N.E.R.A.), 92 – Chatillon, France

Résumé. Un très gros effort est fait à l'heure actuelle pour déterminer la réponse d'un propergol solide à une fluctuation de pression ou de vitesse; il s'agit avant tout d'acquérir des données sur un propergol de manière à prévoir, dans le cadre d'un avant projet, les domaines où le fonctionnement du propulseur sera stable ou instable.

Les méthodes permettant de déterminer la réponse d'un propergol sont décrites. Ces méthodes utilisent soit des propulseurs déjà instables (propulseur L*, propulseur en T) soit des propulseurs à éjection modulée, la fluctuation de pression étant obtenue par variation périodique de la section du col de la tuyère.

Les méthodes basées sur des propulseurs instables ont l'inconvénient d'être limitées à un domaine étroit de fonctionnement; leur développement a cependant permis de préciser les fonctions définissant la réponse du propulseur et en particulier la partie réelle de cette réponse.

La technique de l'éjection modulée permet de travailler dans une large gamme de fréquences et conduit à une évaluation simple de la fonction de transfert.

Les résultats obtenus avec ces techniques et leur domaine de validité sont discutés. Cette discussion a pour base une théorie linéarisée qui montre que la fonction réponse de la surface dépend de deux paramètres liés à la cinétique de pyrolyse et à l'énergie mise en jeu à la surface.

Cette théorie linéarisée, bien qu'étant une approche intéressante, est impuissante à prédire correctement la réponse du propergol.

Prevision of Unstable Operation Domains of Solid Propellant Rocket Motors

Abstract. Important research work is being done at the present time to determine the response of solid propellants to pressure or velocity fluctuations; it aims at acquiring data about the propellant so that, from the design stage, the stable operating domain of a rocket motor could be predicted.

The methods for determining the propellant response function are described; they make use either of already unstable motors (L* or T-shaped), or of modulated exhaust, the pressure fluctuation being obtained through a periodic variation of the nozzle throat section.

The drawback of the methods based on unstable motors is that they are limited to a narrow operating range; their development, however, permitted to know the parameters governing the motor response, and in particular its real part.

The modulated exhaust technique opens the study to a large frequency range and leads to a simple evaluation of the transfer function.

The results obtained with these techniques and their validity domain are discussed. This discussion is based on a linearized theory showing that the response function of the surface depends on two parameters linked to the pyrolysis kinetics and to the energy available at the surface.

This linearized theory, however interesting as a first approach, is unable to predict correctly the propellant response.

1. Introduction

La prévision du fonctionnement stable d'un propulseur est donc un problème difficile, le but de cet article est de déduire les différentes techniques développées à l'heure

* Ce travail a été effectué sous contrat de la Société Nationale des Poudres et avec sa collaboration.

L. G. Napolitano et al. (eds.), Astronautical Research 1971, 265–284. All Rights Reserved.
Copyright © 1973 by D. Reidel Publishing Company, Dordrecht-Holland.

actuelle et qui permettent la détermination de la fonction réponse du propergol [1]. Ces techniques sont développées à partir de propulseurs spéciaux utilisés surtout dans leur domaine de fonctionnement instable (instabilités autoentretenues). L'O.N.E.R.A. a mis au point récemment des propulseurs fonctionnant en vibrations forcées qui permettent de balayer un intervalle plus large de fréquences. Cet article est consacré à une étude de ces différentes méthodes qui, bien qu'imparfaites, permettent cependant d'évaluer la fonction réponse du propergol et dans certains cas, de chiffrer les pertes,

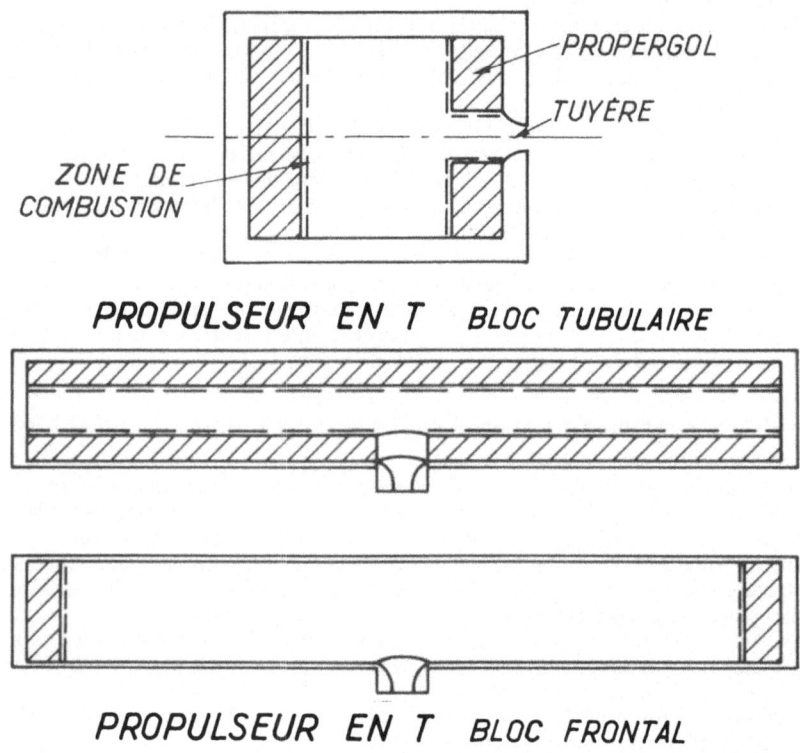

Fig. 1.   Propulseurs d'étude des instabilités de combustion autoentretenues.

il comprend deux parties: la première relative à la description des méthodes et la deuxième concernant l'exploitation des résultats obtenus.

2. Méthodes permettant de déterminer la fonction de transfert du propergol et d'analyser les conditions de stabilité

Nous avons divisé ces méthodes en deux catégories suivant que les oscillations sont autoentretenues (propulseur L^* et propulseur en T, Figure 1) ou qu'elles sont forcées (propulseur à éjection modulée, Figure 2).

2.1. OSCILLATIONS AUTOENTRETENUES

Ce sont les méthodes les plus couramment utilisées au cours de ces dernières années; les paramètres géométriques du propulseur et les paramètres de fonctionnement sont choisis de manière à se placer dans un domaine de combustion instable.

Dans le domaine des basses fréquences, le propulseur mis au point est caractérisé par un fonctionnement à basse pression (~ 5 bars) et par une géométrie variable; il est

PROPULSEUR A BLOC FRONTAL

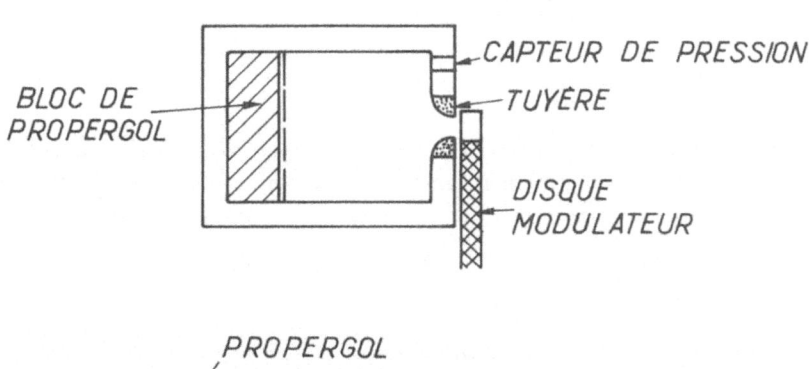

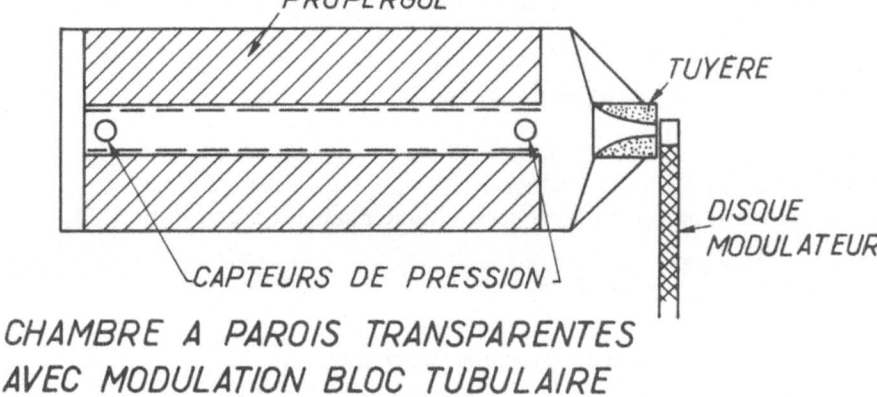

CHAMBRE A PAROIS TRANSPARENTES AVEC MODULATION BLOC TUBULAIRE

Fig. 2. Propulseurs O.N.E.R.A. d'étude des instabilités de combustion.

en effet nécessaire de pouvoir changer d'un essai à l'autre la longueur caractéristique L^*, c'est pourquoi on le désigne par le terme propulseur L^*.

2.1.1. *Propulseur L^**

Il est représenté sur la Figure 1. La combustion est généralement frontale de manière à pouvoir faire varier la longueur de la partie cylindrique et, partant de là, le volume du foyer. Ce propulseur permet de caractériser le propergol dans le domaine des basses fréquences ($N < 100$ Hz) et pour des pressions dans le foyer comprises entre 1 et 10 bars, c'est-à-dire des conditions assez éloignées de celles rencontrées dans la pratique.

Pour des pressions suffisamment basses, le propulseur se met à vibrer, l'allure des enregistrements est alors le suivant: l'amplitude des fluctuations de pression va en augmentant jusqu'à un certain niveau à partir duquel le propulseur s'éteint puis redémarre après un certain délai, lorsque la pression a atteint son niveau nominal, les instabilités réapparaissent, elles sont d'abord de faible amplitude puis celle-ci augmente jusqu'à une nouvelle extinction; on observe donc une combustion par bouffées, au cours de laquelle il est possible de caractériser le régime instable par sa fréquence N et par le coefficient d'amplification α. Une théorie linéarisée va nous permettre de mieux comprendre l'exploitation de ces résultats [2].

La conservation de la masse conduit à l'équation:

$$\frac{\mathrm{d}}{\mathrm{d}t}(M) = \dot{m}_b - \dot{m}_t \tag{1}$$

où M est la masse à l'instant t séjournant dans le foyer, $\dot{m}_b$ le débit brûlé et $\dot{m}_t$ le débit traversant la tuyère. En régime permanent $\dot{m}_b = m_t = \overline{m}$, à partir de M nous définissons un temps de séjour $t_s = \overline{M}/\overline{m}$ de sorte que l'équation précédente devient:

$$\tau_f \frac{\mathrm{d}}{\mathrm{d}t}\left(\frac{p'}{\overline{p}}\right) = \frac{m'}{\overline{m}} - \frac{p'}{\overline{p}} \tag{2}$$

avec comme temps caractéristique du foyer:

$$\tau_f = \frac{1}{\gamma}\, t_s = \frac{1}{\gamma}\, \frac{L^* C^*}{(R/\mathcal{M})T}$$

le signe $'$ indiquant la partie instationnaire $p = \overline{p} + p'$ et $m = \overline{m} + m'$.

Dans le cas d'une évolution isotherme $\tau_f = t_s$.

Introduisons la réponse R du propergol à une fluctuation de pression [3]:

$$R = \frac{m'/\overline{m}}{p'/\overline{p}} = R_r + iR_i \tag{3}$$

comportant une partie réelle et une partie imaginaire.

En supposant une loi d'évolution de la pression de la forme

$$\frac{p'}{\overline{p}} = \frac{\hat{p}}{\overline{p}}\, e^{i(\omega - i\alpha t)},$$

l'égalité des parties réelles et imaginaires permet de calculer R_r et R_i à partir de $N = 2\pi\omega$ et de α, τ_f étant évalué à partir du paramètre géométrique L^* et des paramètres de fonctionnement moyen du propulseur:

$$\begin{aligned} R_r &= 1 + \alpha\tau_f \\ R_i &= \omega\tau_f \end{aligned} \tag{4}$$

$\alpha > 0$ correspondant à une amplification des instabilités.

On peut également caractériser la fonction réponse R par son module $\mathscr{R}$ et le déphasage entre la fluctuation de débit et la fluctuation de pression en introduisant un délai τ_c de sorte que:

$$\mathscr{R} \cos \omega\tau_c = 1 + \alpha\tau_f$$

$$\mathscr{R} \sin \omega\tau_c = \frac{\tau_f}{\tau_c} \, \omega\tau_c. \tag{5}$$

On voit tout de suite la limite de cette méthode. Comme l'indiquent les équations précédentes, le domaine d'instabilités est caractérisé par les faibles valeurs de L^* et de la pression $\bar{p}$; ce domaine de stabilité est défini par l'inégalité

$$1 < \frac{\tau_f}{\tau_c} < \mathscr{R} \tag{6}$$

Comme avec les propergols courants $\alpha \simeq 100 \text{ s}^{-1}$ et par suite des faibles valeurs de L^*, $\tau_f \simeq 10^{-3}$ s, il en résulte que R_r et $\mathscr{R}$ sont voisins de l'unité de sorte que $\tau_f \simeq \tau_c$ égalité qui est bien vérifiée par l'expérience comme l'indique la Figure 3. Le domaine

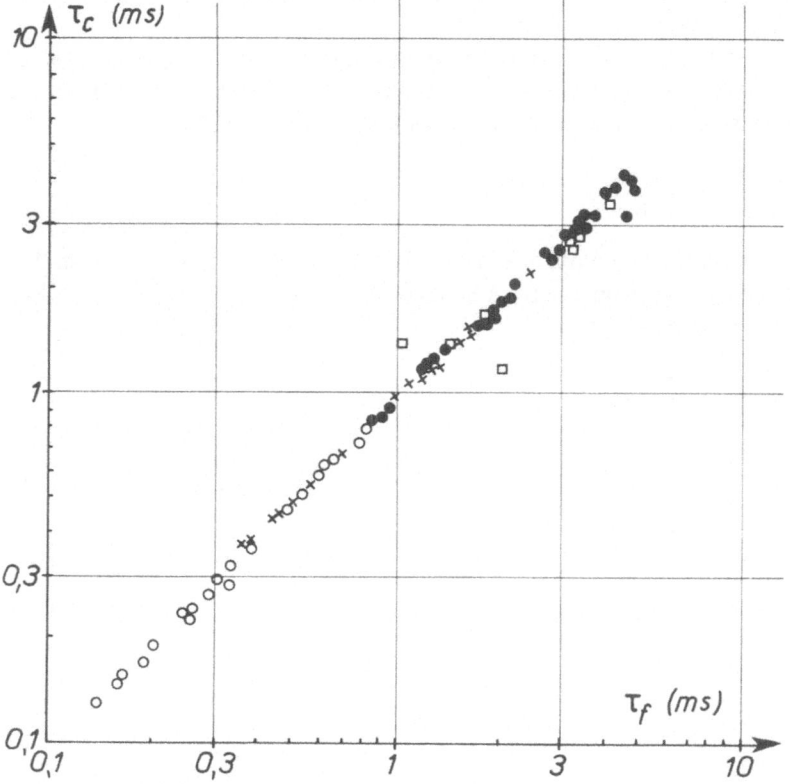

Fig. 3.   Evolution de τ_e en fonction de τ_f (propulseur L^* d'après Price).

de détermination de R est donc très limité. Cette méthode permet cependant de contrôler la validité de la fonction de transfert découlant d'une théorie [1–3].

2.1.2. *Propulseur en T* [4] [5] [6]

Il est constitué par un tube cylindrique fermé aux extrémités (voir Figure 1), la tuyère étant placée à michemin des deux extrémités. Le propergol est placé soit en bout de tube en combustion frontale, soit contre la paroi en combustion radiale, ces deux géométries de bloc pouvant être combinées.

Pour une géométrie de bloc donnée, il est donc possible de faire varier la fréquence en modifiant la longueur L du tube, le mode longitudinal s'établissant, le fondamental a pour fréquence $N = \bar{a}/2L$, $\bar{a}$ étant la vitesse moyenne du son dans la cavité. La pression $\bar{p}$ peut être modifiée par une adaptation du col de la tuyère ou en faisant déboucher la tuyère dans un caisson porté à une certaine pression. Un domaine instable est donc défini dans le plan (fréquence-pression). Les fluctuations de pression enregistrées sont caractérisées par leur fréquence et par le coefficient d'amplification α_c lorsque l'amplitude croît et α_d lorsqu'elle décroît, la croissance et la décroissance ayant une allure exponentielle de la forme $e^{\alpha t}$.

Ce dispositif permet donc de balayer un large domaine qui n'est pas forcément celui d'un autre propulseur utilisant le même propergol.

Des ondes stationnaires s'établissent dans le tube et l'admittance A_b est plus souvent utilisée que la fonction réponse R, on peut passer très facilement d'une notion à l'autre.

Définissons l'admittance A_b comme étant égale à [6] [7] [8]:

$$A_b = \gamma \, \frac{u'/\bar{a}_b}{p'/\bar{p}} \tag{7}$$

où $\bar{a}_b$ est la vitesse du son moyenne près de la surface en combustion et u' la fluctuation de vitesse des gaz perpendiculaire à la surface.

Comme $m = \rho u$ on obtient:

$$A_b = \frac{\gamma \bar{u}}{\bar{a}_b} \left[R - \frac{\rho'/\bar{p}}{\rho'/\bar{\rho}} \right] \tag{8}$$

Pour une onde isentropique à l'interface $p'/\bar{p} = \gamma(\rho'/\bar{\rho})$ de sorte que:

$$A_b + \overline{M}_b = \gamma \overline{M}_b R \tag{9}$$

ici intervient le nombre de Mach $\overline{M}_b = \bar{u}/\bar{a}_b$.

Une étude linéarisée du propulseur dans le cas d'une combustion frontale aux deux extrémités conduit à l'expression du nombre d'onde k [5]:

$$k^2 = k_l^2 - (A_b + \overline{M}_b) \, \frac{4ik_l}{L} \, \frac{S_b}{S_f} \tag{10}$$

k_l correspond à la valeur de k en l'absence d'écoulement et de transfert à la surface; pour le mode longitudinal $k_l = l\pi/L$, l étant égal à l'unité pour le fondamental, S_b/S_f

est le rapport de la surface de combustion S_b à la surface de la section droite du cylindre S_f.

Comme $k = (\omega - i\alpha)/\bar{a}$, en égalant les parties réelles et imaginaires on obtient

$$\left(\frac{\omega L}{\bar{a}}\right)^2 - (l\pi)^2 = 4\left(\frac{\omega L}{\bar{a}}\right)\frac{S_b}{S_f}A_{bi},$$

$$\frac{\alpha L}{a} = 2[A_{br} + \overline{M}_b]\frac{S_b}{S_f},$$

(11)

ou encore, en passant à la fonction R,

$$R_r = \frac{S_f}{S_b} \cdot \frac{1}{2\gamma\overline{M}_b} \cdot \frac{\alpha L}{\bar{a}}$$

$$R_i = \frac{S_f}{S_L} \cdot \frac{1}{2\gamma\overline{M}_b} \cdot \frac{1}{2l\pi}\left[\left(\frac{\omega L}{\bar{a}}\right)^2 - (l\pi)^2\right].$$

(12)

La valeur de α tient compte des pertes et est la somme de la constante α_c correspondant à l'amplification et α_d relatif à l'atténuation

$$\alpha = \alpha_c + \alpha_d.$$

(13)

En ne tenant pas compte de l'évolution de la température le long du tube ($\bar{a}_b = \bar{a}$), la partie réelle de la fonction R s'exprime donc simplement en fonction des données de l'expérience

$$R_r = \frac{1}{4L}\frac{\bar{p}}{\bar{m}}\frac{\alpha}{N^2}\frac{S_f}{S_b},$$

(14)

– pression moyenne $\bar{p}$,
– débit unitaire moyen $\bar{m} = \rho_p\bar{v}_b$ ($\bar{v}_b$ vitesse moyenne de combustion),
– fréquence N,
– constante d'évolution de l'amplitude α.

– La réponse du propergol R a été choisie couplée avec la pression, mais avec les blocs tubulaires un couplage avec la fluctuation de la vitesse v le long du canal c'est-à-dire parallèlement à la surface est également possible; sous forme linéarisée la réponse R comprend alors deux fonctions supplémentaires R_p et R_v de sorte que:

$$\frac{m'}{\bar{m}} = R_p\frac{p'}{\bar{p}} + R_v\frac{v'}{\bar{v}}$$

(15)

la fonction R_v introduite variant tout le long de la surface, sa détermination nécessite des mesures localisées de pression, de vitesse et de température.

– La partie imaginaire fait intervenir $\omega L/a$ qui diffère peu de $l\pi$ puisque la fréquence est assez voisine de la fréquence d'un tube fermé aux deux extrémités de sorte que:

$$R_i = \frac{S_f}{S_c} \cdot \frac{1}{2\gamma\overline{M}_b}\left[\frac{\omega L}{a} - l\pi\right]$$

(16)

Des études systématiques ont été effectuées sur le propulseur en T qui ont permis d'évaluer $R_r(N)$ en fonction de la nature du propergol (composite, double base), de la granulométrie des ingrédients, de la nature du métal, du pourcentage du métal, de la pression de fonctionnement, de la température du bloc, etc....

L'utilisation de ces courbes par le constructeur est difficile car α ou R_r dépendent de la différence entre gains et pertes et les pertes sont fonction de l'organisation de la combustion et de l'écoulement dans le propulseur, donc fonction du propulseur lui-même.

2.2. OSCILLATIONS FORCÉES

Avec les méthodes précédentes l'évaluation de la fonction réponse du propergol ne peut se faire que dans le domaine où le propulseur L^* ou le propulseur en T sont utilisables; ce domaine ne coïncide pas forcément avec le domaine des instabilités rencontrées dans les propulseurs d'application; avec le développement des propulseurs de grande dimension les fréquences observées se situent dans la zone intermédiaire où les fréquences sont comprises entre 100 et 1000 Hz, l'étude de ces fréquences conduit à des propulseurs en T de grande longueur donc avec des pertes aux parois importantes. Pour limiter ces inconvénients, des méthodes ont été développées à l'O.N.E.R.A. permettant d'étudier la gamme de fréquences désirées et principalement le domaine des fréquences intermédiaires pour lequel on dispose encore de peu de résultats.

2.2.1. Propulseur à combustion frontale à éjection modulée

Ce propulseur est schématisé sur la Figure 2. Le bloc de poudre est cylindrique et brûle en cigarette, la tuyère est limitée au col; un disque modulateur placé devant le col permet de faire varier périodiquement sa section. Un piston peut faire avancer le bloc en cours de combustion de manière à garder le volume constant, mais cette possibilité a été encore peu utilisée. La Figure 4 est une vue du dispositif.

Une théorie linéarisée va nous permettre, comme précédemment, d'analyser le fonctionnement du propulseur et de décrire la méthode employée pour déterminer la fonction réponse R.

Le débit traversant le col de la tuyère est proportionnel à la surface du col S_t de la tuyère $\dot{m}_t = pS_t/C^*$, p et S_t variant, l'équation de conservation de la masse devient:

$$\tau_f \frac{d(p'/\bar{p})}{dt} = \frac{m'}{\bar{m}} - \frac{p'}{\bar{p}} - \frac{S_t'}{\bar{S}_t}. \tag{17}$$

Dans le cas d'une modulation sinusoïdale de la section du col de la forme $S'/\bar{S}_t = de^{i\omega t}$ la modulation est de la forme;

$$\frac{p'}{\bar{p}} = \frac{\hat{p}}{\bar{p}} e^{i[(\omega - i\alpha)t - \varphi]} = be^{i[(\omega - i\alpha)t - \varphi]} \tag{18}$$

avec un déphasage φ entre la modulation de section et la modulation de la pression.

Fig. 4. **Vue du propulseur à combustion frontale et à éjection modulée.**

En introduisant dans l'équation différentielle la fonction réponse R et en égalant les parties réelles et imaginaires, nous obtenons dans le cas où $\alpha = 0$:

$$-\tau_f b\omega \sin \varphi + (R_r - 1)b \cos \varphi + bR_i \sin \varphi = d$$
$$\tau_f \omega \cos \varphi - R_i \cos \varphi + (R_r - 1) \sin \varphi = 0$$

(19)

le système étant linéaire en R_r et R_i on explicite R_r et R_i:

$$R_r = 1 + \frac{d}{b} \cos \varphi$$

$$R_i = \tau_f \omega + \frac{d}{b} \sin \varphi.$$

(20)

Lorsqu'à la section maximale correspond la pression minimale, on obtient le système simplifié:

$$R_r = 1 - \frac{d}{b}$$

$$R_i = \tau_f \omega$$

(21)

La mesure de d, b, $\omega = 2\pi N$ et le calcul du temps caractéristique τ_f déterminent donc R_r et R_i.

La technique de modulation est en général la suivante: le disque est divisé en trois secteurs, le premier correspond à une ouverture complète du col, la pression dans le foyer étant p_0, le secteur suivant obture le col de moitié, la pression $\bar{p}_i$ dans le foyer correspond sensiblement à la pression moyenne $\bar{p}$, le passage de p_0 à $\bar{p}_i$ permet d'avoir une idée du temps de séjour τ_f, le troisième secteur module la section du col autour de $\bar{p}$. La Figure 5 est un enregistrement de pression en deux points du foyer, la fréquence est dans ce cas de l'ordre de 600 Hz; on note une différence entre $\bar{p}_i$ et $\bar{p}$ pouvant provenir de la variation de la vitesse de combustion. A partir de ces enregistrements, il est possible de mesurer $\hat{p}/\bar{p}$ ainsi que le déphasage défini à partir d'un repère placé sur le disque.

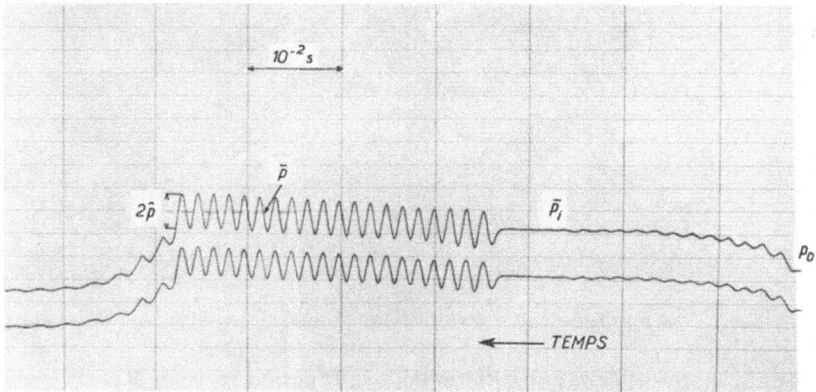

Fig. 5.   Enregistrement des fluctuations de pression pour un tour de disque.

Un exemple de résultats obtenus est donné sur la Figure 6. La gamme de fréquences balayée est celle des fréquences intermédiaires comprises entre 100 et 1000 Hz; le paramètre $b = \hat{p}/\bar{p}$ est porté en fonction de la fréquence et on observe pour ce propergol une décroissance régulière de b en fonction de N lorsque la fréquence croît, le déphasage est constant jusqu'à 600 Hz et égal à 270°, il augmente au delà. Il en résulte que dans ce domaine exploré la partie réelle de R est voisine de l'unité, elle passe à 1.6 à 650 Hz. La mesure du déphasage est donc très importante dans ce cas puisqu'elle conditionne la valeur de R_r et R_i, tg $\varphi = (R_i - \tau_f \omega)/(1 - R_r)$. Bien que cette méthode soit un progrès par rapport aux méthodes précédemment décrites puisqu'elle offre l'avantage de pouvoir balayer en un seul tir un plus large domaine de fréquence, elle n'est cependant pas parfaite et les résultats fournis doivent être examinés avec soin comme le souligne les remarques qui suivent.

– Etudions tout d'abord brièvement, en partant toujours de la théorie linéarisée, la sensibilité de la méthode. Nous considérons deux cas fournis par la fonction réponse de Culick; le premier correspond à une forte variation de R_r avec la fréquence, le maximum atteignant 15 (cas I), le deuxième relatif à un effet plus modéré, le maximum de R_r se situant à 6,5 (cas II), les évolutions de R_r et R_i en fonction de N sont données sur la Figure 7. En portant ces grandeurs dans la théorie linéarisée nous obtenons les valeurs de b et du déphasage φ entre la variation de section et la pression dans le foyer.

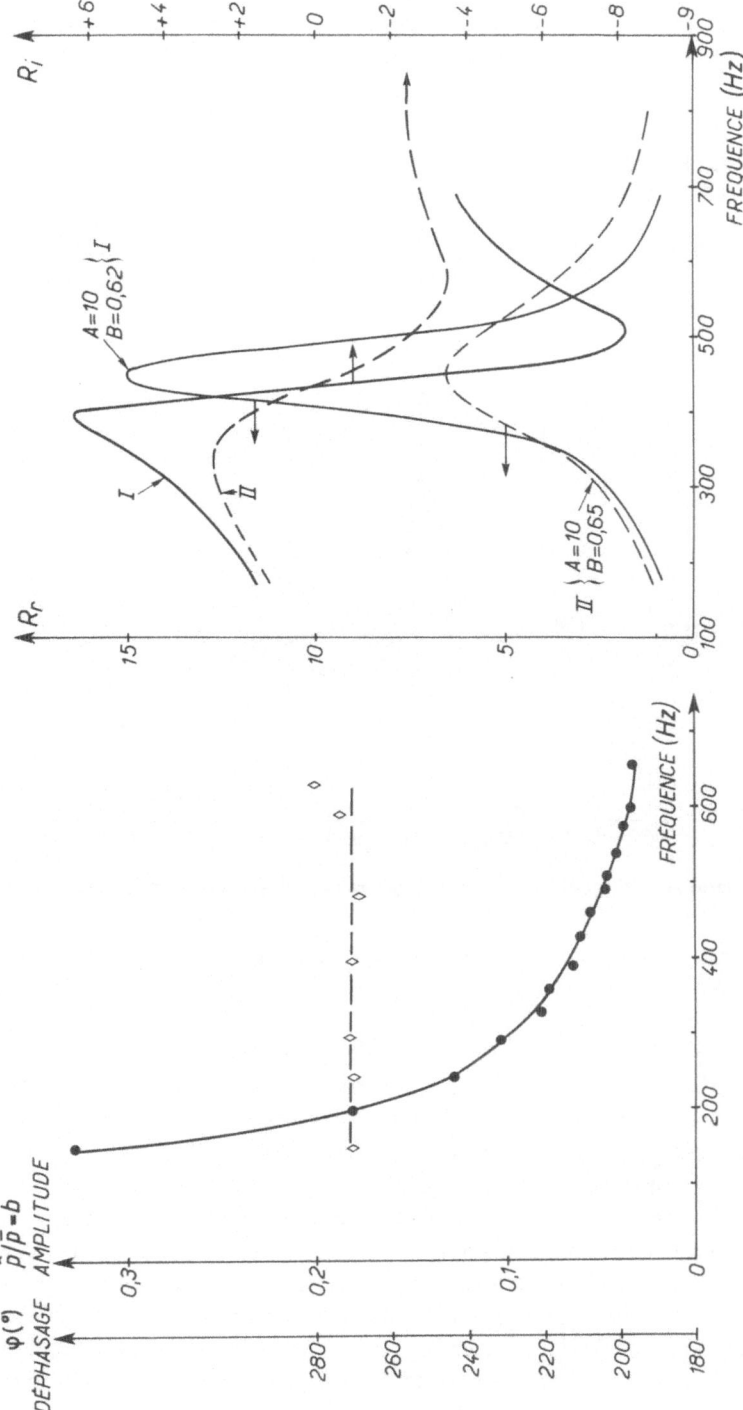

Fig. 7. Variations de R_r et R_i avec la fréquence pour deux fonctions réponse ($\tau_f = 0,004$ s).

Fig. 6. Variation du déphasage et de l'amplitude en fonction de la fréquence.

Les cas I et II sont comparés également à la valeur b lorsque $R_r = 1$. Bien que l'on observe des différences dans les diverses valeurs de b lorsque R_r et R_i varient (Figure 8), la précision de la méthode demeure cependant faible par suite des erreurs pouvant provenir de la mesure de la pression principalement aux faibles valeurs de b; il ne faut

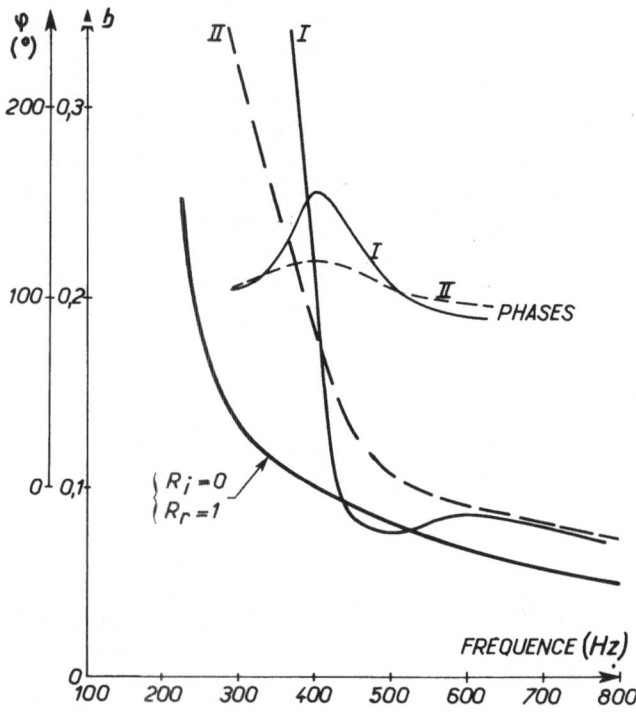

Fig. 8. Variations de l'amplitude et du déphasage avec la fréquence pour les cas I et II.

donc pas s'attendre à une détermination précise de R_r et R_i en particulier lorsque R_r est voisin de l'unité.

2.2.2. *Propulseur à combustion radiale et à éjection modulée*

Pour se rapprocher de la configuration d'un propulseur classique et éviter la forme particulière du propulseur en T, l'O.N.E.R.A. a développé un foyer bidimensionnel à voyant dont la configuration et les cotes sont données sur la Figure 9.

Deux plaques, l'une métallique, l'autre en plexiglass maintiennent deux blocs de poudre cylindriques à section rectangulaire écartés d'une certaine distance de manière à ménager un canal central à travers lequel les gaz brûlés vont circuler. Du côte fond avant, l'extrémité est limitée par un plan et une tuyère unidimensionnelle est fixée au fond arrière; une chambre ménagée à l'extrémité aval du bloc permet de passer de la section rectangulaire du canal à la section circulaire de la tuyère. A la sortie du col, un disque tournant module la section de sortie de la tuyère. Deux capteurs de pression à

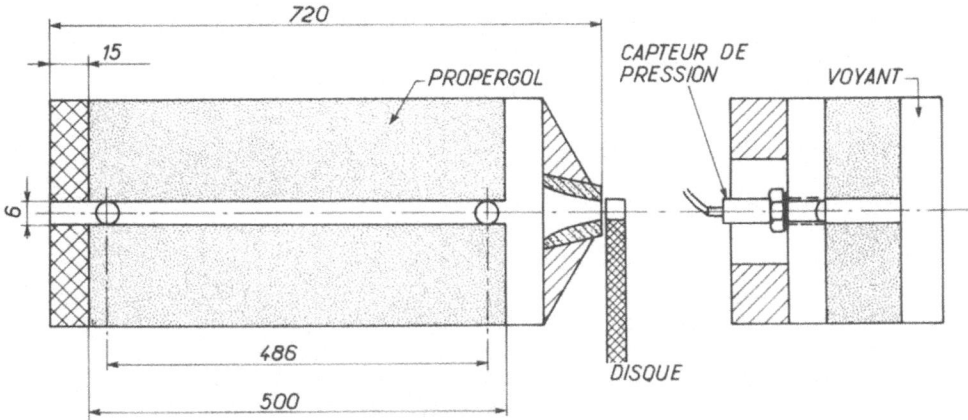

Fig. 9. Coupe du propulseur à voyant avec éjection modulée.

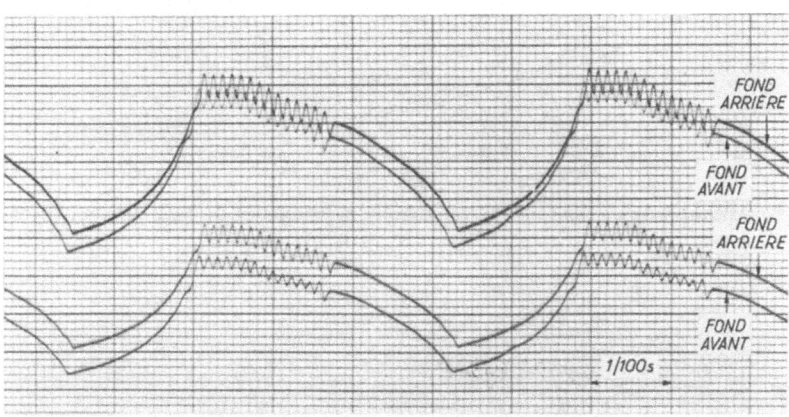

Fig. 10a. Enregistrement des pressions en amont et an aval du bloc pour des fréquences élévées ($N \simeq 880$ Hz).

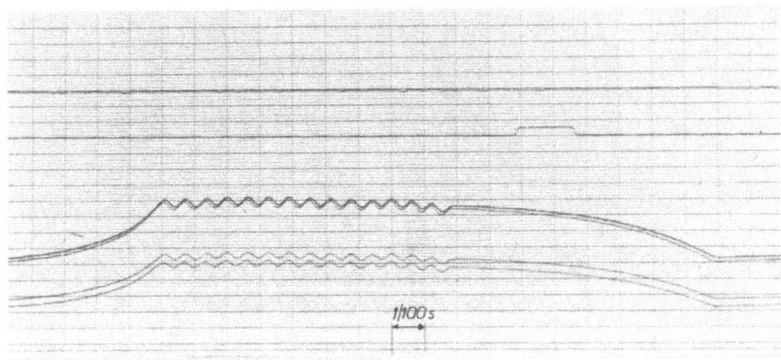

Fig. 10b. Enregistrement des pressions à basse fréquence ($N \sim 160$ Hz).

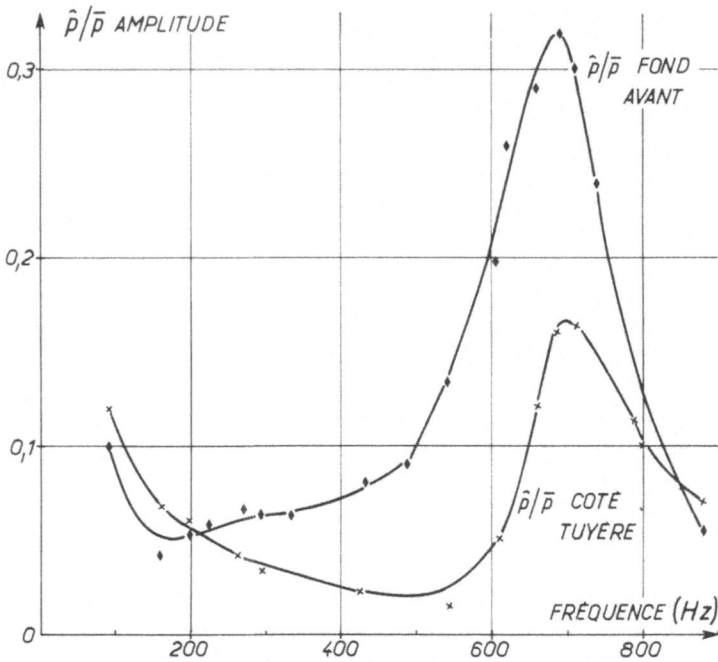

Fig. 11.   Variation de l'amplitude en fonction de la fréquence de modulation ($\bar{p} = 33$ bar).

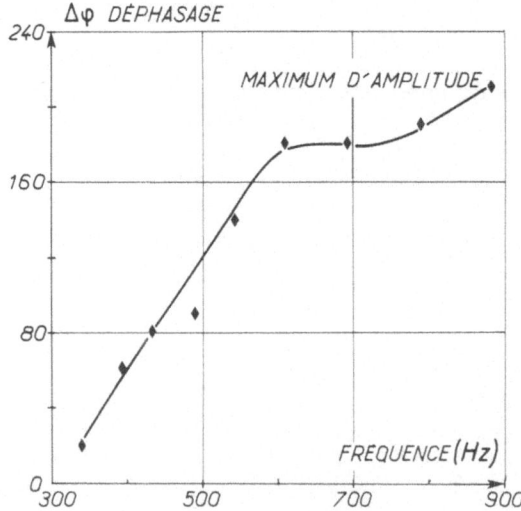

Fig. 12.   Déphasage entre les deux capteurs en fonction de la fréquence. Bloc tubulaire.

membrane affleurante sont placés aux deux extrémités du bloc. Cet ensemble est bien adapté à l'étude des fréquences intermédiaires avec possibilité à la surface de combustion d'un couplage du débit unitaire de propergol brûlé avec les fluctuations de pression et les fluctuations de vitesse le long de la surface. Le disque, comme précédemment, est conçu de manière à avoir trois séquences successives: une séquence laissant un col à pleine ouveture, une séquence à demi-ouverture, le passage de S_t à $S_t/2$ permettant d'avoir une idée de la constante de temps du foyer et enfin une séquence à modulation. Pendant un tour de disque on peut admettre que la fréquence de modulation est constante et que la constante de temps du foyer est également constante, un tour de disque correspond donc à un point de mesure.

Au cours d'un même tir, il est possible, en faisant varier la vitesse du disque, de balayer en fréquences. La Figure 10a est un exemple d'enregistrement de pression obtenu de part et d'autre du bloc à constante de temps du foyer élevée et à une fréquence de l'ordre de 880 Hz, et la Figure 10b à une fréquence beaucoup plus faible de 160 Hz, permettant d'atteindre le régime permanent. L'aspect de la zone de combustion pendant les trois séquences p_0, $\bar{p}_i \sim p_0/2$ et modulation souligne que l'épaisseur de la zone de combustion diminue lorsque la pression augmente, la combustion est plus hétérogène en régime de modulation.

Les résultats obtenus pour un propergol sont portés sur la Figure 11, les variations de l'amplitude des fluctuations de pression en fonction de la fréquence sont différentes suivant la position du capteur. On retrouve bien une fréquence de résonance à 700 Hz correspondant au fondamental du mode longitudinal. Les amplitudes du côté fond avant sont plus importantes que du côté tuyère, à la fréquence de résonance le rapport d'amplitude est de l'ordre de deux. Du côté fond avant, on observe une amplification pour des fréquences allant de 200 à 400 Hz faisant apparaître dans cette zone un couplage "pression" plus important que le couplage "vitesse".

Ces différences enregistrées sur les deux capteurs peuvent trouver une explication dans les conditions d'écoulements très différentes mais aussi dans la structure de la zone de combustion. Le déphasage entre les deux signaux est représenté sur la Figure 12 en fonction de la fréquence. Ce déphasage croît avec la fréquence en passant par 180° à la résonance, ce qui correspond bien à un tube vibrant en $\lambda/2$.

3. Exploitation des résultats

Les résultats obtenus par ces méthodes peuvent être utilisés à deux fins:

(a) comparer avec l'expérience les fonctions élaborées par la théorie définissant la réponse du propergol,

(b) définir des méthodes permettant de prédire la stabilité du propulseur et de définir des procédés stabilisant un propulseur.

3.1. COMPARAISON DES FONCTIONS 'RÉPONSE' THÉORIQUES AVEC L'EXPÉRIENCE

Le schéma le plus classique est celui d'un écoulement unidimensionnel en combustion issu de la surface du propergol transmettant par conduction l'énergie nécessaire au solide pour le transformer en gaz. On admet dans cette théorie que la réponse du gaz à une fluctuation s'effectue en un temps court par rapport à la réponse du solide, toujours en théorie linéaire l'ensemble des résultats est de la forme [7]:

$$R = \frac{m'/\bar{m}}{p'/\bar{p}} = \frac{nAB + n_s(s - 1)}{s + (A/s) - (A + 1) + AB} \tag{22}$$

où n est l'exposant de la pression entrant dans la loi de vitesse de combustion, A caractérise la température d'activation T_{AS} choisie dans la loi de pyrolyse, c'est donc un paramètre lié à la cinétique de la pyrolyse $A \simeq (1 - T_i/\bar{T}_s) \cdot (T_{AS}/T_s)$, T_i est la température initiale du propergol et $\bar{T}_s$ la température moyenne à la surface, B est un paramètre énergétique qui dépend de l'énergie ΔQ_p nécessaire pour transformer le solide en gaz

$$B \simeq \frac{\dfrac{C_p}{C} - \dfrac{T_i}{\bar{T}_s} - \dfrac{\Delta Q_p}{C\bar{T}_s}}{1 - \dfrac{T_i}{\bar{T}_s}},$$

C_p est la chaleur spécifique du gaz et C celle du solide, n_s est l'exposant de la pression intervenant dans la loi de pyrolyse et s solution de l'équation:

$$s(s - 1) = i\omega\tau_{\text{th}} = i\Omega$$

comporte une partie réelle et une partie imaginaire, τ_{th} est le temps thermique dans le solide quotient de la diffusivité thermique par le carré de la vitesse moyenne de combustion $\tau_{\text{th}} = a_{\text{th}}/\bar{v}_b^2$. La Figure 7 donne les variations de R_r et R_i en fonction de la fréquence $N = \omega/2\pi$ pour $A = 10$ et deux valeurs de B 0,62 et 0,65, ces courbes montrent l'influence de B qui peut modifier profondément l'allure des courbes R_r et R_i, l'action de A est moins importante.

3.1.1. *Propulseur L* et propulseur à combustion frontale à éjection modulée*

Avec cette définition de R, les expériences effectuées sur le propulseur L^* et sur le propulseur à combustion frontale à éjection modulée, conduisent à une détermination de A et de B.

Les résultats obtenus sur le propulseur L^* par Beckstead et Culick [2] montrent que pour certains propergols, A est négatif. Ils trouvent également que A et B sont fonction de la fréquence. Des résultats similaires ont été obtenus sur le propulseur à éjection modulée, comme l'indique les Figures 13a et 13b, A et B sont fonctions de la fréquence N, les résultats étant obtenus avec un propergol métallisé. Ce résultat n'a pas de sens car A et B devraient être constants dans le domaine de fréquences où le couplage avec la pression est prédominant, il met donc en cause la validité de la fonction R telle qu'elle est définie par la théorie unidimensionnelle. Ce résultat peut cependant provenir du déphasage voisin de 90°. Toujours en théorie linéaire, des expressions plus complexes ont été élaborées tenant compte de la variation de la vitesse de combustion v_b et de la température de la surface T_s en fonction de la température initiale T_i et de la pression p en régime permanent. Novikov et Riazent-

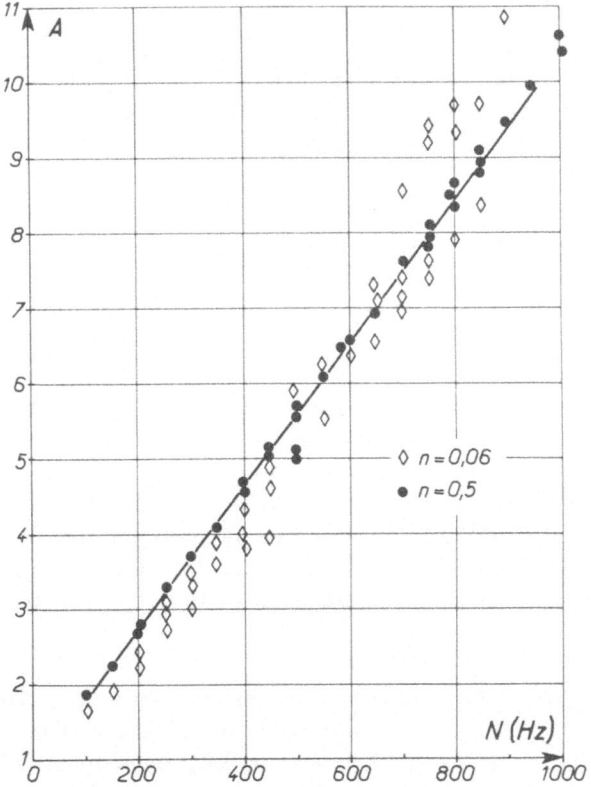

Fig. 13a. Variation de A avec la fréquence pour deux valeurs de l'exposant n.

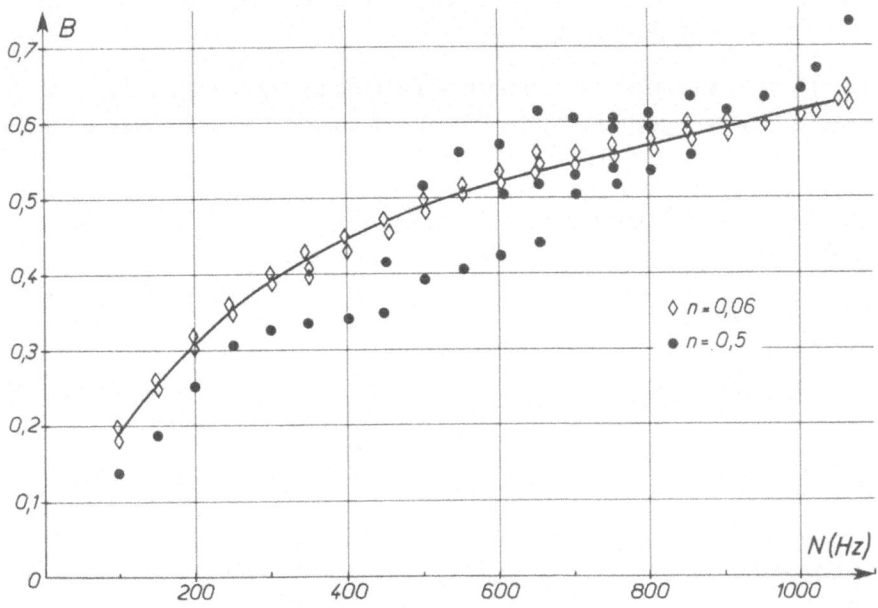

Fig. 13b. Variation de B avec la fréquence.

siev ont déterminé la partie réelle de l'admittance en utilisant la méthode de
Zeldovich [11], devant la complexité des expressions obtenues, la validité de ces
expressions n'a pu encore être vérifiée car nous ne disposions pas en régime permanent des grandeurs

$$e = \left(\frac{\partial \overline{T}_s}{\partial T_i}\right)_{\bar{p}} \text{ et } \mu = \frac{1}{\overline{T}_s - T_0}\left(\frac{\partial \overline{T}_s}{\partial \ln \bar{p}}\right)_{T_i}$$

définissant les variations de la température de surface $\overline{T}_s$ avec $\bar{p}$ et T_i.

3.1.2. *Propulseur en T*

La comparaison de la théorie et de l'expérience se fait principalement à partir de la
première équation (12). En effet, par suite du faible nombre de Mach M_b, la fréquence
observée est sensiblement la fréquence acoustique $(\omega L/a) \simeq l\bar{\pi}$ de sorte qu'il est très
difficile de déterminer R_i avec précision. On peut d'ailleurs montrer que lorsque
$\omega L/a = \pi$, la condition $R_i = 0$ conduit à des valeurs de A et B fonction uniquement de
Ω, de sorte que les variations de $\alpha L/\bar{a}$, si la fonction théorique est correcte, devraient
conduire à une même valeur de Ω, l'expérience montre qu'il n'en est rien.

3.1.3. *Propulseur à combustion radiale et à éjection modulée*

Avec ce propulseur, les conclusions sont sensiblement identiques à celles découlant de
l'expérimentation du propulseur en T, R_i est voisin de zéro à la résonance. Mais on
n'a pas cherché avec ce propulseur à vérifier la fonction réponse [22] mais plutôt à
connaître l'importance du couplage avec la vitesse.

En conclusion, la majorité des résultats expérimentaux obtenus avec les quatre
types de propulseurs, conduisent à des réponses du propergol assez différentes de
celles données par une théorie linéarisée.

3.2. DÉFINITION DE MÉTHODES PERMETTENT DE PRÉVOIR LA STABILITÉ DU PROPULSEUR [9]

La plupart des instabilités rencontrées dans un propulseur étant du type acoustique,
il sera possible de définir la stabilité d'un propulseur à partir du coefficient d'amplification α qui comprend un terme α_g relatif aux gains et un terme α_l correspondant aux
pertes, la stabilité se réduisant à la condition $\alpha_g - \alpha_l \leq 0$. Le problème consiste donc à
pouvoir évaluer correctement α_g et α_l sur le propulseur en projet.

3.2.1. *Evaluation de α_g*

Nous avons vu que α_g dépendait des phénomènes de combustion qui ont lieu à la
surface de combustion car on admet en général que la majorité de l'énergie est libérée
dans cette zone $(\alpha_g = \alpha_b)$. Il nous faut alors résoudre deux problèmes:

(1) A partir des expériences précédemment décrites, évaleur α_b pour un propergol
donné;

(2) α_b étant évalué dans la configuration des expériences précédentes relatives à un
modèle α_{bM}, il faut passer à la valeur de α_b dans la configuration du propulseur en
projet α_{bp}.

Nous avons vu que α_b peut être calculé à partir de la partie réelle de l'admittance A_{br} où à partir de la partie réelle de la fonction réponse R_y qui tiennent compte des pertes intervenant dans le propulseur ayant servi à la mesure. Cette détermination n'est possible que si l'on admet un couplage avec la pression, ce qui n'est pas forcément le cas d'un propulseur classique à canal central.

Avec un propulseur en T par exemple, la détermination directe de α_g peut se faire de deux manières:

(1) α_b est obtenu en faisant la différence des α à la montée en amplitude et à la décroissance en fin de combustion:

$$\alpha_{bM} = \alpha_c - \alpha_d$$

mais, comme nous l'avons signalé, avec des erreurs importantes dues aux différences de condition durant l'évaluation de α_c et α_d.

(2) α_b est déterminé à partir de deux expériences avec des surfaces de combustion différentes par exemple combustion aux deux extrémités α_{2c} et combustion à une extrémité α_{1c} uniquement pendant la montée:

$$\alpha_{bM} = \alpha_{2c} - \alpha_{1c}$$

Oberg et Ryan [10] ont montré qu'il était difficile de comparer les deux expériences, en effet, α_c dépend de la distribution de température dans le propulseur.

3.2.2. *Evaluation* de α_l [6–8]

Il existe peu de méthodes permettant d'évaluer d'une manière satisfaisante les pertes. D'un point de vue théorique, on distingue les pertes par les parois (effet visqueux et thermique), les pertes dans le gaz par relaxation et dans un écoulement à deux phases (particules), les pertes par rayonnement dues à la présence de l'écoulement, les pertes dans la partie solide entourant la cavité.

4. Conclusion

On dispose à l'heure actuelle de très nombreux résultats sur les instabilités de combustion mais qui sont bien souvent difficilement exploitables par le constructeur. Pour remédier à cet état de chose, des progrès devront être réalisés dans les domaines suivants:

(1) Il faut tout d'abord développer des méthodes qui permettent une évaluation plus précise de l'admittance ou de la fonction réponse du propergol, le propulseur en T et le propulseur L^* conduisant à des résultats peu exploitables.

Les propulseurs à éjection modulée constituent un progrès, mais il est nécessaire d'améliorer encore les méthodes de mesure et de dépouillement de manière à bien connaître les pertes et à mieux définir l'apport dû à la combustion, il faut enfin pouvoir caractériser l'importance du couplage avec les fluctuations de pression et les fluctuations de vitesse.

(2) Il est nécessaire pour passer d'un propulseur à un autre de pouvoir déterminer les pertes et vérifier les relations données dans le paragraphe précédent.

(3) Les théories linéarisées donnant la fonction réponse de la surface conduisent à des résultats souvent assez éloignés de l'expérience, un gros effort doit être fait également pour améliorer cette situation.

References

[1] Price, E. W., *Status of Solid Rocket Combustion Instability Research*, NOTS TP 4275 (Feb. 1967).

[2] Beckstead, M. W. and Culick, F. E. C., *AIAA J.* **9** (1971), 147.

[3] Price, E. W., *Recent Advances in Solid Propellant Combustion Instability*, XIIth Symposium (International) on Combustion, Poitiers, 1968. The Combustion Institute, p. 101.

[4] Beckstead, M. W., Mathes, H. B., Price, E. W., and Culick, F. E. C., *Combustion Instability of Solid Propellants*, XIIth Symposium (International) on Combustion, Poitiers, 1968. The Combustion Institute, p. 203.

[5] Culick, F. E. C., *Combustion Sci. Technol.* **2** (1970), 177.

[6] Williams, F. A. Barrère, M., and Huang, N. C., *Fundamental Aspects of Solid Propellant Rockets*, AGARDograph N° 116.

[7] Culick, F. E. C., *Astronaut. Acta* **13** (1970), 221.

[8] Hart, R. W. and McLure, F. T., *Theory of Acoustic Instability in Solid Propellant Rocket Combustion*, Xth Symposium (International) on Combustion, Cambridge (G.B.), 1964. The Combustion Institute, p. 1047.

[9] Coates, R. L. and Horton, M. D., *J. Spacecraft Rockets* **6** (1964), 296.

[10] Oberg, C. L., Ryan, N. W. and Baer, A. D., *AIAA J.* **6** (1968), 1131.

[11] Novikov, C. C., Riazentsiev, I. C. and Tulbckir, B. E., *Žu. Prikl. Meh. Tehničeskoj Fiz.* **5** (1969), 29.

APOLLO PROPULSION SYSTEMS
DEVELOPMENT AND FLIGHT EXPERIENCE

CHARLES H. KING, JR.

Apollo Program Office, NASA Headquarters, Washington, D.C., U.S.A.

Abstract. The design, development, testing and integration experience; and the highly successful operational utilization of the complex array of Apollo/Saturn V propulsion system elements have clearly demonstrated the value of extensive pre-flight analysis, ground testing, and rigorous qualification test, reliability and quality assurance programs to ensure flight vehicle operational integrity with limited all-up vehicle flight testing. This background has also served the program well in providing a substantial data base from which flight anomalies could be rapidly assessed and resolved with minimum program impact.

A brief account of Apollo flights to date and a description of the Apollo/Saturn V propulsion system thrusters is provided as introductory background, followed by a discussion of the design and development philosophy associated with launch vehicle and spacecraft engine selection and development testing. A presentation of propulsion system flight anomalies follows, with emphasis on the assessment process, the value of prior analyses and tests, and the approach to establishing satisfactory and timely fixes.

The paper concludes by calling attention to the value of the exhaustive test, redesign, and retest cycles in the ground development program, the significance of test and operational interaction assessment, and provision for uncovering and correcting early flight propulsion problems while simultaneously achieving mission objectives.

1. Introduction

Over the past three and one half years, the Apollo/Saturn V space vehicle has progressed from the first integrated system development flights to full operational status with an impressive record of performance and demonstrated capability to correct or work around a relatively small number of significant in-flight anomalies.

The flight record shows ten Apollo/Saturn V vehicle flights to date. The first two flights were unmanned development flights in late 1967 and early 1968 to flight-prove the spacecraft elements. The last eight flights over the past two and one-half years have been manned flights, with sufficient confidence demonstrated in the hardware through extensive ground testing and early flight success to schedule and complete a lunar orbiting mission on the first manned Apollo/Saturn V flight in December 1968 and the accomplishment of the first manned lunar landing on the fourth manned Apollo/Saturn V flight in July 1969. The near disastrous service module O_2 system failure on the Apollo 13 flight in April 1970 amply demonstrated how the various features of flight operations flexibility (system redundancy or backup modes), and crew training could effectively be used for satisfactory crew return from an extremely adverse situation.

This paper focuses on Apollo propulsion experiences associated with the Apollo/Saturn V systems developmental and operational flights. Emphasis is placed on the significance of extensive development and qualification ground test programs to establish adequate levels of reliability and quality assurance prior to flight; and the

L. G. Napolitano et al. (eds.), Astronautical Research 1971, 285–303. All Rights Reserved.
Copyright © 1973 by D. Reidel Publishing Company, Dordrecht-Holland.

use of extensive instrumentation in the early flights to measure systems performance in relation to predicted performance. This background of extensive ground testing and early flight test instrumentation is shown to have been instrumental in supporting the rapid and accurate diagnostic assessment of flight malfunctions and has provided an excellent data base for establishing fixes with a minimum of hardware and program schedule impact.

2. Space Vehicle Description

The Apollo/Saturn V space vehicle consists of a Saturn V launch vehicle (LV) and an Apollo spacecraft. The Saturn V launch vehicle consists of three propulsive stages

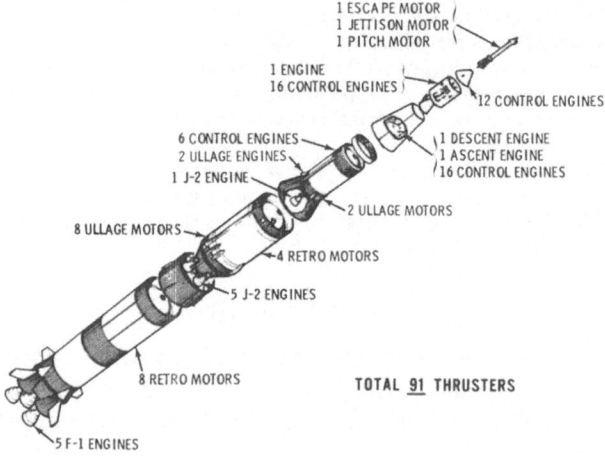

Fig. 1.   Saturn V Apollo propulsion systems.

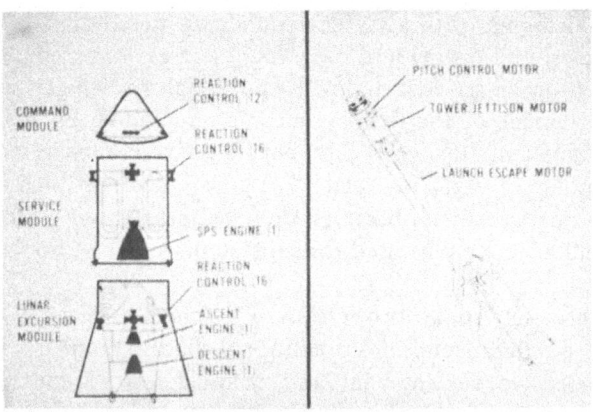

Fig. 2.   Apollo spacecraft propulsion systems.

Fig. 3.   Launch vehicle engines.

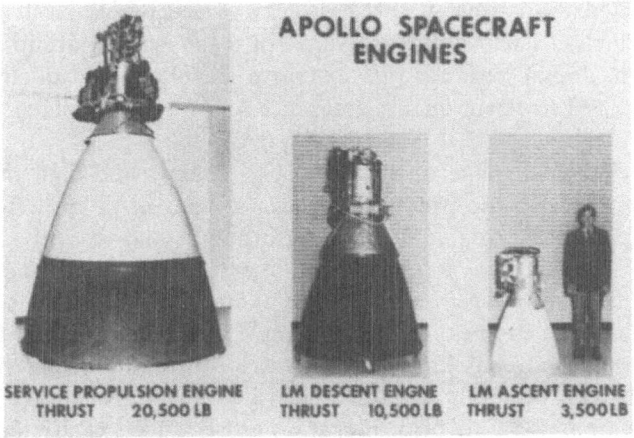

Fig. 4.   Spacecraft engines.

S-IC, S-II, S-IVB with two interstages, and an instrument unit (IU). The Apollo spacecraft consists of the propulsive launch escape system, command module (CM), service module (SM), and lunar module with a non-propulsive spacecraft lunar module adapter (SLA). The CM and SM as a unit are referred to as the command/service module (CSM).

The Apollo/Saturn V configuration consists of a total of 91 thrusters (Figure 1) which vary in size from the 1.5 thrust F-1 engine (five of which power the S-IC stage) to the twelve 90 lb thrust spacecraft control engines on the command module. Of the 41 launch vehicle thrusters, 19 are liquid propellant main propulsion or control thrusters, while 22 are solid propellant retro and ullage motors. Of the 50 spacecraft thrusters, 47 utilize hypergolic liquid propellants and the three housed in launch escape tower are solid propellant motors (Figure 2).

Figures 3 and 4 illustrate the relative size and thrust of the primary propulsion elements of the launch vehicle and spacecraft, respectively, for Apollo with a 6 ft man as a standard for comparison. Shown also in Figure 3 are the H-1 and RL-10 engines used, respectively, in the first and second stages of the Saturn IB. These provided valuable experience in engine design and development and required the same high standards of reliability as the Saturn V engines.

3. Design and Development Philosophy

When the complex Apollo/Saturn V was being designed for the manned lunar mission at the beginning of the decade, there were serious arguments whether a vehicle powered by so many rocket engines could ever be made reliable enough for manned flight. These arguments were accentuated when the all-up concept of a limited number of development flights with complete vehicles was adopted in the face of experience in which many flights had been required to achieve acceptable levels of reliability for unmanned vehicles. The Apollo program approach was to attain the required reliability through design redundancy, extensive development, qualification and acceptance testing, and stringent quality assurance for both development and production hardware.

For the Saturn V launch vehicle, extremely large engines and high energy propellants were almost a necessity to produce a vehicle of reasonable size. Regeneratively cooled turbopump engines capable of high performance and repetitive testing were therefore selected. These engines were subjected to extensive developmental, qualification and acceptance testing which resulted in hundreds of thousands of seconds of ground operation of each engine type before flight. During development the results of all tests were recorded in a reliability system. It was a contractual requirement that the reliability of the launch vehicle engines demonstrated in this system be at least 99% with 50% confidence at the time of qualification test completion. By the time of the lunar landing mission, design improvements, proven by ground testing, had increased the reliability of the F-1 and J-2 engines to 99% reliability with 95% confidence.

For the smaller spacecraft engines, required to operate after long exposures to the space environment, intrinsically high reliability was designed into the engine by simplification, i.e., by employing pressurized hypergolic propellants, ablative thrust chambers, and redundant components such as valves, regulators, actuators, etc.

Solid propellant motors with previously proven high reliability were used in both the launch vehicle and spacecraft for short duration, single operation applications such as ullage, separation and launch escape. Apollo flight experience with these motors has been uniformly good.

Ground and flight test experience for the Apollo/Saturn V propulsion systems at the time of the initial lunar landing is summarized in Tables I and II. In addition to conventional engine testing, these programs include thousands of tests in which most of the injectors were optimized to achieve dynamic combustion stability; i.e., combustion

TABLE I

Spacecraft engine test data

Test phase	Engine	Service module (SPS)	LM descent (DPS)	LM ascent (APS)	Command module RCS	Service module RCS	Lunar module RCS
Ground							
Tests		3 802	3 713	10 370	51 250	1 041 000	300 000
Firing time (s)		100 786	204 300	283 386	24 000	471 300	100 000
Flight							
Starts		46	15	8	6 400	385 000	100 000
Firing time (s)		3 946	2 388	1 833	1 325	15 100	10 000

oscillations would damp out within milliseconds after combustion process was disturbed by an explosive charge. This new engine development required was intended to avoid the unpredictable incidence of failures caused by combustion instability in early unmanned missile flights.

TABLE II

Booster engine test times (s)

Test phase	Engine	J-2	F-1
Ground (s)		459 377	270 600
Flight (s)		20 374	6 220

In spite of the differences in design philosophy, the test experience is not very different in magnitude for the launch vehicle and the spacecraft engines. The extensive testing of the engines formed the background for the successful flight experience of the liquid propellant engines in the Apollo/Saturn V space vehicle.

4. Flight Experience

The Apollo/Saturn V propulsion systems' performance throughout the flight program has been relatively free of major problems, with no mission as yet aborted or significantly degraded due to a propulsion system malfunction or failure. In every case it has been possible to assess the mechanics of the problem, the risks associated with recurrence of the problem, and the steps necessary to provide a satisfactory fix in accord with crew safety and mission success standards, hardware status, and mission schedules.

In general, the systems design analyses, design redundancy in critical areas, rigorous development, reliability and quality assurance, and qualification test programs have assured the satisfactory performance of the majority of the propulsion system elements. This comprehensive background of analysis and testing, along with extensive ground and flight test instrumentation data, has served as a significant data base to understand and resolve the relatively few propulsion system anomalies which occurred in flight.

A selection of some of the more critical Apollo spacecraft and Saturn V launch vehicle propulsion system flight anomalies is presented in the paragraphs which follow. These illustrate the diversity of problems, the manner in which diagnoses were carried out, the value of and dependence upon prior analyses and test results, the building of additional analyses and testing upon this data base, and the generally straightforward manner in which fixes were accomplished with minimum impact on mission operations and schedules.

A. LM DESCENT STAGE PROPELLANT SLOSH

For both the Apollo 11 and 12 lunar landings, the propellant low level sensor (LLS) light was activated earlier then predicted. The data were reviewed and sloshing was indicated as the problem (Figure 5). A test program was initiated at the Langley

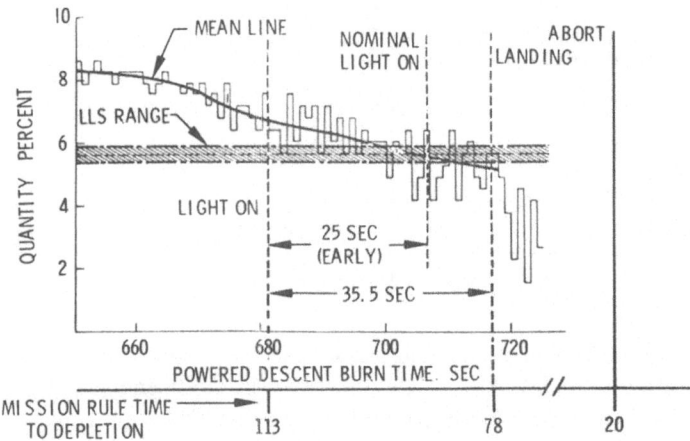

Fig. 5.   DPS quantity vs. burn time (fuel tank #2) LM-6 (Apollo 12).

Research Center (LRC) to determine the slosh modes and the subsequent action of the propellant inside the propellant quantity gauging system (PQGS). The low level sensor is inside the PQGS and was designed to be isolated from sloshing.

Tests at LRC indicated that slosh did affect the low level sensor gauge. The tests provided design parameters for better isolation technique and for baffles to reduce the sloshing. The LM contractor was able to design and qualify the light-weight collapsible baffle (Figure 6), which was installed through the propellant gauge hole for Apollo 14 and subsequent vehicles.

This corrective action proved successful. The Apollo 14 propellant slosh was minimal and the low level sensor was not activated prematurely. By preventing slosh,

Fig. 6.   Slosh baffle installation.

propellant formerly allocated for sloshing was available to provide additional hover time of approximately 20 s for Apollo 14 and subsequent missions.

B. J-2 ENGINE AUGMENTED SPARK IGNITER (ASI) LINE FAILURES

During the second unmanned Apollo/Saturn V launch (Apollo 6), real-time tele-metered data indicated that outboard engine #2 of the S-II stage had cutoff early (412.3 s into the flight) as shown on the timeline in Figure 7. Outboard engine #3 of

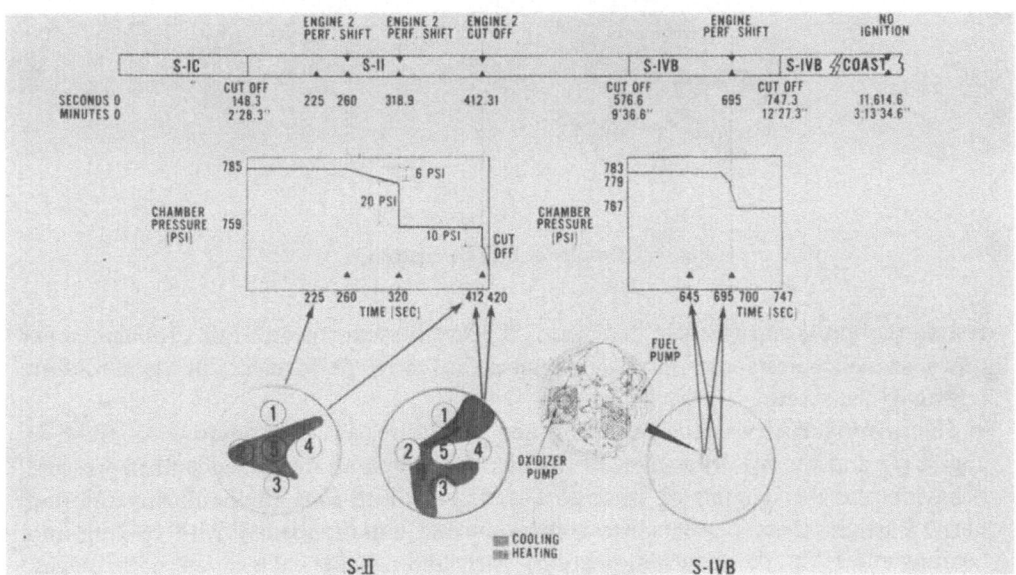

Fig. 7.   J-2 engine ASI anomaly.

the S-II stage also shutdown a few seconds later (not shown). The J-2 engine in the S-IVB also failed to restart as programmed 3 h and 13 min into the flight.

These engine shutdowns were caused by two separate problems. One of which was the ASI line failure on the S-II and S-IVB while the other was a miswired prevalve on the S-II. The ASI line failure on the S-II (engine #2) caused the J-2 engine thrust to drop sufficiently to trigger the thrust okay pressure switches which commanded engine shutdown. The engine shutdown sequence includes closing the stage prevalves for that particular engine. Examination of telemetered S-II stage prevalve position data indicated that engine #2 had been cross-wired to the prevalve of engine #3. When engine #2 attempted to shutdown its prevalve, it instead closed the prevalve on engine #3. Since all engines are commanded shutdown simultaneously during S-II stage static

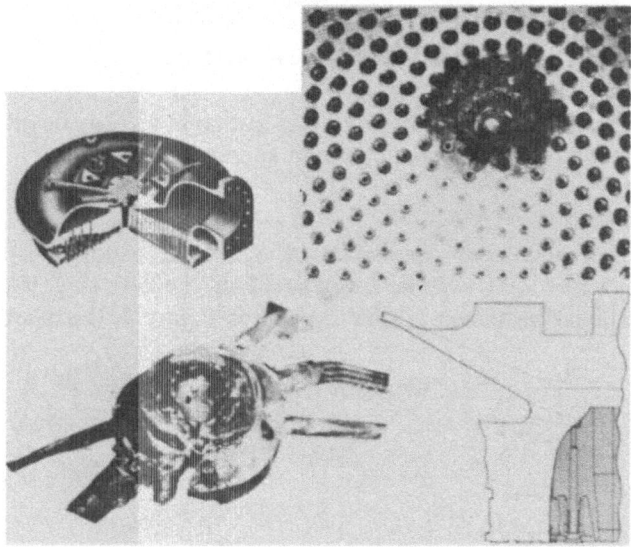

Fig. 8.   Results of S-II simulation.

testing the miswiring was not detected. To prevent recurrence of this problem, shutdown sequence tests of individual engines are now performed during check-out testing of the stage.

The investigation was then concentrated on the reason for shutdown of the S-II engine #2 and the S-IVB engine. Both engines showed anomalous chamber pressure behavior, the S-II engine #2 prior to cutoff and the S-IVB engine during the first burn. Further, these performance shifts occurred simultaneously with cooling and heating effects in the boat-tail sections surrounding the engines on both stages. Chamber pressure plots and boat-tail sketches are keyed to the S-II and S-IVB stage timelines (Figure 7).

The loss in engine chamber pressure, combined with initial cooling followed by heating, indicated gross leakage of hydrogen and oxygen near the engine injector. Examination of the location and construction of the lines and components indicated that the small lines supplying hydrogen and oxygen to the augmented spark igniter (ASI) were a potential source of the leakage. The augmented spark igniter is a small 3-in. diameter copper block located in the center of the J-2 engine injector. In it a hydrogen-oxygen torch provides a constant ignition source for the J-2 engine. ASI injector details are shown above (Figure 8).

Analyses using a mathematical model of the engine system indicated that the engine #2 performance shifts could be duplicated by an approximate 1 lb s^{-1} hydrogen leak from the ASI hydrogen supply line for 60 s, followed by a larger leak of 11 lb s^{-1} for approximately 92 s. This situation was simulated in an engine test at Marshall Space Flight Center, Huntsville, Alabama. Similar tests were conducted at the engine contractor's field laboratory. In this test the ASI, which normally runs at an O_2/H_2 mixture ratio of about 1:1 (1000 °F) ran for approximately 30 s at the hot O_2/H_2 ratio

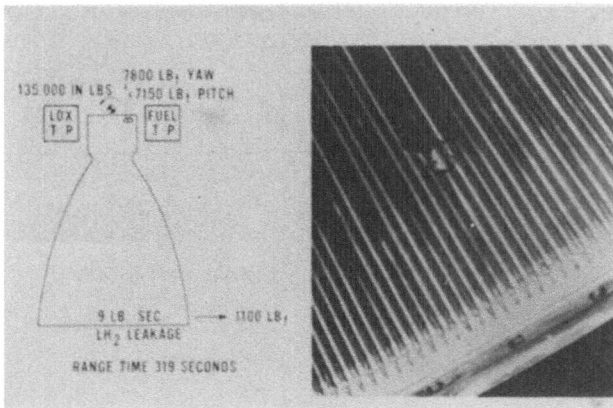

Fig. 9.   S-II/J-2 side load.

of 8:1 (5000 °F). The resulting action was like a cutting torch as shown by Figure 8 detail. The ASI was almost completely eroded away as was the LOX manifold. Had the engine not been shutdown because of the resulting fire, the LOX dome would undoubtedly have failed and caused the chamber pressure switch to trigger automatic shutdown. This sequence of events would explain the Apollo 6 S-II flight results.

At the time of the performance shift on the Apollo 6 S-II stage (319 s), actuator loads were also noted. A leak in the lower portion of the thrust chamber jacket of about 9 lb of hydrogen per second would explain this actuator load (Figure 9). The probable cause of such a leak was revealed in the previously discussed S-II ground test simulation when a small piece of the eroding main injector struck the thrust chamber (Figure 9) and produced a leak of almost the exact size and location required to duplicate the anomalous Apollo 6 S-11 engine flight side load.

The S-IVB performance shifts were also simulated by means of a math model and computer runs, and the results were used to plan a J-2 engine test at the engine contractor's facility. In this test, a simulated leak of approximately 0.5 lb s^{-1} from the ASI hydrogen supply started after 60 s of normal operation. After approximately 100 s, the leak gradually increased to approximately 2 lb s^{-1} at 129 s. The line was then fully ruptured and the run continued to 152 s when normally programmed first burn shutdown occurred. The resulting high O_2/H_2 mixture ratio of 20:1 provided cooler gases than in the S-11 case and caused lower, but major, erosion of the ASI injector and LOX dome. No shutdown occurred, but the engine would not have restarted because of the ruptured ASI feedline. Both conditions agreed with Apollo 6 S-IVB flight results. Having shown that failures of the hydrogen fuel line to the ASI duplicated both the S-II and S-IVB flight anomalies, tests were initiated to determine the cause of the line failures. Component and engine vibration tests and component flow tests at ambient conditions with normal flows of 1.1 lb s^{-1} did not cause failures.

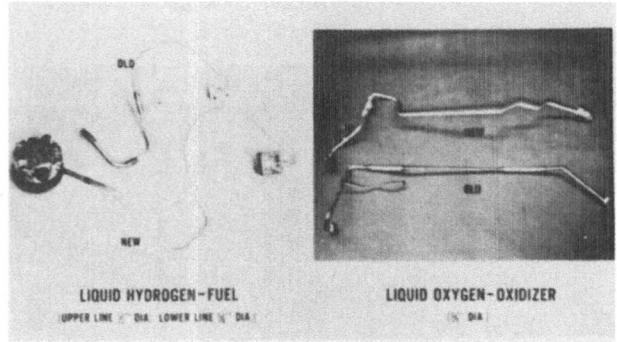

LIQUID HYDROGEN-FUEL   LIQUID OXYGEN-OXIDIZER
(UPPER LINE ¼ DIA. LOWER LINE ⅜ DIA)   (⅝ DIA)

Fig. 10. ASI lines duplication.

High flows of 1.8 lb s^{-1} did cause failures in 55 to 2400 s at ambient conditions. The latter high flow was shown to have been impossible in Apollo 6.

When component tests were run simulating Apollo 6 flows and the altitude (vacuum) environment, a flow induced vibration of about 20 800 Hz occurred and the bellows failed in less than 100 s. Observations and motion pictures indicated that liquid air condensed on the bellows surface under ambient conditions and damped the bellows vibration. Under vacuum conditions, however, no liquid air could condense and the undamped vibrations could produce bellows failures. This explained why the ASI line had survived thousands of component, engine and stage tests on the ground and 14 J-2 engine flights (including four other engines in Apollo 6), but could fail in flight.

The hydrogen feedline to the ASI was then redesigned. By improving tolerance in the unit, two bellows were eliminated without introducing assembly problems. The new line consisted of $\frac{1}{2}$ in. and $\frac{5}{8}$ in. diameter seamless stainless steel with welded joints only at the joining fittings. At the same time, the oxygen feedline to the ASI was also redesigned to eliminate two welded joints and thus provide increased structural integrity. The original and new lines are shown in Figure 10. Both lines were subjected to rigorous component qualification testing consisting of vibration and flow tests in vacuum environment. Extensive qualification tests in engine system firings were also so conducted (9 engines, 11 877 s). Eighteen engine static firings were also conducted on one S-IVB and three S-II stages with the redesigned lines installed on the J-2 engines. The redesigned lines were retrofitted to engines in the Apollo 7 and 8 and in all subsequent launch vehicle S-II and S-IV stages. As a result of discovering this generally unrecognized and obscure phenomenon of bellows vibration under internal gas flows, all bellows lines on the launch vehicle stages and modules were reviewed.

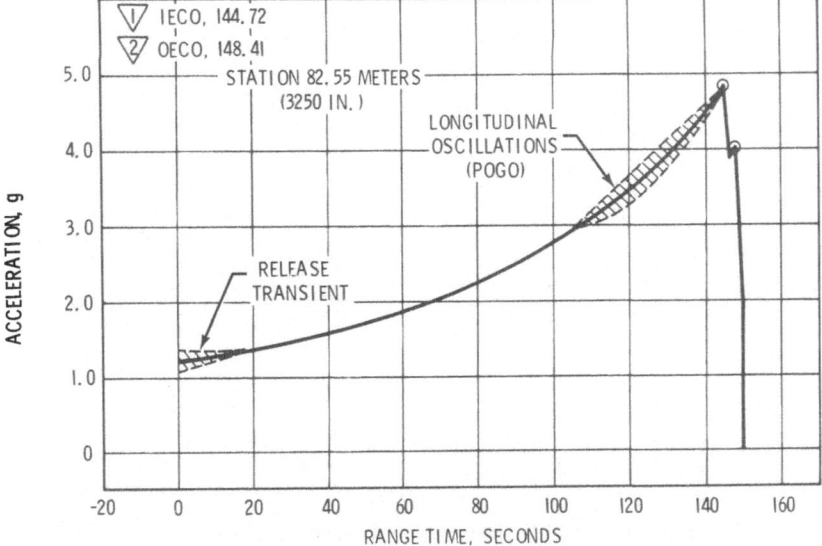

Fig. 11. Apollo 6 longitudinal acceleration – time history.

Analysis indicated that approximately 90% of these lines would not be subject to the phenomenon. In the 10% of the cases where analysis indicated that a potential resonant frequency might occur within the flow range, tests of the lines were conducted. In no case were resonant vibrations encountered. These analyses and tests qualified all Apollo vehicle and engine flexible lines for manned flight.

C. S-IC STAGE POGO

Throughout the design phase of the Saturn V launch vehicle development, extensive analyses were conducted by the Marshall Space Flight Center (MSFC), Saturn V stage and engine contractors, and other consultants to assess the POGO stability characteristics of the vehicle. POGO is the regenerative closed-loop coupling of the structure and propellant feed system. Since POGO only manifests itself fully within an unrestrained vehicle in free flight, the phenomenon could not be completely simulated in the ground environment. The design analyses were therefore based on mathematical modeling techniques developed in part from earlier liquid propellant rocket experience modified to suit Saturn V propulsion/stage peculiarities. These analyses indicated that that Saturn V launch vehicle would have a small but acceptable margin of stability.

In the first unmanned Apollo/Saturn development flight (Apollo 4), the vehicle performed as predicted with no evidence of POGO instability. However, during the second unmanned development flight (Apollo 6), moderately high levels of 5 Hz longitudinal vibrations were experienced during the latter portion of the S-IC stage burn. POGO occurred when a longitudinal resonance of the structure coincided with a resonant frequency of the liquid oxygen (LOX) feedline. The resulting thrust perturbations were of sufficient magnitude to sustain and amplify the oscillations for a short interval near the end of the S-IC first stage burn (Figure 11).

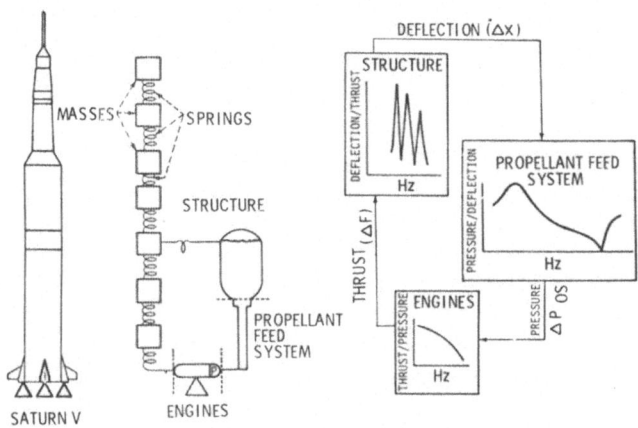

Fig. 12.   POGO phenomenon conceptual models.

Following the Apollo 6 flight, NASA initiated an intensive analysis and test effort to understand and resolve the S-IC POGO problem. The POGO Working Group, which had carried out the early Saturn V launch vehicle studies, was reactivated and enlarged with members and consultants from other organizations having backgrounds in POGO phenomenon assessment. This group conducted an exhaustive study of the S-IC stage, which had exhibited POGO, and a re-examination of all other stages of the Saturn IB and Saturn V vehicles.

The mathematical modeling which was the basic tool for POGO analysis represented the vehicle as a coupled system comprising a mass-spring structure, a propellant feed system, and the engines (Figure 12).

The engine thrust normally exhibits small fluctuations or perturbations above average level. The thrust perturbations cause vibrations in the vehicle structure with acceleration peaks at frequencies corresponding to the structure resonant frequencies. The structural accelerations couple with the propellant system which exhibits fluid pressure peaks at frequencies corresponding to the resonance of the feedlines and tanks. The fluid pressure pulsations at the engine pump inlets modify the propellant mass flow to the engines causing further perturbations in the engine thrust at the frequencies of the pressure pulsations. Coalescence at the structure and propellant feedline resonant frequencies increase the pressure pulsations resulting from the thrust excitation. The system becomes unstable in character if the pressure perturbations produce thrust perturbations larger than those which initially excited the system.

For the POGO assessment, structural data were obtained analytically from a mass-spring model of the vehicle which had been verified by comparison with test data from the full-scale Saturn V dynamic test vehicle. Characteristics of the propellant feed system were derived analytically and were measured in flow and pulsation tests of flight configuration components utilizing water or actual propellants where feasible.

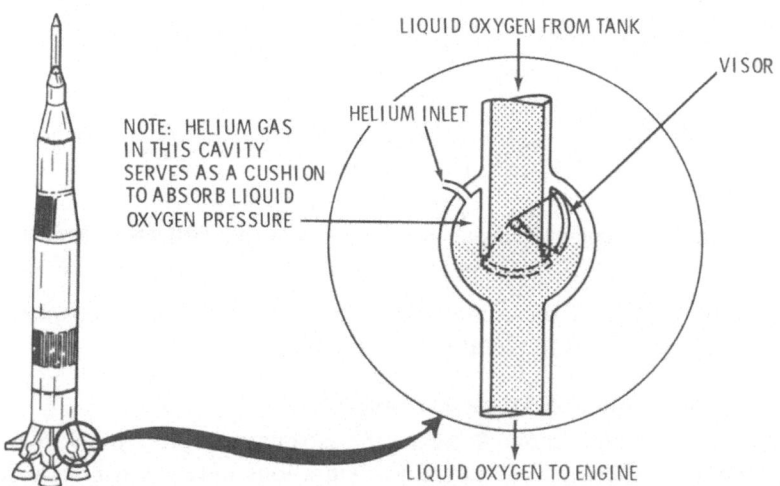

Fig. 13.   Helium gas prevalve accumulator.

The dynamic characteristics of the engine were analytically derived from component test data and verified by end-to-end pulse testing of the complete engine to establish engine transfer functions (thrust perturbations resulting from propellant pressure perturbations) at various frequencies. The analytical work indicated that the stability could be improved most easily by changing the resonant frequency of the LOX feed-line. The desired frequency change was effected by filling the existing cavity in the four outboard prevalves with helium gas (Figure 13). This simple pneumatic spring, or gas accumulator, lowered the resonant frequency of the LOX lines sufficiently to prevent regenerative coupling with the vehicle structure.

The stability of the resulting system was verified by analyses based on extensive mathematical modeling (Figure 14) and the compatibility of the fix with the propulsion system was verified by ground tests of the modified stage, engine and other components of the structure/propulsion system. The accumulator fix was installed on the four outboard F-1 engine LOX feedlines of Apollo 8, the first manned flight of Saturn V, and has proven to be consistently satisfactory in subsequent S-IC stage static firings and manned operational flights. Although POGO is not an engine problem per se, it

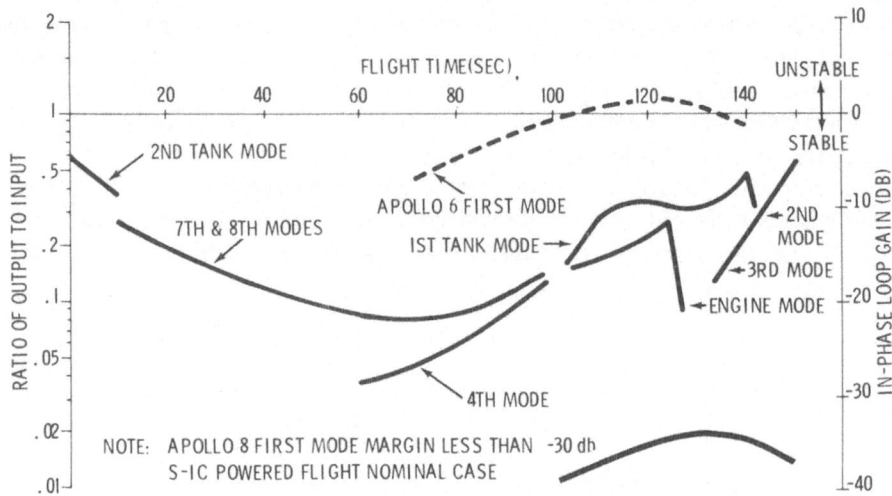

Fig. 14. Apollo 8 minimum stability margins with O/B engine accumulators.

is again illustrative of the complex problems associated with subsystems integration and interaction within the total vehicle system.

D. S-II STAGE LONGITUDINAL OSCILLATIONS (POGO)

On the Apollo 8 and Apollo 9 flights, the astronauts reported low frequency longitudinal vibrations during S-II burn. These vibrations were confirmed by flight measurements as illustrated by the center engine chamber pressure data (Figure 15). The

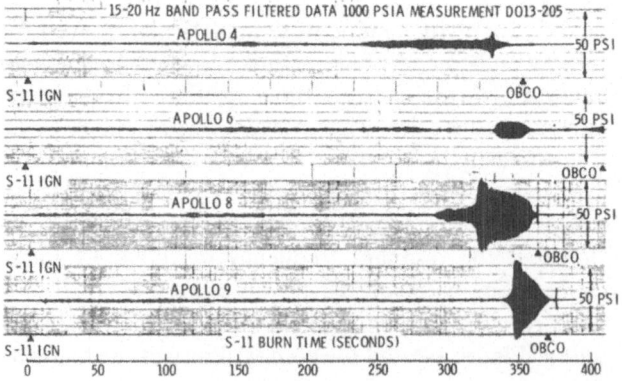

Fig. 15. S-II center engine chamber pressure flight data.

phenomenon was found to be a form of POGO in which the center engine, its beam support, and the oxygen feedline were the primary elements in the feedback loop (Figure 16). Coupling of these elements occurred in the 16–18 Hz range several times during the flight, with the most severe coupling and amplitude buildup occurring near the end of S-II burn.

To resolve this flight anomaly, the POGO Working Group, composed of NASA and industry members, conducted extensive ground tests and analyses to provide an effective interim fix for Apollo 10 and a permanent fix for later flights. Data from ground tests and flight were used as inputs to six mathematical models. Line pulsing tests were conducted by the Marshall Space Flight Center and the engine contractor to establish the dynamic characteristics of the engine and liquid oxygen tank, thrust structure and mass-simulated engines were used to establish structural transfer functions. S-II stage static firings at the Mississippi Test Facility also made important contributions in verifying flight data and qualifying design fixes.

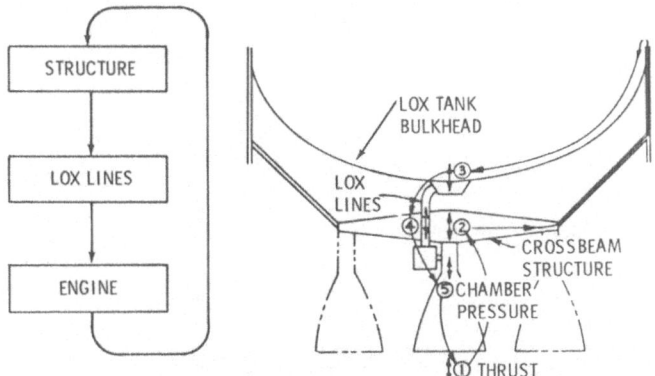

Fig. 16.   S-II center engine-structure feedline, closed loop oscillation.

Since flight experience and analysis indicated that the most severe coupling occurred at a critical tank level near the end of S-II burn, a change was incorporated on Apollo 10 which cut off the center engine early (approximately 100 s prior to S-II burnout). Flight data (Figure 17) confirmed that this change eliminated the high amplitude oscillations near the end of the burn.

Work also continued on a long-range fix supported by additional ground tests and mathematical model improvements. The helium gas accumulator system (Figure 18) resulted from this effort. Both analyses and stage static firings indicated that the accumulator decouples the center engine/line/structure POGO loop by lowering the liquid oxygen line frequency well below the 16 Hz structural frequency (Figure 19). The decision was made to install the accumulator on Apollo 14 after high amplitude oscillations (33 g's) occurred on Apollo 13 at the center engine gimbal point 163 s into the S-II burn (Figure 17). The center engine was automatically shutdown by the thrust OK switches which were activated when the chamber pressure oscillations reached approximately 235 psi (single amplitude) or 60% of operating pressure.

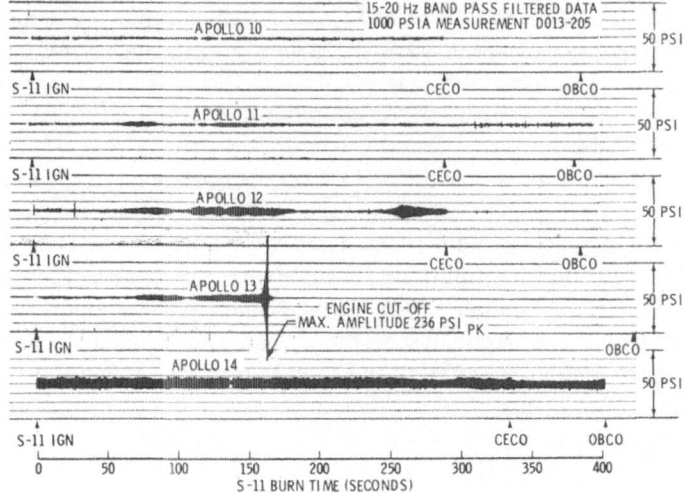

Fig. 17.   S-II center engine chamber pressure flight data (continued).

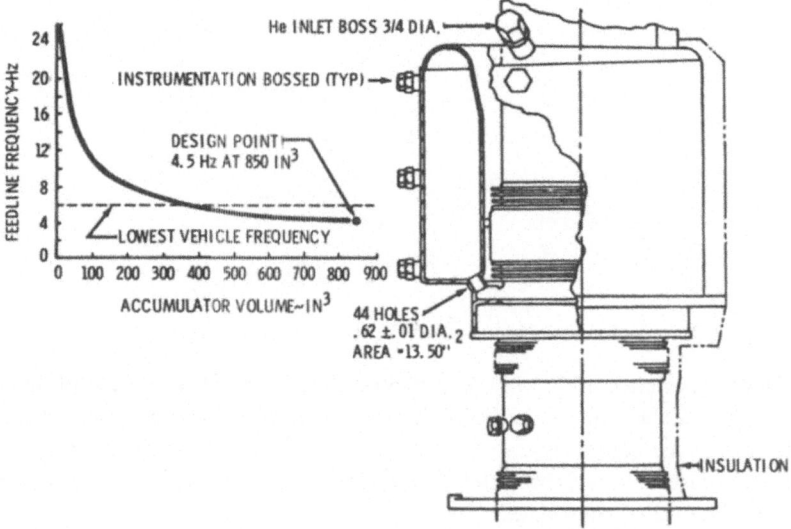

Fig. 18.   S-II accumulator design – center engine LOX feedline.

Because there was some concern that malfunction instability could result from an accumulator during fill, a backup cutoff device was designed, qualified and installed on Apollo 14. This consists of three force-balancing accelerometers or g switches which have the capability to vote and automatically shutdown the center engine should oscillation amplitudes exceed 13.6 g's.

The stability predictions for the Apollo 14 vehicle (Figure 20) indicated that the accumulator would suppress the 16–18 Hz for center engine POGO, but the stability

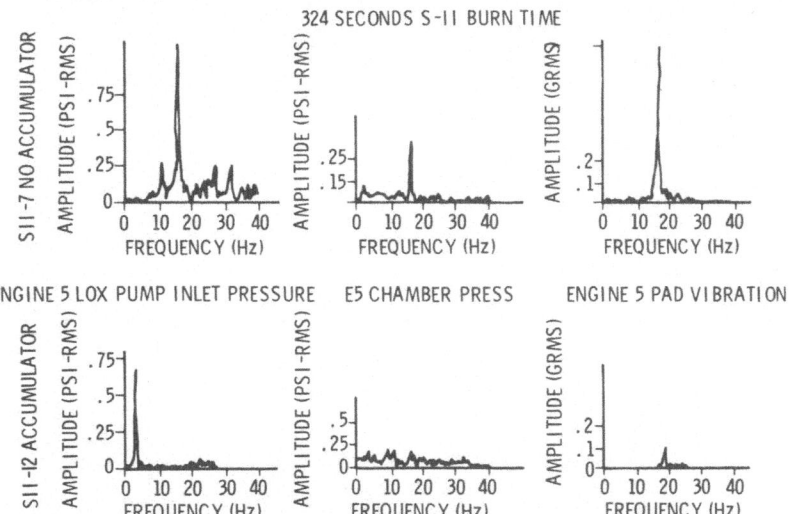

Fig. 19.   S-II center engine static firing frequency response data.

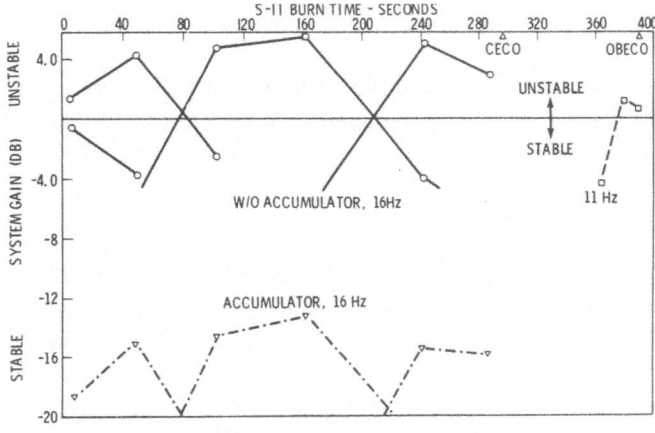

Fig. 20.   Apollo 14 S-II POGO system stability analysis.

of the outboard engines for 11 Hz oscillation would be marginal near the end of burn. These results were confirmed by Apollo 14 flight test data. The accumulator performed excellently in eliminating the 16 Hz oscillations during the critical accumulator fill time (Figure 21) or any other time during S-II burn. The backup cutoff device had no requirement to function. Both accumulator and cutoff device performed equally as well on Apollo 15.

Low level 11 Hz oscillations (0.16 g's) occurred as predicted near the end of the Apollo 14 S-II burn. These low level (0.10–0.22 g) oscillations have occurred on all Apollo flights. Studies to better understand the 11 Hz phenomena are currently being conducted to provide further confidence that oscillations, if any, will continue to be of an acceptable low amplitude.

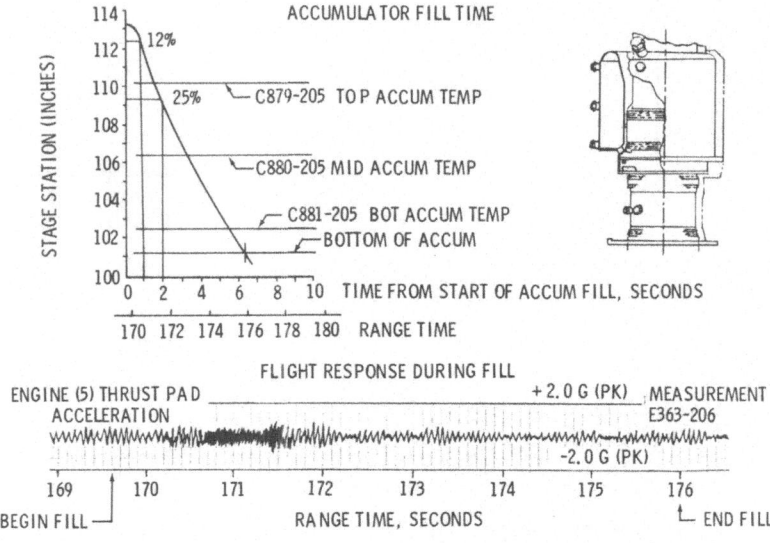

Fig. 21.   S-II Apollo 14 accumulator fill performance.

In addition to the significant propulsion system anomalies discussed above, there have been other minor anomalies encountered, analyzed and resolved. A representative grouping of these less significant problems is discussed below.

E. LM DPS FROZEN FUEL

Shortly after the Apollo 11 LM 'Eagle' successfully landed on the Moon, the pressure in the DPS fuel lines started to rise. Through analysis and tests it was determined that the simultaneous venting of the propellant and the very cold super critical helium (SHe) tanks through the propellant tanks with no fuel flow to the engines caused the fuel to freeze in the heat exchanger. Heat soakback from the descent engine caused the vapor pressure of the trapped fuel to rise. Ground tests were conducted to ensure that monopropellant decomposition of trapped fuel could not occur at soakback temperature. Procedures were also changed to vent the propellant tanks before venting the super critical helium tanks to minimize the fuel freezing problem. A method of releasing pressure from DPS fuel line through a by-pass to the fuel tank was also provided by Apollo 12 and subsequent spacecraft.

F. S-IVB THIRD BURN RESTART – APOLLO 9

It was recognized that translunar injection (TLI) could be prevented by failure of the S-IVB stage hydrogen circulation pump which would prevent cooldown and normal start of the J-2 engine. To demonstrate restart with a failed hydrogen circulation pump, an engine work around and an 80 min coast period, an in-flight demonstration was scheduled after Apollo 9 reached Earth orbit and S-IVB had completed a simulated TLI.

The long overboard hydrogen bleed calculated to provide engine cooldown caused excessively low thrust chamber temperature. Resulting (and predictable) engine roughness caused a pneumatic system failure, subsequent valve closures, and performance degradation. The data obtained, however, were valuable for developing subsequent mission rules which defined an improved backup J-2 starting procedure for a potentially failed hydrogen circulation pump to be used with S-IVB stage.

5. Concluding Remarks

From this 'propulsion eye view' of the Apollo flight program, it is evident that the Apollo spacecraft and Saturn V launch vehicle propulsion systems did their part in placing man on the Moon within the decade of the 1960's with relatively few flight problems. The very low incidence of basic engine design problems can be attributed to the exhaustive test, redesigning and retest cycles in the ground development program. Where design faults were detected in flight, generally the importance of some operational environmental condition had not been fully understood and the ground tests were not made in the critical environment, e.g., high vacuum effects on the J-2 ASI lines. The majority of flight problems resulted from errors in quality control, procedures on the ground, or through systems interactions which were not accurately predicted in the integration process, e.g., LM-DPS early cutoff, S-IC and S-II stage POGO.

The Apollo/Saturn V flight experience indicates that, with a well planned ground and flight test development program, highly reliable flight propulsion systems can be developed. On future NASA programs increased emphasis must be given to understanding the more obscure effects of the operational environment. Even more important will be an intensive pre-flight systems integration program to ensure that the interaction of the vehicle subsystems is completely understood. One of the principal lessons learned from the Apollo program which can have a significant bearing on future programs is that early flight tests must be achieved while simultaneously uncovering and correcting flight propulsion problems. Based on the Apollo program propulsion experience, these problems should be relatively minor in nature if the ground development and test program is sufficiently rigorous in component and integrated systems reliability and qualification testing.

D. STRUCTURE AND MATERIALS

THE POGO PHENOMENON: ITS CAUSES AND CURE

ABNER RASUMOFF and ROBERT A. WINJE

TRW Systems, Houston, Tex., U.S.A.

During the last ten years, a new phenomenon has appeared to plague the designers of space boosters. This phenomenon, nicknamed POGO, is characterized by a low-frequency, longitudinal oscillation of the vehicle structure occurring during discrete phases of the powered flight regime. In addition, oscillations at the same frequency can be observed in the engine thrust and in pressure measurements in the propellant feed system. The nickname POGO is not an acronym but is derived from the similarity of the structural oscillations to the movements of a child's jumping stick (Figure 1).

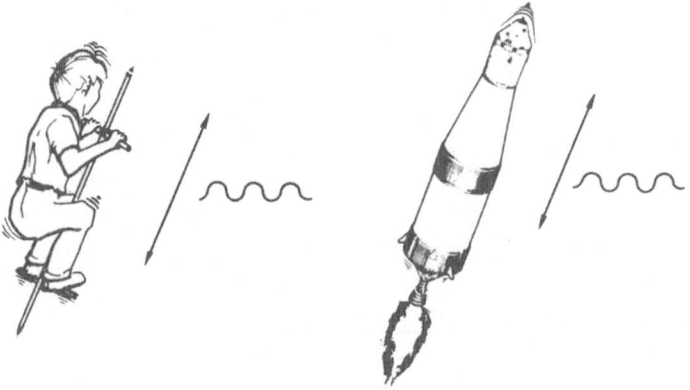

Fig. 1.   Illustration of POGO phenomenon.

Although POGO oscillations were observed on earlier vehicles, it was first recognized as a dynamic stability problem on the Gemini booster vehicle [1]. The fact that this vehicle carried a crew, and that the longitudinal oscillatory 'g' levels at the crew compartment exceeded those considered to be allowable for the crew, required that the phenomenon be understood and that the oscillations be eliminated or reduced to acceptable levels. Since that time, ten years ago, the POGO phenomenon has appeared in other United States launch vehicles, including the Apollo/Saturn V, and in European vehicles such as the French Diamant B [2, 3, and 4].

For purposes of illustrating the phenomenon, Figures 2 and 3 are presented. These figures show a typical time history of POGO oscillations in the structure, engine thrust, and propellant feedline pressures. The oscillations occur at identical frequencies and will occur in all of the 'dynamic' systems that are coupled in the POGO phenomenon.

While the importance of avoiding or suppressing POGO oscillations has been emphasized because of its deleterious effect on manned crews, it is also a significant factor for structural design (several actual and potential failures have been identified);

L. G. Napolitano et al. (eds.), Astronautical Research 1971, 307–322. All Rights Reserved.
Copyright © 1973 by D. Reidel Publishing Company, Dordrecht-Holland.

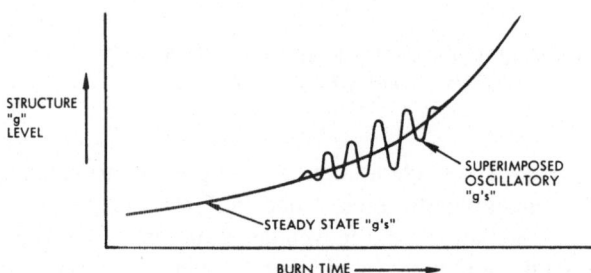

Fig. 2.   Typical POGO oscillations superimposed on steady state acceleration.

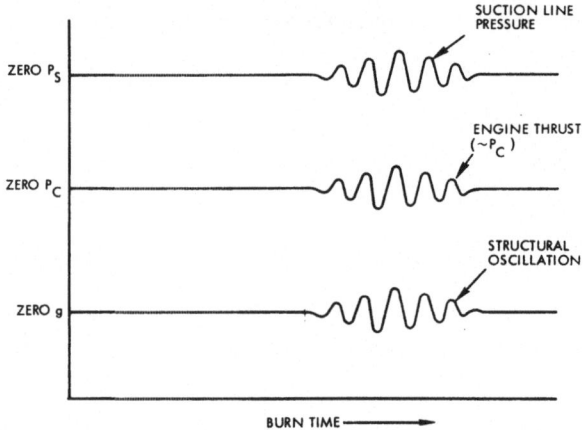

Fig. 3.   Relation between POGO oscillations in structure, engine thrust, and propellant feedlines.

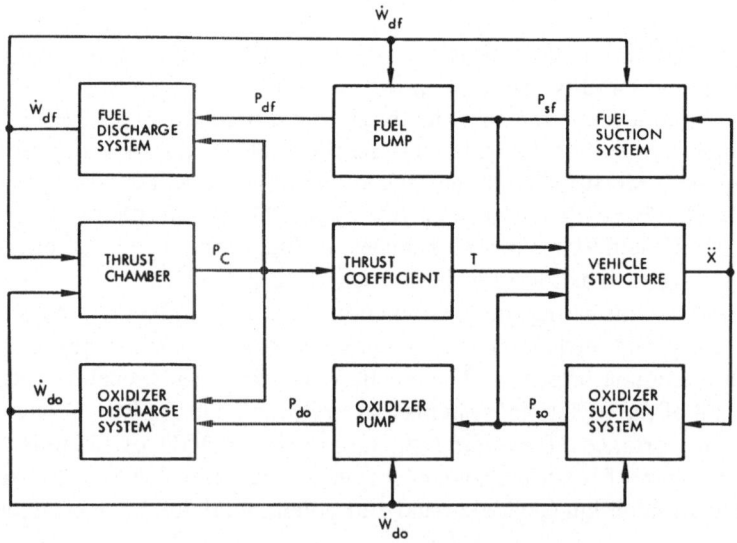

Fig. 4.   Diagram of feedback loops for a typical bi-propellant system including turbopumps.

high pressure oscillations have resulted in propulsion performance penalties and the premature shutdown of liquid rocket engines [3].

1. Technical Basis

The POGO phenomenon has been recognized and studied for many years and has been experienced on many vehicles (including upper stages and pressure-fed systems); however, there has been a reluctance, in the recent past, to accept it as a real possibility. Analytical stability analyses performed prior to flight often were made with inadequate test data and, furthermore, positive stability results were greeted with a minimum of skepticism. Similarly, low level indications of POGO oscillations that showed up in early flight tests were rationalized away on the basis of their being forced responses to vibrations in turbo-pumps, or rocket engines or other conventional dynamic responses. However, it is only with the recognition that the POGO phenomenon is a self-excited oscillation, mathematically similar to control system (autopilot) instability and to aircraft flutter, (such as that resulting from coupling between the bending and torsional motions of aircraft wings), that an understanding of the phenomenon can be attained. To be sure, an energy source is required, which for POGO is the engine thrust, but this does not mean that the rocket engines 'cause' the POGO oscillations. Neither do the aircraft engines cause wing flutter, although they are the energy source providing the aircraft velocity [5, Section 9-2].

Diagrams of the engine-airframe coupling that represent the basis for POGO instability have appeared in many publications [1, 3 and 4]; however, for purposes of discussion, a typical diagram is shown here. Figure 4 shows the basis feedback loops involved.

The symbols in Figure 4 are defined as follows: P_c = engine chamber pressure; T = engine thrust; P_{sf} = pump inlet or suction pressure, fuel; P_{so} = pump inlet or suction pressure, oxidizer; P_{df} = pump discharge pressure, fuel; P_{do} = pump discharge pressure, oxidizer; W_{df} = weight flow rate out of pump, fuel; W_{do} = weight flow rate out of pump, oxidizer; and $\ddot{X}$ = acceleration of vehicle. (It should be noted that the above symbols represent perturbations, i.e., deviations from the quasi-steady state.)

The determination of the stability of this system requires solution of the coupled differential equations representing the various subsystems shown. However, determining the coefficients of these equations requires knowledge of the characteristics of these systems, much of which, unfortunately, cannot be derived analytically but rather requires test data, obtained either from full-scale or scale model hardware.

2. Structural System

For example, the 'Vehicle Structure' box in the diagram refers to data on modal frequencies and mode shapes and on structural damping in the modes. Particular accuracy is required at specific points in the structure of the thrusting stage, such as the engine thrust structure, the pump support structure, propellant feedline support points, and propellant tank bottoms. The modal displacements, however, of all

points of the vehicle are important inasmuch as the structural 'gain' is inversely proportional to the generalized mass. This means that motion of large masses in all
stages (including the propellants in tanks in stages above the thrusting stage) must be
accurately modeled.

The structural damping associated with each mode is a very significant factor in
the stability analyses. Accurate analytical estimates of damping are impossible to
make, and experience has shown that modal damping obtained from full scale modal
ground tests are generally unreliable because of nonlinear effects that occur at relatively low levels of structural oscillation. Based on experience, structural damping
should be limited to one percent of critical damping, and the use of lower values would,
in most cases, not be unnecessarily conservative.

3. Feedline System

The propellant feedlines, both fuel and oxidizer, also play a very significant role in the
POGO phenomenon. In Figure 4 the feedlines are divided into two parts; the part
above the turbopumps (if a pump driven system) is referred to as the suction system,
while the part below the turbopumps is called the discharge system. Experience has
shown that most POGO instabilities have occurred when the suction system natural
frequency is close to the frequency of a high-gain structural mode. Therefore, it is very
important to accurately determine the resonant frequencies of the propellant lines
above the turbopumps. The resonant frequencies are a function of the ordinary
geometric properties of the feedline system, the bulk modulus of the particular propellant, the compliance of the feedline material, and the compliance at the pump entrance,
which is a result of the pump cavitation, and the compliance of an accumulator, if
one is present. These data may be used in the simplified transcendental equation
below to calculate the fundamental feedline frequency and the frequencies of what are
commonly called the organ-pipe modes.

$$\frac{Ag}{C_B\sqrt{B/\mu}} = \omega \tan\left(\frac{\omega l}{\sqrt{B/\mu}}\right) \qquad [11]$$

where A = area of propellant feedline cross-section; B = effective bulk modulus of the
fluid, reduced by pipe flexibility; μ = mass density of fluid; C_B = compliance at pump
inlet; pump cavitation plus accumulator, if present; ω = feedline characteristic
frequencies, radians per second; g = gravitational constant; and l = line length.

While most of the parameters in the equation may be readily calculated, unfortunately the cavitation compliance at the pump entrance, C_B, cannot.

Some attempts have been made to analytically determine pump compliance by
solving the compressible flow equations in the pump, but satisfactory results have not
yet been obtained [6]. In addition, comparison to other pumps is not often informa-
tive. Figure 5 is a plot of non-dimensionalized cavitation compliance versus cavitation
index for several turbo-pumps, derived from test data [6]. As may be seen, other than
for the expected trend lines showing compliance to increase as the cavitation index

decreases, the data could not be used to predict pump cavitation for new pump designs.

Nevertheless, POGO stability estimates should be made early in the design of a new space vehicle, and if the pump compliance cannot be estimated by comparison to turbo-pumps of a similar design, the conservative assumption should be made of

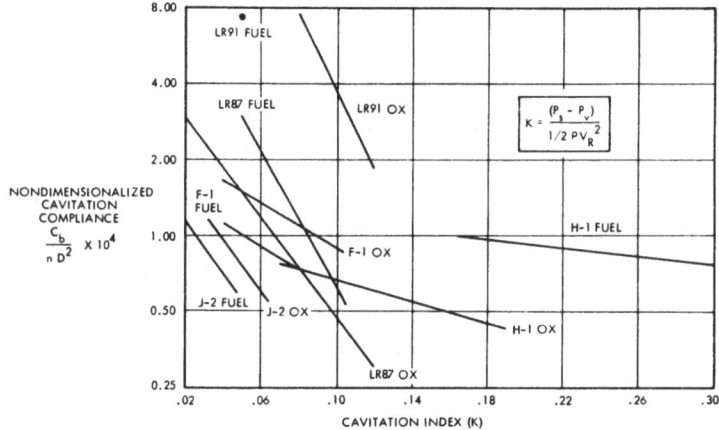

Fig. 5. Typical curves of non-dimensionalized pump cavitation compliance vs. pump cavitation index.

assuming the lowest feedline frequency to be coincident with the first high-gain vehicle structural mode. Naturally, some judgement should be used to avoid unreasonable estimates of the compliance.

4. Propulsion System

A similar problem exists in the modeling of the next element in the POGO loop, that is, the propulsion system. The propulsion system consists, basically, of the turbopump (if a pump-fed system), the discharge line, and the rocket engine itself. One approach to determining the characteristics of the propulsion system is to experimentally determine the propulsion system transfer function from tests that produce pressure and flow perturbations upstream of the pump, with pressure measurements taken at the pump inlet and in the engine chamber. These measurements allow the ratio P_c/P_s to be formed and transfer functions to be developed that model this ratio and the phase between them over a range of P_s oscillation frequencies.

Here again, dependence upon test data for modeling information would require that preliminary POGO stability analyses be delayed until the propulsion system hardware had been built. However, analytical techniques are available that allow reasonable estimates to be made of the propulsion system transfer function [7]. The first item in the propulsion system, the turbopump, can be simply represented by an equation of the form:

$$\sqrt{P_d} = (m + 1)P_s + Q\dot{W}_d,$$

where $(m+1)$ = pump gain; and Q = pump resistance coefficient.

Typical steady state performance curves for turbopumps are shown in Figures 6 and 7. The major parameter, the pump gain, may be estimated for purposes of preliminary analysis from the pump specifications relative to the design operating point of the projected pump. Operating on the relatively flat portion of the head rise curve yields $m=0$; hence, a gain of 1.0. During normal operation, pump gains of 1.5 to 2.0 are not uncommon and may be assumed in preliminary analysis.

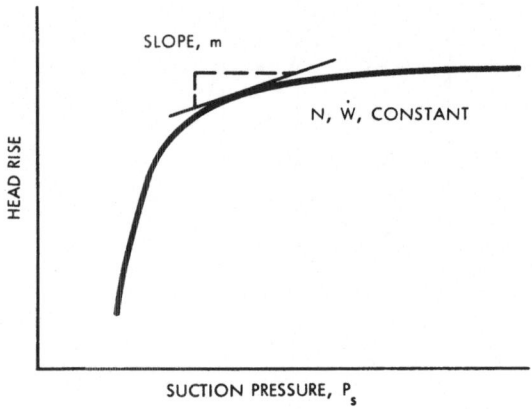

Fig. 6.   Pump performance curve, head rise vs. suction pressure.

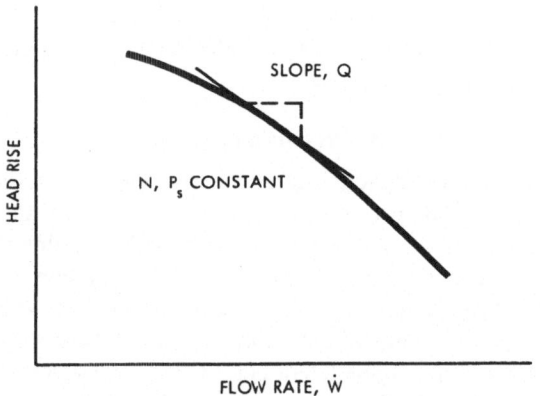

Fig. 7.   Pump performance curve, head rise vs. flow rate.

The remaining portion of the propulsion system, that is, the discharge lines and the engine itself, may be modeled by calculating the inertance of the propellant and the resistance in various parts of the system such as the pump discharge system, the engine cooling lines, the injector head, the injector orifices, and by using the combustion chamber equations. Analytical modeling of a propulsion system has been performed quite successfully for the Apollo/Saturn V first stage engines. Comparison with test-derived transfer functions showed relatively small differences in the phase between P_c and P_s. Furthermore, the availability of an analytical model provides the analyst with a better understanding of the propulsion dynamics and allows the study of parameter changes internal to the system that could not be handled by use of the transfer function only.

The preceding sections have discussed the significance of the various elements in the POGO stability loop and have emphasized assumptions and analyses that can be performed prior to the availability of any hardware components for test purposes. The next section discusses the ground testing required to update the analytical model and emphasizes the importance of prior planning for these tests.

5. Ground Testing

Ground testing is usually a very costly procedure. Therefore, it should not be performed unnecessarily; however, on the other hand, when it is performed, careful planning should serve to maximize the information obtained and minimize the number of tests required. Needless to say, proper test planning and performance will minimize the necessity to do it over again. This fact cannot be emphasized too strongly.

Testing to determine the structural parameters should be confined to that necessary to verify the analytically determined modal frequencies, modal deflections, and damping. High confidence can be had in the capability to analytically predict the frequencies and gross modal deflections of the major longitudinal modes. Furthermore, experience has shown that modal damping obtained from full-scale modal ground tests is generally unreliable because of nonlinear effects that occur at relatively low levels of structural oscillation. Therefore, full scale modal testing to support POGO in terms of *gross vehicle* modal frequencies, deflections, and damping is not considered necessary or worthwhile.

The area in which structural testing has the largest payoff is in obtaining data on the deflection characteristics of those locations that directly participate in the POGO stability equations; that is, the engine motions, pump motions, motion of propellant line attach points, and tank bottom motions. These data can be obtained from static testing to determine influence coefficients, and also from dynamic testing of the *local subsystems*, preferably full scale. In some cases, an insight into the localized damping may be obtained thereby; however, if the determination of damping is an objective, then the testing should include excitation at several amplitude levels in order to enable assessment of nonlinearities usually associated with structural damping.

An example of the usefulness of subsystem testing occurred during the Apollo/Saturn program when such testing was used on the aft end of the second stage to more accurately define the influence coefficients of the engine thrust structure and the lower tank bottom dome.

The most significant ground testing required in order to be able to predict the existence of POGO is dynamic testing of the propulsion system, especially of the propellant feedlines. The important parameters to be determined are those that are not particularly amenable to prediction by analysis. If a turbo-pump is in the system, an accurate determination of the compliance at the pump entrance cannot be obtained without ground testing. Other factors that are difficult to model analytically, such as the effect on compliance of various types of bellows designs, may also be determined more accurately by testing.

The type of testing recommended by the authors is pulsing upstream of the pump, wherein the feedline system is excited sinusoidally by an external pulser, and frequency response information is found directly. This type of testing should be performed on the basic propulsion/propellant feed system, in order to determine the lowest frequency and the higher, organ-pipe frequencies. It may also be performed on a system incorporating suppression devices as a check on the predicted lowest frequency and the higher, organ-pipe frequencies and also to define any extra feedline modes and frequencies introduced by the suppression device. A sketch of a typical test setup using a piston to produce the upstream pulsing is shown in Figure 8.

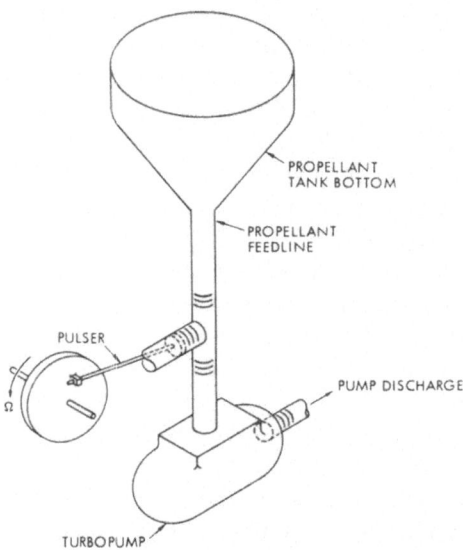

Fig. 8.   Sketch of test set-up showing external piston-type sinusoidal pulser.

The prime instrumentation required for these tests are pressure transducers. These should have good dynamic response (40–50 cps) and be accurate to within 1%. In addition to the usual pressure measurements at the pump inlet and discharge, and at the engine, flush-mounted transducers should be installed along the entire length of the propellant feedline in order to obtain the mode shape of the standing wave. Because of the 'noise' at the pump inlet, the standing wave generated in the line can be used to assist in the determination of pump inlet pressure for both ground and flight test. It is important that pressure instrumentation provisions be made an integral part of suction line design.

In the past, propellant feedline testing has been of two types: cold-flow and hot-firing. In the cold-flow tests, the engine resistance and other characteristics are simulated through the use of a cavitating venturi down the line from the turbo-pump (which is included in the test set-up). However, no propellants are burned. In the hot-firing tests, as the name implies, the actual rocket engine is included in the test set-up and the propellants are burned. Needless to say, the cold-flow tests are much cheaper and, in our opinion, can provide almost as much useful data as the hot-firing tests. However, no dynamic data are obtained on the engines themselves. Therefore, it is important to be able to model the engines sufficiently accurately to eliminate the requirement to determine their characteristics from dynamic tests. Inasmuch as steady-state performance testing of engines is always required, much of the data necessary to model the engines dynamically, such as impedance, can be obtained from the engine steady-state tests.

If, in the final analysis, dynamic hot-firing tests are considered essential, their extent can be minimized and cold-flow pulsing tests be employed for most of the data.

With reference to the turbo-pumps, while the pump gain can be determined from normal steady-state pump tests, and while elementary considerations and most experience lead to the conclusion that the static pump gain will be equal to the 'dynamic' pump gain, measurement of pump inlet and discharge pressure during the pulsing tests (either cold-flow or hot-firing) can provide the necessary data to confirm or deny that fact.

6. Analysis

The POGO analysis of a vehicle system to determine its stability or instability involves the simultaneous solution of the several differential equations describing the dynamic behavior of the various vehicle elements shown in Figure 4. These equations have been described in several publications including [1, 7, and 8]. While not wishing to repeat these equations, some representative equations are shown below for the purpose of illustrating the fundamental data required.

7. Structure Equations

A typical equation relating the structural modes to the engine thrust and the pressure force at the turbo-pump inlet is

$$(S^2 + 2\zeta_i\omega_i S + \omega_i^2)q_i = \frac{1}{M_i}\left[\sum_j T_j\varphi_{ij}^e - \sum_k P_k A_k \varphi_{ik}^p\right]$$

$$X_j^e = \sum_i \varphi_{ij}^e q_i$$

$$X_k^p = \sum_i \varphi_{ik}^p q_i$$

where S = Laplace operator; q_i = generalized coordinate i; φ_{ji}^e = modal displacement of location $_j^e$ (jth engine) of mode i; φ_{ik}^p = modal displacement of location $_k^p$ (kth pump) of mode i; M_i = generalized mass of mode i; ω_i = circular frequency of mode i; ζ_i = fraction of critical damping of mode i; P_k = inlet pressure at pump k; A_k = pump inlet area at pump k; T_j = thrust of engine j; X_j^e = displacement of engine j relative to system c.g.; and X_k^p = displacement of pump k relative to system c.g.

The forces acting on the structure are the thrust forces, summation of $T_j\varphi_j$, and the force due to suction pressure times suction area acting at the pump location, summation of $P_k A_k \varphi_k$. The latter term is included because the fluid column is not included in the structural modes. A corresponding term acting on the tank bottom is not included because of its negligible effect.

8. Feedline Equations

Of prime significance in the POGO phenomenon are the characteristic frequencies of the usually long propellant feedlines. These frequencies are the result of two mechanisms: (1) the compressibility of the fluid and ductility of the line itself and (2) a lumped compliance at the pump inlet due to local inlet cavitation of the pump. Solution of the one-dimension lossless wave equation for a uniform line with a lumped compliance at one end and free at the other results in the following transcendental expression which gives the characteristic frequencies of the line.

$$\frac{Ag}{C_B \sqrt{B/\mu}} = \omega \tan \left(\frac{\omega l}{\sqrt{B/\mu}} \right)$$

where A=line area; g=acceleration of gravity; C_B=lumped compliance at line termination; l=effective line length; μ=propellant mass density; and B=effective modulus of propellant and line.

The values of ω which satisfy the above expression are the characteristic frequencies. Solutions to the above for the limiting cases $C_B \to 0$, $C_B \to \infty$, corresponding to a fixed-free and free-free uniform column are

$$C_B = 0 \qquad \omega_n = (2n - 1) \frac{\pi}{2} \frac{a}{l} \qquad n = 1, 2, 3, \ldots$$

$$C_B = \infty \qquad \omega_n = (n - 1)\pi \frac{a}{l} \qquad n = 1, 2, 3, \ldots$$

where $a = \sqrt{B/\mu}$, the acoustic velocity.

For a linear stability analysis, the transcendental equation relating pressure and flow must be replaced by an equivalent representation. An appropriate number of the open-open or open-closed organ-pipe modes, or a combination of both, can be used, depending on the value of the lumped compliance and on the convergence rate. On the Titan II and Apollo/Saturn Stage I, it was adequate to use only one open-open mode with the lumped compliance to represent the first two characteristic frequencies.

The pressure/flow equations are

$$P_t - P_s = LS\dot{W}_s$$

$$(S^2 + 2\xi_l \omega_l S + \omega_l^2)q_l = \frac{1}{M_l} (P_t \eta_t - P_s \eta_s)$$

where P_s=pressure at termination of line (pump suction pressure); P_t=tank bottom pressure; $\dot{W}_s$=weight flow rate in suction line; L=inertance of line l/Ag; q_l=generalized modal coordinate of the open-open organ pipe mode; M_l=generalized mass of the above mode; η_t=modal displacement of the tank end of the line mode; η_s=modal displacement of the pump suction end of the line mode; ξ_l=assumed dumping fraction of that mode; and ω_l=frequency of that mode.

Those two equations along with the following flow continuity equation giving flow through the pump (no storage capacity is assumed in the pump) relate the feedline variables. The continuity equation also includes the effects of relative pump motion.

$$\dot{W}_d = \dot{W}_s + \gamma A \eta_s \dot{q}_l + \gamma A \dot{X}_p - C_B S P_s,$$

where $\dot{W}_d$=weight flow rate through pump; $\gamma A \eta_s \dot{q}_l$=additional flow term into pump due to organ-pipe mode (or 2nd line resonance); $\gamma A \dot{X}_p$=additional flow term into pump due to pump relative motion; and $C_B S P_s$=additional flow term into pump due to inlet cavitation 'bubble' or vapor volume at pump inlet. A, γ, and C_B are the line area, propellant weight density, and bubble compliance, respectively.

The inclusion of a branch suppression device requires additional equations relating pressures and flows in the device to the above described flows. The tank bottom pressure P_t is generated by the complex motion of the propellant/tankage in terms of the structural modes.

9. Pump Equations

The equations representing the pumps are of two types: one representing the pressure rise across the pump and the other relating the continuity of flow through the pump. The pressure rise is given by

$$P_d = (m + 1)P_s + Q\dot{W}_d + K_p N,$$

where P_d=pump discharge pressure; P_s=pump inlet or suction pressure; $\dot{W}_d$=weight flow rate out of pump; N=pump speed; $(m+1)$=effective 'gain' of pump; Q=pump resistance coefficient; and K_p=pump speed coefficient.

The continuity equation has been given previously.

10. Engine Equations

The pertinent equations for the engine, which is defined as that part from the pump discharge pressure and flow, through engine thrust, are given below. They contain only resistive and inertial terms. The relatively high pressure short line lengths have obviated the need for including compliant effects on the discharge side of the pumps.

$$P_d - P_c = (L_d S + R_d)\dot{W}_d,$$

where P_c=chamber pressure; L_d=total discharge inertance, including fluid in the cooling jackets; and R_d=total discharge linearized resistance coefficient, including cooling jackets, and injector.

The thrust chamber performance is given by

$$(\tau S + 1)P_c = \frac{C_o^*}{A_t g} \dot{W}_{do} + \frac{C_f^*}{A_t g} \dot{W}_{df}$$

$$T = C_f A_t P_c$$

where $\dot{W}_{do}$ = oxidizer weight flow rate into chamber; $\dot{W}_{df}$ = fuel weight flow rate into chamber; A_t = nozzle throat area; g = acceleration of gravity; C_o^* = 'characteristic velocity' associated with oxidizer flow variations; C_f^* = 'characteristic velocity' associated with fuel flow variations; τ = chamber equivalent time constant to account for transport time and combustion delay; T = thrust per engine; and C_f = thrust coefficient.

The preceding collection of structural and fuel and oxidizer propellant equations represents the dynamical system in terms of the variables of pressure, P; flow rates, $\dot{W}_i$; and structural motions, X. The equations are written in terms of the physical elements of the system, and the values of the coefficients must be determined by analysis or test as discussed previously.

11. Stability Determination

The stability of the coupled system is determined by examination of the set of equations previously discussed. The equations are 1st and 2nd order linear differential equations written in terms of the Laplace variable S. Several analytical methods are available and have been used to assess the stability of the set of equations. For relatively simple systems (i.e., those with only one structural mode), open-loop frequency response analyses can be performed yielding qualitative and quantitative assessments of system stability. Referring to Figure 4, if the vehicle structure consists of only one mode, the loop can be opened at the structural motion X (or $\dot{X}$). The transfer function X_{out}/X_{in} can then be examined as a function of frequency. If the ratio is *greater* than 1 at zero phase, the system is *unstable*. If the ratio is *less* than 1 at *zero phase*, but greater than 1 at some *non-zero phase*, it is said to be *phase stabilized*. Ratios of less than 1 at any phase indicate a stable system.

The inclusion of multiple structural modes and engine groups results in multiple parallel loops where one loop opening does not 'open' the total loop, although a transfer function can be defined and results calculated. However, the magnitude and phase of the ratio will be dependent on where the loop is opened.

In addition to open-loop techniques, stability may also be assessed by examination of the roots of the closed-loop characteristic equation. Here, *qualitative* stability can be determined immediately; however, additional computations are required to determine the degree of stability. A 'damping gain margin', as suggested in [10] can be employed to assess the *quantitative* measure of stability. This involves reducing the structural damping assumed in each mode by the same percentage amount until neutral stability exists. The amount of change is then used as a quantitative expression of stability. For example, if 1% modal damping is assumed in all modes and the system is stable and a reduction to 0.5% is required to produce at least one neutrally stable root, the 'gain margin' is a factor of 2.

The gain margins using either open loop or closed loop are expressed in decibels (db) as follows:

Open Loop   *Closed Loop*

Gain (db) = 20 $\log_{10}$ (X_{out}/X_{in})   Gain (db) = 20 $\log_{10}$ (ζ/ζ_N)

where X_{out}/X_{in} is the open-loop transfer function as previously discussed. For the closed loop, ζ is the assumed modal damping and ζ_N is that required to produce neutral stability. Thus, a gain margin of 6 db is a factor of 2.

12. Stability Criteria

The determination of a suitable criterion to ensure POGO stability has not been firmly established. The reason is that experience in the aerospace community with the POGO phenomenon is still too limited to verify that any chosen criterion is valid. In addition, specific tolerances in test generated data are difficult to quantify. Nevertheless, criteria have been generated and used. These, in general, consist of requiring a 6-db gain margin and a 30-deg phase margin. This corresponds to a factor of 2 used in the sense of structural factors of safety. In the same vein, these margins should not be used to account for uncertainties in component basic data, such as structural damping, pump gain, etc. By illustration, in comparison to the philosophy behind factors of safety used in structural analysis – which by now is a well established discipline – the design loads applied to structural components are not 'nominal', but are the so-called '3σ' values. When vehicle loads are a result of winds aloft, critical trajectories are chosen, and critical wind profiles are used. In addition, the material properties used for structural analysis purposes are not nominal values, but are usually 'minimum-guaranteed' values. In this manner, the loads used for structural analysis are high probability loads, and the material properties used are high-probability, minimum-guaranteed. Nevertheless, in addition to the preceding conservative procedures, structural factors of safety are still used, primarily to provide a cushion – dictated by experience to be needed – for variations in structural capability from vehicle to vehicle that result from variations in the manufacturing process and from the in-service history, and for possible variations in structural load distribution that cannot be accounted for by analysis.

Similarly, the stability margins of safety for POGO should be used to account for *unknown* variations that occur from vehicle to vehicle and in recognition of the fact that deficiencies still exist in our ability to model the problem and consider all the factors that could affect the POGO stability of the vehicle. The basic data used in the stability analysis calculation, whether obtained analytically or experimentally, should be conservative, high confidence data – especially when it is realized that verification of the stability analysis can only occur during flight test.

13. POGO Prevention

The preceding sections have discussed the factors that produce the POGO phenomenon, have provided some information on how these factors may be determined analytically or from test, and reviewed the dynamic equations involved. This section discusses the question: what to do if the stability analysis shows a POGO instability to be present; that is, the cure.

If a POGO instability exists, it is probably the result of the near coincidence of a feedline propellant natural frequency with a high-gain structural mode.

If the structural gain is high, the mode should be examined to determine if it is one involving participation of the entire vehicle or if it primarily consists of a local deflection. In the former case, the only recourse would be to stiffen the entire vehicle by increasing the vehicle skin thickness. This would be generally unacceptable because of the weight penalty involved. Even so, the net effect would not be to change the structural gain of the mode but would only result in an increase in the modal frequency because the dominant mass is in the weight of the propellants. The actual effect would be to shift the time of coincidence of the structural mode and the feedline propellant natural frequency to an earlier time of flight when, perhaps, the structural gain would be lower. On the other hand, if the mode consists, primarily, of a local deflection, the weight penalty involved in stiffening the particular structure may be acceptable and would result in a decrease in the structural gain. A case in point is the POGO instability that occurred during the second stage burn of the Apollo/Saturn V. In this case, the high structural gain mode primarily involved large deflections of the center engine support structure. If this POGO instability had been recognized during the stage design phase, it might have been possible to 'design-out' the high-gain mode.

The near coincidence of a feedline propellant natural frequency with a structural mode occurs often because, while the feedline propellant natural frequency varies very little during flight, the structural natural frequencies have large variations with flight time. This is the result of propellant usage and the corresponding reduction in vehicle weight during flight, while the vehicle stiffness remains constant. In many cases, a vehicle structural mode frequency will cross (and become coincident with) the feedline propellant natural frequency sometime during the flight. If this occurs when the structural gain is high, then POGO instability is very likely. This effect is illustrated in Figure 9. (A definition of 'high' structural gain may be found in the Appendix of [10].)

In this case, the simplest and most effective cure is to employ a device to reduce the feedline propellant natural frequency below that of the structure and thereby decouple the systems. This is most easily accomplished by designing an accumulator to be placed upstream of the turbo-pump. Several accumulator designs that have been used successfully in launch vehicles are shown in Figure 10 (reproduced from [10]). The use of a helium bleed as a device to reduce the feedline propellant natural frequency is shown but not recommended. The helium bubbles, while effective in reducing the fundamental frequency, also lower the organ-pipe modes, which are usually high with respect to the important structural frequencies. This raises more problems than it cures.

For best results, the accumulator should be placed reasonably close to the turbo-pump entrance. The reason for this is that introducing an accumulator into the feedline adds another mode to the propellant line, which can most simply be characterized as that due to the mass of fluid between the accumulator and the turbo-

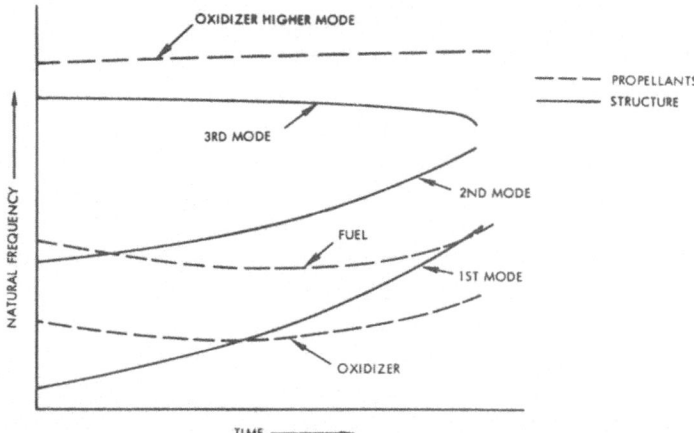

Fig. 9. Plot of typical natural frequencies of vehicle structure and propellant feedlines showing structural frequencies crossing propellant line frequencies.

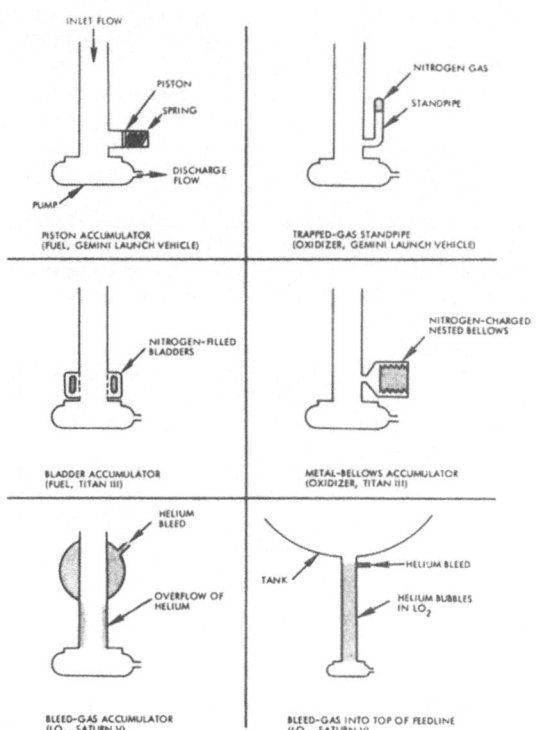

Fig. 10. Corrective devices for engine-coupled POGO.

pump, acting on the turbo-pump cavitation compliance. This mode, which has a relatively high frequency, may interact with the higher structural modes. However, care must be exercised that the accumulator is not so close as to adversely affect the flow field at the pump inlet.

In summary, the cure to POGO instability involves the following:

(1) Early analysis should be made using as many analytically derived parameters as possible, plus estimates of other parameters based on comparison to similar vehicles.

(2) Early ground testing should be performed for the major components that may contribute to POGO instability; emphasizing careful planning to minimize costs, and extensive and accurate instrumentation to maximize data return.

(3) If dynamic analysis indicates that POGO instability exists, and if structural re-design is not possible, then corrective devices, such as accumulators, should be designed and included as early as possible in the ground testing to insure the use of the proper data in POGO analyses.

References

[1] McKenna, K. J., Walker, J. H., and Winje, R. A., 'Engine-Airframe Coupling in Liquid Rocket Systems', AIAA Paper 64–81, January 1964; also *J. Spacecraft Rockets* **2** (1965), 254.
[2] Rose, R. G. and Harris, R., 'Dynamic Analysis of a Coupled Structural Pneumatic System Longitudinal Oscillation for Atlas Vehicles', AIAA Paper 64–483, July 1964.
[3] Sterett, James B. and Riley, G. F., 'Saturn V/Apollo Vehicle POGO Stability Problems and Solutions', AIAA Paper 70–1236, October 1970.
[4] Barrère, M. and Bouttes, J. (ONERA) and Lemonie, J. (LRBA), 'Influence on the POGO Effect of the Aerodynamic Phenomena', Space Transportation System Technology Symposium, NASA Lewis Research Center, NASA TM X-52876, Vol. II, July 1970.
[5] Bisplinghoff, R. L., Ashley, H., and Halfman, R. L., *Aeroelasticity*, Addison-Wesley Publishing Company, Inc.
[6] Vaage, R. D., 'Characteristics of Feed System Instabilities', Transportation System Propulsion Technology Conference, MSFC, April 1971.
[7] Goldman, R. L. and Reis, G. C., 'A Method for Determining the POGO Stability of Large Launch Vehicles', Report TR-69-7C, Research Institute for Advanced Studies, The Martin-Marietta Company, June 1969.
[8] Rubin, S., *J. Spacecraft Rockets* **3** (1966), 1188.
[9] Worlund, A. L., Hill, R. D., and Murphy, G. L., 'Saturn V Longitudinal Oscillation (POGO) Solution', AIAA Paper 69–548, AIAA 5th Propulsion Joint Specialist Conference, June 1969.
[10] 'Prevention of Coupled Structure-Propulsion Instability (POGO)', NASA Space Vehicle Design Criteria, NASA SP 8055, October 1970.
[11] Den Hartog, J. P., *Mechanical Vibrations*, 2nd ed., McGraw-Hill Book Company, 1940, p. 164.

GASEOUS HYDROGEN EMBRITTLEMENT OF
AEROSPACE MATERIALS

D. P. WILLIAMS and H. G. NELSON

National Aeronautics and Space Administration, Ames Research Center,
Moffett Field, Calif., U.S.A.

Abstract. With the increasing use of hydrogen in the aerospace industries, there has been an increasing concern with the problem of gaseous hydrogen embrittlement. The problems have led to a rather large amount of research on the phenomenon. This paper reviews some of the results of this research and discusses possible mechanisms for the phenomenon.

The advent of the space age brought with it an increasing demand for the use of high purity hydrogen and, consequently, an increasing incidence of gaseous hydrogen embrittlement. Advanced chemical engines use high purity hydrogen as a fuel; the need for relatively large amounts of space power requires the use of hydrogen–oxygen fuel cells; the nuclear rocket engine uses hydrogen as a propellant. For these and for other applications, the hydrogen is stored, transferred, reacted, and expelled at temperatures which may range from liquid hydrogen temperature ($-260\,°C$) to temperatures above $2000\,°C$. For any and all of this temperature range the hydrogen is contained by and in contact with various materials. Such a variety of material-hydrogen environment combination has resulted in a great deal of concern about, and a number of failures due to, gaseous hydrogen embrittlement.

Many different phenomena are included in the descriptive term 'hydrogen embrittlement' [1]. All of these phenomena result in a ductility reduction of the material and most are caused by some effect of super-saturating the bulk of the material with hydrogen. These phenomena have been studied since before 1940 and although there is still some controversy concerning the responsible mechanisms, they were thought to be sufficiently well understood that hydrogen-embrittlement could be predicted and prevented by proper selection of materials and/or processing techniques. However, the gaseous-hydrogen embrittlement problems which began to plague the U.S. Aerospace Industry 10 yr ago were generally unexpected. They occurred with hydrogen–material combinations and conditions which were not expected to produce any conventional form of hydrogen embrittlement. Since the gaseous hydrogen problems initially occurred at relatively high pressures (greater than $3 \times 10^7\,N\,m^{-2}$), the phenomenon was initially called 'high pressure hydrogen embrittlement', and was thought to be a new high pressure phenomenon.

Most of the early failures encountered by NASA and its contractors were related to ground support equipment used for fueling of liquid hydrogen fueled rocket engines. For this application, high pressure, high purity gaseous hydrogen was used to force liquid hydrogen from ground based storage bottles into the vehicle fuel tanks. In this gaseous hydrogen system which operated at pressure levels near $3 \times 10^7\,N\,m^{-2}$, various storage tanks, transfer lines, gauges, valves, etc. began to leak or fail catas-

L. G. Napolitano et al. (eds.), Astronautical Research 1971, 323–330. All Rights Reserved.
Copyright © 1973 by D. Reidel Publishing Company, Dordrecht-Holland.

trophically. The frequency of these failures and the variety of the material affected
led NASA to initiate an exhaustive literature search and to simultaneously initiate
some contract research in an effort to characterize the phenomenon. The literature
survey indicated that such embrittlement had in fact been observed previously in
high strength steel and nickel alloys in high pressure gaseous hydrogen environments [2, 3]. However, these studies had not explored or defined the mechanisms of
the embrittlement. The NASA sponsored work [4] gave the first real indication of
the magnitude of the problem. A partial listing of the materials found to be affected
is shown in Table I.

TABLE I

Materials tested in gaseous hydrogen at 7×10^7 N m^{-2}

Extremely embrittled	Severely embrittled	Slightly embrittled	Negligibly embrittled
18 Nickel maraging steel	Ti-6Al-4V (STA)	AISI 304L stainless	AISI 310 stainless
AISI 410 stainless	A-212	AISI 305 stainless	A-286 stainless
AISI 1042 (Quenched and tempered)	AISI 430F stainless	Be-Cu alloy 25	7075 T-73 Al alloy
17-7 pH stainless	Nickel 270	Titanium (commerci-	6061 T-6 Al alloy
H-11 tool steel	ASTM A-515	ally pure)	1100-0 Al
9Ni-40Co-0.20C	HY-100		OFHC copper
Rene 41	ASTM A-372		AISI 316 stainless
AISI 4140	Ti-6Al-4V (annealed)		
Inconel 718	AISI 1042 (normal-		
AISI 440C stainless	ized		
Electroformed nickel	ASTM A-302		
	4Y-80		
	AISI 1020		
	Ti–5Al–2.5Sn		
	ARMCO iron		
	ASTM A-517		

The table shown summarizes the results of these initial studies and gives some
idea of severity of the problem. The definitions of 'extreme', 'severe', 'slight', and
'negligible' are those of the contractor [4] and are somewhat subjective. However,
the classifications 'extreme' and 'severe' both represent material which suffered
substantial reductions in notched strength and ductility when tested in high-pressure
gaseous hydrogen as compared to these same properties when tested in high-pressure
gaseous helium. As can be seen, high strength and medium strength steels, nickel
base alloys, and titanium alloys are all seriously embrittled by gaseous hydrogen and
only pure copper, stable austenitic steels, and aluminium alloys appear to be definitely
immune from embrittlement. (Excluded from this table are alloys of those metal
systems for which chemical reactions at room temperature are known to cause
hydride formation – this, however, is also accompanied by embrittlement. This
includes alloys of tantalum, zirconium, niobium, etc.)

From these studies several characteristics of this embrittlement were noted. They are: (1) that the embrittlement results in a loss of mechanical strength and ductility; (2) maximum embrittlement for most of the materials investigated occurs near ambient temperature; (3) the gaseous hydrogen causes a change in the fracture mode from ductile tearing to intergranular and quasi-cleavage cracking; (4) embrittlement occurs only in very pure hydrogen; (5) embrittlement occurs at pressures less than 7×10^7 N m^{-2}; (6) embrittlement is most severe at very slow strain rates.

Based somewhat on these early observations and on our experience in similar research fields [5, 6], we postulated that these unexpected gaseous hydrogen embrittlement problems were not a unique function of the high pressures involved. We postulated that the phenomenon which was being called 'high-pressure hydrogen embrittlement' could also occur at low gaseous hydrogen pressures. We postulated that the fact that the problem had not often been observed previously was due to the higher

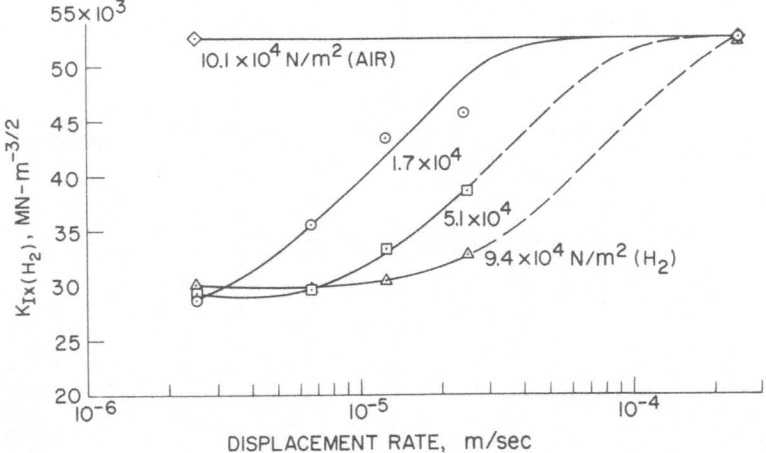

Fig. 1. The effect of hydrogen pressure and displacement rate on the embrittlement of 4130 steel.

purity of the hydrogen involved in the NASA programs rather than the higher pressures involved in these programs. We initiated research programs to investigate the postulate and some of the initial results of this program are shown in Figure 1 [7].

The data shown on this slide were obtained using fracture toughness specimens made from a high strength steel and the data reported are in terms of the stress intensity parameter (K). The values $K_{IX(H_2)}$ represent the stress intensities (i.e., proportional to the stress at the tip of a propagating crack) at which measurable crack growth is first observed in a hydrogen environment. From this figure, it can be seen that the stress-intensity for cracking in hydrogen is considerably lower than that for failure in air at pressures as low as about 1.7×10^4 N m^{-2} (one-sixth of normal atmospheric pressure). Also, the stress intensity for observable cracking in hydrogen was found to be a function of both testing speed and hydrogen pressure – suggesting a dynamically controlled cracking mechanism. Further tests on a titanium alloy indi-

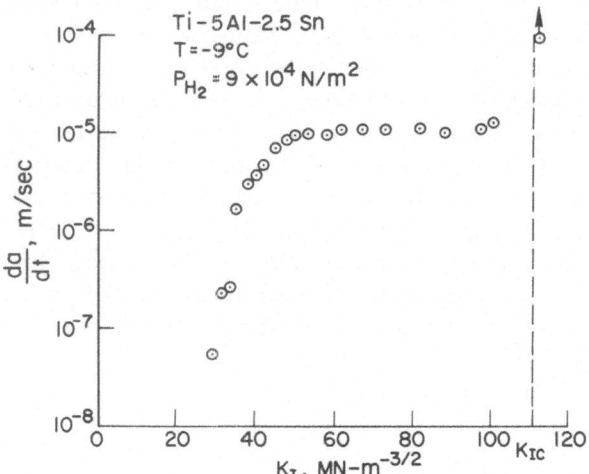

Fig. 2. The relationship between hydrogen-induced crack-growth rate and applied stress intensity for Ti-5Al-2.5Sn.

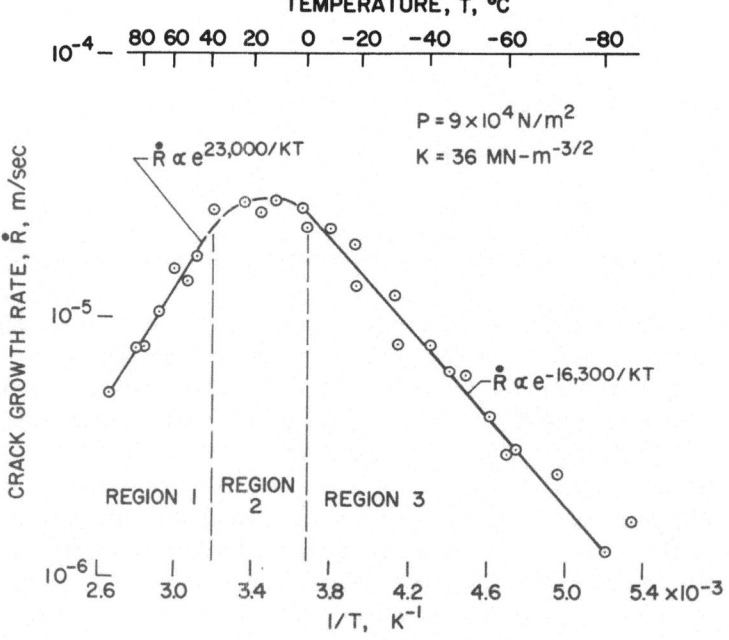

Fig. 3. The temperature dependence of the slow crack-growth rate of 4130 steel.

cated the significant reduction in useful strength that could be caused by the hydrogen environment [8]. Figure 2 shows the measured relationship between the hydrogen induced crack-growth rate and the stress-intensity. As can be seen from this figure the hydrogen environment at a pressure less than normal atmospheric pressure caused cracking at stress-intensities of the order of 25% of those which would cause cracking (and failure) in air. These observations certainly confirmed the hypothesis that gaseous hydrogen embrittlement could (and did) occur at low as well as high pressures.

In our work we confirmed the earlier findings that embrittlement could be inhibited by small amounts of impurity gases with as little as 100 parts per million oxygen inhibiting cracking of steel and even smaller amounts inhibiting the cracking of

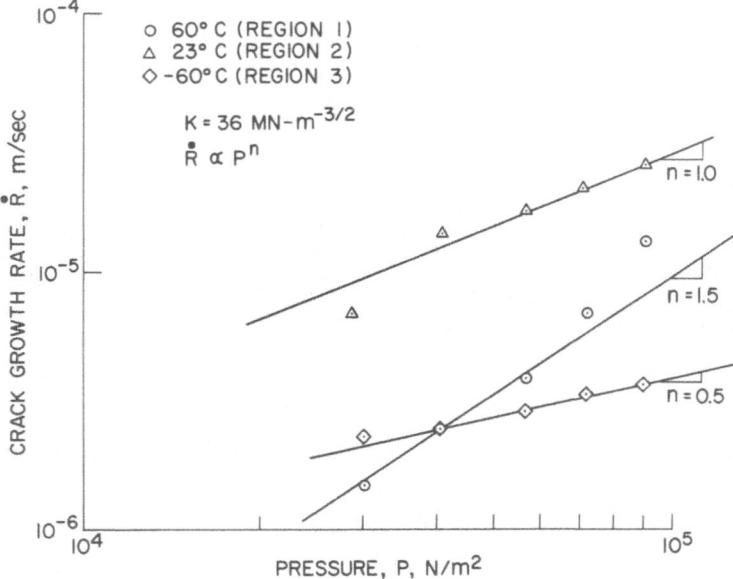

Fig. 4. The effect of hydrogen pressure on the slow crack-growth rate of 4130 steel.

titanium. The use of such inhibitors was not, however, a practical solution to the aerospace industries' problems.

In an attempt to define the mechanisms of this hydrogen induced cracking, we measured the temperature and pressure dependencies of the crack growth rate for a high-strength steel [9]. Figure 3 shows the effect of temperature on the hydrogen induced crack growth rate at a pressure of 9×10^4 N m^{-2} and over the temperature range from $+80$ to -80 °C. In the temperature range from 80 to 25 °C, the crack growth rate increased with decreasing temperature; in the range from 0 to -80 °C, the rate decreased with temperature. In both regions the rate varied with the exponential of reciprocal temperature with apparent heats of reaction of $+23\,000$ J mol^{-1} and $-16\,300$ J mol^{-1} for the high and low temperature regions, respectively. The behavior at temperatures between $+25$ and $0\,°C$ appeared typical of a transition region.

Figure 4 shows the effect of hydrogen pressure on the crack growth rate in all three of the temperature regions identified on the previous slide. As can be seen, the growth rate was proportional to hydrogen pressure to the 1.5 power and to hydrogen pressure to the 0.5 power at low pressure at the high and the low temperatures, respectively. The pressure dependence in the transition region was 1.0.

The observations of the temperature and pressure dependence of the crack growth rates gave considerable insight into the mechanism of the gaseous embrittlement

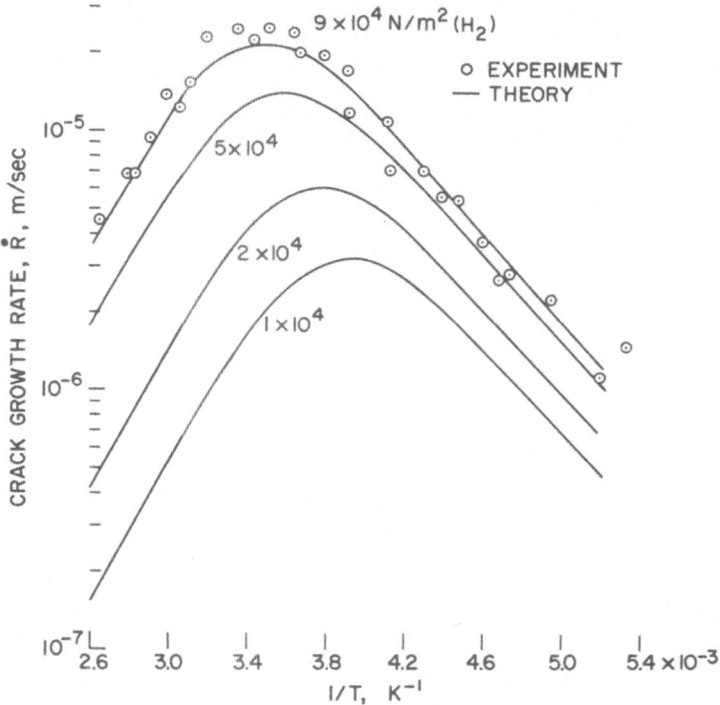

Fig. 5.   Calculated crack-growth rate in hydrogen-embrittled steel as predicted from absorption rate.

phenomenon. We were able to directly relate the behavior demonstrated in the previous two slides to an equation for activated adsorption of hydrogen on iron to give the quality of agreement indicated in Figure 5 [1]. On this figure the experimental data are shown superimposed on the theoretically predicted curves. Although the exact positioning of the theoretical curves required selection of constants by use of the experimental data, the close agreement between the slopes, and the transitional behavior indicates that the activated adsorption of hydrogen is likely the rate controlling mechanism in the hydrogen-induced cracking of steels. This same mechanism appears to control cracking in nickel base alloys and may control cracking in titanium, zirconium, tantalum, and niobium for certain conditions of temperature, hydrogen pressure, and alloy structure [9].

For iron and nickel base alloys, calculations can show that not only is the surface reaction of adsorption the rate controlling step but that diffusion to an appreciable depth beneath the surface occurs too slowly to be involved in the embrittlement process. For alloys of the other metal systems, this may or may not be the case, depending on the particular environmental conditions. This indicates that some surface or very near-surface reaction can cause embrittlement of alloys of most metal systems and that diffusion to any appreciable depth beneath the surface is *not* a requirement for hydrogen-induced cracking. The fact that hydrogen-induced cracking could be caused by surface (or near-surface) interactions has several implications. It means that since no driving force is required for large scale *ab*sorption and diffusion of hydrogen within the metal, any application involving hydrogen in contact with metals at any pressure and temperature must be approached with caution. Given sufficient time, hydrogen-induced crack growth can cause embrittlement of a great many alloys over a wide range of pressures, temperatures, and stress levels.

To present, several methods have been defined for reducing the threat by hydrogen. It appears that any method which prevents the hydrogen molecule from dissociating to its constituent atoms on the clean metal surface will prevent the necessary hydrogen–metal interaction and prevent cracking. The fact that hydrogen does not readily dissociate on aluminium or copper appears to, in fact, explain their immunity from cracking. Some methods used to reduce the hydrogen threat include: selection, when possible, of non-susceptible materials; contaminating the hydrogen gas with an impurity which will preferentially adsorb on the metal surface; the use of surface liners or coatings on which hydrogen will not adsorb and dissociate. However, all of these methods have limited applications and until such a time that the basic hydrogen–metal interaction mechanisms are defined, some design situations will continue to require the use of material–hydrogen combinations which are known to produce cracking. For this reason it has been necessary to develop a design criterion and experimental test technique which is suitable for predicting conservative design conditions [10].

It is beyond the scope of this paper to discuss the work done in developing, or the theory behind, the failure criterion which has been developed. It is sufficient to say that the criterion is based on well established kinetic relationships between the hydrogen-induced crack-growth rate and such design conditions as temperature, hydrogen pressure, gross stress, fracture toughness, etc. These parameters have been combined in a quite general analytical expression which relates these parameters to a conservative design stress-intensity which can be used to predict failure under various conditions. An example of the sort of information available by use of the criterion is shown in Figure 6. This figure shows how the useful design stress-intensity varies with necessary service life for three different temperatures. As can be seen, for infinitesimally short life-times the full fracture toughness of the material could be utilized at any of the temperatures. Conversely, the material could not be used at all in a hydrogen environment at any of the temperatures if required to last for an infinite time. The utility of the criterion is in predicting useful design conditions for situations between these two extremes.

We recognize the limitations of a semi-empirical design approach of the type used to develop this criterion. However, until we learn more about this extremely severe

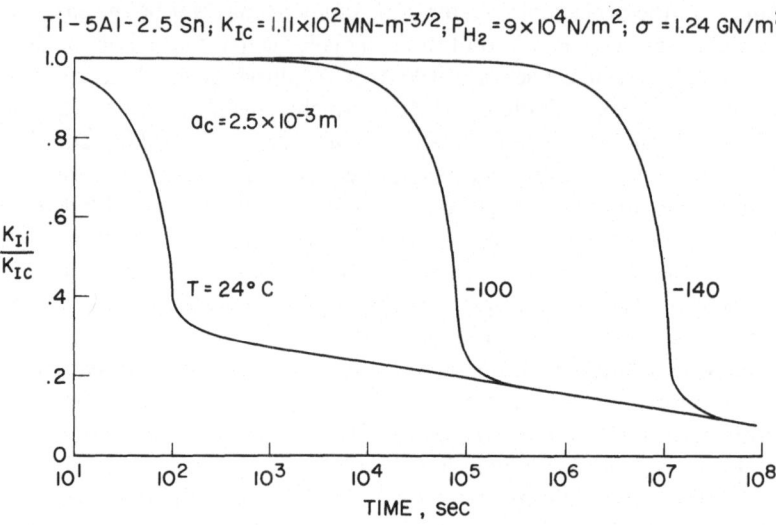

Fig. 6.   Analytically predicted relationship between the normalized stress intensity and failure time for Ti-5Al-2.5Sn at three temperatures.

problem of gaseous hydrogen embrittlement and until we develop the necessary understanding of the basic interaction responsible for the problem, we must use such approaches to design the required aerospace hardware.

References

[1] Rogers, H. C., *Science* **159** (1968), 1057.
[2] Cavett, R. A. and Can Ness, H. C., *Weld J. Res. Supp.* **42** (1963), 316S.
[3] Hofmann, W. and Rauls, W., *Weld J. Res. Supp.* **44** (1965), 225S.
[4] Walter, R. J. and Chandler, W. T., Rocketdyne Research Report, R6851, January 1967.
[5] Williams, D. P. and Nelson, H. G., *Trans. AIME* **233** (1965), 1339.
[6] Nelson, H. G. and Williams, D. P., *Environment-Sensitive Mechanical Behavior*, Gordon & Breach, New York, 1966, p. 107.
[7] Williams, D. P. and Nelson, H. G., *Met. Trans.* **1** (1970), 63.
[8] Williams, D. P. and Nelson, H. G., *Met. Trans.* **3** (1972), 2107.
[9] Nelson, H. G., Williams, D. P. and Stein, J. E., *Met. Trans.* **3** (1972), 469.
[9] Nelson, H. G., Williams, D. P. and Stein, J. E., 'Environmental Hydrogen Embrittlement of an Alpha-Beta Titanium Alloy' (to be published in *Met. Trans.*).
[10] Williams, D. P., 'A New Criterion for Failure of Materials by Environment-Induced Cracking', (submitted for publication in *Int. J. Fract. Mech.*).

FATIGUE FAILURE OF MATERIALS
UNDER NARROW BAND RANDOM VIBRATIONS
PART 2

T. C. HUANG

Dept. of Engineering Mechanics, University of Wisconsin, Madison, Wis., U.S.A.

RALPH B. HUBBARD

Dept. of General Engineering, University of Hawaii, H.I., U.S.A.

and

R. W. LANZ

College of Environment Sciences, University of Wisconsin, Green Bay, Wis., U.S.A.

1. Introduction

The present paper is a continuation of a previous paper (Part 1) on the fatigue failure of materials under the multi-factor influence of narrow band random vibrations. In Part 1 the basic principle of response surface design has been described, and a first order fatigue life predicting equation had been developed. The research was based on a 2^3 factorial design with 4 added centers and an experiment of 12 tests. It was shown that the analysis was possible with a 1/2 replicate, i.e., an experiment of 6 tests. Based on the results of statistical analysis of variance, these linear models have been accepted. However, only the 95% confidence intervals which are based on 12 tests are of practical use.

In this paper a more comprehensive second order design that can handle the increasing operating ranges will be explored. It will be shown that a composite design of 18 tests, i.e., 8 corner points, 4 centers and 6 axial points, can be used to estimate the coefficients of a second order fatigue life predicting equation. These 18 tests can be obtained by simply adding 6 new tests at axial points to the 12 tests which were conducted for the first order design. For increasing precision a 24-test design can be developed sequentially by adding another 6 new tests at the same axial points to the 18 tests which we already have. With more tests conducted the confidence intervals may be improved. The procedure of the composite design and the associated statistical analyses are illustrated. From the analyses of these particular experiments of 18 or 24 tests, the lack of fit is not important and the models are acceptable. However, the second order effect has been found to have no statistical significance for these particular ranges of test conditions.

2. A Second Order Model

The first order model which was developed in Part 1 of this investigation, i.e.,

$$y = c_0 + c_1 x_1 + c_2 x_2 + c_3 x_3 + \varepsilon \tag{1}$$

L. G. Napolitano et al. (eds.), Astronautical Research 1971, 331–342. All Rights Reserved.

in which x_1, x_2, x_3 are the coded variables, c_0, c_1, c_2, c_3 are the four estimated coefficients, and ε is the experimental error, can be extended to the second order model as

$$y = c_0 + c_1x_1 + c_2x_2 + c_3x_3 + c_{11}x_1^2 + c_{22}x_2^2 + c_{33}x_3^2$$
$$+ c_{12}x_1x_2 + c_{13}x_1x_3 + c_{23}x_2x_3 + \varepsilon. \tag{2}$$

In this equation the square terms x_1^2, x_2^2, x_3^2, represent the quadratic effect of three variables, and the product terms, x_1x_2, x_1x_3, x_2x_3, represent their interaction effects. A total of ten coefficients, i.e., c_0, c_1, c_2, c_3, c_{11}, c_{22}, c_{33}, c_{12}, c_{13}, c_{23} are to be estimated. Each coded variable x_i is obtained by the same equation in Part 1 which is

$$x_i = \frac{2(\log V - \log H)}{\log H - \log L} + 1. \tag{3}$$

Before proceeding to illustrate the design and analysis of second order equations with 18 and 24 tests, it might be mentioned that the previous 12 tests in Part 1 can be used to estimate a simplified second order equation

$$y = c_0 + c_1x_1 + c_2x_2 + c_3x_3 + cx^2 + c_{12}x_1x_2 + c_{13}x_1x_3 + c_{23}x_2x_3 \tag{4}$$

in which only one term, x^2, represents the quadratic effects of three variables.

3. Design and Analysis of Second-Order Equation with 18 Tests

A. COMPOSITE DESIGN AND EXPERIMENTATION

In order to estimate all the coefficients in Equation (2) for the second order model, a central composite design is planned. This design consists of 18 tests: 6 new tests

TABLE I

Factors and coded variables

Factor			Code		
R	S	F	x_1	x_2	x_3
27.2	10.8	43.2	2	2	2
23.3	7.2	36.0	1	1	1
20.0	4.8	30.0	0	0	0
17.0	3.2	25.0	−1	−1	−1
14.7	2.13	20.8	−2	−2	−2

added to the 12 previous tests. These added tests are at the 6 axial points numbered 13 through 18 as shown in Figure 1. Figure 1 also includes the previous 12 tests. Table I shows the complete test conditions and the coded variables. A distance of 2 units between each axial point and the center is chosen instead of 1.682 units. This gives a somewhat wider range of operating conditions. The test conditions and results for the 6 added tests are shown in Table II.

TABLE II

Test conditions, codings and fatigue life – Test No. 13–18

Test No.	Desired test conditions			Code			Actual test conditions			Life	
	R	S	F	x_1	x_2	x_3	R	S	F	T	y
13	27.2	4.8	30.0	2	0	0	26.5	5.0	30.0	10.43	2.3447
14	14.7	4.8	30.0	−2	0	0	14.8	5.0	30.0	318.40	5.7633
15	20.0	10.8	30.0	0	2	0	23.2	11.2	30.0	13.13	2.5749
16	20.0	2.13	30.0	0	−2	0	19.6	2.1	30.0	102.35	4.6284
17	20.0	4.8	43.2	0	0	2	20.5	5.0	43.2	37.38	3.6211
18	20.0	4.8	20.8	0	0	−2	19.6	4.6	20.8	87.00	4.4659

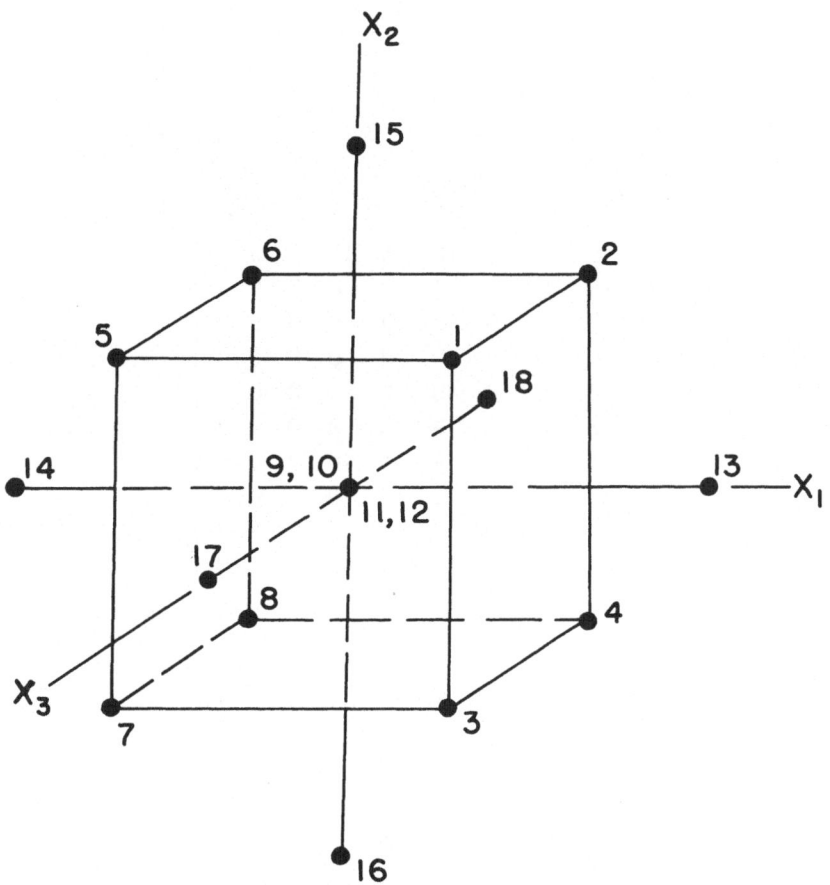

Fig. 1. A three-factor composite design.

For the complete second order model with 18 tests the design matrix X is

$$X = \begin{array}{c} \begin{array}{ccccccccccc} x_0 & x_1 & x_2 & x_3 & x_1^2 & x_2^2 & x_3^2 & x_1x_2 & x_1x_3 & x_2x_3 \end{array} \\ \left[\begin{array}{cccccccccc} 1 & 1 & 1 & 1 & 1 & 1 & 1 & 1 & 1 & 1 \\ 1 & 1 & 1 & -1 & 1 & 1 & 1 & 1 & -1 & -1 \\ 1 & 1 & -1 & 1 & 1 & 1 & 1 & -1 & 1 & -1 \\ 1 & 1 & -1 & -1 & 1 & 1 & 1 & -1 & -1 & 1 \\ 1 & -1 & 1 & 1 & 1 & 1 & 1 & -1 & -1 & 1 \\ 1 & -1 & 1 & -1 & 1 & 1 & 1 & -1 & 1 & -1 \\ 1 & -1 & -1 & 1 & 1 & 1 & 1 & 1 & -1 & -1 \\ 1 & -1 & -1 & -1 & 1 & 1 & 1 & 1 & 1 & 1 \\ 1 & 0 & 0 & 0 & 0 & 0 & 0 & 0 & 0 & 0 \\ 1 & 0 & 0 & 0 & 0 & 0 & 0 & 0 & 0 & 0 \\ 1 & 0 & 0 & 0 & 0 & 0 & 0 & 0 & 0 & 0 \\ 1 & 0 & 0 & 0 & 0 & 0 & 0 & 0 & 0 & 0 \\ 1 & 2 & 0 & 0 & 4 & 0 & 0 & 0 & 0 & 0 \\ 1 & -2 & 0 & 0 & 4 & 0 & 0 & 0 & 0 & 0 \\ 1 & 0 & 2 & 0 & 0 & 4 & 0 & 0 & 0 & 0 \\ 1 & 0 & -2 & 0 & 0 & 4 & 0 & 0 & 0 & 0 \\ 1 & 0 & 0 & 2 & 0 & 0 & 4 & 0 & 0 & 0 \\ 1 & 0 & 0 & -2 & 0 & 0 & 4 & 0 & 0 & 0 \end{array}\right] \begin{array}{c} \text{Test No.} \\ 1 \\ 2 \\ 3 \\ 4 \\ 5 \\ 6 \\ 7 \\ 8 \\ 9 \\ 10 \\ 11 \\ 12 \\ 13 \\ 14 \\ 15 \\ 16 \\ 17 \\ 18 \end{array} \end{array} \tag{5}$$

from which

$$X'X = \begin{bmatrix} 18 & 0 & 0 & 0 & 16 & 16 & 16 & 0 & 0 & 0 \\ 0 & 16 & 0 & 0 & 0 & 0 & 0 & 0 & 0 & 0 \\ 0 & 0 & 16 & 0 & 0 & 0 & 0 & 0 & 0 & 0 \\ 0 & 0 & 0 & 16 & 0 & 0 & 0 & 0 & 0 & 0 \\ 16 & 0 & 0 & 0 & 40 & 8 & 8 & 0 & 0 & 0 \\ 16 & 0 & 0 & 0 & 8 & 40 & 8 & 0 & 0 & 0 \\ 16 & 0 & 0 & 0 & 8 & 8 & 40 & 0 & 0 & 0 \\ 0 & 0 & 0 & 0 & 0 & 0 & 0 & 8 & 0 & 0 \\ 0 & 0 & 0 & 0 & 0 & 0 & 0 & 0 & 8 & 0 \\ 0 & 0 & 0 & 0 & 0 & 0 & 0 & 0 & 0 & 8 \end{bmatrix} \tag{6}$$

and

$$(X'X)^{-1} = \frac{1}{480} \begin{bmatrix} 112 & 0 & 0 & 0 & -32 & -32 & -32 & 0 & 0 & 0 \\ 0 & 30 & 0 & 0 & 0 & 0 & 0 & 0 & 0 & 0 \\ 0 & 0 & 30 & 0 & 0 & 0 & 0 & 0 & 0 & 0 \\ 0 & 0 & 0 & 30 & 0 & 0 & 0 & 0 & 0 & 0 \\ -32 & 0 & 0 & 0 & 22 & 7 & 7 & 0 & 0 & 0 \\ -32 & 0 & 0 & 0 & 7 & 22 & 7 & 0 & 0 & 0 \\ -32 & 0 & 0 & 0 & 7 & 7 & 22 & 0 & 0 & 0 \\ 0 & 0 & 0 & 0 & 0 & 0 & 0 & 60 & 0 & 0 \\ 0 & 0 & 0 & 0 & 0 & 0 & 0 & 0 & 60 & 0 \\ 0 & 0 & 0 & 0 & 0 & 0 & 0 & 0 & 0 & 60 \end{bmatrix} \tag{7}$$

B. PREDICTING EQUATION

By the method of least squares the coefficients $\bar{c}$ can be estimated by the equation

$$\bar{c} = (X'X)^{-1}X'\bar{y}. \tag{8}$$

Using the results of 18 tests Equation (8) gives the ten coefficients as follows:

$$\begin{Bmatrix} c_0 \\ c_1 \\ c_2 \\ c_3 \\ c_{11} \\ c_{22} \\ c_{33} \\ c_{12} \\ c_{13} \\ c_{23} \end{Bmatrix} = \begin{Bmatrix} 3.9153 \\ -0.8416 \\ -0.3483 \\ -0.1520 \\ 0.0306 \\ -0.0825 \\ 0.0280 \\ -0.0648 \\ 0.0885 \\ 0.0171 \end{Bmatrix} \tag{9}$$

Thus the predicting equation based on 18 tests is

$$\hat{y} = 3.9153 - 0.8416x_1 - 0.3483x_2 - 0.1520x_3$$
$$+ 0.0306x_1^2 - 0.0825x_2^2 + 0.0280x_3^2$$
$$- 0.0648x_1x_2 + 0.0171x_1x_3 + 0.0885x_2x_3. \tag{10}$$

C. ANALYSIS OF VARIANCE

Table III shows the analysis of variance with 18 tests. The F-ratio indicates [3] that the lack of fit is not significant. Thus the model is acceptable. Other F-ratios show that the first order terms are significant and the second order terms are not.

TABLE III

Analysis of variance of 18 tests

Source	Sum of squares	Degree of freedom	Mean square	Ratio
Zero-order terms	272.961403	1	272.961403	
First-order terms	13.643871	3	4.547957	221.39
Second-order terms	0.367733	6	0.061289	2.98
Lack of fit	0.498569	5	0.099714	4.85
Error	0.061628	3	0.020543	
Total	287.533203	18		

D. CONFIDENCE INTERVALS

Table IV shows the complete test and estimated results together with the residuals and 95% confidence intervals. It is seen that the precision of these intervals is not enough to be of practical use.

TABLE IV

Results of 18 tests

$$\hat{y} = 3.9153 - 0.8416x_1 - 0.3483x_2 - 0.1520x_3$$
$$+ 0.0306x_1^2 - 0.0825x_2^2 + 0.0820x_3^2$$
$$- 0.0648x_1x_2 + 0.0171x_1x_3 + 0.0885x_2x_3$$

No	T	y	$\hat{y}$	$\hat{\bar{T}}$	$y - \hat{y}$	$(y - \hat{y})^2$	95% confidence levels			
							$\hat{y}$		$\hat{T}$	
1	16.62	2.8106	2.5904	13.33	0.2202	0.048499	2.1096	3.0712	8.2447	21.5675
2	16.23	2.7869	2.6832	14.63	0.1037	0.010746	2.2024	3.1640	9.0466	23.6652
3	26.43	3.2745	3.3825	29.44	−0.1080	0.011659	2.9017	3.8633	18.2045	47.6214
4	27.53	3.3153	3.5436	34.59	−0.2283	0.052121	3.0628	4.0244	21.3867	55.9458
5	83.28	4.4222	4.2262	68.46	0.1960	0.038423	3.7454	4.7070	42.3252	110.7192
6	115.43	4.7487	4.6730	107.02	0.0757	0.005729	4.1922	5.1538	66.1658	173.0843
7	101.82	4.6232	4.7591	116.65	−0.1359	0.018481	4.2783	5.2400	72.1208	188.6621
8	151.67	5.0217	5.2742	195.24	−0.2525	0.063760	4.7934	5.7550	120.7118	315.7724
9	44.25	3.7899	3.9153	50.17	−0.1255	0.015744	3.6206	4.2101	37.3588	67.3629
10	56.38	4.0321	3.9153	50.17	0.1168	0.013638	3.6206	4.2101	37.3588	67.3629
11	45.15	3.8100	3.9153	50.17	−0.1053	0.011097	3.6206	4.2101	37.3588	67.3629
12	58.07	4.0616	3.9153	50.17	0.1463	0.021409	3.6206	4.2101	37.3588	67.3629
13	10.43	2.3447	2.3546	10.53	−0.0099	0.000099	1.8502	2.8591	6.3611	17.4452
14	318.40	5.7633	5.7211	305.23	0.0422	0.001783	5.2167	6.2255	184.3158	505.4803
15	13.13	2.5749	2.8888	17.97	−0.3139	0.098559	2.3844	3.3933	10.8527	29.7631
16	102.35	4.6284	4.2822	72.40	0.3462	0.119870	3.7777	4.7866	43.7174	119.8937
17	37.38	3.6211	3.7234	41.41	−0.1023	0.010466	3.2190	4.2279	25.0034	68.5709
18	87.00	4.4659	4.3313	76.04	0.1346	0.018113	3.8269	4.8358	45.9197	125.9332

In arriving at the 95% confidence intervals from

$$\hat{y} \pm t_{\text{df0. }05/2} \sqrt{V(\hat{y})}$$

the calculation of the variance of estimate fatigue life $V(\hat{y})$ is now based on the equation

$$V(\hat{y}) = V(c_0x_0 + c_1x_1 + c_2x_2 + c_3x_3 + c_{11}x_1^2 + c_{22}x_2^2 + c_{33}x_3^2$$
$$+ c_{12}x_1x_2 + c_{13}x_1x_3 + c_{23}x_2x_3)$$
$$= V(c_0) + x_1^2V(c_1) + x_2^2V(c_2) + x_3^2V(c_3) + x_1^4V(c_{11}) + x_2^4V(c_{22})$$
$$+ x_3^4V(c_{33}) + (x_1x_2)^2V(c_{12}) + (x_1x_3)^2V(c_{13}) + (x_2x_3)^2V(c_{23})$$
$$+ 2(x_0x_1)\,\text{cov}\,(c_0c_1) + 2(x_0x_2)\,\text{cov}\,(c_0c_2) + \cdots$$
$$+ 2(x_1x_2)\,\text{cov}\,(c_1c_2) + 2(x_1x_3)\,\text{cov}\,(c_1c_3) + \cdots$$
$$+ 2(x_2x_3)\,\text{cov}\,(c_2c_3) + \cdots$$
$$+ \cdots \tag{11}$$

where the new notation cov is for covariance. Covariance is evaluated by

$$\text{cov}\,(c_i c_j) = C^{ij} s^2 \tag{12}$$

in which C^{ij} are the elements of $(X'X)^{-1}$ and s is the estimated standard error. The superscripts i and j of C correspond to the subscripts of c in the order of 0, 1, 2, 3, 11, 22, 33, 12, 13 and 23.

4. Design and Analysis of Second Order Equation with 24 Tests

A. COMPOSITE DESIGN AND EXPERIMENTATION

In order to increase the precision of confidence intervals and to further test the effect of added axial points in quadratic coefficients, a composite design of 24-tests is in order. The additional six tests (19–24) are performed under identical conditions as the previous six tests (13–18). Table V shows the test results.

The design matrix for the second order model with 24 tests is

	x_0	x_1	x_2	x_3	x_1^2	x_2^2	x_3^2	$x_1 x_2$	$x_1 x_3$	$x_2 x_3$	Test No.
	1	1	1	1	1	1	1	1	1	1	1
	1	1	1	−1	1	1	1	1	−1	−1	2
	1	1	−1	1	1	1	1	−1	1	−1	3
	1	1	−1	−1	1	1	1	−1	−1	1	4
	1	−1	1	1	1	1	1	−1	−1	1	5
	1	−1	1	−1	1	1	1	−1	1	−1	6
	1	−1	−1	1	1	1	1	1	−1	−1	7
	1	−1	−1	−1	1	1	1	1	1	1	8
	1	0	0	0	0	0	0	0	0	0	9
	1	0	0	0	0	0	0	0	0	0	10
	1	0	0	0	0	0	0	0	0	0	11
$X =$	1	0	0	0	0	0	0	0	0	0	12
	1	2	0	0	4	0	0	0	0	0	13
	1	−2	0	0	4	0	0	0	0	0	14
	1	0	2	0	0	4	0	0	0	0	15
	1	0	−2	0	0	4	0	0	0	0	16
	1	0	0	2	0	0	4	0	0	0	17
	1	0	0	−2	0	0	4	0	0	0	18
	1	2	0	0	4	0	0	0	0	0	19
	1	−2	0	0	4	0	0	0	0	0	20
	1	0	2	0	0	4	0	0	0	0	21
	1	0	−2	0	0	4	0	0	0	0	22
	1	0	0	2	0	0	4	0	0	0	23
	1	0	0	−2	0	0	4	0	0	0	24

$$\tag{13}$$

TABLE V

Test conditions, codings and fatigue life – Test No. 19–24

Test No.	Desired test conditions			Code			Actual test conditions			Life	
	R	S	F	x_1	x_2	x_3	R	S	F	T	y
19	27.2	4.8	30.0	2	0	0	24.95	4.86	30.0	5.17	1.6429
20	14.7	4.8	30.0	−2	0	0	15.11	4.79	30.0	247.82	5.5127
21	20.0	10.8	30.0	0	2	0	22.03	8.65	30.0	12.92	2.5588
22	20.0	2.13	30.0	0	−2	0	19.61	2.21	30.0	60.20	4.0977
23	20.0	4.8	43.2	0	0	2	19.48	4.78	43.2	39.92	3.6869
24	20.0	4.8	20.8	0	0	−2	20.76	5.06	20.8	53.52	3.9801

from which

$$X'X = \begin{bmatrix} 24 & 0 & 0 & 0 & 24 & 24 & 24 & 0 & 0 & 0 \\ 0 & 24 & 0 & 0 & 0 & 0 & 0 & 0 & 0 & 0 \\ 0 & 0 & 24 & 0 & 0 & 0 & 0 & 0 & 0 & 0 \\ 0 & 0 & 0 & 24 & 0 & 0 & 0 & 0 & 0 & 0 \\ 24 & 0 & 0 & 0 & 72 & 8 & 8 & 0 & 0 & 0 \\ 24 & 0 & 0 & 0 & 8 & 72 & 8 & 0 & 0 & 0 \\ 24 & 0 & 0 & 0 & 8 & 8 & 72 & 0 & 0 & 0 \\ 0 & 0 & 0 & 0 & 0 & 0 & 0 & 8 & 0 & 0 \\ 0 & 0 & 0 & 0 & 0 & 0 & 0 & 0 & 8 & 0 \\ 0 & 0 & 0 & 0 & 0 & 0 & 0 & 0 & 0 & 8 \end{bmatrix} \tag{14}$$

and

$$(X'X)^{-1} = \frac{1}{192} \begin{bmatrix} 44 & 0 & 0 & 0 & -12 & -12 & -12 & 0 & 0 & 0 \\ 0 & 8 & 0 & 0 & 0 & 0 & 0 & 0 & 0 & 0 \\ 0 & 0 & 8 & 0 & 0 & 0 & 0 & 0 & 0 & 0 \\ 0 & 0 & 0 & 8 & 0 & 0 & 0 & 0 & 0 & 0 \\ -12 & 0 & 0 & 0 & 6 & 3 & 3 & 0 & 0 & 0 \\ -12 & 0 & 0 & 0 & 3 & 6 & 3 & 0 & 0 & 0 \\ -12 & 0 & 0 & 0 & 3 & 3 & 6 & 0 & 0 & 0 \\ 0 & 0 & 0 & 0 & 0 & 0 & 0 & 24 & 0 & 0 \\ 0 & 0 & 0 & 0 & 0 & 0 & 0 & 0 & 24 & 0 \\ 0 & 0 & 0 & 0 & 0 & 0 & 0 & 0 & 0 & 24 \end{bmatrix} \tag{15}$$

B. PREDICTING EQUATION

Applying Equation (8) to the results of 24 tests gives the ten coefficients as follows:

$$
\begin{Bmatrix} c_0 \\ c_1 \\ c_2 \\ c_3 \\ c_{11} \\ c_{22} \\ c_{33} \\ c_{12} \\ c_{13} \\ c_{23} \end{Bmatrix} = \begin{Bmatrix} 3.9533 \\ -0.8836 \\ -0.3605 \\ -0.1257 \\ -0.0269 \\ -0.1146 \\ 0.0038 \\ -0.0648 \\ 0.0885 \\ 0.0171 \end{Bmatrix} \tag{16}
$$

Thus the predicting equation based on 24 tests is

$$
\begin{aligned}
\hat{y} = {} & 3.9533 - 0.8836x_1 - 0.3605x_2 - 0.1257x_3 \\
& - 0.0269x_1^2 - 0.1146x_2^2 + 0.0038x_3^2 \\
& - 0.0648x_1x_2 + 0.0885x_1x_3 + 0.0171x_2x_3
\end{aligned} \tag{17}
$$

C. ANALYSIS OF VARIANCE

Table VI shows the analysis of variance with 24 tests. An investigation of F-ratios for lack of fit and for the first and second order terms yields the same conclusion as did the similar investigation with the 18 tests.

TABLE VI

Analysis of variance – 24 tests

Source	Sum of squares	Degree of freedom	Mean square	Ratio
Zero-order terms	349.407670	1	349.407670	
First-order terms	22.234199	3	7.411400	360.78
Second-order terms	0.683006	6	0.113834	5.54
Lack of fit	1.007774	11	0.091616	4.45
Error	0.061628	3	0.020543	
Total	373.394280	24		

D. CONFIDENCE INTERVALS

The complete test data and the estimated results, together with the residuals and 95% confidence intervals are tabulated in Table VII. Each of the confidence intervals is smaller than the corresponding intervals from the 18 tests. This was as expected because there were more tests used in the evaluation.

TABLE VII

Results of 24 tests

$$\hat{y} = 3.9533 - 0.8836x_1 - 0.3605x_2 - 0.1257x_3$$
$$-0.0269x_1^2 - 0.1146x_2^2 - 0.0038x_3^2$$
$$-0.0648x_1x_2 - 0.0885x_1x_3 - 0.0171x_2x_3$$

							95% confidence levels			
No.	T	y	$\hat{y}$	$\hat{T}$	$y-\hat{y}$	$(y-\hat{y})^2$	$\hat{y}$		$\hat{T}$	
1	16.62	2.8106	2.4866	12.02	0.3240	0.104988	2.0503	2.9229	7.7700	18.5952
2	16.23	2.7869	2.5270	12.52	0.2599	0.067550	2.0906	2.9633	8.0901	19.3612
3	26.43	3.2745	3.3029	27.19	-0.0284	0.000809	2.8666	3.7393	17.5777	42.0667
4	27.53	3.3153	3.4116	30.31	-0.0963	0.009277	2.9753	3.8479	19.5951	46.8948
5	83.28	4.4222	4.2063	67.11	0.2159	0.046619	3.7700	4.6426	43.3792	103.8148
6	115.43	4.7487	4.6006	99.55	0.1480	0.021916	4.1643	5.0369	64.3484	153.9979
7	101.82	4.6232	4.7635	117.16	-0.1403	0.019686	4.3272	5.1998	75.7320	181.2412
8	151.67	5.0217	5.2261	186.07	-0.2044	0.041788	4.7898	5.6624	120.2790	287.8507
9	44.25	3.7899	3.9533	52.11	-0.1634	0.026715	3.6695	4.2371	39.2324	69.2068
10	56.38	4.0321	3.9533	52.11	0.0788	0.006211	3.6695	4.2371	39.2324	69.2068
11	45.15	3.8100	3.9533	52.11	-0.1433	0.020538	3.6695	4.2371	39.2324	69.2068
12	58.07	4.0616	3.9533	52.11	0.1083	0.011739	3.6695	4.2371	39.2324	69.2068
13	10.43	2.3447	2.0787	7.99	0.2660	0.070764	1.7057	2.4517	5.5052	11.6075
14	318.40	5.7633	5.6129	273.94	0.1504	0.022619	5.2399	5.9859	188.6567	397.7784
15	13.13	2.5749	2.7739	16.02	-0.1990	0.039604	2.4009	3.1469	11.0334	23.2636
16	102.35	4.6284	4.2158	67.75	0.4126	0.170267	3.8428	4.5887	46.6550	98.3711
17	37.38	3.6211	3.7169	41.14	-0.0958	0.009171	3.3439	4.0899	28.3299	59.7331
18	87.00	4.4659	4.2199	68.03	0.2460	0.060528	3.8469	4.5929	46.8477	98.7772
19	5.17	1.6429	2.0787	7.99	-0.4358	0.189920	1.7057	2.4517	5.5052	11.6075
20	247.82	5.5127	5.6129	273.94	-0.1002	0.010042	5.2399	5.9859	188.6567	397.7784
21	12.92	2.5588	2.7739	16.02	-0.2151	0.046281	2.4009	3.1469	11.0334	23.2636
22	60.20	4.0977	4.2158	67.75	-0.1181	0.013946	3.8428	4.5887	46.6550	98.3711
23	39.92	3.6869	3.7169	41.14	-0.0300	0.000902	3.3439	4.0899	28.3299	59.7331
24	53.52	3.9801	4.2199	68.03	-0.2398	0.057518	3.8469	4.5929	46.8477	98.7772

5. Response Surface

Figures 2 and 3 show the response surfaces for R, S, F in the logarithmic scale based on Equations (10) and (17). These surfaces give an overall graphical view of the fatigue life under narrow band random loading based on the second order design and within the prescribed ranges of the operating conditions.

Fig. 2.   Response surfaces showing relations of fatigue failure time T and three factors, R, S and F for a second order composite design of 18 tests.

Fig. 3.   Response surfaces showing relations of fatigue failure time T and three factors R, S and F for a second order composite design of 24 tests.

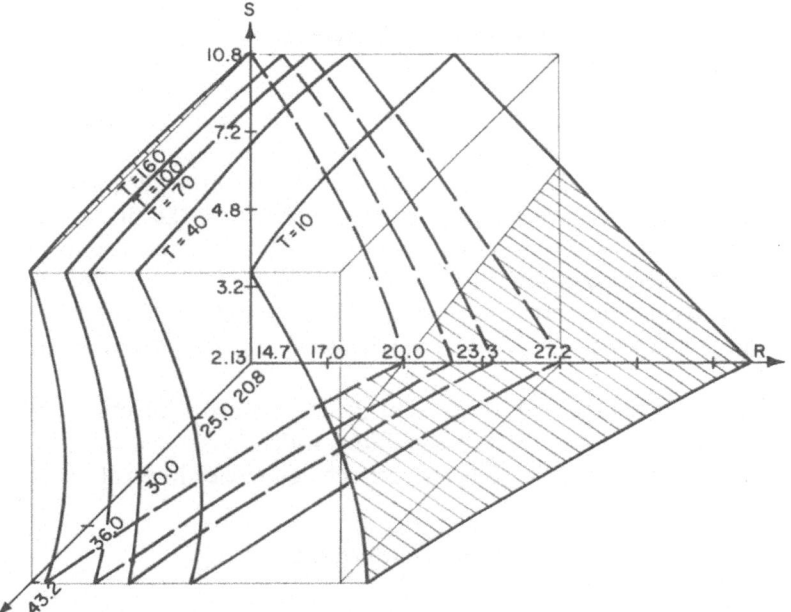

Fig. 2.

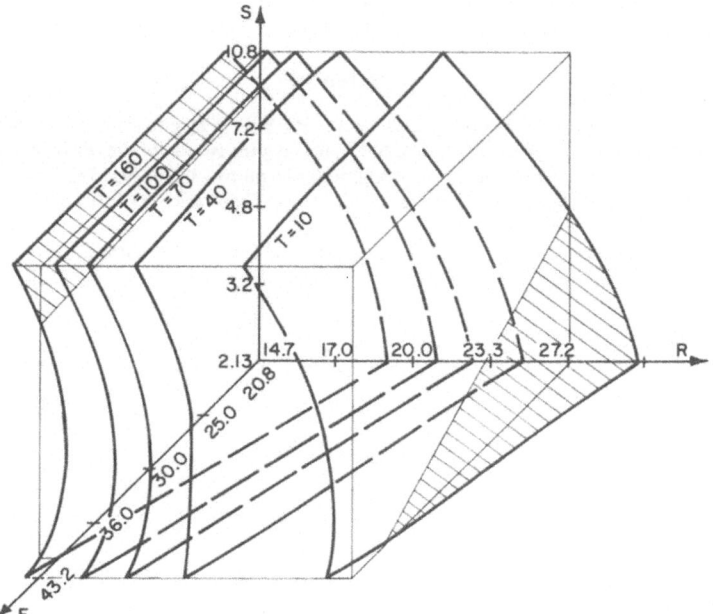

Fig. 3.

6. Conclusion

(1) It is possible to derive the second order predicting equation for the fatigue life of materials under random vibrations by the method of response surface with a small number of tests, i.e., 18 or 24 tests.

(2) In the analysis of variance, the lack of fit tests show the adequacy of the postulated second order model. The same analysis reveals that the first order terms are dominant and the second order terms have no statistical significance. In other words the first order model is adequate.

(3) Due to experimental error in general and the deviation of test conditions from desired conditions used in the analysis in particular, the second order effect might be obscured.

(4) Further investigation may reveal that the second order effect might be increasingly important as the ranges of test conditions widen.

(5) Increasing the number of tests reduces the confidence intervals as expected.

Acknowledgements

This investigation was carried out to near completion during 1965–68 with partial support from the NASA Institutional Grant to the University of Wisconsin. In completing the study, further support was also received from the Wisconsin Alumni Research Foundation and the Engineering Experiment Station of the College of Engineering, the University of Wisconsin.

References

[1] Huang, T. C., Hubbard, Ralph B., and Lanz, R. W., 'Fatigue Failure of Materials under Narrow Band Random Vibrations – Part 1', to be published in the *Proceedings of the 21st International Astronautical Congress*, Constance, Germany, 1970, pp. 593–610.
[2] Davis, O. L. (ed.), *Design and Analysis of Industrial Experiments*, 2nd ed., Oliver and Boyd, London, 1963.
[3] Croxton, F. E., Cowden, D. J., and Klein, S., *Applied General Statistics*, 3rd ed., Prentice Hall, 1967.

E. TELEMETERING AND DATA MANAGEMENT

DATA MANAGEMENT FOR THE SKYLAB PROGRAM

JOHN H. DISHER*

National Aeronautics and Space Administration, Washington, D.C., U.S.A.

Skylab is an experimental manned space station, scheduled for initial launch in 1973. The objectives of the Skylab Program may be summarized as follows:

(1) To learn more about the long duration effects of the essentially zero gravity environment in conjunction with other space environmental factors on man, and to evaluate the effectiveness of the selected design features of Skylab in providing the three-man crew for effective working and comfortable living conditions for Earth orbit missions up to 56 days duration.

(2) To carry out specific scientific and engineering investigations principally in the areas of biomedicine, solar astronomy, Earth resource surveys and space manufacturing processes.

In order to meet these objectives, the Skylab is designed to afford a large living and working volume for its crew, necessary habitability and life support equipment, and necessary experimental equipment and sensors to conduct the more than 50 experiments included in the program. A sketch of the Skylab in its orbital configuration is shown in Figure 1. The primary habitability and living quarters consist of a 12000 ft³ 22-ft diameter cylinder called the Orbital Workshop Module. This module includes a kitchen, bedroom, bathroom and two levels of laboratory work space. One work level has a 7-ft height so that the crew can walk around relatively normally by pressing at the 'ceiling' with the hands. The other work level is approximately 20 ft high and affords a large free volume for those experiments requiring free astronaut movement in the zero gravity field. Other units or modules of the Skylab orbital assembly are identified in Figure 2. These include the airlock module (AM) which is the 'engineering and control room' and which provides free access to space via a hatch; the multiple docking adapter (MDA), which carries the Earth resource sensors, controls and displays for the solar astronomy and Earth resource sensors and which provides the docking interface with the man-carrying taxi vehicle (e.g., the command and service module); the Apollo telescope mount (ATM) which contains the solar astronomy instruments; and the command and service modules (CSM) which provides for launch and recovery of the three-man crews. The workshop, MDA, airlock and ATM are launched as an assembly, unmanned, by the first two stages of the Saturn V launch vehicle. The man-carrying spacecraft is a modification of the standard Apollo lunar command and service module.

The total nominal mission duration is eight months with a total manned occupancy of the Skylab for up to five months.

---

* Deputy Director, Skylab Program.

L. G. Napolitano et al. (eds.), Astronautical Research 1971, 345–363. All Rights Reserved.
Copyright © 1973 by D. Reidel Publishing Company, Dordrecht-Holland.

Fig. 1.   Skylab.

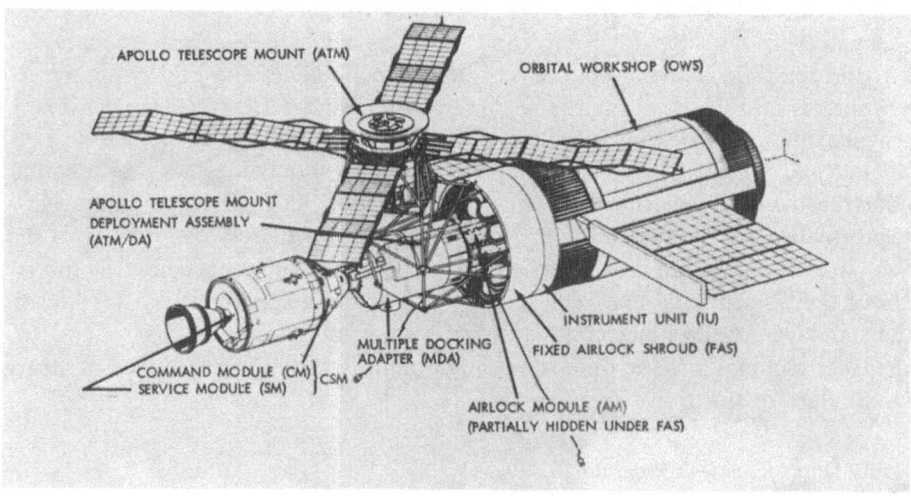

Fig. 2.   Skylab modules.

Principal to the Skylab mission are the over fifty scientific and engineering experiments carried onboard. These experiments, of course, also are major contributors to the data management load on Skylab. Summaries of experiment objectives in each of the principal areas of experimentation are as follows:

Skylab Medical Experiment Objectives: Provide quantitative data on man's ability to withstand weightless flight up to 56 days.

Skylab Solar Astronomy Experiment Objectives: Acquire high-resolution observations of the structure and behavior of the Sun from above the Earth's atmosphere and to evaluate the ability, effectiveness and support requirements of man operating complex scientific instruments in the space environment.

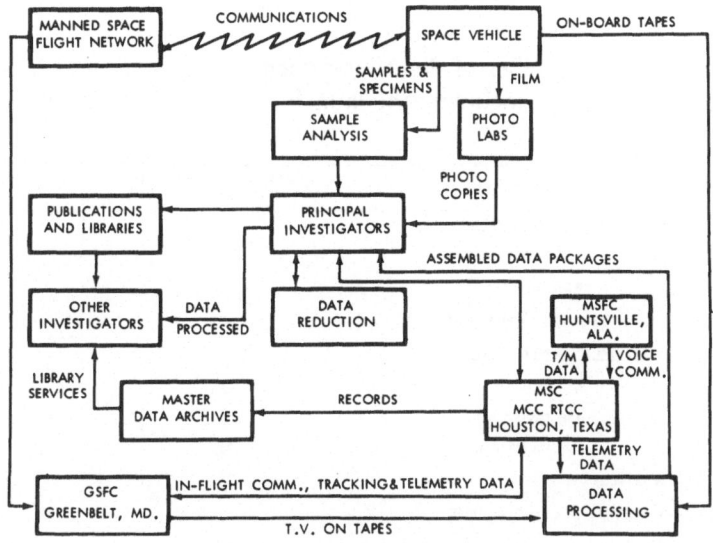

Fig. 3.   Skylab data flow concept.

Skylab Earth Resource Experiment Objectives: Expand investigation of remote sensing of the Earth from orbit by carrying relatively large, flexible, and high-performance sensors and by utilizing the crew to operate them in laboratory fashion.

Skylab Independent Experiment Objectives: 'Independent experiments' is terminology applied to the variety of engineering, technology, operations and miscellaneous science experiments outside the three major categories. Objectives are to acquire data on a variety of engineering, scientific and operationally oriented experiments which require the space environment for effective conduct.

The remainder of the paper will be devoted to a discussion of the Skylab data system. The discussion will cover: the data to be collected; the systems onboard the Skylab for data collection and return to Earth; the system on the ground for receiving, processing, and relaying the data to the Mission Control Center at Houston,

Texas; Mission Control Center processing of the data; and the software for the various space and ground located computers used throughout the program.

An overall schematic of Skylab data flow is shown in Figure 3.

The data to be gathered in Skylab include telemetry, voice recordings, television, samples and specimens, logs and records, magnetic tapes, and photographic film.

1. Data to be Collected

In describing the data to be collected, a fifth category of data, 'system status and operation', will be added to the four categories of experiment data already mentioned. A description of the data to be gathered for each of the five categories follows:

A. SYSTEM STATUS AND OPERATION DATA

Figure 4 summarizes the number of flight instrumentation system measurements for the major systems of Skylab, including the launch vehicles. Telemetry data provides

MODULE/LAUNCH VEHICLE	NUMBER OF MEASUREMENTS (TELEMETRY)
● SATURN 1B LAUNCH VEHICLE	
● FIRST STAGE	250
● SECOND STAGE	209
● INSTRUMENT UNIT	435
● SATURN V LAUNCH VEHICLE	
● FIRST STAGE	296
● SECOND STAGE	554
● INSTRUMENT UNIT	250
● AIRLOCK	552
● MULTIPLE DOCKING ADAPTER	67
● ORBITAL WORKSHOP	398
● APOLLO TELESCOPE MOUNT	919
● COMMAND AND SERVICE MODULE	374

Fig. 4.   Skylab summary of flight instrumentation.

flight control information and system status to the launch and mission control teams. The data include temperatures, pressures, voltages, currents, frequencies and rates, accelerations, inertial positions and attitudes, rotational velocities, quantities of consumables and event indications. Approximately 1100 launch vehicle measurements will be made during the launch of the Skylab on the Saturn V launch vehicle. After attaining orbit, and separating from the launch vehicle, about 1900 orbital assembly measurements will be made for flight control and system monitoring. Critical systems measurements will also be displayed onboard the Skylab for use by the crew. During the launch of the manned Saturn IB vehicle, a total of about 1300 telemetry measurements will be monitored. Of this total, about 400 relate to the spacecraft and 900 to the Saturn IB launch vehicle. After the manned spacecraft and the Workshop are joined together in orbit, a total of about 2300 measurements will be made on a continuing basis.

Telemetry is transmitted in both digital and analog form from the launch vehicles. However, only digital data are transmitted from the orbital assembly even though many measurements are originally analog. Sampling rates for these data range from one sample every 15 s for slowly changing temperatures and voltages up to 320 samples per second for physiological data such as astronaut electrocardiogram data. Motion picture coverage of operational tasks and log book entries also provide data for post-flight analyses of system performance.

TYPE OF MEASUREMENT	SENSOR	HOW RECORDED
BLOOD PRESSURE	PRESSURE TRANSDUCER & MICROPHONE	TELEMETRY & RECORDED VOICE
LEG VOLUME	CAPACITIVE PLETHYSMOGRAPH	"
HEART RATE	ELECTRODES	"
HEART FUNCTION	ELECTRODES	"
BODY TEMPERATURE	THERMISTER	"
RESPIRATION RATE	SPIROMETERS	"
RESPIRATORY FUNCTION	"	"
OXYGEN CONSUMPTION	MASS SPECTROMETER	"
CARBON DIOXIDE PRODUCTION	"	"
METABOLIC RATE	BICYCLE ERGOMETER	"
MOTION SENSITIVITY THRESHOLD	OBSERVER MATRIX CODING	"
OCULOGYRAL ILLUSION THRESHOLD	" "	"
EEG/EOG (SLEEP)	ELECTRODES	TELEMETRY & TAPE
BODY MASS	SPRING PENDULUM/TIMER	LOG BOOK & RECORDED VOICE
WATER INTAKE	CALIBRATED DISPENSERS	LOG BOOK
FOOD INTAKE	PREDETERMINED MENUS	LOG BOOK
FECES MASS	SPRING PENDULUM/TIMER	LOG BOOK
URINE VOLUME	POSITIVE DISPLACEMENT DEVICE	LOG BOOK

Fig. 5.   Skylab in-flight medical data.

B. MEDICAL EXPERIMENTS

The in-flight medical data are summarized in Figure 5. Daily samples of urine and all feces in dehydrated form from each of the three crewmen will be returned for ground analysis after in-flight weighing. Blood pressure, heart rate and function, body temperature, respiration rate and function; food, water and oxygen intake; brain wave activity during sleep, limb volume, energy expenditure, and other physical parameters will be continuously or periodically recorded during the flight. Most of these data will be telemetered to the ground stations and selected data will be voice tape recorded and read to the ground as well. Onboard photographs and motion pictures will also be used for medical post-flight analyses.

C. SOLAR ASTRONOMY EXPERIMENTS

A sketch of the Skylab solar experiment assembly is shown in Figure 6. Data return for the solar astronomy experiments, summarized in Figure 7, will be primarily in the form of still and motion picture photographs and TV images of solar phenomena in the white light, ultraviolet and X-ray regions of the spectra. A total of 115 000 photographic frames, equivalent to 111×10^{12} bits of data will be obtained photographically. This photographic bit equivalent is based on specific characteristics of

Fig. 6.   Solar astronomy.

TYPE	DESCRIPTION	QUANTITY	DATA BIT EQUIVALENT
FILM	35mm B & W	93,000 FRAMES	57×10^{12}
	35mm B & W 35mm X 258mm STRIP	1,600 STRIPS	4.5×10^{12}
	70mm B & W	20,300 FRAMES	49×10^{12}
		TOTAL FILM DATA	(111×10^{12})
TELEVISION	4 CAMERAS 525 LINES, 30 FRAMES/SEC.	70 HOURS	12.4×10^{12}
TELEMETRY	DIGITAL DATA	24 MEASUREMENTS	0.06×10^{12}
VOICE	ASTRONAUT COMMENTS	140 DAYS	0.16×10^{12}
FLIGHT LOGS	HANDWRITTEN NOTES	--	--
		TOTAL DATA	124×10^{12} BITS

Fig. 7.   Skylab solar physics data.

the ATM film which ranges from 110 to 225 line pairs per mm in resolution (at high contrast) and in size from 35 mm to 70 mm. In addition, a total of 12.5×10^{12} bits will be recorded electronically during the eight-month Skylab mission for an overall total of about 124×10^{12} equivalent bits of data.

D. EARTH RESOURCE EXPERIMENTS

The Earth resource experiment Earth coverage is shown in Figure 8. These data consist of photographs of selected areas on Earth in the visual and near infrared regions of the spectrum, electronic scanning and digital recording of the spectral characteristics of selected areas on Earth in the visible through the thermal infrared regions, digital recording of radar backscattering cross-section at a frequency of 13.9 GHz, and digital recording of surface brightness temperature at the L-band frequency of

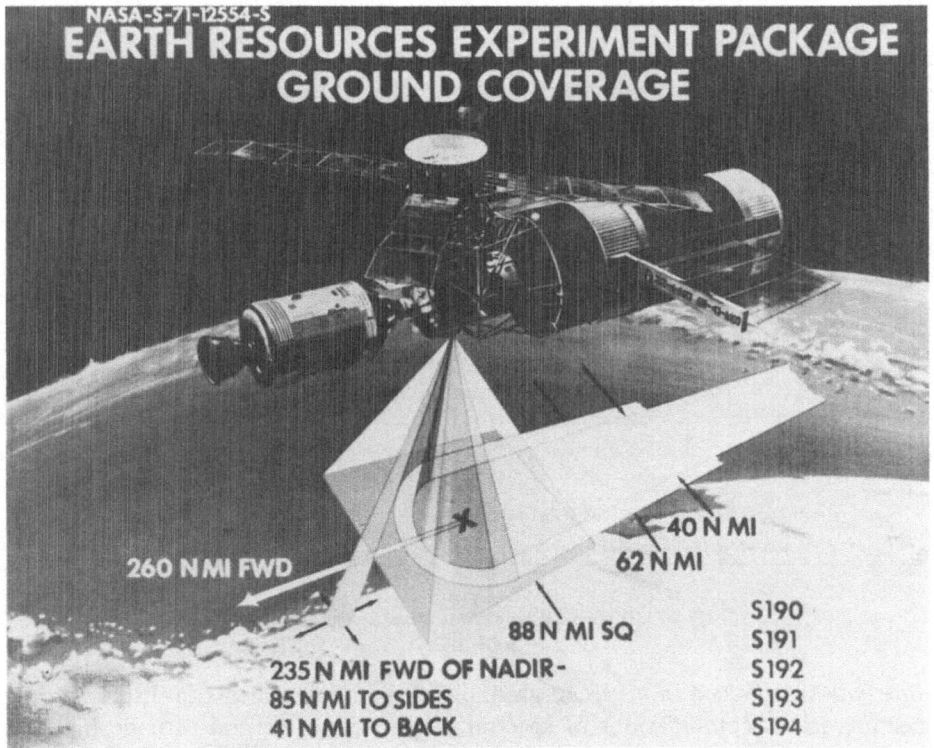

Fig. 8.

1.4 GHz. Earth resource data are all recorded onboard the Skylab with no capability for radio frequency transmission to the ground other than voice description and commentary. The total in-flight data recorded for Earth resource experiments are listed in Figure 9. A total of 3870 frames of 125 mm color, 14 400 frames of 70 mm black and white, 7200 frames of 70 mm color and 16 800 frames of 16 mm black and white film will be carried. The total data bit equivalent of these photographs is about 10^{13} bits, assuming six bits per element, which is equivalent to a 64 level gray scale discrimination. The Earth resource non-photographic data are returned on special high-density, 28-channel magnetic tape with a recording density of 20 000 bits per inch per channel. A total of 86 400 ft of tape with a resulting bit capacity of about 6.00×10^{11} bits are carried. However, due to the multiple use of the tape recorder, only about 25% of the total capacity of the tape is usable and actual data recorded total about 1.5×10^{11} bits. Most of the taped data come from the S192 multi-spectral scanner which produces data at the rate of about 27×10^6 bps. These data are placed on 24 of the 28 tracks of the tape recorder which runs at a rate of 152 cm s^{-1}. When not recording scanner data, the recorder is run at 19 cm s^{-1}. After return to Earth the original 28-track magnetic tapes will be reproduced and reformatted onto 14-track, 2.5 cm tapes. The reformatting process will reduce tape skew, convert the PCM code from the special high-density code to standard bi-phase L code, and reduce the maximum packing density from 20 000 to 10 000 bits per 2.5 cm. This processing will result in computer compatible tapes suitable for further data reduction and analysis by investigators using general-purpose computers. Data from the

MEDIA	DESCRIPTION	QUANTITY	DATA BIT EQUIVALENT
FILM	125mm COLOR	3,870 FRAMES	7.30×10^{12}
	70mm BLACK & WHITE	14,400 FRAMES	2.28×10^{12}
	70mm COLOR	7,200 FRAMES	0.99×10^{12}
	16mm BLACK & WHITE	16,800 FRAMES	0.06×10^{12}
		TOTAL FILM DATA	10.63×10^{12}
MAGNETIC TAPE	2.5CM WIDE, 28 CHANNEL	12 TAPES	
	20,000 BITS/2.5CM/CHANNEL	7,200 FT/TAPE	0.58×10^{12}
		TOTAL DATA	11.21×10^{12}
FLIGHT LOGS	SPECIFIED ASTRONAUT RECORDS		
VOICE RECORDINGS	ASTRONAUT COMMENTS		

Fig. 9.   Skylab Earth resources experiments data.

scanner will be reduced in a special ground-based station which can display ground scenes in various combinations of spectral bands on an integral cathode ray tube. Color photographs of these scenes will be available for further analysis.

E. INDEPENDENT EXPERIMENTS

The independent experiments run the gamut from stellar astronomical observations to an evaluation of crew maneuvering ability with a cold gas propelled, 3-axis-controlled backpack maneuvering unit. The majority of data for the independent experiments will be recorded on still and motion picture film with some TV, telemetering, log book, and voice data. In addition, several of the experiments require physical return of samples or specimens. Data from these experiments are summarized in Figure 10.

A summary of the overall data return to Earth is shown in Figures 11 and 12. For discussion, the data return can be conveniently divided into two classes – data return

TYPE OF MEASUREMENT	SENSOR OR VARIABLE SENSED	HOW RECORDED
RADIATION IN SPACECRAFT	DOSE RATE, PROTON ENERGY, TEMPERATURE	TELEMETRY
SURFACE EMISSIVITY/ABSORBTIVITY	TEMPERATURE AND POSITION	TELEMETRY AND 16mm CAMERA
ASTRONAUT MOTION	EVENTS, VELOCITY, ACCELERATION	TELEMETRY, 16mm CAMERA, 70mm CAMERA
FLAMMABILITY IN ZERO GRAVITY	FLAME PROPAGATION	16mm CAMERA
MATERIALS BEHAVIOR IN ZERO GRAVITY	MATERIALS PHENOMENA	16mm CAMERA
SINGLE HUMAN CELLS GROWTH IN ZERO GRAVITY	CELL GROWTH	2 MICROSCOPE CAMERAS
STELLAR ULTRAVIOLET SPECTRA	SPECTROGRAPH	UV CAMERA
HORIZON UV AIRGLOW	UV AND VISIBLE LIGHT	ULTRAVIOLET CAMERA VISIBLE LIGHT CAMERA
CIRCADIAN RHYTHM	TIME, TEMPERATURE, PRESSURE, ANIMAL ACTIVITY	TELEMETRY
PARTICLE COLLECTION	EVENTS, TEMPERATURE	TELEMETRY
GALACTIC X-RAY MAPPING	PULSE HEIGHT ANALYZER, VOLTAGE, TEMPERATURE	TELEMETRY
ULTRAVIOLET PANORAMA	EVENTS, POSITIONS, SPECTROGRAPH	TELEMETRY, UV SPECTROGRAPHIC CAMERA, VOICE
CREW/VEHICLE MOTION INTERACTION	LIMB POSITION ANGLES, FORCE, CREW/VEHICLE MOTIONS	TELEMETRY, TWO 16mm CAMERAS
EXTERNAL CONTAMINATION	EVENTS, PARTICLE SIZE, MASS, BRIGHTNESS, FREQUENCY	TELEMETRY, 16mm CAMERA

Fig. 10.   Skylab independent experiment data.

SOURCE	FORM	TOTAL EQUIVALENT BITS TRANSMITTED TO EARTH
● SYSTEM STATUS AND OPERATIONS	● TELEMETRY	600×10^9
	● TELEVISION	7200×10^9
	● VOICE	30×10^9
● MEDICAL EXPERIMENTS	● TELEMETRY	3×10^9
	● TELEVISION	4800×10^9
	● VOICE	60×10^9
● SOLAR ASTRONOMY EXPERIMENTS	● TELEMETRY	60×10^9
	● TELEVISION	$12,000 \times 10^9$
	● VOICE	150×10^9
● EARTH RESOURCES EXPERIMENTS	● VOICE	12×10^9
● INDEPENDENT EXPERIMENTS	● TELEMETRY	4×10^9
	● TELEVISION	4800×10^9
	● VOICE	30×10^9
	TOTAL	30×10^{12}

Fig. 11.   Skylab radio frequency data return.

by radio frequency communication with the Earth and data physically returned to Earth.

In the interest of completeness, the analog TV and voice data have been quantified in terms of equivalent bits and are included in the tabulation of RF return data. The relative bits of telemetry, voice and TV are, of course, not necessarily indicative of relative information content, considering the various degrees of redundancy in the several types of data. The RF data return totals about 30×10^{12} bits for the eight-month mission, of which about 7×10^{11} bits are digitized telemetry data.

	FORM	WEIGHT*(Kg)	VOLUME (cc X 10⁻³)**	QUANTITY	EQUIVALENT BITS OF DATA
● SOLAR ASTRONOMY EXPERIMENTS	● FILM	324	589	5000 METERS OF 35MM & 70MM FILM	111 X 10¹²
● EARTH RESOURCES EXPERIMENTS	● FILM	48	71.3	7210 METERS OF 16, 70 AND 125MM FILM	11 X 10¹²
	● MAGNETIC TAPE	46	7	26,330 METERS OF 2.4 CM TAPE	.58 X 10¹²
● MEDICAL AND INDEPENDENT EXPERIMENTS AND OPERATIONAL DATA	● FROZEN URINE AND DRIED FECES AND VOMITUS	114	391	—	—
	● MAGNETIC TAPE	1	1	4 REELS	—
	● MATERIALS SAMPLES AND SPECIMENS	44	60	96 ITEMS	—
	● FILM	69	71	10,130 METERS	5.9 X 10¹²
● ALL EXPERIMENT AND OPERATIONAL DATA	● LOGS & REPORTS	54	100	—	—
TOTAL		700 Kg*	1290 X 10³cc	—	128 X 10¹²
* INCLUDES CONTAINER WEIGHT		** PROVIDED STORAGE VOLUME			

Fig. 12.   Skylab physical data return.

The physically returned data include frozen urine samples and dried feces for chemical analysis, film for all experiments and for post-flight operational analyses, and magnetic tape, logs and reports for all experiments. Physical samples and specimens are also returned to Earth for selected independent experiments. Examples are materials samples exposed to the space environment and solid-state crystals grown in the zero gravity environment. Returned film has been evaluated in terms of equivalent bits by use of appropriate resolution and gray scale or color scale discrimination factors which vary with the specific film types used.

The returned tapes all represent recording of data in digital form.

The total weight and volume of physically returned data are 700 kg and 1.3×10^6 cm³, respectively, and the total bit content of all the returned film and tapes are estimated to be about 128×10^{12} bits.

2. Onboard Systems for Data Collection and Transmission

A. ONBOARD RECORDERS

The Skylab uses a total of eight onboard tape recorders (plus two spares) for data collection. With the exception of the recorder for the Earth resource experiments, previously discussed, and two recorders for selected medical data, the recorders are all for the purpose of recording data when out of contact with the ground and subsequently telemetering the data to the ground when over an appropriate station. The characteristics of the onboard recorders are summarized in Figure 13. The various recorders are located in several places in the orbital assembly; there being one in the command module, three in the airlock module, two in the telescope mount, one in

RECORDER CHARACTERISTIC	LOCATION				
	CSM	AM	ATM	MDA	OWS
NO. OF RECORDERS	1	3	2	1 + SPARE	1 + SPARE
NO. OF CHANNELS/RECORDER	5 PCM, 1 VOICE, 8 SPARE	1 PCM, 1 VOICE	1 PCM	28	7
RECORD SPEED (ips)	3.75 - 15	1 $7/8$	4	3.75 - 60	.03
DUMP SPEED (ips)	120 - 15	41 $1/4$	72	N/A	N/A
DUMP RATIO	32:1 - 1:1	22:1	18:1	N/A	N/A
RECORD DATA RATE (KBPS)	1.6 - 51.2	5.12 — 5.76	4	2100 - 33,600	--
DUMP DATA RATE (KBPS)	51.2 - 51.2	112.64 - 126.7	72	N/A	N/A
MAXIMUM RECORD TIME	120 MIN - 30 MIN	4 HOURS	90 MIN	384 MIN/24 MIN	100 HOURS
MAXIMUM DUMP TIME	3.75 MIN - 30 MIN	10.9 MIN	5 MIN	N/A	N/A
DATA CONTENT	OPERATIONAL AND EXPERIMENT DATA, CSM VOICE	OPERATIONAL OR EXPERIMENT DATA, VOICE	ATM OPERATIONAL AND EXPERIMENT TM	EARTH RESOURCE SENSOR SCIENTIFIC & SUPPORT DATA	MEDICAL EXPERIMENT DATA

Fig. 13.   Skylab tape recorders.

the MDA for the Earth resource experiments, and two in the workshop for medical data. The telemetry recorders have a total recording capacity of about 72 kbps, a total capacity of 385×10^3 kbps and a maximum data dump rate of 503 kbps.

B. SKYLAB ONBOARD FILM CAMERAS

Film remains one of the best recording mediums for obtaining large quantities of data over a short period. In some instances, no currently suitable alternative exists to photographic film. Not including TV, a total of 37 cameras of varying types, sizes, and capacities are carried on Skylab for data recording purposes. Some of these are built into experimental equipment, with access only for film reloading. Others are hand-held or bracket-mounted at various locations within the orbital assembly. Several are accessible only by EVA and film exchange is made by the astronaut outside the spacecraft. The specific locations, purpose and sizes of the several cameras are indicated in Figure 14. Thirteen different types of film are carried on Skylab and they cover a wide range of speed (from ASA 40 to ASA 6000) and resolution (30 line pairs per mm to 225 line pairs per mm). Collectively, the 22 300 m of film on Skylab will store about 127×10^{12} equivalent bits of data for ground analysis.

One of the considerations relating to photographic film usage for space data recording is environmental protection of the film prior to, during and after use. Much of the film for the eight-month Skylab mission will be carried up on the initial launch and thus must be protected from radiation for an eight-month period. Film storage vaults are provided onboard Skylab for this purpose. The film vault in the workshop

PURPOSE	NUMBER AND TYPE OF CAMERAS	LOCATION	FILM QUANTITY	EQUIVALENT BITS OF DATA
SOLAR ASTRONOMY	3 - 35mm FRAME	ATM	3350 METERS	57×10^{12}
	2 - 35mm STRIP		415 METERS	4.5×10^{12}
	1 - 70mm FRAME		1220 METERS	49×10^{12}
EARTH RESOURCES	6 - 70mm FRAME	MDA	6585 METERS	3.3×10^{12}
	1 - 125mm FRAME		495 METERS	7.3×10^{12}
	1 - 16mm MOTION		130 METERS	$.06 \times 10^{12}$
MEDICAL AND INDEPENDENT	5 - 70mm MOTION	OWS, MDA, CM	75 METERS	03×10^{12}
EXPERIMENTS AND	11 - 16mm MOTION	OWS, MDA, CM	9750 METERS	5.6×10^{12}
OPERATIONAL DATA	7 - VARIOUS TYPES	OWS, MDA, AM	305 METERS	$.25 \times 10^{12}$
TOTAL	37 CAMERAS		22,300 METERS	127×10^{12} BITS

Fig. 14.   Skylab camera locations and film.

weighs 2400 lb and will have a maximum shielding thickness of 3.4 in. of aluminium. The film vaults also must maintain temperature and humidity within acceptable limits.

C. DATA TRANSMISSION TO THE GROUND

The system for relaying Skylab radio frequency data to and from Earth is shown schematically in Figure 15. Location, frequency and use of the several links are shown.

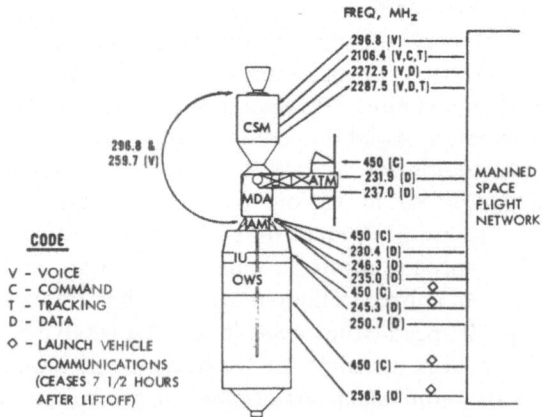

Fig. 15.   Skylab to ground communications.

Telemetry data can only be sent when the orbital assembly is in view of the ground network which is about 28% of the time. The rest of the time the telemetry signals are recorded onboard and subsequently dumped. Similarly, commands can be sent from the ground to Skylab to turn equipment on and off and to supply data to the onboard computers and the crew. An onboard teleprinter will permit hard copies of this information to be available to the crew. Voice contact with the crew is also possible whenever the Skylab is in view of a network station.

A TV camera onboard will be used to transmit pictures to Earth via the USB link in its frequency modulation mode from the CSM. Similarly, the video signals from the closed circuit TV on the ATM will be sent to earth over the same link at selected times.

Maximum PCM bit rate capacity to the ground is 173 kbps for real time data and 377 kbps for recorded data. Total PCM data quantity transmitted for a typical day will be about 4×10^9 bits.

3. Ground-Based Systems

A. GROUND TRACKING AND DATA ACQUISITION

The stations of the manned space flight network (MSFN) which will be operational for Skylab and the Earth coverage of these stations relative to selected Skylab orbits are shown in Figure 16.

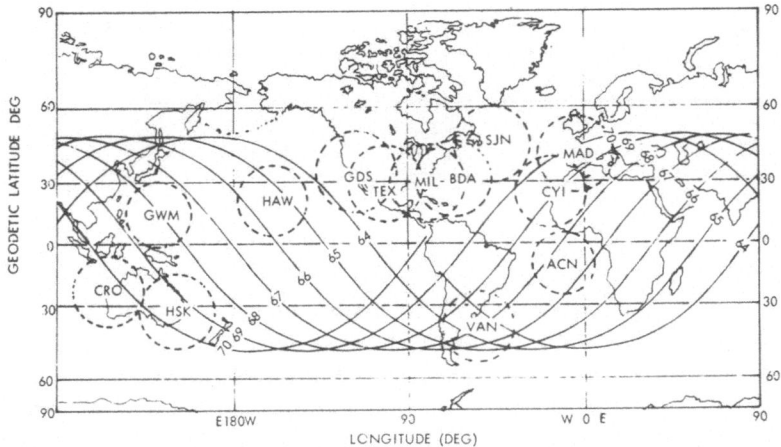

Fig. 16. Typical Skylab ground network coverage. Revolutions by trough 70.

The manned space flight network was initiated for Project Mercury, adapted for the Gemini program, and further modified and augmented for Apollo. The earlier programs relied upon VHF telemetry in the 225–260 MHz band; and the Skylab launch vehicles use these links to transmit information at rates up to 72 kbps. The

Skylab Program will also use the Unified S-Band (USB) system developed for Apollo in addition to the VHF links. The carriers on the S-band 'down' links can be frequency modulated over a 3 MHz band when high data rates are required. The USB signal margins are sized to supply a signal-to-noise ratio in excess of 30 db over a 500 KHz information bandwidth at lunar ranges. The VHF telemetry transmitter bandwidth allocations are 500 KHz each.

Pulse Code Modulation (PCM) decommutation systems at the MSFN stations have been added for increased data handling needs of Skylab missions. PCM decommutators are used in the network stations to prepare data for input to the on-site telemetry data processing systems where parameters are formatted for transmission to the mission control center (MCC) at Houston.

B. DATA PROCESSING AT REMOTE SITES

Early in the Skylab program, a goal was established to handle the Skylab RF data without major changes to the existing network used for Apollo. Using present Apollo techniques, only a small fraction of the approximately 4×10^9 bits per day of Skylab

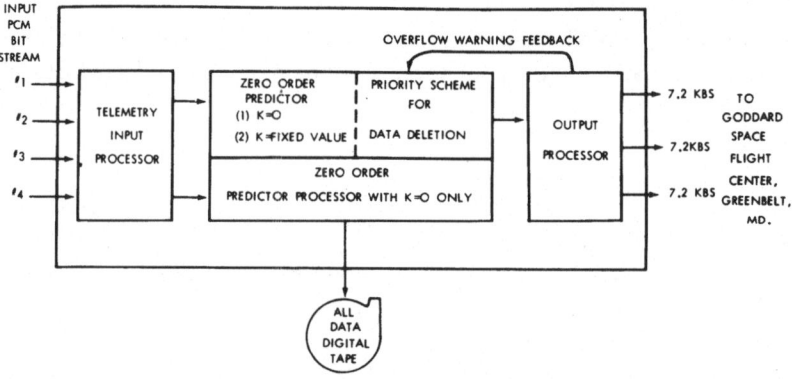

Fig. 17.   Skylab remote site telemetry data processing.

data generated could be returned to the mission control center in Houston in real or near-real time. Therefore, it was decided to use data compression and redundancy removal at the remote MSFN sites for Skylab. The selected approach is shown schematically in Figure 17. This scheme will permit transmission of essentially all telemetry data from remote sites to MCC in real time or nearly real time. The system will make use of hardware already existing at MSFN sites but will require increases in the number of high-speed data lines linking these sites to the MCC and the bit rates transmitted on these lines. The compression system takes advantage of the high redundancy inherent in the spacecraft-to-ground data. It will filter out redundant data, transmitting to MCC only the values of parameters that have changed more than a specified amount since they last were sampled, along with appropriate parameter identification and time information.

The data will be protected by an error correcting code. Based on careful consideration of past experience and expected variation with time of Skylab measurements, a 20 to 1 compression ratio was selected for design of the Skylab system. Thus, if the downlinked data is at least 95% redundant (excluding real time biomedical data, which will not be compressed) all the information content in the data can be transmitted to MCC in real time. If the downlink redundancy falls below 95%, adaptive program control will be used to preclude overload. The following steps will be taken as necessary in the order indicated:

(1) The magnitude of change in parameter values (corridor width) considered to be non-redundant would be increased.

(2) The next step would be to reduce the sampling rate within the computer on selected parameters with lower priorities.

(3) If still further reduction were needed, lower priority parameters would be deleted entirely from real time processing.

The more extreme adaptive control measures to avoid overflow will most likely only be required in the event of loss of one or more of the output lines. The adaptive processing will apply only to the output of data for real time transmission to the mission control center. Concurrently, an 'all data digital tape' will be prepared containing all non-redundant data without regard to the adaptive control procedures needed to prevent real time overflow. This tape may then be played back to MCC post-pass with the tape speed controlled to match the output line capacity.

Biomedical data has a low redundancy factor, so at least initially the real time biomedical data will not be subjected to compression. The biomedical data (EKG's and respiration waveforms) will be stripped from the downlinked data prior to entering the computer, and will be transmitted to the MCC on a dedicated high-speed data line.

Three additional high-speed data lines from each MSFN site, each operating at a rate of 7.2 kbps will be used to transmit the non-medical data to the Goddard Space Flight Center for relay to the MCC.

C. UTILIZATION OF TELEMETRY AND VOICE DATA

Telemetry and voice data are used at the mission control center for monitoring of crew and system status, for real time management of the mission, and for providing data to the experimenters, engineers, and operations analysts. A flow diagram of the telemetry data is shown in Figure 18. The remote site data processing and data relay to MCC were previously discussed and are shown on the left hand portion of this figure. Upon receipt at the mission control center, compressed data will be processed in real time and provided for display and real time mission control.

Concurrently, the real time data plus delayed transmission data ('all data') will be fed into a special system for mission management. This system is called the 'Mission operations planning system (MOPS)' and is shown schematically within the heavy-dashed lines of Figure 18. The MOPS is a terminal controlled, multi-computer system which contains a large portion of the Skylab data processing capability and which provides special routines and programs for:

(1) Rescheduling inflight activities as the inevitable real time deviations from the nominal flight plan occur (called activities scheduling program). The activities scheduling program will provide computer assistance for scheduling crew activities and experiments and is intended to provide a rapid and efficient means of utilizing a continually updated large data base for flight plan development and updating.

(2) Data storage and retrieval of all data, and processing for a variety of purposes (called mission data retrieval system or MDRS). The mission data retrieval system receives and processes all digital data from the network and maintains a 48-h data base of all data. It processes data for flight control in near real time and processes data for experimenters and engineering and operations analyses in accord with a pre-determined plan or upon demand. The prime output devices are microfilm, cathode ray tube displays, and printers.

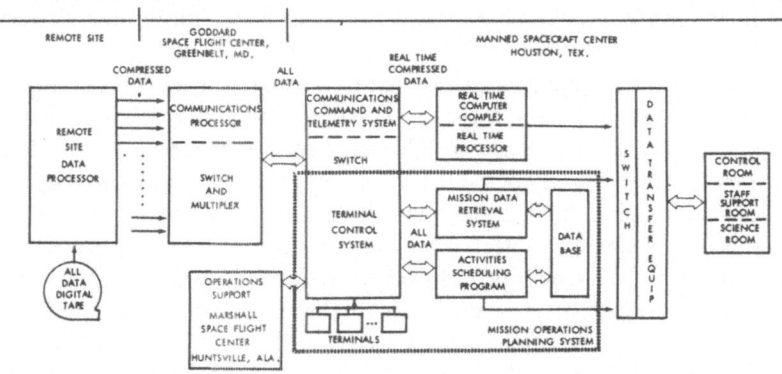

Fig. 18.   Skylab telemetry data flow and utilization.

(3) Management of complex spacecraft onboard subsystems – specifically the attitude control system and the electrical power system, both of which have significant constraints and interactions with the activities scheduled onboard the spacecraft.

The MOPS uses local and remote cathode ray tube (CRT) terminals for display and control. The communications, command and telemetry system (CCATS) (Univac 494 Computer) provides switching between CRT terminals and the various applications which are programmed in IBM 360-75 and Univac 1108 computers. Major portions of 'all data' are also relayed to the Marshall Space Flight Center in Huntsville, Alabama for analyses in support of the mission. The Marshall center has engineering responsibility for much of the Skylab hardware and for design integration of the Skylab system. The activities at the Marshall center include systems and hardware assessment for the orbital assembly and for experiments; analyses of flight problems and operation of flight simulation facilities and breadboards for problem solving and subsystem management.

4. Skylab Software

The total computer software for Skylab can be conveniently divided into two categories – that carried in the space vehicle and that located on the ground. A summary of the space-borne programs is presented in Figure 19. The computers for the launch vehicles and the command module are unchanged from the Apollo program and the software has been modified as required by the Skylab mission. These computers have 32 768 and 38 912 28-bit and 16-bit word capacity, respectively. The ATM computer, on the other hand, is being developed specifically for Skylab and has a

COMPUTER	COMPUTER SIZE	PRIMARY PROGRAM FUNCTIONS	REMARKS
LAUNCH VEHICLE DIGITAL COMPUTER (LVDC)	32768 28-BIT WORDS	BOOSTER GUIDANCE, NAVIGATION & CONTROL	ONE LVDC PER LAUNCH VEHICLE INTERNALLY TMR
		SEQUENCING	MODIFIED APOLLO SOFTWARE
		COMMUNICATIONS	
COMMAND MODULE COMPUTER (CMC)	38912 16-BIT WORDS	CSM GUIDANCE, NAVIGATION & CONTROL	ONE CMC PER CSM
		COMMUNICATIONS	BACKUP BY GROUND COMPUTERS AND CREW CHARTS/PROCEDURES
			MODIFIED APOLLO SOFTWARE
APOLLO TELESCOPE MOUNT DIGITAL COMPUTER (ATMDC)	16384 16-BIT WORDS	SUBSYSTEM REDUNDANCY MANAGEMENT	TWO ATMDCs PER SWS ONE IN STAND-BY REDUNDANCY
		SWS ATTITUDE CONTROL	SYSTEM AND SOFTWARE ARE UNIQUE TO SKYLAB
		ATM EXPERIMENT SUPPORT	
		COMMUNCATIONS	

Fig. 19. Skylab flight software.

capacity of 16 384 16-bit words. The ATM software provides for sub-system redundancy management, attitude control management for fine pointing of the solar experiments as well as pointing of the whole Skylab, and support for the astronomy experiment operation and communications. One unique software aspect for this computer is the programming to cyclically unload the control moment gyro momentum accumulation, working against the gravity gradient torque during the dark side of each Earth orbit. Two computers are carried – one in standby redundancy.

The ground-based computer facilities and associated software required for Skylab are so extensive that only a cursory overview of their scope can be mentioned here. Such a summary is provided in Figure 20. The ground-based computers and related software can be further divided into two categories – those used for general purpose applications and which are required for part of their operating time to support Skylab, and those that are fully dedicated to Skylab requirements for a specific period

- COMPUTERS USED IN GENERAL-PURPOSE APPLICATIONS
  - 41 "LARGE" COMPUTERS
    - SOME ARE STAND-ALONE
    - SOME ARE PART OF A MULTI-COMPUTER SYSTEM

  - 31 "SMALL" COMPUTERS
    - SOME ARE STAND-ALONE
    - SOME ARE USED AS PERIPHERAL PROCESSORS IN LARGE SYSTEMS

  - TOTAL CAPABILITY APPROXIMATELY 165 X 10^{12} OPERATIONS/YEAR

  - FUNCTIONS PERFORMED:
    - SCIENTIFIC AND ENGINEERING ANALYSIS
    - SYSTEMS DESIGN AND VERIFICATION
    - PRELAUNCH VEHICLE CHECKOUT SUPPORT
    - MISSION CONTROL
    - DATA PROCESSING AND ANALYSIS
    - ADMINISTRATIVE

- COMPUTERS USED IN DEDICATED, SINGLE-TASK APPLICATIONS
  - 25 "LARGE" COMPUTERS

  - 323* "SMALL" COMPUTERS

  - FUNCTIONS PERFORMED:
  - AUTOMATED SYSTEMS TESTS
  - TEST DATA PROCESSING
  - SYSTEMS SIMULATORS
  - COMMUNICATIONS PROCESSORS
  - CREW TRAINING SIMULATORS
  - MISSION CONTROL SUPPORT
  - REMOTE SITE PROCESSORS

                                        * APPROXIMATE INVENTORY AS OF
                                          JULY 31, 1971

Fig. 20.   Skylab ground software.

- SKYLAB SOFTWARE BOARD

  - CHAIRED BY PROGRAM DIRECTOR
  - PERIODICALLY REVIEWS OVERALL SKYLAB FLIGHT AND GROUND SOFTWARE STATUS
  - REVIEWS SELECTED AREAS IN DEPTH
  - PROVIDES OPEN CHANNEL OF COMMUNICATIONS FOR ALL SKYLAB SOFTWARE
  - PROVIDES MANAGEMENT VISIBILITY AND INSIGHT

- FLIGHT SOFTWARE IS UNDER FORMAL CONFIGURATION CONTROL BOARD CONTROL

  - LEVEL I: PROGRAM DIRECTOR
  - LEVEL II: CENTER PROGRAM MANAGERS OR DELEGATES
  - LEVEL III: REPORT TO LEVEL II; HANDLE ROUTINE CHANGES

- GROUND SOFTWARE MANAGEMENT SYSTEM IS A FUNCTION OF CRITICALITY

  - FORMAL CONTROL OF MISSION CRITICAL GROUND SOFTWARE
  - INFORMAL, BUT RIGOROUS, CONTROL OF OTHER GROUND SOFTWARE

Fig. 21.   Skylab software management.

of time during either the developmental, test and checkout or operational phases of the Skylab program. In the general purpose category are 41 'large' and 31 'small' computers. The total capacity of these computers is about 165×10^{12} operations per year. Typical functions performed for Skylab are pre-launch checkout support and data processing and analysis. In the dedicated category are 25 'large' and about 325 'small' computers of a variety of types. Typical functions are automated systems tests, system simulators, crew training simulators and remote site data processing.

The collective software for all of these space and ground located computers of course requires effective coordination and control in the same way as does the mission hardware. As summarized in Figure 21, all flight software and mission critical ground software are placed under formal configuration review and control at appropriate levels after the baseline configuration has been approved. The software for each application of course goes through its own development, qualification and acceptance program to verify mission compatibility.

5. Concluding Remarks

In summary and conclusion, the Skylab program presents a large and complex task of data management – both as regards data physically returned and data electronically returned to Earth. This task will be carried out through:

(1) Combined use of a variety of data collection devices onboard Skylab.

(2) Selected augmentation of the vast ground system capabilities developed for the Mercury, Gemini and Apollo programs.

(3) Selective use of new ground-based capabilities in adaptive-controlled data compression, computer-aided mission planning and computer-aided systems management of the spacecraft.

(4) Formal management and control of program software.

SPACE SHUTTLE DATA MANAGEMENT SYSTEM

JERRY C. McCALL

International Business Machines Corporation, Huntsville, Ala., U.S.A.

The report presented herein is derived from IBM's activity over the past two years as a member of the North American Rockwell, General Dynamics, Honeywell, American Airlines, British Aircraft Corporation Phase B shuttle team and from IBM's investment program and other various contracts. This is the approach that IBM has suggested to NASA through North American Rockwell and should not be construed as NASA's position.

During the past two years, an intensive effort has been made to select and define a configuration for a two-stage fully reusable Space Shuttle. This effort has included detailed analyses of the airframe, propulsive systems, all internal subsystems, ground operations, flight operations, and, finally, the program management of all these elements. The data management system for the orbiter is addressed in this paper. It will be addressed based upon this fully reusable concept rather than upon one of the partially reusable concepts currently under investigation in the extensions to the Phase B contracts. Whereas changing to a partially reusable vehicle significantly affects the airframe design, it need not and probably will not have a major impact on the avionics subsystem. It is intended that the data management systems for the reusable booster and orbiter be substantially identical.

The data management system to be described should be viewed as the eventual operational system rather than the initial system for the early horizontal and vertical test flights. Consistent with the concept of a phased development program for the vehicle, these early flights will utilize a subset of the operational avionics system, such that by adding to this subset the final operational system will be obtained. In addition to this subset, the test flights will utilize substantial quantities of development flight instrumentation (DFI) for extensive monitoring of all vehicle elements. This instrumentation will be excluded later on the operational vehicles.

Some consideration is being given to using avionics equipment on the early shuttle flights that would be taken from older programs. Such equipment would not be a subset of the eventual operational shuttle system. Using such equipment would have several disadvantages: namely, very little experience with and confidence in the ultimate operational avionics system would be obtained; there would be a physical impact on the vehicles containing this older equipment when it was stripped out and replaced by the operational equipment; ground testing, flight testing, and astronaut training would still be required for the operational equipment in addition to the older equipment; more current engineering expertise is available on the new systems now coming off the production lines than on older equipment no longer in production; and, lastly, current technology provides advantages in weight, power, reliability, and capability not available in the older technologies. Since these new systems currently in

L. G. Napolitano et al. (eds.), *Astronautical Research 1971*, 365–374. *All Rights Reserved.*

production will be adequate, with some modification, for the final operational shuttle, we believe they should be used for the early flights, on a subset basis.

Perhaps more than any other single element, the data management system (DMS) directly interacts with virtually all other elements of the vehicle. In the course of normal flight, if an element of the vehicle is expected to experience change in position or temperature or in electrical or chemical condition, it is a candidate to be sensed, analyzed and possibly controlled through the DMS. The following is a partial listing of elements that will interact in some way with the DMS:

The DMS itself (self-checking)
Astronauts and cockpit controls
Communications system
Aircraft landing aids
Main propulsion system
Orbital maneuvering propulsion system
Attitude control propulsion system
Air breathing engines
Fuel and oxidizer tanks and distribution systems
Guidance, navigation and control system
Environmental control and life support systems
Aerodynamic surfaces
Electrical power system
Hydraulic power system
Vehicle structure
Thermal protection system
Landing gear
Booster stage
Ground systems.

The DMS is the principal system through which the astronauts sense, analyze and control these elements; however, much of the vehicle management is done automatically to minimize the workload on the astronauts and to handle those responses which exceed the capability of a human being.

The onboard Data Management System (DMS) consists of the following elements:

Central processing/input-output units (CPU)
Main memory units and buses to CPUs
Drums and bases to CPUs
Interlaced avionics system data bus (5 line pairs)
Acquisition, control and test (ACT) units
CRT malfunction displays
Display electronic units (DEU)
Display select keyboards
Alphanumeric displays and computer entry keyboards
Caution and warning unit

Flight software

Data handling equipment within the DFI.

In addition, the ground portion of the data management system will interface with the onboard system to perform the following functions:

Centralized data processing for the launch facility

Automated ground-based checkout of vehicle (preflight and postflight)

Computer support of mission control

Simulation and training support.

The ACT unit is the standard interface between the DMS and other subsystems or vehicle elements and contains a 32-channel multiplexer and a 32-channel demultiplexer. Beyond this, some subsystems may utilize submultiplexers. In general, there is one ACT for each unit that is coupled to the data bus. Redundant units are separately coupled to the data bus; hence, one ACT is assigned to each redundant unit. When a unit or its assigned ACT fails to perform properly, both are electronically switched out by the computer and replaced by one of the redundant units and its assigned ACT.

Many elements of the vehicle, of course, have no redundant units yet will have ACTs assigned to them to determine such things as fuel levels, flow rates, skin temperatures, landing gear positions, etc. The assignment of ACTs to such units is based upon the number and proximity of the points to be sensed or controlled and the similarity of the data to be gathered.

Each ACT is separately, transformer-coupled to each of the five twisted, shielded line pairs that constitute the data bus system. These five lines run longitudinally and are located in different sectors near the exterior of the vehicle so that a major failure in one sector of the vehicle will not destroy all data bus transmission lines simultaneously, but would permit continued access to any undamaged ACTs and their associated subsystems. The transformer coupling obviates the necessity to physically tap into the transmission lines and provides DC isolation of subsystems.

Five lines are chosen for the data bus system so that during normal operation two will be in use: one for transmitting data from the computers and one for receiving data from the ACTs. The other three are redundant and any one of the three can be used to replace either the sending or receiving line that may have failed. The input/output logic in the computers has been designed to perform the function of controlling the operation of the bus system.

Within the ACTs the required A/D and D/A conversion resolution is eight bits, expandable to 10 or 12 bits if required. The data bus transmission rate of one megabit per second appears to be comfortably adequate to handle the expected data load of 200 000 bits per second, which arises from about 2000 analog signals, 2000 discretes, and 1500 digital signals.

The principal physical characteristics of the bus system are as follows: Each ACT will weigh 4.5 lb, occupy 0.07 ft^3, and will consume power based upon its use. In the standby mode it uses 1.5 W and during transmission, 25 W. Based upon expected utilization rates, the average power consumption per ACT is under 10 W. Only one ACT will transmit at any time; therefore, average power for the data bus/ACT system will be low. Each of the twisted, shielded lines can be up to 600 ft in length.

The data transmitted over the bus system must flow to and from the central computer complex, consisting of three basic sets of equipment: (1) the central processing units, which contain the input/output modules and bus controller; (2) the main storage units; and (3) the mass memory devices. Each operating element within one of these three sets of equipment is backed up by three redundant elements, satisfying the FO/FO/FS rule for electronics. Data are transferred between each CPU and the other two sets of equipment by two additional bus systems internal to the central computer complex and not connected to the vehicle data bus system. Each of the two internal buses is interlaced so that any CPU can directly access any of the twelve main storage units and any of the four mass memory devices.

An analysis of the software programs and data storage requirements indicates that 41 000 32-bit words of main storage will be adequate for all phases of the mission profile, including boost, abort, rendezvous and landing, and this includes a comfortable design contingency margin of 50% over the actual requirements of 27 000 words. These 41 000 words are contained in a bank of five main storage modules, 8192 32-bit words per module. (One CPU can address any of the twelve 8K modules, constituting the total main storage, but in normal operation it will only address the five assigned to it.)

A second CPU is in operation during critical mission phases and is addressing a different bank of five 8K main storage modules. The remaining two 8K modules are spares. This second CPU is running the same program simultaneously with the first CPU. A comparison technique has been devised to determine when one of these ten main storage units or one of these two operating CPUs has failed, which one has failed, and how to switch in one of the two remaining spare CPUs or one of the two spare main storage modules without degrading the vehicle performance. When the spares are exhausted through multiple failures, the vehicle will be returned safely to its base operating in the simplex mode with one CPU and a sufficient number of 8K modules to contain the programs necessary for this safe return – it could even be less than the normal five modules.

An alternate configuration consisting of five CPUs is still being analyzed wherein during these critical phases of the mission, three CPUs would be performing the same program simultaneously and continuous voting would be used to detect and eliminate a failed CPU.

The main storage modules are made up of low power, miniature cores with internal and external diameters of 0.008 and 0.013 in., respectively. The words are addressable as 32-bit full words or as 16-bit half words. Thus both 32- and 16-bit data and both long and short instructions are available. For each 16 bits of usable memory, there is additionally one partly bit and one memory protect bit. The operation is full parallel using binary, fixed point, 2's complement arithmetic. Floating point can be added. Sixteen general registers are built into the hardware. The 1 μs memory results in approximately a 2 μs add, 6 μs multiply and 10 μs divide. With typical instruction mixes, this results in an average instruction execution rate of 430 000 equivalent adds per second. Circuitry for the CPU and memory modules is at the MSI level. The system is designed to exceed an MTBF of 1325 hr. One CPU will weigh 34 lb and will consume 136 W when in use. The two redundant CPUs will not consume power until brought on line. The twelve 8K main storage units are packaged in four groups of three units each. One such group would weigh 71 lb. Power consumption of the memory modules varies with usage. The five memory units assigned to one CPU could use 238 W, but since the CPU only addresses one 8K unit at a time, methods are available and are being investigated to reduce the power utilization of the other four units not being addressed. The same would be true, of course, for the second CPU and its five memory units running in parallel with the first set.

The third set of elements in the central computer complex consists of the mass storage units. The use of mass storage devices is not absolutely mandatory since the CPU can address an enlarged set of main storage units sufficient to hold all programs and data required for the entire mission; however, we have chosen to limit the main storage to the programs and data needed for the CPU to execute the particular phase of the mission in which the vehicle is currently operating and overlay operational programs between mission phases. Under software control, data will be loaded from mass memory into protected areas of main memory through the same procedures and circuits used for initial loading on the ground prior to a mission. Historical data from previously executed phases of the mission as well as the programs, data and display formats for yet to be executed phases would reside in mass storage to be brought in only when needed. The programs and data needed for computer restarts would be in mass storage. Such restarts could become necessary due to a number of reasons, including radiation exposures, unaccounted for power transients, unpredicted modes of failure from internal or external causes, astronaut requests, ground checkout operations, and normal system shutdown when docked to the space station.

From the three classes of currently available mass storage devices – tapes, disks and drums – we have chosen the drum as the best compromise of weight, power, and access time. Only one drum would be required for the entire mission under normal conditions. The remaining three would be spares, but would be loaded with programs and data prior to launch. Any CPU can access any drum through the internal bus system. Current drum technology provides, on one drum, all the mass storage space required with a comfortable 100% growth factor over the actual storage requirements of 180 000 words. Thus, the higher storage capacity of tapes is not needed and the limitation of the lower access time of tape systems is avoided.

The capacity of one drum is 15 megabits on the surface of a 10.8 in. long and 6.5 in. diameter cylinder. The overall dimensions of the drum subassembly are $8.75 \times 12 \times 19$ in and it weighs 45 lb. The power supply for each drum has dimensions of $8.75 \times 4 \times 12$ in and weighs 25 lb. The data are recorded on 392 data tracks, plus spares. In addition, there are two timing tracks, plus spares. Of the fifty head assemblies, nine can be used to read or write. The heads are floated above the drum surface with gas bearings to achieve positive control of head spacing without the need for a mechanism to lower the heads to the proper position. During startup the heads rest on lands between the recording tracks on the drum surface. All internal signal distribution is accomplished through printed circuits and tape cables. The drum turns at 4800 RPMs which yields an average access time to the first addressed location of 6.25 μs. Worst case is 12.5 μs. The data transfer rate is 3.2 megabits per second. Average power consumption is 215 W and the MTBF is predicted to exceed 2000 hr.

The last major complement of operational flight hardware in the DMS is the set of integrated displays and controls in the cockpit. Each piece of this equipment is connected to the remainder of the DMS through an ACT unit, which is connected in turn to the vehicle data bus. The computer thus has access to the displays and controls and can send processed data to the displays or can pick up any data that has become available from the cockpit controls.

To simplify the software and to increase confidence in the integrity of proper sequencing of all data flowing in the system, no external interrupts are permitted to the computer. Instead, the computer is programmed to sample each data source at a rate that is adequate to receive, analyze and implement actions associated with the data from that source without unacceptable delays. A prescribed priority sequence is followed to resolve conflicts that will arise when normal sequencing is changed.

The necessity to handle spacecraft, aircraft, and some payload requirements from one cockpit led to the acceptance of the CRT, multifunction, computer-driven display as the fundamental element of the cockpit display system. There will be four identical CRTs in the cockpit, driven by two display electronic units (DEUs), one unit for two CRTs. A third DEU will be available as a spare. All three DEUs will be interconnected, however, so that any DEU may drive any CRT.

The face of the CRT is 6.125 by 5.25 in. and the tube is 17 in. long. The face is covered by a filter which will permit usage of the display under ambient lighting conditions as high as 10 000 ft lamberts. Data and background formats are presented using the random position stroke technique.

The DEU may be considered a preprocessor in that it holds 25 600 bits of data which are used to store formats, refresh the display, rotate and translate symbols, and control the deflection and blanking signals. The DEU has its own power supply. The memory in the DEU is composed of a mixture of read-write and read-only storage. The read-only portion is used to store standard format controls and symbols not subject to change while the read-write portion is used for input data and refresh buffering, and for changeable formats and symbols. The DEU storage consists of monolithic circuits which can be reloaded from mass memory, via the CPU, if necessary during the flight. The size of the monolithic DEU memory permits the storage of 800 characters, 500 vectors or 250 conics, and 128 alphanumeric characters.

There is a keyboard which the astronaut will use to select the flight mode he wishes to initiate. The appropriate sequence of actions will then be taken by the computer and the corresponding sequence of displays will automatically appear on the CRTs. In addition, however, the astronaut can use a second keyboard to request that the formats and data currently visible on the CRTs be replaced by others stored in a DEU or in the computer. Each key on these two keyboards is lighted to indicate what the astronaut has requested and is coded for multiple use.

A third, unlighted keyboard is available for the astronaut for inserting data and commands to the computer. A light emitting diode type of alphanumeric display is used to make visible the data associated with this keyboard prior to entering the computer. This display is also used as an output from the computer to indicate position and velocity, status of consumables, subsystem status, mission event times, and other selected parameters.

All three of the keyboards and associated displays are replicated for the second astronaut. Other elements of the cockpit, such as the hand controllers, pedal assembly, caution and warning lights, various switches, some dedicated displays, etc., are connected to the computers through the ACT/data bus system but are not considered part of the data management system, hence, will not be discussed in this paper.

In addition to the DEU preprocessor for the CRTs, there is a set of computers internal to the main thrust engines. These computers are considered part of the engine system rather than the DMS; however, they interact with the central computer via the vehicle data bus. For example, thrust level and thrust vector commands are generated by the central computer and are transmitted to the main thrust engine system via the data bus for execution.

The operational hardware previously described will be used to launch, operate and land the vehicle. It will also be used to determine the state of health of the vehicle before, during, and after the mission. In the event that trouble is encountered during the flight, the DMS must assist in detecting the trouble, isolating it to the malfunctioning unit, and electronically replacing the malfunctioning unit with a redundant unit. If the trouble is within the DMS itself, the self-diagnosis and self-replacement capability of the DMS must be automatically invoked by the DMS. In addition, the astronaut monitors the performance of the DMS and can initiate checkout routines. If the trouble is outside the DMS, then a different diagnostic and switching technique is used. When time is available for diagnosis, decision and implementation of remedial action, the astronaut will participate in this process as observer and decision maker; however, during time critical portions of the flight the DMS will have to handle certain critical checkout and fault isolation (COFI) functions automatically.

The DMS performs its self-diagnosis by running two CPUs simultaneously on the same program and continuously comparing results at each step of the process. In addition, there is built-in test hardware in the computers and software diagnostic routines. Prior to initiating these, however, the first indication of an error is treated as a transient and the program is immediately rerun. If the data still does not compare, then memory and CPU pattern-sensitivity checks are performed and other checking routines are initiated to determine which CPU or memory module should be replaced. The combination of these procedures provides 100% detection coverage. Even after two CPUs have failed and two memory modules have failed, the system remains fully operational with virtually 100% detection coverage.

The capability to run two elements of the vehicle simultaneously and compare the digital results to the precision of the calculation is unique to the CPUs and main storage, giving us a higher level of confidence in our knowledge of the state of health of the DMS than of the subsystems being monitored by the DMS. The astronauts will assist in identifying and diagnosing trouble with the cockpit displays and controls by observing vehicle responses to their commands. The DEU contains built-in-test equipment for self checking. The drums are periodically tested by reading and sum-checking and by writing and verifying the data. During passive portions of orbital flight, only one CPU will be used to save power. It will be checked during that time by self-test diagnostics and by astronaut observation of the displays and the failure of the system to implement commands. Data bus failures are indicated by the failure of any ACT units, and in particular those furthermost down the line, to respond to test or command signals. When a data bus failure is indicated, further testing is performed by the computer to determine whether a transmit or receive line has failed, then one of the redundant lines is switched in to replace it. The ACT units will auto-matically utilize built-in-test circuits each time a command is transmitted. Parity and status bits are also checked for validity.

To check out and monitor the performance of the other subsystems, the DMS is necessarily limited to: (1) limit testing, (2) reasonableness testing, and (3) short term trending. Each subsystem has been examined and estimates have been made as to the number and type of checks that should be made for that subsystem to monitor its per-formance. These total to about 1600, with about two-thirds of them being limit tests and one-third reasonableness tests. The 1600 figure does not include those internal to the DMS or the main thrust engine computers.

There is a question as to how much checking should be done by the central computer versus subsystem built-in-test equipment or manual checking. As previously indicated, some elements of the DMS utilize built-in-test equipment. An analysis of 18 sub-systems containing 66 simplex functional paths was made to determine how they should be checked. Of these 66 paths, 34 could be checked by any of the three methods at equal cost; 32 had a cost effective selection – 16 for manual, 7 for built-in-test, and 7 for checking by the central computer. Time critical functions cannot be checked manually. Furthermore, for convenience to the astronauts and ground crews, manual checking will be discouraged. When the central computer performs the checking function it does so over the same ACT/data bus line that is used to communicate with that subsystem during normal operation. Some subsystems, for example, com-munications or electrical power, will routinely be checked by more than one of the three methods.

6700 32-bit words are needed to perform the subsystem testing by the central computer. Two-thirds of this is devoted to checking the power and GN & C sub-systems. In addition, the ability of the astronaut and ground crews to detect trouble by observing the displays and responses of the vehicle to commands constitutes a sub-stantial, independent check on the performance of the vehicle subsystems.

The DMS hardware is, of course, under the control of the software, all of which always resides on the drum so that main memory and DEU memory can be loaded as needed. Of the 180 000 words utilized from the drum's 400 000 word capacity, 50 000 are for recording historical flight data, 30 000 are for COFI diagnostics, 18 000 for GN & C, and 12 000 for all other programs, including the executive. The remaining 70 000 words are used for tables, 30 000 of which are for display formats.

The 60 000 words of programs are separated into individual application modules and an executive. Only 27 000 of the 60 000 are needed in main storage at one time. Since there are no external interrupts, the executive remains in positive control at all times. The application modules will be written so that the module being executed can communicate with another module only through the executive. The application modules use a set of general registers that are different from those used by the executive to ensure that the application modules will never interfere with executive control. By isolating each application module in this way, it can be written, checked out and modified with a minimum impact on the remainder of the modules, thus reducing the exposure to interacting errors and simplifying the isolation of errors when they are detected. This isolation also improves management interfaces and control during the development of the modules.

All program instructions will reside in protected (hardware or software) main memory locations. Protected memory is defined as read-only while unprotected is defined as read-write. Protected and unprotected areas are predetermined. Normal program operation cannot change the protected-unprotected status of main memory. All unused memory locations will be assigned as protected so that if a program erroneously attempts to write into protected memory, the computer will enter a failure isolation program.

When the computer is used to present dynamic data on the CRT, augmenting the capability of the DEU, it does so by loading the input data buffers of the display unit with the continuously changing data at the required rate. This data is then repeatedly fetched and processed by the DEU at a sufficiently high rate so that the dynamic displays appear as continuous motion to the human eye. The background formats required to generate these dynamic displays are stored in mass memory and are transferred to the DEU as required, which minimizes the main storage required for the display processing function.

All data taken from the computer for operational use will be checked by limit or reasonableness (or both) to ensure that the data are valid prior to use in critical computations. Both reasonableness and limit values may vary with time or mission phase. For example, expected velocity increments from the inertial sensors will vary widely for different mission phases and will vary with time within a mission phase. If input data fails to meet the reasonableness or limit criteria, diagnostics are invoked as necessary to determine the failed element which is then replaced by component switching.

To speed the process of implementing the COFI function and to minimize main memory requirements, the COFI programs are kept in mass storage; and, when needed, the executive loads the first COFI program into main storage and initiates execution. While it is being executed, the next COFI program is loaded from mass memory into another section of main memory. This second program is then executed immediately upon completion of the first.

In addition to the operational flight hardware and software, the early shuttle flights will be equipped with development flight instrumentation (DFI). It is intended that the DFI will make maximum use of the operational data bus technology, documentation and avionics subsystem components to minimize DFI costs. By following this plan, it may also be possible to utilize some of the DFI stripped from early vehicles as operational equipment in later vehicles or as logistic spares. The DFI will be distributed throughout the vehicle to minimize cable length and will be attached to the vehicle data bus through an ACT/select buffer, so that it can have access to that data without interfering with its flow. In general, the physical interface between the DFI and the operational systems will be minimized to reduce the rework associated with removing the DFI out when it is no longer needed.

Various types of DFI recorders will be used onboard and on the ground while the vehicle umbilical is still attached. The magnetic tapes from these recorders will be removed and processed on ground at the central computing complex. Onboard data processing by the DFI will be limited to such things as format generation, sequencing, limit checking, etc., as opposed to data analysis.

The shuttle data management systems on the ground can rather arbitrarily be separated into two categories – those associated directly with the vehicle and those only indirectly associated. The latter category is important but of less interest in this discussion. It would include such things as payload support, fleet management, logistics, general launch facility operation, contingency control, mission control, system development, factory support, etc. The first category would include facilities to develop the onboard software, train astronauts in the use of this software, perform vehicle checkout and launch on the pad, and perform postflight analysis for rapid turnaround.

The more nearly the hardware and software in the first category can be made identical to the onboard hardware and software, the easier it will be to manage the interfaces. Even where larger computers and tape drives will be used on the ground, keeping the instruction sets, data formats, executives, etc., similar to those used onboard will be effective in simplifying the processing. When changes are introduced to the vehicles and missions or when trouble develops, the work load associated with solving these problems will be substantially reduced if the ground and flight hardware and software are closely related.

During turnaround between flights, most of the time will be consumed by the manual effort associated with replacing faulty equipment, checking the system, loading the next payload, mating and fueling the vehicles, and performing the final launch checkout. Therefore, to achieve a short turnaround, it is extremely important that the postflight data analysis, including recommending the fixes, be limited to the shortest possible time period. This demands automated analysis of simplified designs.

In summary, we believe the data management system can be developed from current technology and that it should be. Early flights should utilize a subset of the ultimate operational equipment. The ground data management hardware and software used directly with the onboard DMS should be nearly identical to the onboard DMS. It will be necessary to implement the development of this onboard and ground data management system – specifically software – concurrently with that of other vehicle subsystems to achieve a balanced design that will result in the production of a cost effective space transportation system.

PART III

UTILIZATION AND APPLICATIONS
OF SPACE TECHNOLOGY

A. EARTH RESOURCES SATELLITES

GLOBAL MONITORING AND
REMOTE SENSING FROM SATELLITES

BENGT G. LUNDHOLM

Ecological Research Committee, Swedish Natural Science Research Council,
Sveavägen 166, XV, S-113 46 Stockholm, Sweden

Abstract. *Air monitoring*. At present there is no existing technology for immediate satellite-borne monitoring of air pollutants. The instrumentation now being developed seems well adapted to use for ground observations or to be carried by aircraft. The very special conditions in the *lower* atmosphere indicate that in the near future this medium will not be monitored for information on the global environment, but the new techniques will be important for local monitoring. On the other hand the technology apparently exists for global monitoring of the *upper* atmosphere to follow large-scale climatic changes.

Water monitoring. Fresh water monitoring is mostly of local and regional interest and remote sensing from aircraft may be of future importance. The planned Earth resource satellites have too long intervals between observations to be really useful. In the future special ocean watching satellites, placed in Earth-synchronous equatorial orbit or in low polar orbits, could be used.

To follow the changes in the ocean ecosystems only a limited number of variables should be necessary. The most basic and important variable is primary production. Primary production is partly related to the number of green cells present – or the biomass. Since there is a fairly constant relation between the amount of chlorophyll and the biomass, it might be possible to get an estimate of the biomass by measuring the chlorophyll from the air by aircraft or satellites. For development of instrumentation and for coordination of the global approach special international pilot projects have to be started.

Terrestrial monitoring. The existing methods offer great possibilities for the use of remote sensing to survey terrestrial ecosystems and to record changes. A recording of the vegetation changes in marginal areas as in deserts and tundras would be good variables for monitoring from satellites. The technology is available. Ground truth research has to be organized as pilot projects on an international level.

1. Definitions

'Monitoring' is defined here as measurements of variables or properties in relation to time. For measurements in relation to space, the word 'surveying' is used. Thus monitoring describes temporal variation and surveying the geographical or spatial distribution.

This paper will deal with environmental monitoring. The motivation for this special monitoring is that we fear that we are confronted with an environmental crisis. There are a multitude of problems in relation to the changing environment, but here we will only deal with problems of importance from a global point of view.

The main problem is the threat that pollution will affect the whole biosphere. Biosphere refers here to the global ecosystems including air, water, soil and biota. This pollution affects the biological systems including man and also the geophysical systems that determine global climate. The monitoring must be connected with these problems.

L. G. Napolitano et al. (eds.), Astronautical Research 1971, 379–387. All Rights Reserved.

The main task for global environmental monitoring is to establish secular trends. If it is possible to establish such trends, e.g., a change in the global albedo or an increase in the acidity of the rainwater or the disappearance of certain ecosystems, the question will be if these changes are natural or if they are caused by man. From the human point of view, the consequences of these changes are independent of the basic cause. It is, however, of importance to be able to predict these changes with highest possible certainty, to make counter-measures possible. If it is possible to establish the cause-effect relationship, e.g., that the change is caused by human activities or that it is natural, it might – in the first case – be possible to break the trend by special actions such as prohibiting ocean dumping or banning leaded gasoline. For this reason it is important to be able to separate the 'natural' changes from the 'human made' changes. To be able to do that the natural baselines have to be established. These natural baseline-values are sometimes very complicated with different temporal cycles superimposed on each other, each with a specific statistical variation.

It is thus possible to measure both causes and effects and in many cases it is impossible to separate these two categories of variables as they are very interlinked. From a practical point the variables can be divided into two groups.

(1) Physico-chemical or abiotic variables.
(2) Biological variables.

When the proper variables have been chosen, different operational systems can be used for the actual measuring. A global network of reference stations covering both oceans and continents is now being considered. It is also clear that for certain variables a station network is not well suited, for example to register changes in distribution of vegetation. In these cases other systems of measurement have to be considered and here remote sensing from satellites is a possibility and offers a new tool for global monitoring. I will now discuss the advantages and disadvantages of this new technique.

There is a very marked difference between different types of environmental monitoring. As pointed out earlier, the aim of global monitoring is to establish secular trends. For that we need an integration both in time and space. To take one example, we want to know if the DDT content in the oceans has changed from one year to the next, or if the timberline has changed. In regional monitoring, the area may have very different size, e.g., a city, a country or an international region. The aim for the monitoring in such an area might be to continuously follow pollution levels. It might be important to get information on short periods with extremely high levels of pollution, as they have biological significance. In many cases a high spatial resolution is wanted in order to map the single emissions. It must be stressed that instrumentation to be used in satellites in many cases might be very well adapted to such regional recording. The instruments, however, have in this case to be placed on the ground or in air-crafts. It is thus important to have a cross-fertilization between regional monitoring and the emerging technique for satellite instrumentation.

We now have lines of development within the technique for satellite instrumentation which are very promising from a regional monitoring point of view.

For larger regions it is also possible that satellite-borne instrumentation can be used where a continuous surveillance is needed, e.g., to watch oil spill in an ocean area. In such a case a geo-synchronous satellite might be a useful watchdog.

The great advantage of a satellite system for global monitoring is the high degree of geographical coverage, which might be continuous or repeated. Depending on the technique used for measuring the variables, this coverage has very marked restrictions, e.g., a cloud cover may stop the recording for extensive periods. The importance of these disturbances is, however, to a great extent dependent upon the variable which is recorded. From an information point of view, extensive coverage increases the amount of information in such a way that it may be difficult to handle. This may be partly balanced by different degrees of resolution. High resolution increases the amount of information, but in global monitoring high resolution may not be wanted, as geographical integration is needed. That means that the demands on the instruments for global monitoring in relation to resolution can be rather moderate. This is one of the main reasons why satellites are especially suited for global monitoring. It is, however, hardly necessary to point out that data handling is as important as proper instrumentation.

The fact that through satellite monitoring it is possible to use the instrumentation over large areas and that this instrumentation – even the same instrument – may be continuously used is very essential. This will give results comparable both in space and time. It is, of course, obvious that such criteria must be the base for all monitoring. A review of the present activities and the present attempts to set up monitoring systems shows, however, how extremely difficult it is to fulfil these criteria. Some claim, even, that much of the monitoring activities now planned may be useless because of difficulties with intercalibration in space and time. In this field the application of satellite-borne instrumentation may serve both as a good example and also as a tool for intercalibration for other monitoring systems.

It is of the nature of measurements taken from satellites that they reflect the momentary conditions. This is a disadvantage of global monitoring, where measurements integrated in time are needed in order to establish the secular trends. The value of these instant measurements, are, however, very different in the different media such as water, air, soil, etc.

2. Changes in the Atmosphere

Air pollutants are mostly confined to the thin layer near the Earth surface. At the same time, it has to be kept in mind that there are very extensive variations between different areas depending on local emissions and local differences in air mixing. There are also very rapid changes in the concentrations of the same point from time to time. That means that instruments have to cover a large range of concentrations from very low 'back-ground' levels to 'high exposure' levels. Satellites look down at pollution through the thinnest dimension, and here the extremely variable surface concentrations will dominate the results. Under such conditions, instant measurement has, from a global monitoring point of view, little value. This can be compensated for by short intervals between the measurements or by continuous measuring.

The present instrumentation related to air monitoring and remote sensing is either based on optical absorption or on different kinds of back-scattering of energy in the ultra-violet, visible and infrared parts of the electromagnetic spectrum.

The most useful part of the spectrum is the middle infrared, where most pollutants have their absorption bands or 'finger-prints'. Here we also have windows for the Earth radiation. Day and night observations are thus possible. Quantitative measurements of the pollutants are possible only when the surface temperature as well as the atmospheric temperature profiles are known. It is now possible to get such profiles from existing satellites. A very serious handicap is, however, that a cloud cover makes detection impossible.

These techniques have now been tested from the ground and from aircraft, and they will be of importance for local and regional monitoring. It is also clear that the presence of air pollutants have been registered from satellites, but it is rather doubtful if these recordings have enough sensitivity to be used for global monitoring.

Measurements from down-looking satellites are thus dominated by conditions in the lower atmosphere. Conditions in the upper atmosphere (above 5 km) are quite different. Here we have a high degree of mixing and integration, and very small variations. At the same time it has been pointed out that conditions in the upper atmosphere are of great importance to the global climate. It is thus necessary to follow any changes in the upper atmosphere. This has been stressed by the two summer studies on global changes arranged by Professor Carroll Wilson of Massachusetts Institute of Technology. The first was held last year in Williamstown and the second this year in Stockholm, the report from which will be published within a few days. Satellites offer now the only possible technique to monitor the upper atmosphere, and it is probable that in the future we may have co-orbiting satellites monitoring pollutants along horizontal paths in the upper atmosphere in order to avoid the disturbances from the 'surface layer'. This is a technical challenge.

It has, however, to be stressed that we already have suitable techniques to follow certain large-scale climatic changes from satellites: that is changes in the global cloud patterns and cloud properties. Of importance also are recordings of differences in the relation to the heat budget of the Earth atmosphere system as changes in albedo etc. We have an international scientific program, GARP (Global Atmospheric Research Program) especially designed to deal with these conditions. This program is organized by International Council of Scientific Unions (ICSU) and the World Meteorological Organization (WMO).

There are also other variables related to human activities which are important for the assessment of climatic changes and can be measured from satellites in special censuses. Such variables are the arctic ice cover, areas under irrigation, artificial lakes, and the extent of urbanization. The frequency with which information should be gathered will vary according to the variable, but in most cases once a year is more than sufficient.

3. Changes in the Waters

Fresh water monitoring is mostly of national and regional interest and remote sensing from aircrafts may in the future be of importance. Monitoring of ocean conditions from satellites may in the future be of great global importance. Water variables suited for satellite monitorings are for the present rather limited. We can separate them in two groups:

(1) Variables directly related to pollution.
(2) Variables related to biological productivity of the waters.

The pollution variables can be grouped in three categories.

A. THERMAL POLLUTION

There are excellent methods of recording the surface temperature, where the infrared band is used. We have here a well developed technology which already has been used in satellites. A very important limitation is, however, that the energy emitted is coming from a very thin water layer. For the strongest emission at a wavelength of 10 μ the emitting layer is only 0.04 mm. Thermal pollution is a local and regional problem – but as far as can be estimated – hardly a global problem.

B. POLLUTANTS IN WATER

In this area the possibility of getting information is very limited by the fact that solar radiation has a very restricted ability to penetrate water. Only certain parts of the visible light have this penetration ability. This light is the energy source for the life in the oceans. The colour of water, especially in the oceans, is a good general index of the pollution load, but it is not yet possible to determine which pollutant is present. It is doubtful if it ever could be possible to record water pollutants from satellites with an accuracy needed for global monitoring.

C. POLLUTANTS ON THE SURFACE

Aircraft instrumentation and techniques are now ready to record surface pollutants as oil spills. This will be very important for local and regional monitoring. The same technique can probably also be used in satellites. Because of the importance of keeping a continuous watch on pollution of the oceans, special geostationary satellites may be needed for this purpose.

Variables related to the biological productivity are of global significance. From the human point of view, the oceans are a source of protein – a critical food stuff in a hungry world. It has been indicated that it would be possible to increase the protein output from the oceans, and great expectations are now directed towards the oceans. However, these prospects are not very good, as the productive areas are very limited in the oceans and these areas are the same ones which are used as recipients of human wastes and as highways for oil-spilling tankers. The human wastes may cause eutrophication, that is 'run-away' productivity, but these wastes also contain poisons and biologically active substances, which may decrease the biological productivity or make seafood unsuitable for human consumption. It is thus of utmost importance to follow changes in the oceanic biological systems.

Our knowledge of the oceans is not very good. There are great difficulties in grasping the oceans as a totality and it has been very hard to get synoptic views of ocean conditions. For that reason we know little about the geographical and time variations. However, through the satellite technique, we have for the first time a tool suited to surveying large areas. This new 'horizontal' approach to the oceans will increase our knowledge in a revolutionary way.

Our problem is, however, that for the present moment it is impossible to follow the long-term biological changes in the marine ecosystems. We think now, however, that only a limited number of variables should be necessary, but the recording must take place at critical points and at critical times. These critical points and times are not yet known but have to be found by special pilot studies. A short discussion on the possible variables may, however, be useful. The most basic and important variable is primary production – how much organic material is produced per time unit. This is, however, difficult to measure. Primary production is partly related to the number of green cells present, the biomass. This biomass is easy to measure, but to get a synoptic picture of large areas will be extremely expensive with traditional methods.

Since there is a fairly constant relation between the biomass and the amount of chlorophyll, it might be possible to get a good estimate of the biomass by measuring the chlorophyll content in the water. Here it is possible that remote sensing instrumentation in combination with satellites will offer ways of cheap synoptic mapping. Methods can be spectrographic and based on the visible light which is backscattered from the water. This new instrumentation has now been flown in aircraft with success. Other methods are based on the blue-green laser which penetrates deep into the water. Here fluorescence might be used for identification of the chlorophyll.

There are also other types of monitoring of the oceans which have indirect interest for large-scale changes as monitoring of fish resources in order to increase the catches. I will, however, not deal here with this type of monitoring that will require – among other things – real time data and thus require a quite different organization.

4. Changes on the Continents

In the lower atmosphere and in waters the temporal changes are very rapid. If we thus want to have an integrated value, the measurements have to be taken with high frequency or with short time intervals. The conditions on the continents are quite different. Here the changes are – for instance in the soils and in certain types of vegetation – very slow, and a single measurement in itself represents a time integration. This means that variables can be measured at long time intervals and in some cases at irregular intervals. For many variables one recording a year is sufficient and there are also examples of variables that can be measured for monitoring even with longer time intervals. These favour the use of satellite monitoring as there can be long intervals and disturbances caused by cloud cover are not so important or can be completely ignored.

Another important difference between the continents and the oceans is that there are very marked local differences on the continents. This is for instance the case with variables related to the soil or the vegetation. Even between points that are very near each other, remarkable differences may occur. This makes monitoring in general difficult and special sampling methods have to be used. Here, however, satellite-borne instruments may offer advantages, compared with other monitoring systems, as we have a certain spatial integration because of the low resolution. Not only the operational system but also existing instrumentation offer great possibilities for remote sensing. Here the multiband spectral scanning system is especially promising to record variables related to terrestrial ecosystems. And there is no doubt that this method is a very valuable tool for ecological research and mapping of biological resources. This technique has been tested in aircrafts with very good results. With the launching of ERTS A the space-craft stage is reached.

By using satellite-borne multiband spectral scanning it is possible to map, classify and evaluate the different terrestrial ecosystems. Here the man-made ecosystems, crop lands and forests, are much easier to record and understand than the complicated 'natural ecosystems'. Several projects with the International Biological Programme have shown that this technique will get good results for 'natural ecosystems'. An important difference between man-made ecosystems and natural, is that the latter are more stable. In man-made ecosystems instability appears as plant diseases and pest outbreaks. The main reason for this difference is that man-made ecosystems are simpler, and with fewer stabilizing feed backs. We have also a general tendency that the human impact on the 'natural' ecosystems increases the instability. It might be possible to register outbreaks of diseases and pests by satellite-borne remote sensing. Changes in the frequency and extension of these outbreaks might be a good variable for global monitoring of important changes in biological systems.

Even if we know more about the functioning of the terrestrial ecosystems than about the marine conditions, our knowledge will not yet allow us to select variables for a proper recording of the changes. The structure of the ecosystem has evolved to resist changes. As pointed out earlier the stability is less in simple ecosystems with a few species and few feed backs. In marginal areas, where few species can exist and they often live on the border line of survival, the changes are very pronounced. Examples of such unstable ecosystems are found in tundra and desert areas, where population explosions and marked changes in the distribution of species are common.

This general stress in the marginal areas also affects the vegetation. Changes in vegetation boundaries are thus very pronounced. A very slight change in the environmental variables will result in large areal differences in the vegetation distribution. The impact of man will have a similar effect. The history of man is characterized by increasing desert areas, with examples from Central Asia and North Africa. Also today there are many signs that as a result of man's activities the deserts are encroaching on cultivated land. Both the Sahara and Kalahari are examples of this phenomenon. This is a very serious problem for the less developed countries, which are forced to push new cultivation out into the marginal zones. With the new technology in agriculture supported by national and international aid-organizations, the speed of destruction may increase if mistakes cannot be avoided. These aspects are for the moment very critical both in North Africa and Southern Africa (Botswana) where large schemes for the use of marginal areas are planned. A recording of changes in the vegetation boundaries will thus here be of importance from a global monitoring point of view and will also assist the LDC:s and give background information about critical resources to avoid mistakes.

Such a recording is possible from satellites. The necessary technique is available and ready to be satellite-borne. Different types of vegetation can be registered both by visible and infra-red bands. Both the U.S. and the U.S.S.R. have proved that it is possible to register vegetation from satellites. The Soviet scientists have worked on 'desert problems' in relation to the flights of the 'Soyuz – 6.7.8.' and Leningrad University has prepared geobotanical maps in Africa from satellite 'Zond 5'. The American scientists have worked especially within the deserts in the Western U.S., and there are also bilateral projects between the U.S.A. and South American states dealing with 'dry areas' in South America. An interesting example here is the joint U.S.-Argentine IBP project. These projects are using aircraft-borne instrumentation. A recording of the global vegetation boundaries from satellites will be possible with the launching of ERTS A.

I think it is very important that we now take this opportunity to use the information from satellites to record changes in vegetation boundaries as a variable for global monitoring. The first step will be to form an international organization for ground truth. Since political complications may occur, it is possible to avoid this by using non-governmental organizations for the pioneer research. To cover the ground truth aspects, bodies from the ICSU-family can be used such as the Special Committee for International Biological Programme or the newly formed Scientific Committee on the Problems of the Environment (SCOPE) which is covering the total environment.

The ground truth work in itself is not too complicated. It will be a recording of the main vegetation cover and composition and also phenology for the different main species. Some physical variables have also to be recorded as precipitation, surface moisture, soil and plant temperature registered with infra-red thermometers. It is also necessary to have an initial step with recording of the variables from aircraft. At a later stage, monthly or even less frequent records would be sufficient. Satellites in polar orbits with a 17-day repeated coverage can be used even in the northern areas, which have an extensive cloud cover. Here in the northern areas the vegetation changes are slower and more regular than in the hotter desert areas, where vegetation explodes after the very irregular rains. In desert areas a more frequent coverage is needed, but this is also possible as the cloud cover is less. In the northern areas there is already intensive circumpolar scientific co-operation within the framework of the Tundra Biome within the International Biological Programme. In this project participate the U.S.A., Canada, the Scandinavian Countries, Finland and Soviet Union. Within this biome study are plans to use remote sensing from aircraft and space craft, which can be used as a pilot project for global monitoring. The main thing is now to establish co-operation between the tundrabiome, responsible for the ground truth, and the two space powers in possession of the space technology.

IBP is for the present moment building up a biome study for deserts. This is, however, not yet extensive enough to cover the main deserts on the globe. It should, however, be possible to develop pilot projects between the international scientific community, represented by the ICSU-family, the space powers and the countries with aid organizations interested in the use of the semiarid areas. Such projects will have two objectives: global monitoring and inventory of resources, and both these objectives can be met by a recording of vegetation changes.

5. Telemetry

It was earlier stressed that the measurements taken from satellites are instant values and that this had many disadvantages. A possibility to avoid these difficulties is to place the sensor on the ground and use the satellite to collect the information and to store it for later transfer to ground stations. In such a case it is possible for the sensor to give integrated information for any time period. This system has many advantages as it will cover large areas with the same standardized method. The *in situ* measurements are in general more accurate than the satellite measurements. Even remote and distant areas will be easily available for continuous recording. I have, however, no time to penetrate this technique which has many other advantages but also very special problems.

6. Evaluation

To sum up, it is quite clear that global monitoring from satellites offers many possibilities. In some cases the technology is available, in other cases it is developing, and in some cases extensive pilot studies are needed. It is important that before any variable is accepted and included in a monitoring system for satellites, it must be evaluated in relation to its feasibility and relevance. One example will illustrate this. As has been pointed out earlier, the chlorophyll content in ocean water may be a very relevant variable. It is probable that it can be recorded from satellites. The critical question is whether it is possible to use this variable to establish changes in the chlorophyll content from year to year to establish secular trends. Standard deviation and confidence limits are necessary in assessing the usefulness of a variable.

A comparative evaluation must also be done between remote sensing from satellites and other systems for monitoring. From a technical point of view the present methods in use for satellite measuring are less accurate than the *in situ* methods. We must, however, consider that we have a rapid development towards improved methods. The crucial question, however, is what degree of accuracy is needed for the different variables. The selection of systems depends on these answers.

In this comparative evaluation, a cost-benefit analysis must play an important role. It has to be underlined that in many cases it should be possible to get information of importance to global monitoring from already existing satellites and from planned projects with small additional cost, as a recording of vegetational changes.

THE GROWTH OF REMOTE SENSING
THROUGH THE NIMBUS AND ERTS SPACECRAFT

I. SHELDON HAAS and JOHN J. HORAN

General Electric Co., P.O. Box 8555, Philadelphia, Pa. 19101, U.S.A.

1. Introduction

Less than fourteen years ago, most of us experienced a great thrill in listening to the radio signals sent from the first man-made Earth-orbiting satellite. In fact, it will be exactly fourteen years two weeks from today that Sputnik I was launched and achieved its electrifying orbit around the Earth. Since that time many other achievements have added to this new technology of space flight. One of these happened about a decade ago when we had first look at our Earth from space taken by a television camera aboard another man-made satellite. Since that date literally hundreds of thousands of

Fig. 1. Nimbus 4 satellite.

L. G. Napolitano et al. (eds.), *Astronautical Research 1971, 389–401. All Rights Reserved.*
Copyright © 1973 by D. Reidel Publishing Company, Dordrecht-Holland.

photographs, both in the IR and visible spectrum, have been produced by a variety of sensors carried by a variety of automated and manned earth-orbiting satellites.

Considering that it has been centuries since man first had the opportunity to view remote objects through the eyepiece of a telescope, the last ten years have been a veritable explosion in the technology of remote sensing.

A principal contributor to this technological revolution has been the Nimbus satellite program (Figure 1), initiated by the National Aeronautics and Space Administration in the early 1960's to produce an advanced meteorological satellite. We are looking at a photograph of the fourth in the series of Nimbus observatories taken prior to its launch on April 17, 1970. Along with its predecessors, this satellite contributed significant stimulus to the potentials of remote sensing technology with the successful performance of the British and American vertical temperature sounders and the improved infrared scanner techniques. This satellite is still operating and supplying data both to experimental scientists and to environmentally oriented agencies of our government.

The Nimbus program has matured into the United States' principal satellite remote sensing research program. Soon it will be augmented by the Earth resources technology satellite (ERTS) program.

Combined, these programs encompass a wide variety of remote sensing techniques, operating in the entire spectrum from ultraviolet through infrared to the microwave region. Today, I shall briefly review the growth of satellite remote sensing through these two projects.

2. Remote Sensing Considerations

Remote sensors, such as those evident below the satellite structure, measure the electromagnetic energy impinging on them from a remote point, in our case the Earth and its atmosphere. Reflected solar illumination, self-emitted thermal radiation, reflected lunar and stellar illumination, radioactivity, and sensor-generation illumination beamed to illuminate a particular target may all be received at a sensor. Because of their relative magnitude, the reflected solar and self-emitted thermal energy sources have been of prime significance for most remote sensing applications to date, although illumination from spacecraft-borne sensors shows great promise for the future.

We see these two principal energy sources pictured in Figure 2. The 300 °K blackbody curve is representative of the Earth as a thermal source and is approximately the same magnitude as the reflected solar energy, but fortunately shifted in the electromagnetic spectrum. The relative independence of these two sources of energy and the known atmospheric transmission characteristics represented on Figure 3 have been the basis for much of the work in satellite remote sensing up until now. The areas of low transmission indicated here are due to absorption by various mixtures of atmospheric components, such as carbon dioxide, ozone and water vapor.

These high emission (low transmission) segments of the spectrum and the atmospheric windows at 8 and 10 μm continue to be explored by scientists in an effort to improve our detailed knowledge of the world we live in. Remote sensors have already

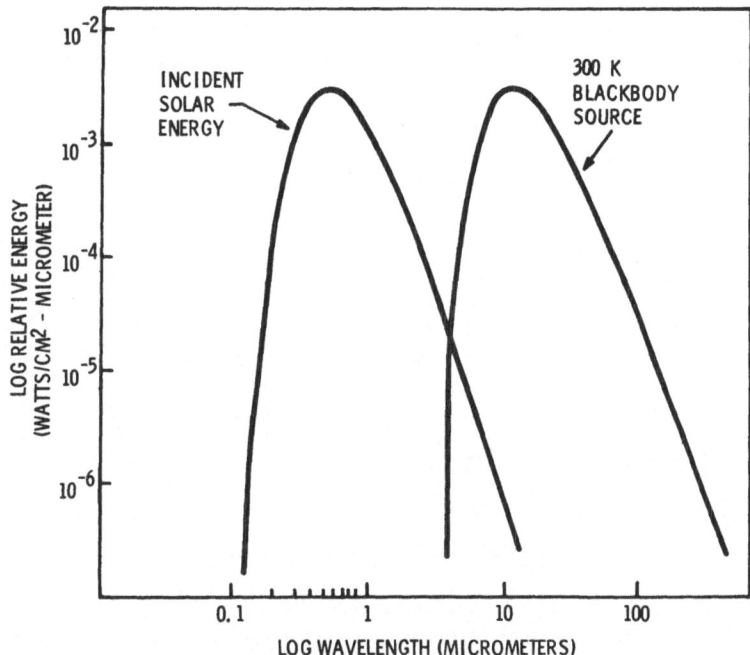

Fig. 2.   Spectral distribution of principal energy sources.

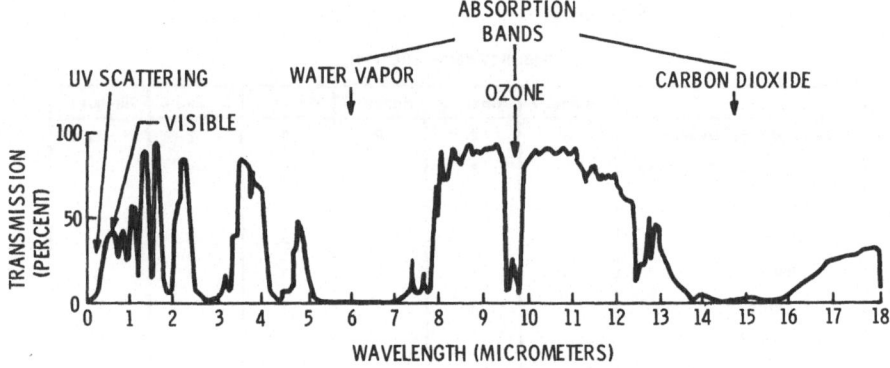

Fig. 3.   Transmission characteristics of the atmosphere.

been designed to measure such things as the Earth's radiation budget, surface and cloud top temperature, and the vertical temperature and structure of the atmosphere. Key atmospheric constituents such as water vapor and ozone have been analyzed.

We did not start out with this much understanding a decade ago. The early remote sensors such as the high resolution infrared radiometer flown on Nimbus 1 and first used to study the cloud patterns during the night portion of the satellite orbit was a rather uncomplicated broadband radiometer. In the simplified sensor schematic shown in Figure 4, the instrument represented the lower extreme of complexity in terms of the number of channels, detector temperature, dynamic range and data bandwidths.

These items are all interrelated, for the noise equivalent detectable signal or minimum detectable change in radiance, referred to as $NE\Delta N$ on Figure 4, is inversely related to

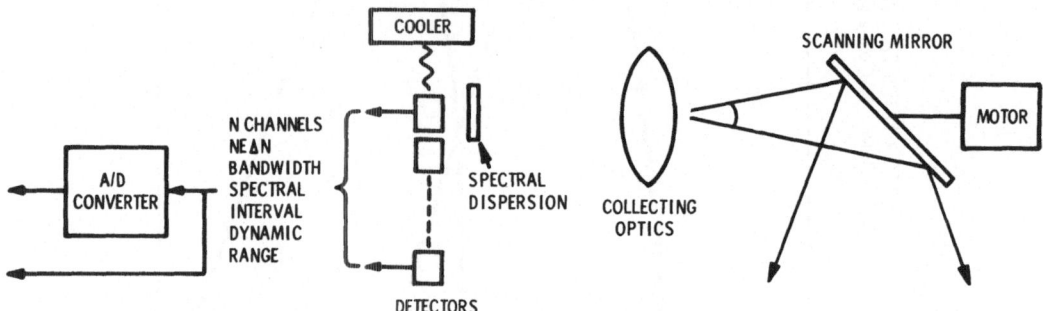

Fig. 4.   Schematic of typical spaceborne remote sensor.

signal bandwidth and the width of the spectral interval. Increasing the number of channels, narrowing the spectral interval or increasing the optics tend to widen the bandwidth, which in turn can have significant impact on the data rate and add complexity to the ground data processing. Other factors such as scan efficiency, which is the ratio of useful data time to total scan time, can also affect transmission bandwidth and ultimately the data processing.

NIMBUS SENSOR GROWTH

	Nimbus 1	Nimbus 2	Nimbus 3	Nimbus 4	Nimbus E	Nimbus F
Number of Experiments	3	4	9	9	7	12
Number of Spectral Channels	3	8	28	43	34	~50
Spectral Regions						
Visible	X	X	X	X	X	X
Infrared	X	X	X	X	X	X
Far Infrared		X	X	X	X	X
Ultraviolet			X	X		X
Microwave					X	X
Total Spacecraft Weight (lbs)	820	916	1268	1366	~1800	~2000

Fig. 5.   Growth of Nimbus sensors.

All of these factors have been considered in the development of each sensor and solutions to all of these trade-offs have contributed to the growth of remote sensors and remote sensing.

Figure 5 identifies the growth of sensor technology from the relatively straightforward visible and infrared payload of Nimbus 1 to the sophisticated complement of sensors to be flown on Nimbus E and F – sensors which encompass virtually the entire electromagnetic spectrum. The key factor in this growth is indicated on the second line, number of spectral channels. This indicates almost double an order of magnitude increase in spectral channels that need to be individually identified within a sensor, detected, rectified, multiplexed, transmitted and ultimately processed by ground equipment for intelligent utilization by the using community.

3. Imaging Systems

The first spaceborne candidate for remotely sensing the Earth and its atmospheric environment was television imagery in the visible range. Using reflected sunlight, excellent images (Figure 6) of Earth's cloud cover were obtained. This was due to the

Fig. 6.   Annotated APT photograph.

difference in reflectivity between clouds and ground. The images were obtained by such cameras as the advanced vidicon camera system (AVCS), the automatic picture transmission (APT) system, and later the image disector camera system (IDCS). The APT photograph shown here, taken more than seven years ago, is not very much different in quality and usefulness than the photography utilized by many meteorologists in their nightly routine television weather reports.

The technology represented by these cameras forms the basis for the evolutionary development of the return beam vidicon camera system (Figure 7) to be flown on the first Earth resources technology satellite. This picture shows the three-camera system in test at our Valley Forge, Pennsylvania Facility. These high resolution, spectrally selective cameras will be geometrically and radiometrically calibrated to provide color photography of the Earth scene in 100 nautical mile square frames.

Fig. 7.   RBV subsystem during test.

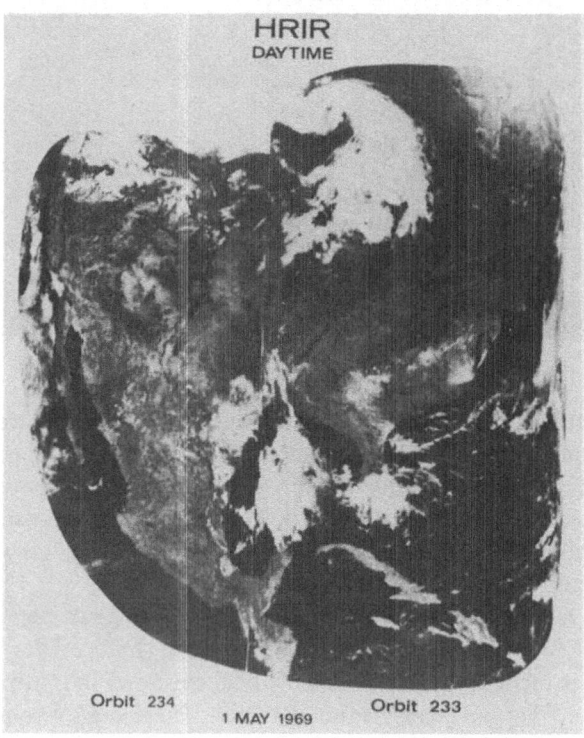

Fig. 8.   Composite of two HRIR pictures from consecutive orbits.

As an example of the evolution involved, consider some key parameters of the APT and the RBV. Where the APT covered the entire visible spectrum, each RBV is limited to 0.1 μm spectral interval. APT had 800 scanning lines, RBV has 4150 with attendant increase in resolution. APT transmitted data to ground receivers at 5 kHz rate and the RBV will return data at a 3.5 MHz rate offering a real challenge to the data processing facility design which will be covered during this session another day.

To achieve cloud photography during the night portion of the orbit, it was necessary to develop an imaging system which used elements of the second source of electromagnetic energy, namely self-emitted thermal radiation. This source of energy was first sensed by the high resolution infrared radiometer (HRIR) on Nimbus 1. Figure 8 is an example of the radiometer output made by aligning two consecutive passes processed from the satellite's tape-recorded data. The outlines of clouds are sharply differentiated from the darker earth because of the large temperature differences.

HRIR produced a cloud map image during the daylight portion of the satellite orbit but the large amount of reflected sunlight prevented relating the radiometric feature to cloud temperature. To overcome this and to produce the same type of data

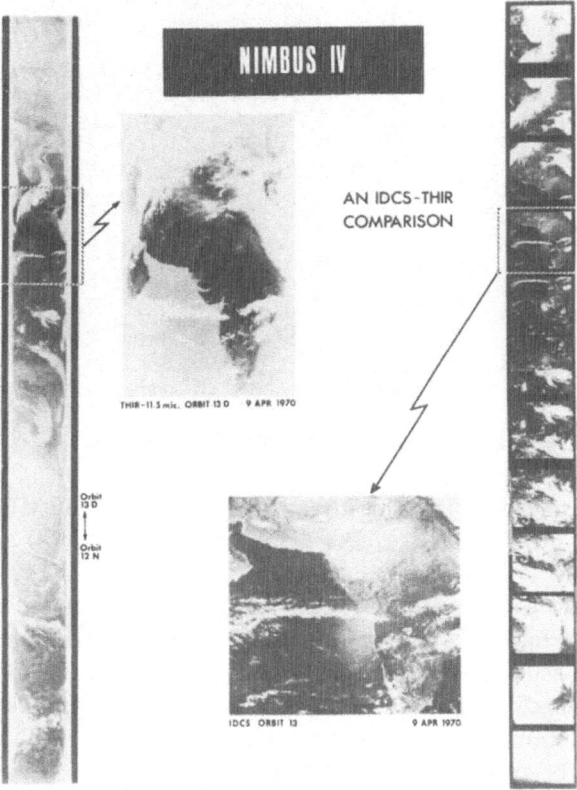

Fig. 9.   Comparison of IDCS and THIR pictures.

both day and night, the next generation of radiometer, the temperature humidity infrared radiometer (THIR) was developed.

This instrument used the long wave window at 10–12 μm, where there is virtually no reflected sunlight. Figure 9 is an example of an IDCS TV picture and a picture taken by the THIR infrared imager. We can note the presence of clouds in both photographs of approximately the same brightness where the ocean appears light in the high infrared, indicating its true temperature colder than the land mass. Another change from the first HRIR was the addition of multiple simultaneous channels. Although not pictured here, the THIR is a 2-channel instrument, with the second channel at 6.3 μm devoted to mapping water-vapor distribution. An even further

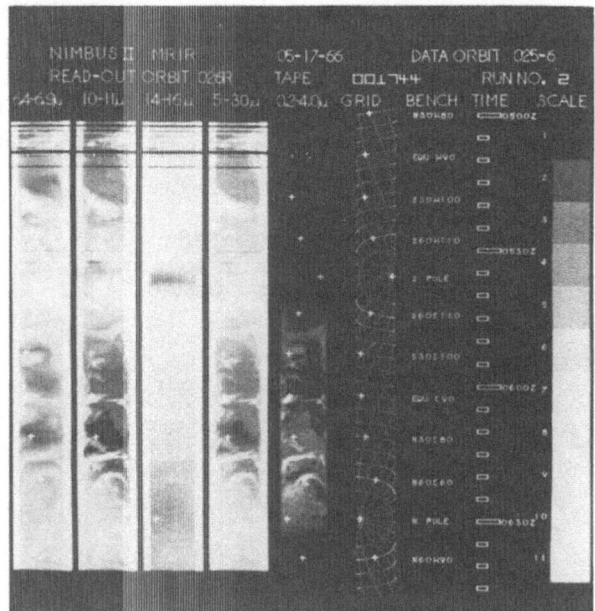

Fig. 10.   Typical example of MRIR data.

development, the surface composition mapping radiometer (SCMR) to be flown on Nimbus E, is a three-channel, higher resolution radiometer. I shall discuss this instrument later.

All of these advances have contributed to the development of the multispectral scanner to be flown on ERTS A next year. Like the advances in technology contributing to the high resolution RBV camera system, significant developments in detector thermal control, space calibration, sensitivity and optics design have been achieved for this high-resolution, 4-channel scanner which provides us with a 15-megobit per second challenge for ground data processing.

One of the earliest multispectral devices was the medium resolution infrared radiometer flown on Nimbus 2 in May 1966 and again on Nimbus 3 three years later. This instrument had five spectral channels in the visible and infrared (Figure 10) and provided scientists with repetitive global data, permitting the detailed study of the effect of water vapor, CO_2 and ozone on the heat balance of the Earth. A data composite of all five spectral channels together with spacecraft annotated data and a calibrated grey scale generated by the sensor is shown here. One such chart is generated for each orbit and permits easy cataloging. Magnetic tapes storing the data are available for detailed analytical use.

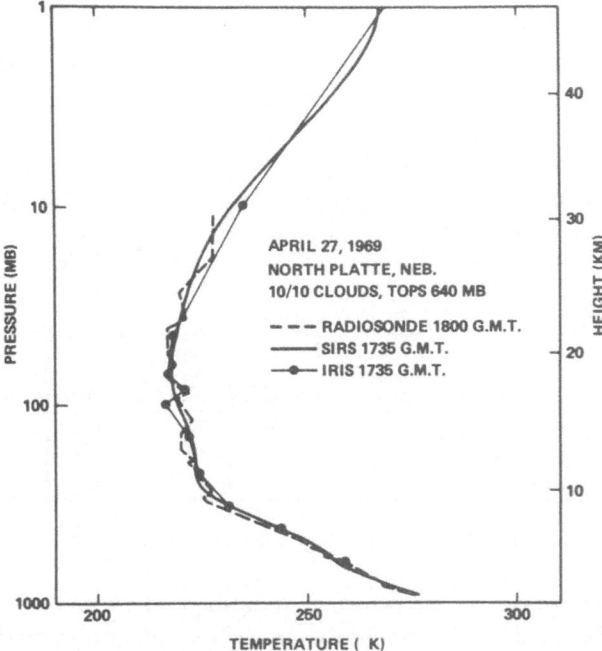

Fig. 11.   Comparison of radiosonde, SIRS and IRIS temperature profiles.

Multispectral radiometry can also be used to remotely sense vertical profiles of atmospheric temperature and the vertical distribution of certain atmospheric constituents. This technique requires detailed narrow spectral band measurements centered near absorption bands of selected atmospheric gases. With a gas such as CO_2, whose vertical distribution in the 15 μm region is well known and constant, multispectral radiance data can be related to vertical temperature profile. Once vertical temperature profile is known, similar measurements centered near absorption bands of gases, such as water vapor and ozone, permit derivation of their vertical concentrations.

A number of instruments have evolved to make these profile measurements. The infrared interferometer spectrometer (IRIS) and the satellite infrared spectrometer (SIRS), both flown on Nimbus vehicles, use different means of dispersing the incoming energy into its spectral components but achieve the same result. Figure 11 provides a comparison of radiosonde data, SIRS data, and IRIS data taken over North Platte, Nebraska, showing extremely good correlation.

Another sensor experiment, the selective chopper radiometer (SCR), uses the absorption and pressure broadening of CO_2 gas itself in varying densities to act as a very narrow-band optical filter providing temperature profile measurements to very high altitudes.

These measurements have been the most significant meteorological data provided by space platforms, in that such data have been directly incorporated into advanced numerical weather analysis and prediction products. For the first time in history, upper air data have been made routinely from such areas of the world as the Southern Hemisphere oceans.

It should be noted that logical extension of these multispectral techniques should permit routine and continuous global environmental monitoring from space in the very near future.

The combination of multispectral sensing for profile measurements and for imaging was used to solve a specific profile problem. This problem, initially encountered with SIRS data, was the presence of clouds in the field of view. The clouds limited the achievable altitude of the temperature profile measurements. One solution is to work

RELATIVE RADIANCE

Material	(A) Black and White 0.5-0.75μ	(B) Multispectral		
		0.5-0.6μ	0.6-0.7μ	0.7-0.8μ
Concrete	30	30	28	20
Green Grass	15	20	12	35
Brown Earth	15	15	17	16

Fig. 12.   Relative radiance reflected from several materials.

on developing a profile instrument which could also produce an image. By doing so, clouds could be identified and measurements made through breaks on the cloud cover.

The infrared temperature profile radiometer (ITPR), which will be flown on Nimbus E, and the high resolution temperature sounder (HIRS), planned for Nimbus F, combine moderate spatial resolution imagery with profile radiometry and appropriate statistical analysis and are expected to achieve a 'clear air profile'.

Another application of multispectral sensing is the fundamental approach to be used in the ERTS mission. Figure 12 shows the relative radiance reflected from several materials on the Earth's surface. The ability to identify features using multispectral imaging can be seen by comparing the relative responses in column 'A', which is typical of the spectral response of such devices as the APT, to the responses shown in the spectral intervals of 'B', which are representative of the ERTS sensors. The average reflectivity of green grass shown in 'A' is considerably lower than that of concrete, but the average of brown Earth and grass is the same and could, therefore, not be discriminated by a broad-banded sensor. However, as indicated under the different spectral intervals in 'B', there appears to be some real hope that all three materials can be differentiated, opening a subject that will be discussed in much detail during the coming sessions of this meeting.

4. Sensors and Their Science

I have indicated that sensor technology has grown significantly since the first Nimbus. I will now discuss some of the sensors to be flown on Nimbus E in that context (Figure 13). Shown here is the Nimbus E spacecraft with the electrically scanned microwave radiometer antenna (ESMR) deployed in front toward the velocity vector. The ESMR data will be reconstructed into images similar to APT in the visible, or THIR in the infrared. The instrument – consisting of an electrically scanned phased array antenna with appropriate electronics – is a relatively large one, three feet by three feet by six inches. The ESMR antenna face must be pointed toward the Earth's surface and is deployed only after the spacecraft achieves orbit.

The experiment's objective is to globally map the thermal radiation emitted from the Earth's surface and atmosphere at a wavelength of approximately 1.5 cm or 19 GHz.

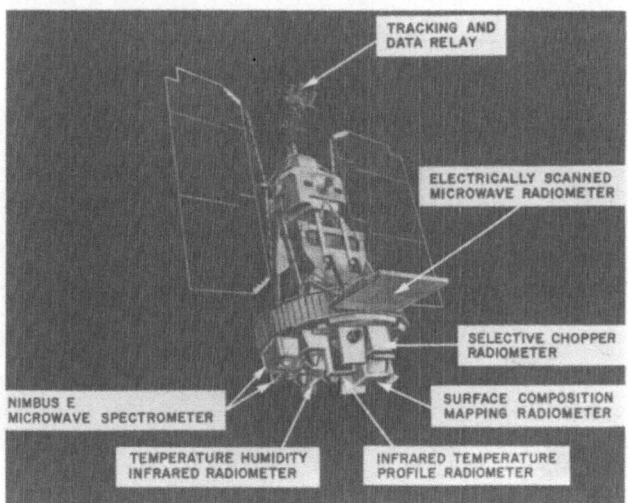

Fig. 13. Nimbus E satellite.

As an example of the ESMR mission, some clouds, such as cirrus and stratus, are relatively transparent in the microwave region, but opaque in the infrared. Thus, where these clouds predominate, such as in the polar areas, it will be possible to map surface objects.

Clouds with high water content will be identified with this new microwave instrument. Areas of precipitation, intense frontal and convective activity, severe storms, and other similar meteorological features will be mapped and the data used to annotate the more conventional IR or visible cloud cover maps. Over land areas the amount of vegetation covering the ground or the water content of the soil may be differentiated.

Another Nimbus E instrument, the surface composition mapping radiometer

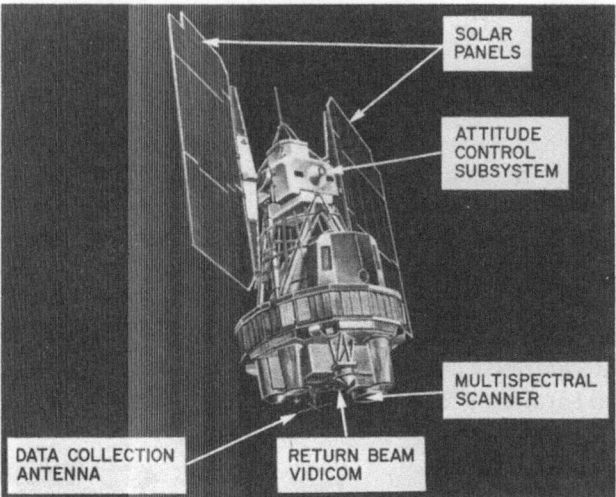

Fig. 14.   Earth resources technology satellite.

Fig. 15.   View of Earth from Apollo 11.

(SCMR), is a scanning IR imaging radiometer, with a nadir ground resolution of about 700 m, and with two spectral channels centered in the 8 and 10 μm atmospheric window.

Thermal emission of a surface object is a function of the temperature of the object and its emissivity, which can be a function of wavelength. In the case of quartz, for example, there is a dramatic decrease in emissivity around 9 μm. The effect can be used to identify the nature of various surface compositions.

I have already briefly mentioned the multispectral scanner (MSS) and the return beam vidicon (RBV) to be flown on ERTS (Figure 14). These sensors will provide a starting place for the growth of multispectral imaging sensors over the next few years. Tradeoffs between spectral bandwidth, number of channels, area coverage, and spatial resolution will yield instruments whose parameters are markedly different from the current devices and could perform new missions that will map ocean surface characteristics, upper atmospheric properties, ice/water distribution in clouds, and most importantly the distribution of atmospheric pollutants.

We have seen how remote sensors and their technology have grown and become increasingly sophisticated in less than a decade (Figure 15). The synoptic sensor data provided by today's satellites are helping man to learn more about his environment and to further his knowledge in many scientific fields. The Nimbus satellites and the ERTS satellites to be launched in the next few years may provide a quantum jump in that knowledge. And the experience gained from assessing sensor operation and data during that time period will be of vital importance to the scientists designing the next generation of remote sensors, and will definitely be applicable to future operational missions which are certain to evolve.

THE ROLE OF APPLICATIONS SATELLITES IN THE
MANAGEMENT OF THE HUMAN ENVIRONMENT

P. A. CASTRUCCIO

Federal Systems Division, International Business Machines Corporation,
Gaithersburg, Maryland, U.S.A.

Abstract. The development of Earth resources satellites originated from the prospect of improving methods of discovery, utilization, and management of the Earth's natural resources. The growing concern over ecological problems suggests an expanded role for such satellites: uncovering, quantifying, and assisting in mitigating the problems generated by interaction between man and environment.

The information desired from such satellite systems must meet two requirements: (1) tactical, uncovering the causes and cause-effect relationships of environmental problems, and (2) strategic, i.e., the assessment of trends to predict the future evolution of the problems.

An essential tool for implementing both tactical and strategic analyses is the art and science of modeling.

Examples of existing models in oceanography and hydrology are presented. Their objectives, mode of operation, and types of input dated requirements are described. Avenues for significant improvement in the model's predictive accuracy and for reduction of the cost of gathering the necessary data via remote sensing are analyzed.

1. General

When a region of the world is still virgin – such as the United States 400 yr ago – people are few, the land is vast, and the resources are large. Thus begins the era of exploration. As time goes on, population grows, but resources remain constant. The human groups exploit the resources of the region to the maximum: agricultural resources, mining, rivers, etc. This is the era of exploitation, which began for the U.S. in the early 19th century. As resources are increasingly exploited, and the population continues to grow, we begin to notice a phenomenon, well known in physical systems: the phenomenon of coupling between human activities.

Experience shows that the effects of 'coupling' between diverse human endeavors are by and large deleterious. Is this necessarily so, or is it caused by our ignorance of the underlying mechanisms? Can the technology which has caused the problem also show the way to the cure? The answer is: very probably yes.

The economic consequences appear to be that increasing portions of the GNP will be devoted to evading the ill effects of coupling. Unfortunately, these particular portions of GNP are non-productive. An $100 000 SO_2 filter in an electric coal-burning plant produces nothing in return except cleaner air. To be productive in the conventional sense, the $100 000 should be spent in more furnaces or in improving the efficiency of the process.

The net effect is to reduce the 'measured' GNP to the lesser 'real' GNP, as depicted in Figure 1. If careful management of the coupling problem were not to be undertaken

L. G. Napolitano et al. (eds.), Astronautical Research 1971, 403–417. All Rights Reserved.
Copyright © 1973 by D. Reidel Publishing Company, Dordrecht-Holland.

soon, a rather 'catastrophic' reduction in GNP may occur, as shown in the lower curve of Figure 1.

There are fundamental differences between the discovery and exploitation of natural resources, and ecological management. In the former, economic return is

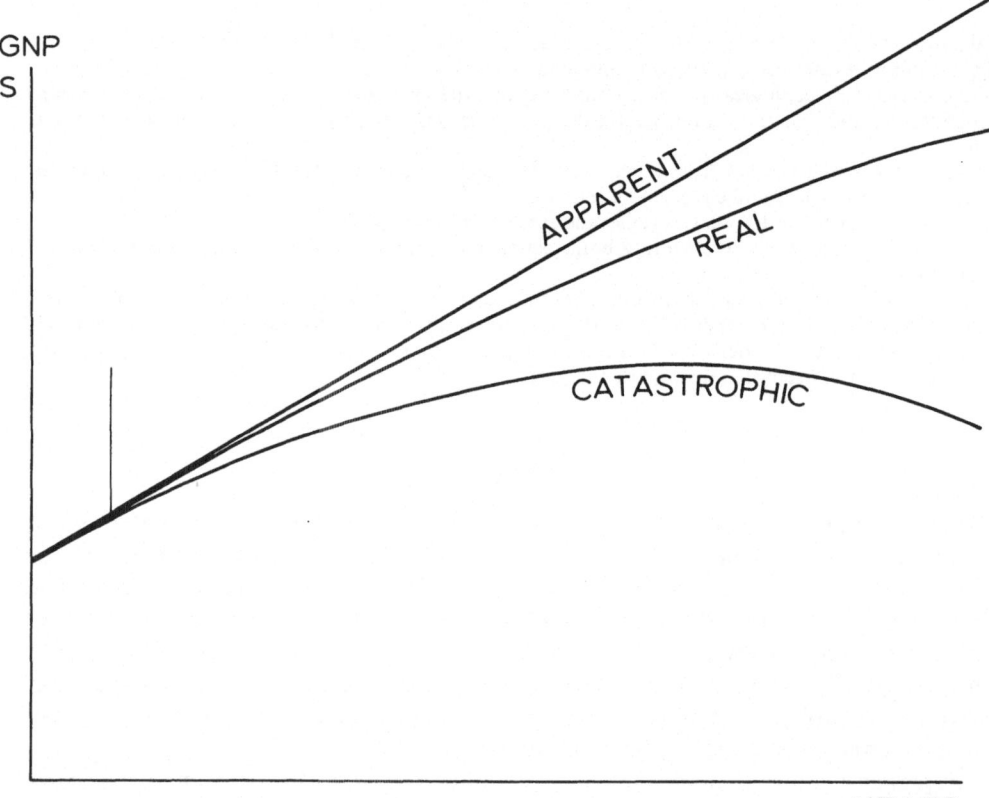

Fig. 1.   Pay this price.

the paramount criterion. In the latter, economic payoffs vie with other, less tangible criteria – such as aesthetic motivations. In the former, the discovery and location of resources is of paramount importance. In the latter we are much more concerned with resource conservation dynamics.

2. The Need for Models

A fundamental requisite of environmental management is ability to predict, to answer such questions as: what improvements will a specific policy of correction yield, and when? Is there a better policy?

Prediction requires the ability to formulate the cause-effect relationships between inputs (environmental variables) and outputs (environmental goals, desired effects). The essential tool of environmental management is the art and science of modeling.

Ecological models can vary in complexity from pencil and paper formulations; to physical, scaled analogues of a region; to complex, computerized mathematical formulations.

The basic tools to achieve environmental modeling exist in the mathematical techniques of optimization and simulation. Some of the principal mathematical techniques currently available and applicable to modeling are listed below: Monte Carlo method; linear programming; dynamic programming; non-linear programming; game theory; decision theory.

TABLE I

Typical environmental models

Simulation of a complex river system on an electronic digital computer
Simulation in plant ecology
Program of river forecasting
Mathematical model of the Delaware estuary
Groundwater precipitation model for streamflow analysis
Simulation of water quality management
Mathematical model for dissolved oxygen

Table I shows typical environmental models currently available. Models of the physical world can be constructed by two fundamental methods (with variations and combinations):

(a) Deterministic: from known cause-effect relationships, which can be either physical laws or empirical relationships. Examples: atmospheric circulation and physical behavior of the ocean.

(b) Heuristic: from correlation of historical behavior of output, in response to variations in input. Examples: models of river flood prediction, behavior of fish. Heuristic models are necessary when the inputs are imperfectly known, or their relationships excessively complex.

Let us look at some examples of existing models, and at the potential contributions that remote sensing can add to their performance.

3. Fishing Models

One-half of the world's population experiences protein shortages. Today's fishery harvest could supply the protein needs of over one billion people at low cost. The long-term sustainable protein yield of the sea has been estimated to be sufficient to satisfy the protein requirements of 6 billion people.

The immediate needs can be taken care of by improving methods of fish location. Many factors affect fishery yield: types of vessels and fishing gear, modernity of supporting infrastructure such as docks, transportation, and distribution networks. A most significant factor is information, which is rather meagre at the present time. Vessels and fleets mostly ply 'historically fertile' fishing grounds. Seasonal variations in migration patterns of important species, and existence and potential of alternative fishing areas, are not well known.

Ideally, a rational approach to fishing requires two classes of information:

(a) A strategic, long-term forecast, to provide information on what total catch is available when, and what portion can be caught without depleting the resource.

(b) A tactical, short-term prediction, to plan individual fishing operations most effectively.

The need for strategic forecasts arises from the large variation – as much as 10 to 1 – in fish availability from season to season in a given region. Long-term information would significantly assist the planning of business and ecological strategy.

Essential to the tactical forecast is the prediction of fish migration characteristics and of environmental conditions affecting fishing operations. Operators intent on maximizing their catch need to know where, when, and for how long to send their ships. This decision is a function of several variables, including:

(a) the location, extent, and average motion of the region to be fished;
(b) density of fish in the region;
(c) depth of fish;
(d) 'mean free path' between fish schools;
(e) composition of fish by species;
(f) size of specimen and life status (spawnings, etc.);
(g) other characteristics such as propensity to 'bite';
(h) sea state;
(i) weather affecting operations – wind, clouds, or ice; and
(j) Potential inhibiting factors – i.e., the presence of oil pollution can affect the taste of the fish and make it unmarketable.

For a typical fishing operation, without benefit of prediction, the ratio of search-to-fishing time can be as high as 10:1.

Even a few items of this information of good reliability, could greatly improve current fishery operations.

The objective of fishing models is to predict the abundance of a certain species of fish in a given area. Because the relationships that exist between fish and their environment are complex and poorly known, present models are constructed by regression analysis from empirical data, rather than from theory.

Typically, the dependent variable is quantity of a species caught; the input variables are related to environmental factors such as salinity and water temperature. The regression analysis is developed by correlating catch reports with measurements of aquatic environmental parameters, and then fitting a polynomial function to the data generally using the least squares criteria. The resulting polynomial equation becomes the 'fishing forecast model'. During the last decade, a number of fishing forecast models have been developed in the U.S. and other fishing nations. The principal among these are shown in Table II. A model which has achieved significant operational success is the tuna forecast model of the Bureau of Commercial Fisheries, Tuna Resources Laboratory, La Jolla, California. It is used during the summer–fall season migration of prime albacore and bluefin along the Oregon and California coasts. The area encompassed is approximately 600 000 square miles. Input parameters to the Tuna forecast model are surface temperature, vertical temperature profiles, and salinity. As output, it yields where, when, and how much albacore and bluefin tuna can be expected to be taken throughout the season.

Gathering the three input parameters for the model is a major undertaking. Surface temperature is recorded by ships at sea, and radioed back. The vertical temperature profile and salinity data are collected by specially equipped research vessels.

Soviet research in the Barents and North Sea has yielded a series of regional cod and herring migration models, and three very effective models for determining optimal herring fishing depth.

The Konstantinov Cod Model is representative of the regional regression fishing models for the Kilden and Rybachi banks. The model predicts total catch as a percentage of a fixed base value. The input is water temperature at 150–200 m depth along the Kola meridian during November and December. The derived regression model for these predictions is $y = 24.3x - 86$, where $y =$ percentage of a fixed-base measurement of catch, and $x =$ subsurface temperature reading.

The vertical migration patterns of herring in the Norwegian Trough are more important to fishermen than their horizontal wanderings within the fishery. The reason is that fishermen must set their lines to correspond to the depth of herring schooling activity in the various regions of the trough. One of the regression models used to predict this data is G. D. Vasil'ev's Herring Distribution Model. The model's forecasts are based upon the vertical position of the layer of maximum vertical stability. The fishing area is subdivided into 34 equal squares; a projection for optimal fishing depth is made for each of the subareas. The regression equation used is $y = 1.2x$, where $y =$ top of fishing layer, $x =$ vertical portion of the layer of maximum vertical stability.

Existing fishing models suffer from the limited extent of data bases. Even though several oceanographic factors correlate rather well with fishing productivity, they are presently excluded from the models because of the inability to regularly and reliably gather pertinent data.

The more meaningful oceanographic variables that warrant inclusion as inputs in fishing forecast models are:

(1) Subsurface temperature profile – defines the location of the thermocline to depths of several hundred feet. Until a stable thermal layer has developed, the food chain and migration of temperature-sensitive fish such as tuna is inhibited. For example, too late a warming trend in Hawaiian waters tends to alter the Skipjack migration pattern enough to severely reduce total tuna catch for the season.

(2) Surface temperature variation – The temporal and spatial variation in surface

TABLE II

Principal fishing models

Model	Agency	Predicted output	Input parameters	Remarks
Kildin and Rybachi Banks cod	Polar Scientific Research Institute of Marine Fisheries and Oceanography – U.S.S.R.	Abundance of cod during spring migration	Water temperature along Kola meridian at 150–200 m depth	Experimenters suggest possible other parameters such as light intensity, chemical content. Similar Soviet models have been developed for cod migrations in Murmansk Shoals, Gusinaya Banks, Western Spitzbergen, and Southern Bering Sea
Lofoten shoals herring fisheries	Northern Fisheries Administration – U.S.S.R.	Productivity of spawning period on Lofoten Shoals	Average catch per net during the foraging period north of 68° 30′ N for the preceding year	Used for long range forecast of commercial herring catch
Vertical distribution of Norwegian trough herring (A)	Atlantic Scientific Institute of Marine Fisheries and Oceanography – U.S.S.R.	Depth below which the bulk of the herring stock will be found	Specific density of the sea water	Predictions are regularly given for 34 different subareas. This data is invaluable for determining optical fishing procedures
Vertical distribution of Norwegian trough herring (B)	Atlantic Scientific Institute of Marine Fisheries and Oceanography – U.S.S.R.	Depth below which the bulk of the herring stock will be found	Depth of upper layer of salinity and temperature discontinuity	Another approach to the above problem. No direct comparison of accuracy available
Vertical distribution of Norwegian trough herring (C)	Atlantic Scientific Institute of Marine Fisheries and Oceanography – U.S.S.R.	Depth below which the bulk of the herring stock will be found	Wind speed at surface	A hydrometeorological approach to predicting depth of herring schools. Results considerably more accurate than Method (B)
Year class distribution of bank (Downs) herring in the North Sea	Atlantic Scientific Institute of Marine Fisheries and Oceanography – U.S.S.R.	Survival rate of spawned bank herring	Rate of change of surface temperature between February and March	This form of model prediction is commercially and ecologically significant

Table II (continued)

Model	Agency	Predicted output	Input parameters	Remarks
Albacore-bluefin Pacific coast seasonal distribution	Bureau of Commercial Fisheries – Tuna Resources Laboratory, La Jolla, California, U.S.A.	Location and quantity of Albacore and Bluefin tuna	Temperature profile, salinity measurement	Has increased average tuna landing by 3×10^6 lb. Starting this year, surface water temperature profiles will be directly gathered by the La Jolla laboratory by analysis of Nimbus weather satellite IR data
Skipjack Hawaiian migration	Bureau of Commercial Fisheries, Biological Laboratory – Dr Gunther Seckel, Chief Experimentor	Qualitative measure of skipjack migration: average, above average, below average	Rate of temperature change and salinity in the waters directly off Koko Point, Oahu	A more advanced model for predicting the location of skipjack schooling activity has been proposed
Groundfish herring	British White Fish Authority – United Kingdom	Program in formulation	—	Presently in formative stage. Two separate models are envisioned: one for groundfish and one for herring
Pacific coast herring	Fisheries Research Board of Canada, Pacific Ocean Group; Namiamo, British Columbia	Program in formulation	—	Model is in formative stage

temperature is the most significant factor in the surface thermal profile. A discrete change in the prevailing surface temperature pattern correlates very strongly with the occurrence of upwelling. This is the ocean phenomenon whereby deep ocean water, rich with nutrients, is brought to the surface. Schools of fish such as tuna are usually found in the vicinity of upwelling activities.

(3) Salinity – Many species of commercial fish are sensitive to this parameter. Skipjack tuna, for example, appear to be most plentiful in areas where dissolved salt content is between 32–35 parts per thousand. The occurrence of this level of salinity in the vicinity of an area of upwelling would tend to have a high correlation with tuna schooling activity.

(4) Chlorophyl content – This parameter correlates very strongly with the degree of primary production. The location of some species of fish relative to these areas can be statistically well predicted.

(5) History of cloud cover – Cloud cover, may correlate strongly with the development of aquatic patterns such as upwellings. Cloud patterns, by a sort of image circulation, bring water up to the surface, sweep the surface water away and therefore cause cold, nutrient-rich water to well up along the continental shelf. Such indirect effects of the meteorological oceanographic interface were theorized to have been responsible for the conditions which led to the biggest congregation of tuna ever recorded off the coast of Peru in 1967. Occurrence of these conditions were detected by the ATS satellite.

(6) Chemical content – Nutrient-rich waters are a catalyst for a whole series of biological events. The nitrates, silicates, and phosphates present are conducive to aquatic productivity.

(7) Surface wind conditions – The degree and direction of surface wind is a prime factor in the mechanics of the upper layer of the ocean's heat budget. Surface convection and subsurface advection are governed directly by the wind. Soviet regression models have indicated a strong correlation between wind velocity and the depth of herring schooling activity.

The direct or indirect detectability of a significant number of these input parameters, is currently under research, development and test – in particular the use of combined buoy–satellite systems appears attractive.

4. Hydrological Models

Water resources constitute a major environmental problem. The current consumption of water is six tons per capita per day in the U.S.; less, but still quite high, elsewhere.

How much water is available? Due to economic reasons, only the water that falls from the sky by precipitation which is on the average 850 mm per year. As a gross figure for the U.S., the coefficient of utilization of rain water is only about $7\frac{1}{2}\%$.

Extrapolating the growth of water demand to the year 2000 and multiplying by the Earth's estimated population – approximately 6 to 7 billion – one computes a total demand. If this is matched to the total availability of precipitation water, assuming a global efficiency of utilization of 4% in AD 2000, it is easy to determine that the available water will have to be recirculated on the average every 2000 h; in highly developed regions, approximately every 500 h.

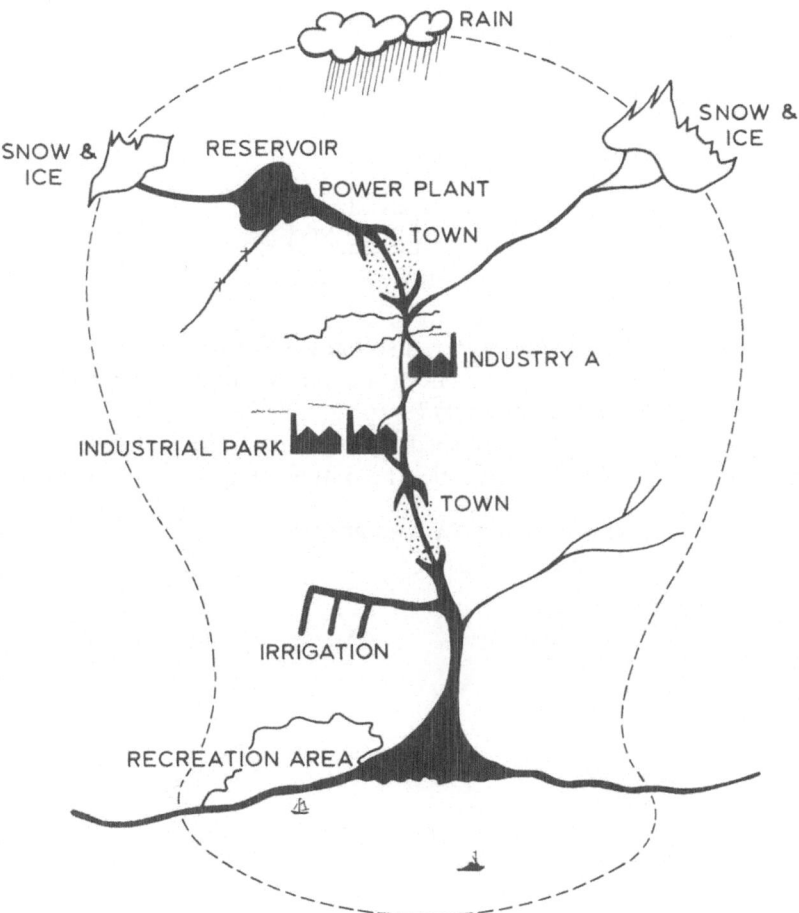

INTERACTIONS WITHIN
TYPICAL WATER BASIN

Fig. 2. Typical watershed interactions.

There is a lot of work going on to find new sources of water. 97% of the world's water is saline. Of the remaining 3%, approximately 95% is locked in ice, mostly in polar caps.

The best price today at which large quantities of water can be desalinated practically is approximately $1 per thousand gallons. A price at which a city is willing to buy is perhaps half of this; for agricultural water, of the order 5 cents per thousand gallons. The price for industrial water ranges from 10 to 20 cents.

Studies to determine the economics of transporting Arctic ice via supertankers found that the transportation rates are too high. Therefore, at the moment, we are

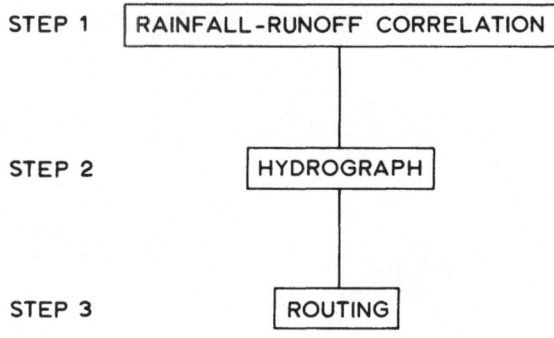

Fig. 3.   River forecasting.

confined to utilizing only the rain water, also known as surface fresh water, which is about one ten-thousandth of one percent of the total water available on Earth. What can be done to utilize it more efficiently?

The problem boils down to watershed management. The watershed is a system in which the input (rainfall) is stochastic, the output requirements are deterministic.

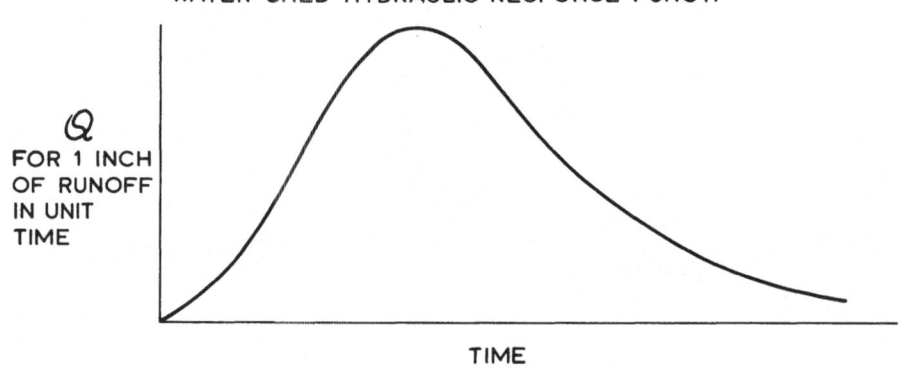

Fig. 4.   Unit hydrograph.

The watershed, as shown in Figure 2, has to provide consumers with power; municipalities with water; the farmers with irrigation water; and perhaps even recreation water. The consumption schedules are relatively fixed, the rainfall is stochastic. The problem is: how do we match the two?

To do this, the environmental science services administration (ESSA) has developed a model to predict how much water will be available in a watershed as a function of rainfall. This model is considered the best available in practice. The ESSA model, as shown in Figure 3, does three things. First it correlates how much rain falls with how much water flows out of the watershed – the utilization coefficient. Second, it predicts the time behavior of the flow. The flow-time behavior is called a hydrograph. Third, it combines the first two parts in the channel flow to give the overall prediction.

The first part of the model, called 'correlation between rainfall and runoff', is based upon four inputs: the quantity of rain; the duration of the rainfall; the season

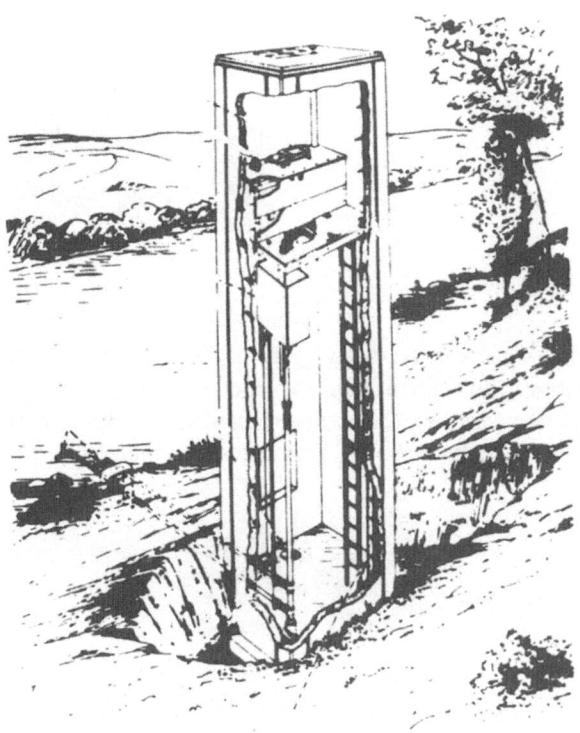

Fig. 5.   A typical water state recording installation.

of the year; the humidity (dry soil absorbs water faster and therefore yields less runoff from a given rain, whereas wet soil tends to become impermeable and therefore a given amount of rain yields more runoff). It is not practical to actually go in the field and measure how wet it is. The model computes something called the antecedent precipitation index which is based upon the rainfall of the preceding weeks.

The second piece of the model is the time-flowcurve or hydrograph, Figure 4. From historical measurements one constructs the so-called unit hydrograph, which is the response of the watershed system to a runoff of one inch, assumed constant over the whole watershed area. Once this unit hydrograph is available, then one can 'multiply' this, by convolution, by the actual runoff and obtain the flow-time output.

The data are collected by three basic types of tools. The river gauge, Figure 5, measures the height of the river. The simpler ones are just sticks with numbers painted

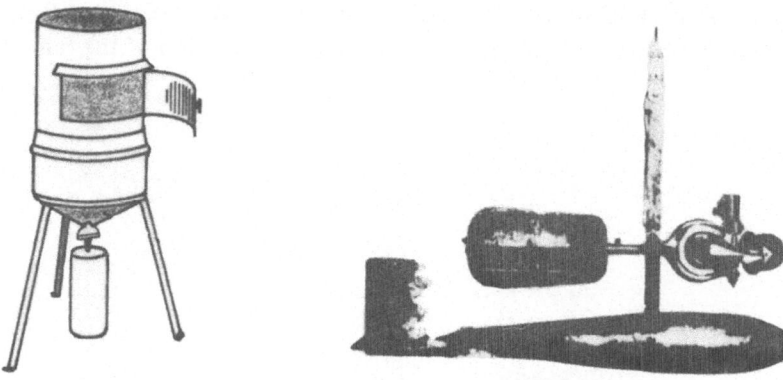

Fig. 6.   Tipping bucket rain gauge.   Fig. 7.   Price current meter.

upon them, which a field worker reads periodically. In the U.S. and Europe, field workers are expensive: thus the manual method is becoming rather costly.

Also, one would like to make this measurement frequently, at least once a day, or oftener during river activity periods. The trend is, therefore, to install automatic

Fig. 8.   Micro hydrology – analytic flow forecasting.

stations. Many use analog reporting; they write·continuously on a strip of paper. The cost of such a unit can be as large as $30 000. Several are needed in a river, depending upon its length, uniformity and other characteristics.

The second tool is the rain gauge, Figure 6. The manual version costs about $300. The trend is to automate them, by attaching them to telephone lines or providing them with radio transmitters.

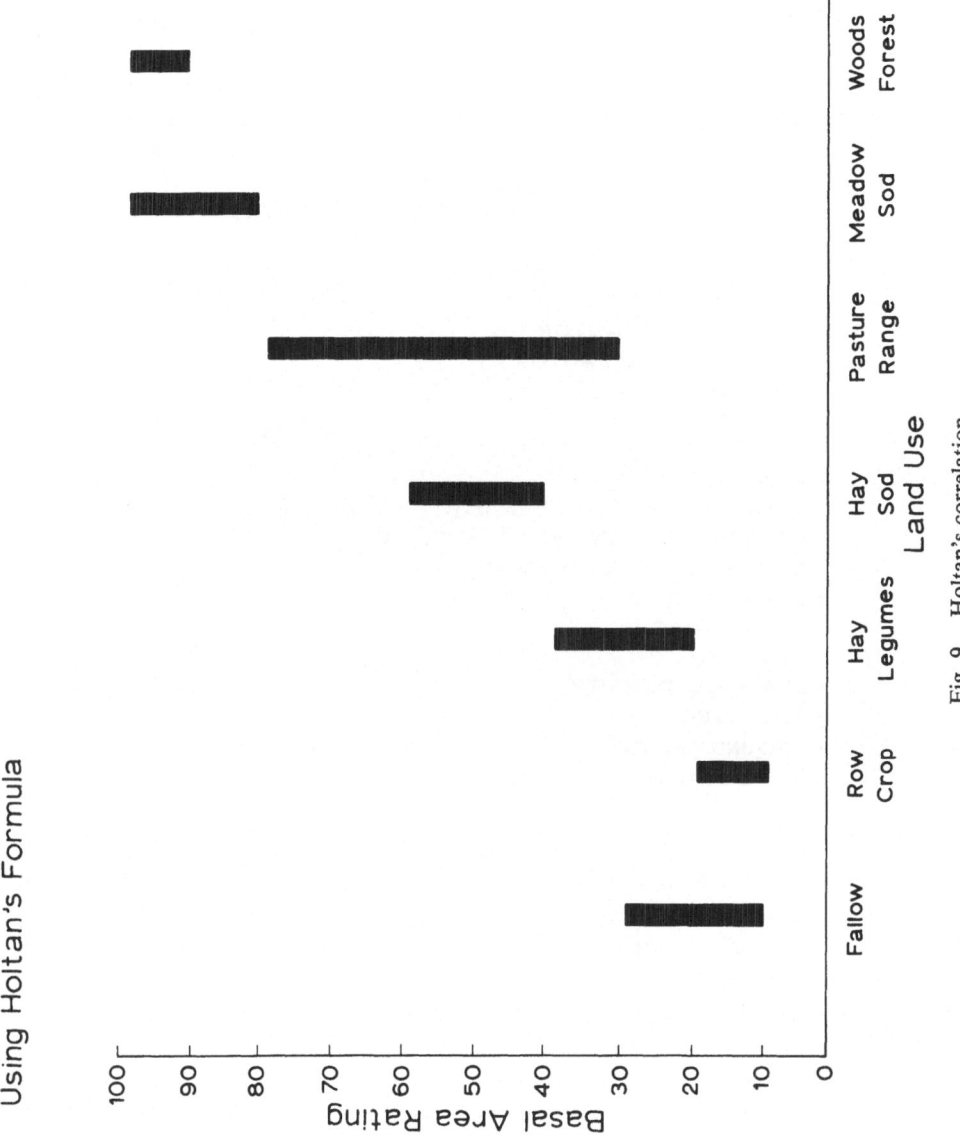

Fig. 9. Holtan's correlation.

The third tool, Figure 7, measures the speed of the water. To compute the flow, one has to measure the area of the river, find the average velocity, and multiply by two. The speed is not the same throughout all sections of the river; so it is measured at different points; the measurements are then correlated and an average speed is calculated.

The cost, labor and time consumption which the present methods entail calls for improved systems of data collection.

How do we accomplish this? The ESSA model assumes the watershed to be a 'black box'. If one understands what is inside the 'black box', one can get a better insight and better prediction. Much laboratory work has been performed in hydrology;

SOME BASIN CHARACTERIZING PARAMETERS

- BASIN AREA
- BASIN SHAPE FACTOR
- VALLEY SHAPE FACTOR
- STREAM ORDER
- STREAM LENGTH
- STREAM SLOPE
- STREAM WIDTH
- VARIANCE OF STREAM WIDTH
- LENGTH OF OVERLAND FLOW
- DEGREE OF URBANIZATION
- LAND COVER
- MEANDERING
- TEXTURE
- DRAINAGE DENSITY
- SOIL TYPE
- GROUND WATER
- SUB BASIN ORIENTATION

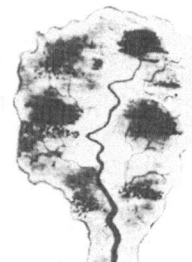

many empirical and theoretical results are available. The problem is to extrapolate results from laboratory to the field. It is too costly to send large amounts of people into the great outdoors to gather data; it is also costly to install permanent, remote measuring instrumentation.

Remote sensing appears to hold the potential for a major step forward in cost-performance.

From imagery we can divide the watershed into areas of homogeneity, as shown in Figure 8. For example, for parking lots which have runoff coefficients of 0.9–0.95, practically all the rain runs off. Forests can have runoff coefficients very close to zero. One can, therefore, label an area which is all forest as, for example, type one; an area which is all parking lots type two, and so forth. For each such homogeneous area one can create a micromodel whose coefficients can be assessed by knowing what lies inside the area. Then tie them all together and come up with a prediction which can be far more refined than the simple 'black box' model used today.

For example, three things which are ignored in the present models are easily recognizable in even the poorest aerial pictures. First is the phenomenon of interception. Anywhere from a hundredth to two-tenths of an inch of rain remains attached to the plants, depending on the type of plant. What we have to do is to recognize how much area is covered by forest, and identify the type of forest: from photography, or at other parts of the spectrum – infra red, for example. Second is the phenomenon of evapo-transpiration, the sweating of the plants. This is a function of the area coverage and of the type of plant. We can also measure this from remote sensing. The third is infiltration. Every soil has different characteristics of water absorption. In parking lots, almost all the water will run off, in sandy soil little will run off. There are under research experimental methods to measure from remote sensing the type of soil. We can also measure its vegetation cover, which has been shown by Holtan and others, Figure 9, to be connected to the absorption coefficient because certain plants grow better or grow only in certain kinds of soil.

Figure 10 shows additional parameters whose knowledge, for a particular watershed, could still further improve its model. All of these parameters are eminently amenable to aerial remote sensing. They are, by the way, difficult if not impossible to gather from maps – because in maps many of the significant features are edited out.

The point of applying remote sensing techniques to the determination of the hydrologic regime of watersheds is twofold: first, the improvement in predictive accuracy of already instrumented and modeled watersheds; two, the determination of the hydrologic regimes of as yet unknown watersheds, with potentially significant reductions in time, labor, and cost over present methods. Such a determination is an essential prerequisite for the planning of flood control and water resource utilization works within the watershed.

B. SCIENTIFIC SPACECRAFTS

THE GERMAN AERONOMY SATELLITE AEROS

(*Special Design Features*)

K. J. GLUITZ
Dornier System GmbH, F.R.G.

AEROS is the designation of the second German Aeronomy Satellite to be launched in 1972 by an improved Scout launch vehicle. Its weight of 120 kg and dimensions of approximately 1000 mm diameter and 800 mm height require the improved Scout booster stage Algol III and a new heatshield diameter of 42 in.

The AEROS scientific mission objective is to conduct Aeronomy research in the upper atmosphere over a period of six months. Mission requirements such as obtaining data at different altitudes, different geographic latitudes and longitudes and different times of day and year led to a 3^{00}–15^{00} Sun synchronous orbit with an inclination of 97.2°, whereby an exact perigee of 230 km and apogee of 800 km has to be achieved. After 4–5 months orbital operation the apogee will be increased again above 600 km in order to repeat scientific measurements as a function of seasonal changes.

The scientific payload will consist of five experiments.

A mass spectrometer (MS) will study the chemical composition of the ambient atmosphere by measuring the density spectrum of ions and neutral particles.

A retarding potential analyser (RPA) will measure the energy distribution of ions and electrons.

An impedance probe (IP) will measure the electron density and a XUV-spectrometer (XUV) the solar radiation in the wave length from 100 to 1000 Å. The above mentioned four German experiments are supplemented by a neutral atmosphere temperature experiment (NATE) from NASA–GSFC.

This experiment will measure the temperature of neutral nitrogen and total atmosphere density. The five experiments will always operate simultaneously during a measuring period. This way the decisive aspects of the physical processes will be covered and their cross correlation can be studied, while the Sun's radiation as their energy source will be included. AEROS will be one of the first aeronomy satellites carrying a XUV-spectrometer.

The different experiment requirements also lead to a major difficulty in designing the satellite. The XUV-spectrometer must point to the Sun, while the sensing axis of the mass spectrometer and the retarding potential analyser may not deviate by more than 60° from the direction of flight during the measuring phases. In a rigid, spin-stabilized satellite these requirements cannot be met simultaneously at all points in the orbit. The solution to the problem was found by matching the satellite configuration to the orbit.

The orbital plane lies at an angle of 45° to the Sun. The satellite's body is basically a cylinder, whose axis of symmetry is also its spin axis. The spin axis is aligned to the

L. G. Napolitano et al. (eds.), *Astronautical Research 1971*, 421–441. *All Rights Reserved.*
Copyright © 1973 by D. Reidel Publishing Company, Dordrecht-Holland.

Sun. The XUV-spectrometer's window is in the cylinder's frontal surfaces; the mass spectrometer and retarding potential analyser are on different meridians on the cylindrical skin. Thus the angle between the direction of flight and the sensor axis of the mass spectrometer and the retarding potential analyser passes through a minimum during each spin revolution. This minimum angle will never be greater than 60° because of the orbital plane's 45° inclination to the Sun's vector. The mass spectrometer and the retarding potential analyser are controlled in such a way that measurements are taken only in a small angular range which is symmetrical to the given minimum angle.

The mass spectrometer raises a further major problem in constructing the satellite.

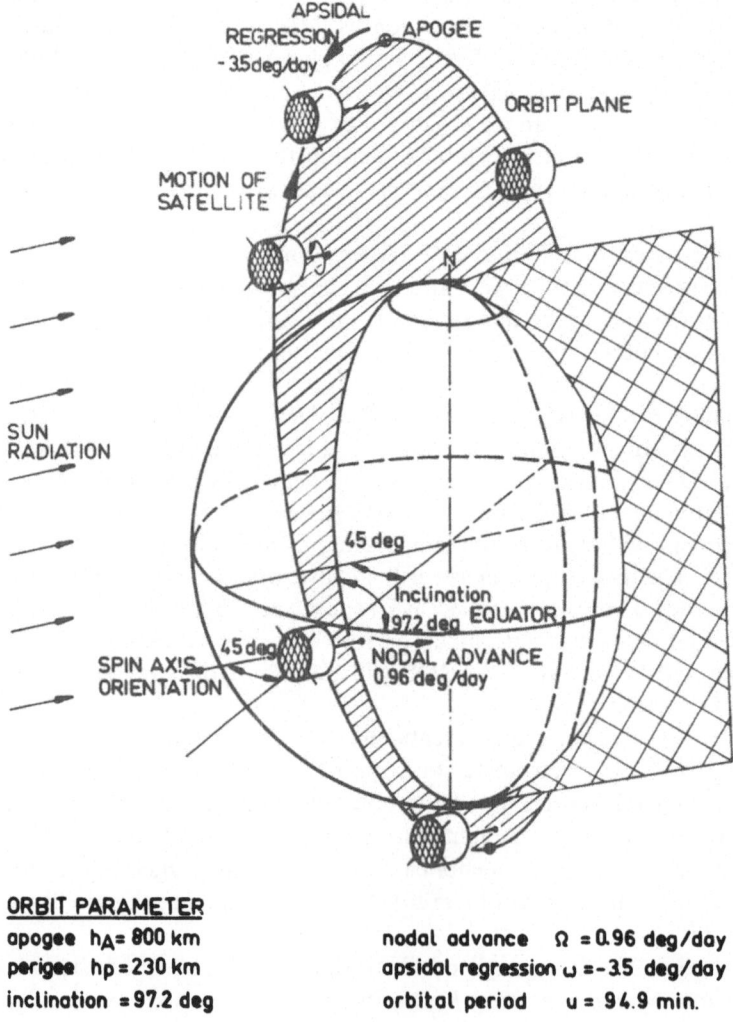

ORBIT PARAMETER

apogee h_A = 800 km	nodal advance Ω = 0.96 deg/day
perigee h_p = 230 km	apsidal regression ω = -3.5 deg/day
inclination = 97.2 deg	orbital period u = 94.9 min.

Fig. 1.

The objective of this experiment is, as already mentioned, to study the chemical composition of the ambient atmosphere. These measurements may not be falsified by any out-gassing products escaping from the satellite itself. The following measures have therefore been taken: All external parts of the satellite are made of materials which emit as little gas as possible. The geometric shape of the satellite is such that no gas-emitting parts are in the mass spectrometer's field of view. Finally, attitude control is achieved not by means of gas nozzles, but by an active magnetic system.

Two scientific measuring programmes have been incorporated: a normal programme and a special programme.

Fig. 2.

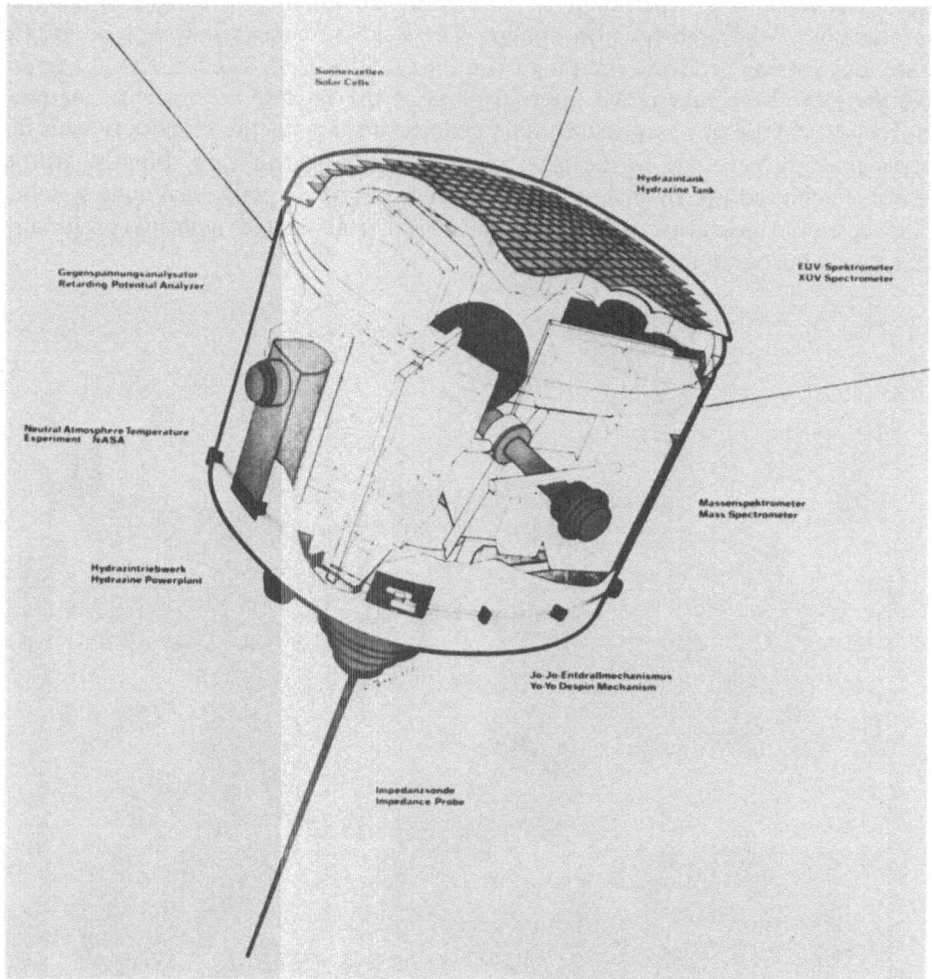

Fig. 3.

In the normal programme the retarding potential analyser and mass spectrometer will measure only in the most favourable angular range, while in the special programme they will sample data over complete spin revolutions. The latter data will be used for empirical calibration, determination of background interference and detection of multi-ionized particles. Both the normal and the special programme will be switched on for complete measuring orbits. If certain functions of the satellite should fail, various emergency programmes can be selected, to ensure partial continuation of scientific data acquisition.

The above defined experiments and their measuring modes established the following requirements for the design of the spacecraft configuration:

– The XUV-spectrometer pointing towards the Sun and thus limiting the maximum solar cell assembly to a 1 m diameter surface.

– Complete gas tight one piece skin with a high electrical conductivity to perform accurate RPA measurements.

– Installation of the MS, RPA and NATE such that measurements can be performed at minimum angle of attack during each spin period.

– Minimum structural dynamic loading of the sensitive experiments thus resulting in a design requirement for all experiments having an input in the spacecraft structure of less than four times that of the launch vehicle interface over the entire frequency spectrum.

These requirements led to a spacecraft configuration with

– A centre structure being a star shaped highly damped system where the XUV-spectrometer, all experiment electronic components and the spacecraft subsystems are installed (Figure 2).

The outer skin as a cylindrical one piece structural component, conically shaped

Fig. 4.

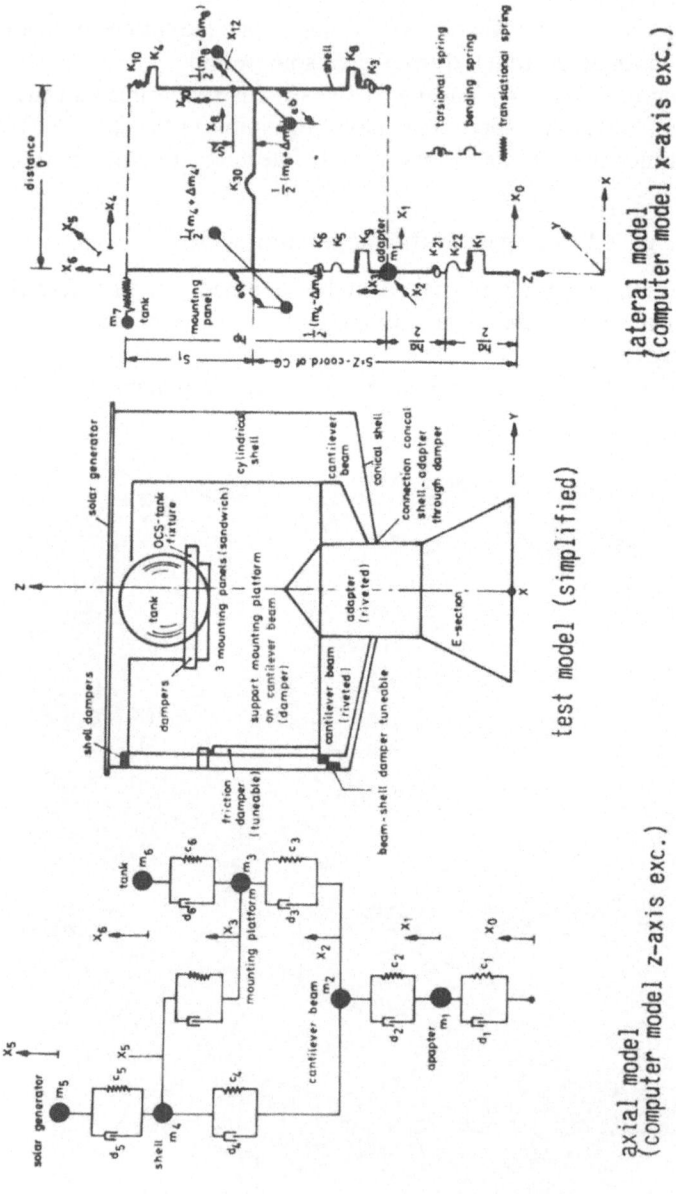

Fig. 5.

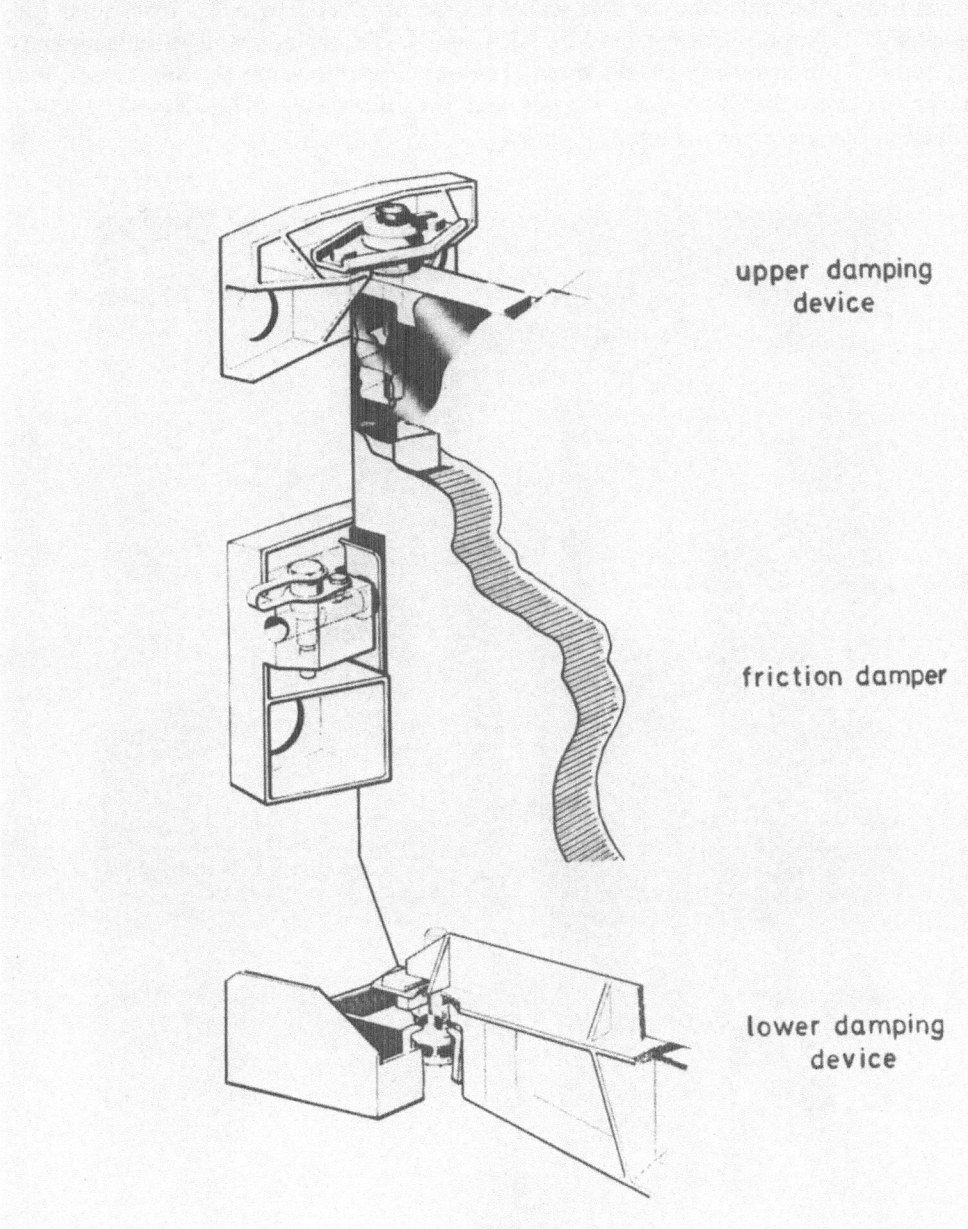

upper damping
device

friction damper

lower damping
device

Fig. 6.   Aeros damping system.

toward the spacecraft adapter and sealed on the other end by a flat plate solar cell assembly. The experiment sensors MS, RPA and NATE, including all attitude measuring sensors, are installed in this section. The only exceptions are the Sun sensor fine, integrated with the XUV-spectrometer and the impedance probe experiment, installed in the spacecraft adapter (Figure 4).

Fig. 7.

In order to obtain the required design parameters for the spacecraft structure with respect to optimized damping characteristics and sufficient structural stiffness, a mathematical model was developed and analysed by means of parameter variation. Preliminary results made it evident that the resonant frequencies of the control structure and the spacecraft skin had to be separated such that the resulting out of phase motion can be utilized to absorb energy. This led to the design of a friction damping system to be installed at the interface between the control structure mounting platforms and the skin (Figure 6). The final configuration of this system consists of a silicon base damper on the top and bottom of each mounting panel and a tunable friction damper assembly in the centre.

Test programmes conducted on a structural test model verified that

– The spacecraft damping system is becoming more efficient as the vibratory input level is being increased. At the flight vibration environment the amplification values at the various experiment and subsystem components stay below a value of $Q=4$ (Figure 8).

– The first resonant frequency is decreasing at the different component locations as the vibratory input is being increased.

– The comparison between theoretical and actually measured vibration response coincide very well when subjected to a 7g launch vehicle input amplitude (Figure 10).

Limited solar cell area – one spacecraft surface always being oriented towards the Sun – and the prevention of solar paddle usage (experiment requirements with respect to out-gassing and 180° field of view) made it necessary to develop an optimized electrical power system. The AEROS spacecraft uses a parallel power system with a

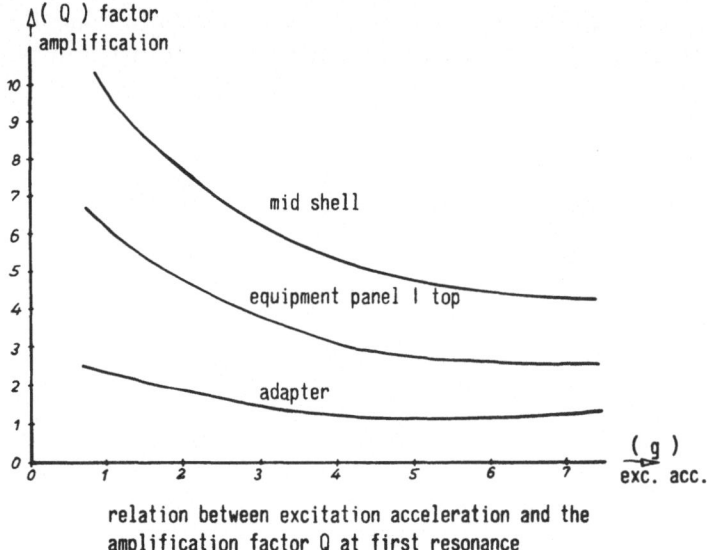

relation between excitation acceleration and the amplification factor Q at first resonance

Fig. 8.   Aeros structure.

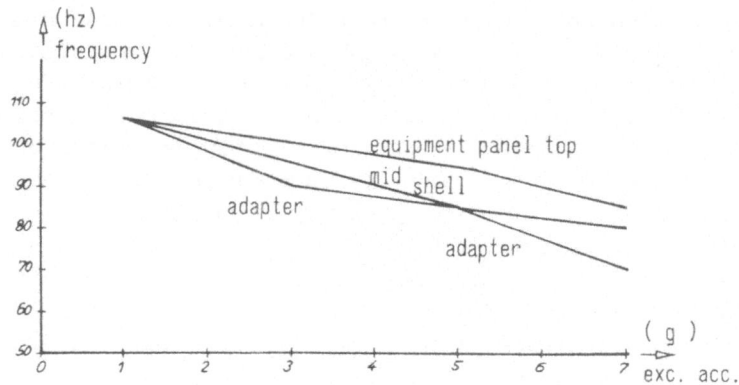

relation between excitation acceleration and the
occurance of the first resonance frequency

Fig. 9.

maximum power point tracker. The adaption is obtained by adjusting the input
impedance of the power control and conditioning system to the differential output
impedance of the solar array. This input impedance results from the parallel impedances of the reflected load (through the direct regulator) and the impedance of the
charging regulator. The first is defined by the load, the latter determined by the
tracker, such that the total impedance is adapted by the array. By steady adaption,
the maximum available array power at all possible panel temperatures can be utilized.
The next advantage will be evident in case of a full battery charge, when the power
demand is a minimum. A corresponding full charge signal overriding the tracker reduces or stops charging, so that the total impedance is increased, leading to a desired

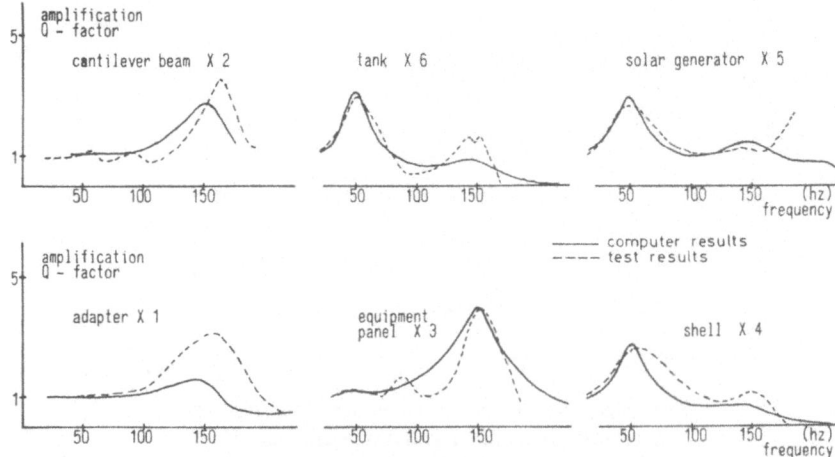

Fig. 10.   Aeros structure dynamic response.

direction of the array. Thus the usual shunt regulator causing increased weight and thermal problems is avoided.

The AEROS power system (Figure 11) feeds solar power directly through the 'direct regulator' to the 28 V main bus paralleled by a 16 V converter.

The input impedance is controlled by the 'overvoltage control', keeping the main bus voltage at 28 V $\pm$ 2%.

The 'maximum power point tracker' determines the input impedance of the 'charging regulator' such that the total impedance of the power control and conditioning system is adapted to the array. The excessive solar power is stored in one of the batteries, selected by the 'battery control'. If the battery charge is stopped or

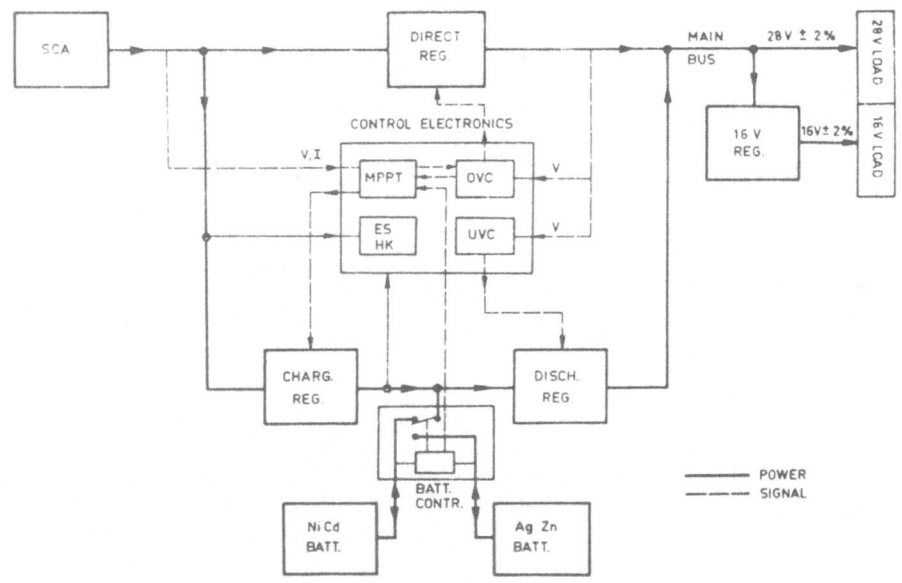

Fig. 11. Block diagram of Aeros power system.

reduced to trickle charge, the battery control overrides the tracker signal, thus increasing the input impedance of the 'charging regulator' and causing a desired misdirection of the array.

If the main bus power demand exceeds the available solar power, the main bus voltage drops to a value of 28 V $-$ 1%. This is controlled by the 'under voltage control' (UVC) and permits an additional power flow from the battery through the 'discharging regulator' to the main bus. In this case battery charging will be stopped. The tracker continues adaption by influencing now the impedance of the 'direct regulator'. The switch-over process of the tracker does not effect remarkably the main bus voltage regulation, so that $\pm$ 2% accuracy of nominal main bus voltage can easily be guaranteed (overall accuracy).

All control circuits are supplied by a regulated, partly redundant supply, getting

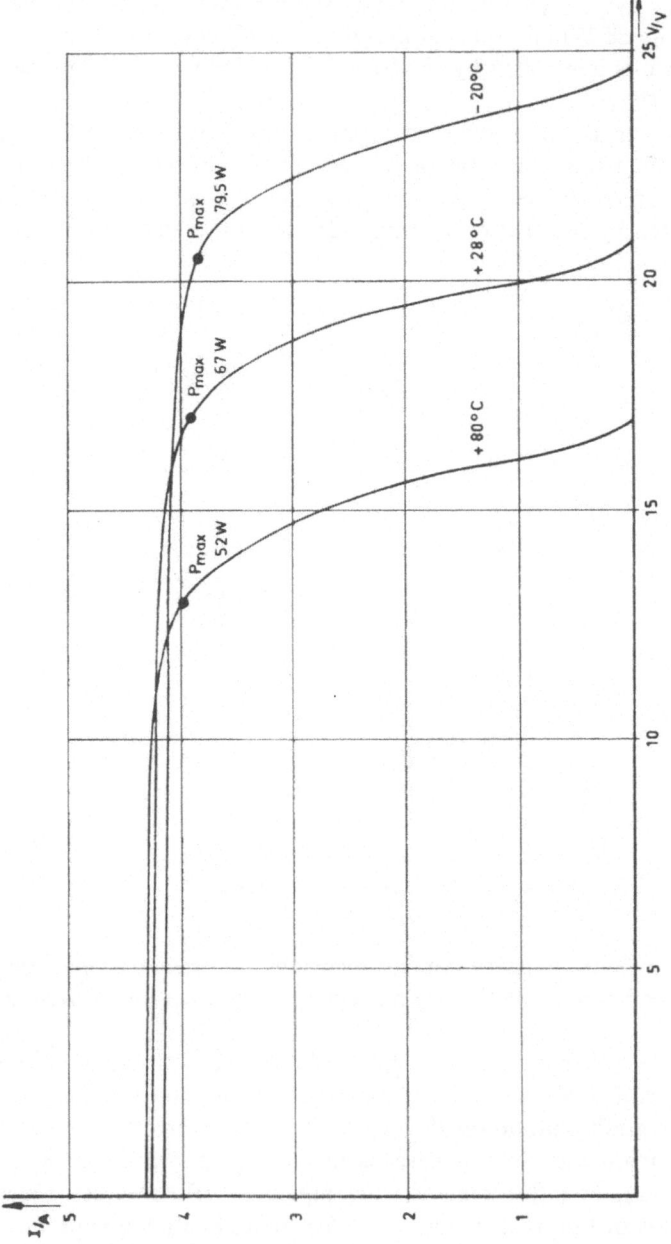

Fig. 12.   Theoretical IV characteristics.

its power from the array in the Sun phase and from the batteries in the shadow phase during each orbit.

The AgZn battery was chosen for its high energy capacity, necessary during the satellite's acquisition phase. It is conditionally rechargeable and can be regarded as redundancy for the NiCd battery, which has been selected for normal mission operation. The voltage of both batteries can vary between 32 and 44 V.

The 'direct-' and 'charging regulator' work on the booster principle from a 10–25 V Sun bus voltage, the one inline the 28 V main bus, the other with the batteries. Their input impedances are frequency-controlled.

The 'discharging regulator' working from the batteries to the main bus is controlled by a frequency controlled buck regulator.

Simulating the I–V characteristics of the solar array at various temperatures (Figure 12) the measured efficiency values are summarized as follows:

Direct regulator	$\eta_{\mathrm{DIR}} = 0.92$	
Charging regulator	$\eta_{\mathrm{CR}} = 0.94$	
Discharging regulator	$\eta_{\mathrm{DR}} = 0.94$	
NiCd battery	$\eta_{\mathrm{Wh}} = 0.71$	

(Figures 13, 14 and 15).

The overall efficiencies of the two power paths leading from Sun bus to the main bus loads can be defined as follows:

Direct path	$\eta_{\mathrm{D}} = 0.92$	through direct regulator
Battery storage	$\eta_{\mathrm{BS}} = \eta_{\mathrm{CR}}\eta_{\mathrm{Wh}}\eta_{\mathrm{DR}} = 0.63$	through charge regulator, battery, discharge regulator
Battery path	$\eta_{\mathrm{BNS}} = \eta_{\mathrm{CR}}\eta_{\mathrm{DR}} = 0.88$	through charge regulator, discharge regulator

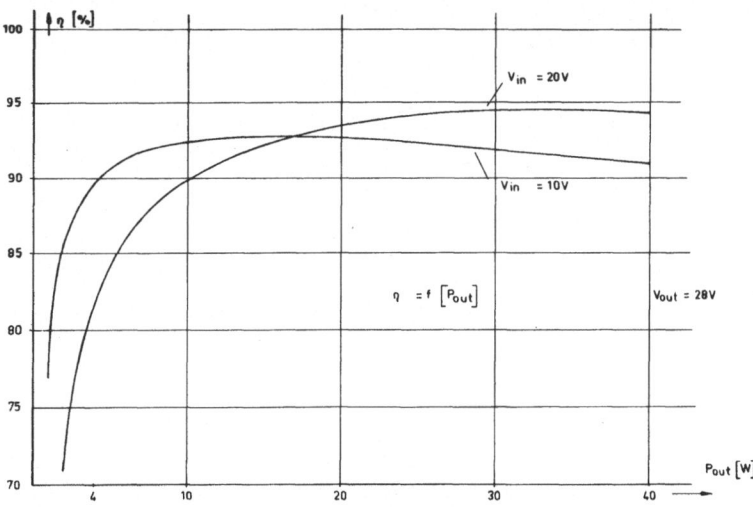

Fig. 13. Efficiency of direct regulator.

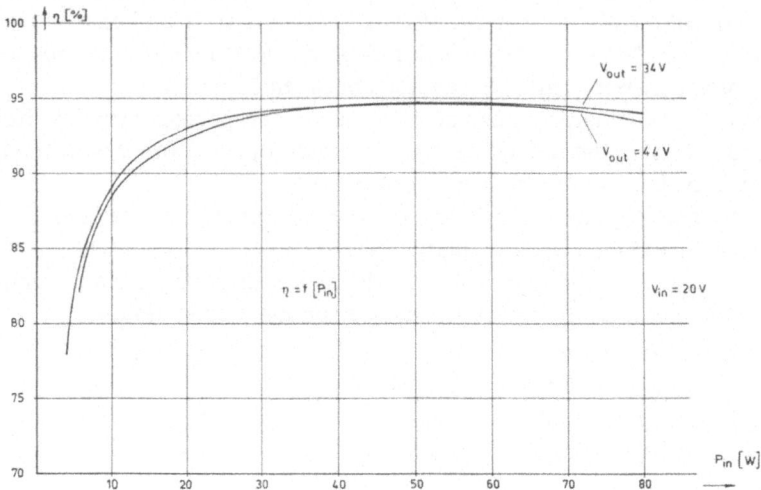

Fig. 14.   Efficiency of charging regulator.

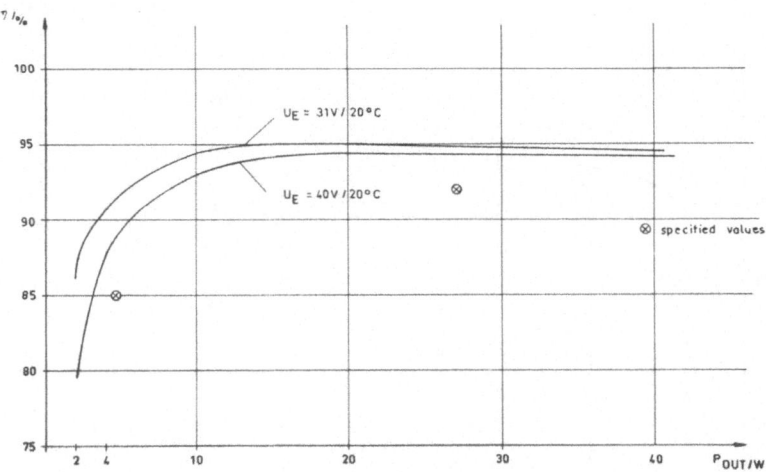

Fig. 15.   Efficiency of discharging regulator.

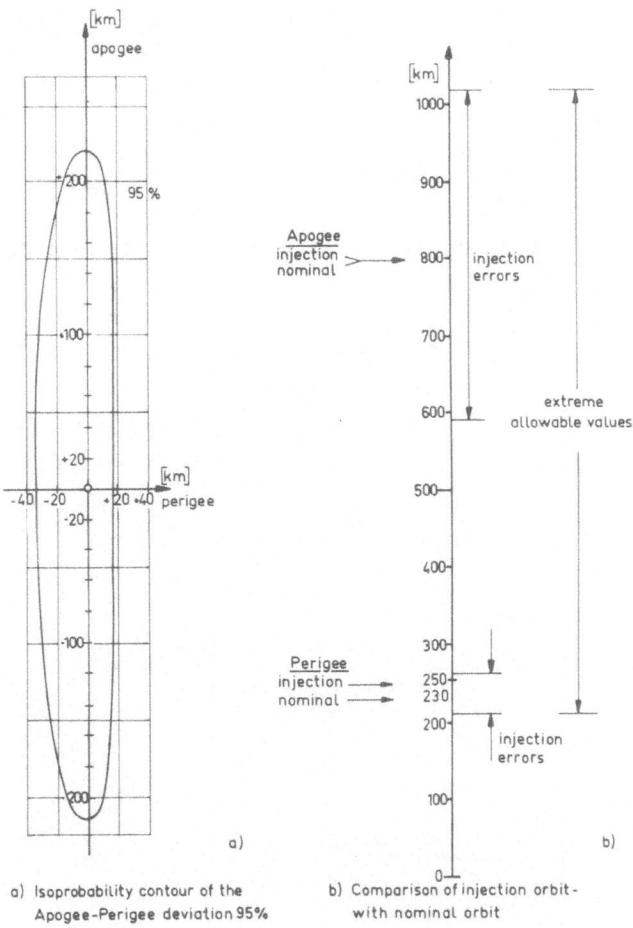

a) Isoprobability contour of the
   Apogee-Perigee deviation 95%

b) Comparison of injection orbit-
   with nominal orbit

Fig. 16.

An additional benefit is evident in case of failure in the direct path, then the solar power can be transferred in the parallel path without being stored in the batteries.

By using the maximum power point tracker the power gain calculated on preliminary solar panel temperatures curves is about 12% compared to a conventional system. This corresponds to about 6 W/orbit in case of AEROS.

The 2σ injection errors of the Scout vehicle required an orbital correction system within the spacecraft in order to obtain the specified nominal orbit parameters:

$$230 \text{ km} - \text{perigee}$$
$$800 \text{ km} - \text{apogee}$$
and $\qquad 97.2° - \text{inclination}$

The system is of the monopropellant type using hydrazine as fuel. Decomposition is effected by a catalyst (Shell 405) in the thruster chamber (Figure 17). The system consists of a pressurized fuel tank centrally located in the spacecraft and two thrusters located on the adapter side of the spacecraft. The system is capable of generating a maximum thrust of 3.7 kg (42 kg cm^{-2} fuel pressure) and a minimum of 1.5 kg (14 kg cm^{-2}). Maximum fuel capacity is 4.7 kg, capable of generating of total ΔV of 83.5 m s^{-1}.

After spacecraft injection into orbit the following manoeuvre can be executed during the initial acquisition phase:

Lowering of perigee by 20 km maximum (6 m s^{-1})
Lifting of apogee by 200 km maximum (50 m s^{-1})
Correction of inclination by $\pm 0.3°$ (40 m s^{-1})

The fourth manoeuvre will lift the apogee after 4–5 months life time with the remaining fuel. The system can be activated by retarding command with a minimum

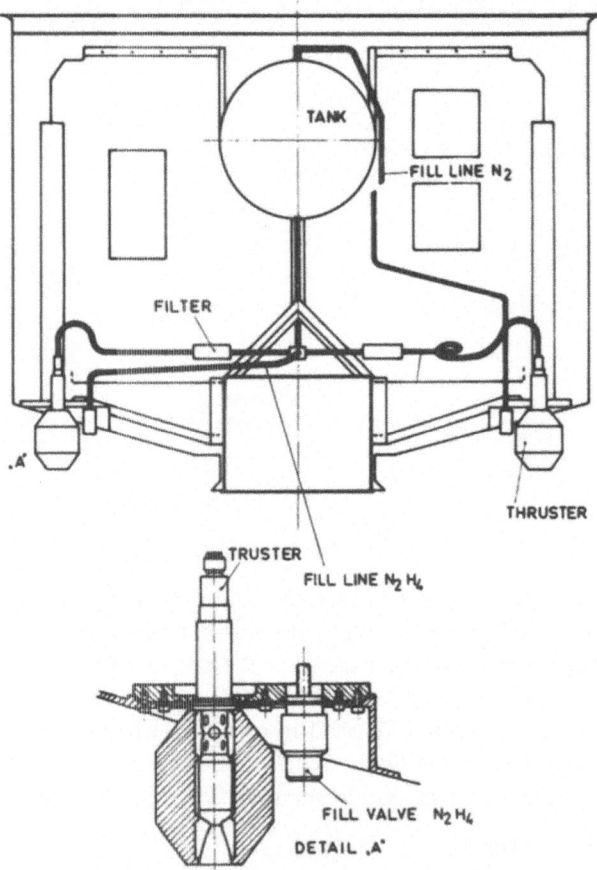

Fig. 17. Aeros orbital correction system.

· burning time of 4 s corresponding to an altitude change of 4 km. If larger changes are required, the basic burning time is increased by a commanded multiplier.

Out-gassing and weight limitation requirements led again to the development of an attitude control system using the principle of active magnetic control.

The attitude measuring system provides data for the operation and evaluation of the experiments and activation of the attitude control and orbital correction system. It consists of

– A coarse Sun sensor covering a field of view over 180° with an accuracy of 1°, primarily activated for spacecraft positioning during orbital correction manoeuvres.
– a fine Sun sensor covering a field of view over ± 5° with an accuracy of 0.2°;
– 2 infrared horizon crossing indicators measuring with an accuracy of 0.3°;
– a three axis magnetometer defining the Earth's magnetic field vector within 1%;
– an ion probe measuring the minimum angle of attack within 3° to be used for the experiment phase control circuit.

The position of the Sun direction in reference to the XUV-spectrometer, the angle between the Earth magnetic field vector and the spacecraft spin axis as well as the angle of attack of the ion sensor with regard to mass spectrometer and retarding potential analyser are required for experiment data analysis on the ground. For on-board attitude determination to activate the attitude control system the combination Sun sensor-magnetometer or Sun sensor-Earth sensor can be used.

The attitude control system will perform the following functions:

– alignment of the satellite's spin axis for orbital correction manoeuvres;
– maintaining the spin axis within ± 5° of the Sun direction; and
– maintaining a constant spin rate of 10 rpm with an accuracy of 1%.

The system consists primarily of

– 2 axis control coils, capable of producing a magnetic dipole moment of 1.8×10^{-4} Vsm each;
– 1 spin control coil of the same type;
– 2 pendulum type nutation dampers, capable of reducing the nutation angle to less than 0.05°;
– and control electronics to activate the various control functions (Figures 18 and 19).

The two axis control coils, installed parallel to the spacecraft spin axis, will be energized, when the plane between spacecraft spin axis and Sun direction is nearly perpendicular to the plane between spin axis and the Earth magnetic field vector. The spin axis control coil, installed perpendicular to the spacecraft spin axis, will be energized, when the Earth's magnetic field vector is located in a plane nearly perpendicular to the spacecraft spin axis.

For both control functions this condition is met 4 times in one orbit, at the so-called switching points, whereby two of each are always located in the Sun-lit portion of the

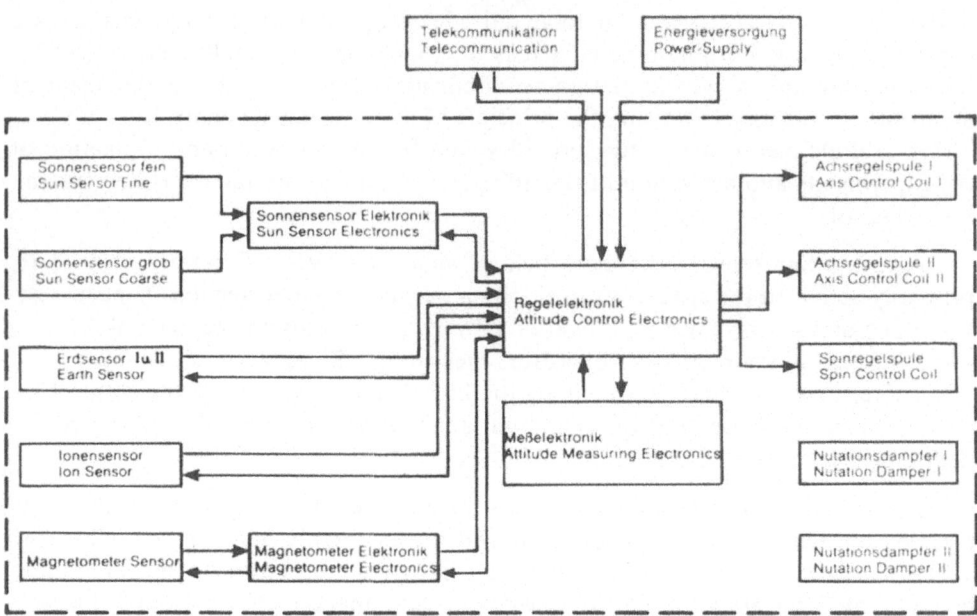

Fig. 18.   Aeros attitude control system (ACS).

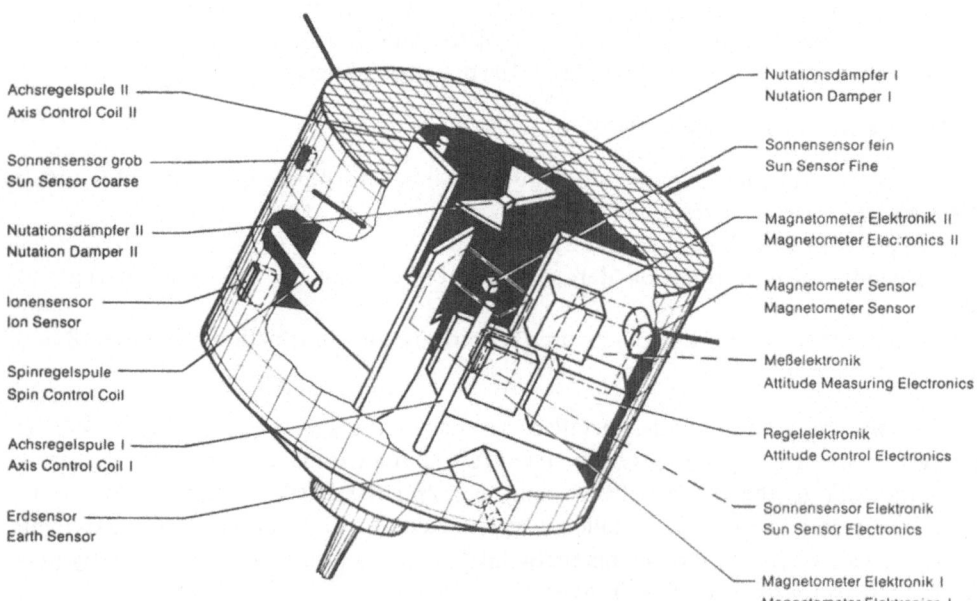

Fig. 19.

orbit. The error signals for the axis control system are generated by the fine or coarse Sun sensors and for the spin control by the Earth sensors or Sun sensors.

If the spin axis deviates more than 4° from the Sun direction, the magnetometer will determine the required switching points in orbit and energize the control coils for approximately 2.5 min each, until the deviation has been reduced to less than 0.7°. For the spin deviation the control system will be energized, when the nominal value of 10 rpm has changed by 0.8%. Based on disturbances from gravity, atmosphere and spacecraft's own magnetic dipole, the spin axis will for example deviate approximately 1°/orbit and therefore requires correction only about every fourth orbit.

As mentioned previously the MS and RPA experiments are only measuring during each spin in the range of the minimum angle of attack. It was therefore necessary to develop a phase control system which synchronizes the data frame clock signal with that of the minimum angle of attack generated by an ion sensor.

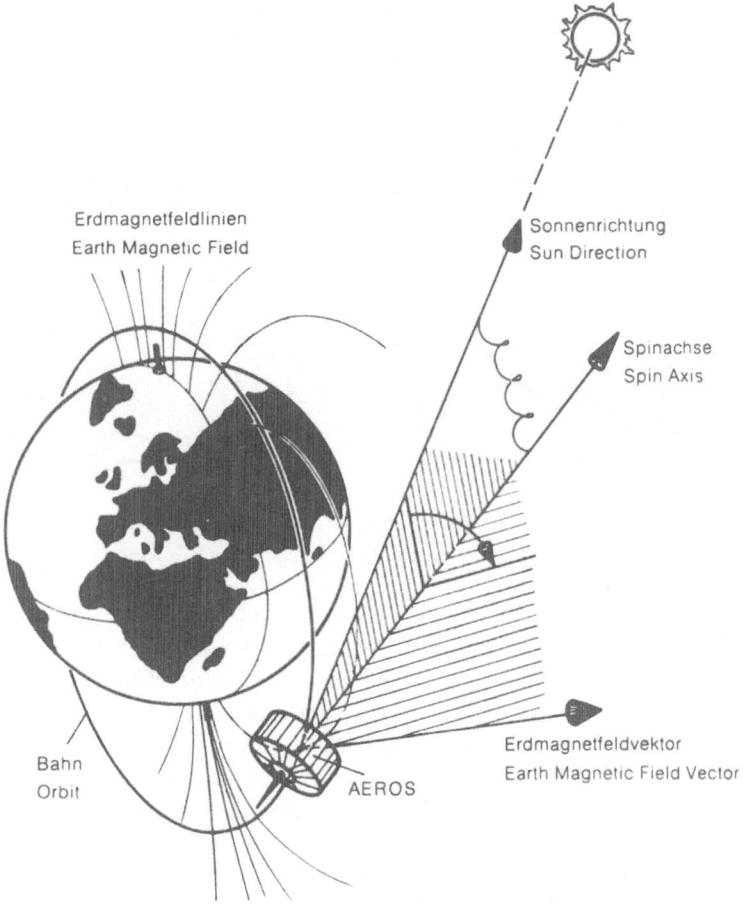

Fig. 20.

The experiment phase control compares the ion pulse, generated once every spin (every 6 s) with the data pulse, controlling the data flow from the buffer store and accordingly the data frame. By comparing the time intervals between the data pulses and the actual ion sensor pulses, the phase difference is determined resulting in a bit rate increase or decrease. Bit rate change however will only be activated at the end of each data frame corresponding to one spin.

However, the bit rate is utilized to control the recording speed of the magnetic tape recorders, such that a constant bit density can be recorded on the tape.

This technique will guarantee that the bit rate errors during the tape recorder play-back mode at constant speed can be kept at a minimum.

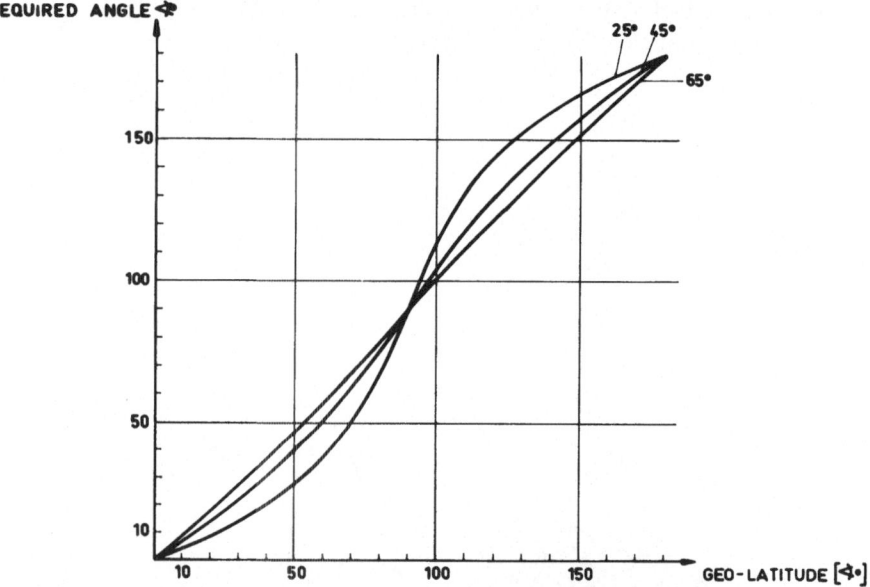

Fig. 21.   Aeros angle of attack change.

The experiment phase control requires a mean value for the bit rate as a reference for the control system. This reference is generated once during every orbit over 128 spins in the equatorial range, where the ion sensor is generating the most accurate pulse period. Since the rate of change in the prescribed orbit is non-linear – attitude of the ion sensor axis with respect to the orbital plane – (Figure 21), one can see that in the equatorial range the rate of change is a minimum. The operational range of the control loop has been limited to a maximum correction capability of $\pm 1\%$, whereby this deviation can only be corrected in steps of 0.3% from one data frame to the next. The control parameters chosen were defined to primarily cover the non-linearity of the angle of attack and suppress excessive disturbances of the ion sensor while the spacecraft is acquiring data over the polar regions. In this region the capability will however exist, to determine on a limited basis the maximum rate of change of the ion sensor disturbance existing there.

Acknowledgements

The information presented in this paper was sponsored by the 'Bundesministerium für Bildung und Wissenschaft' of the Federal Republic of Germany represented by 'Gesellschaft für Weltraumforschung'.

GEOS – THE FIRST SCIENTIFIC GEOSTATIONARY
SATELLITE SPONSORED BY ESRO

K. BERGE and R. KRIEGER

GEOS – Project Department, ERNO Raumfahrttechnik GmBH, Bremen, F.R. of Germany

Abstract. The following paper discusses the scientific objectives and satellite design aspects of this first geostationary scientific mission, sponsored by ESRO. It aims to present the scientific payload in a more popular way and discusses special spacecraft design features in the area of electric and magnetic cleanliness.

The GEOS-satellite is one of the most interesting and fastidious scientific undertakings in space. It carries nine experiments for studying distribution of low, medium and high energy particles, thermal and subthermal plasma, variation and magnitude of magnetic and electric fields.

It is an integrated payload such that certain measurement areas of experiments overlap but different methods are always used. In this way differences in events or in the way of measurement can be detected and instruments calibrated. This kind of payload especially is important in view of the International Magnetospheric Study (IMS) to be carried out between mid 1975 and 1977.

The GEOS-satellite will be launched in 1976 by a Europa II launcher from Kourou, French Guyana. The overall spacecraft mass will be about 360 kg and most of the experiments will be mounted on eight booms in various directions. Data transmission will be done by an S-band Telecommunication system, transmitting 10 to 12 kbit s^{-1}. The spacecraft will also operate with an advanced propulsion system (Hydrazine) and will have an apogee motor for transfer from a parking orbit into the geostationary orbit.

1. Mission Objective and Overall Spacecraft Design

The mission of the GEOS-satellite is to perform a series of integrated and correlated scientific investigations of the distribution of low, medium and high energy particles, thermal and sub-thermal plasma, variation and magnitude of magnetic and electrical fields.

The position of the GEOS-satellite in a geostationary orbit offers, for the first time, the possibility to investigate these phenomena while spatial variations of the satellite relative to the phenomena remains relatively constant. Although similar investigations have been made by previous missions, none has yet been conducted in this orbit.

To attain the scientific aims of the mission the payload is composed of nine experiments, which are listed in Table I.

Five experiments are devoted to measurements of magnetic and electric fields and of waves in the magnetosphere, covering together the frequency range from zero or slowly varying to about 54 kHz. Three of these experiments have passive antennae as sensors to detect natural electromagnetic waves or fields. A triaxial fluxgate magnetometer measures the magnetic vector and its variations from DC to about 5 Hz (S-321). Electric fields from DC up to at least 12 kHz are investigated by the doubles sphere technique (S-328). Several electric and magnetic antennae are used

L. G. Napolitano et al. (eds.), *Astronautical Research 1971*, 443–453. *All Rights Reserved.*

TABLE I
List of GEOS experiments

Experiment	Task	Institute
S-301/304	The study of thermal plasma through resonance and mutual impedance techniques.	Centre National d'Études des Télécommunications, Issy-les-Moulineaux, France.
S-302	Study of the very low energy plasma using an electrostatic analyser in the range of 2–500 eV.	Mullard Space Science Lab., Holmbury St Mary, England
S-303	Ion composition, energy spectra and angular distribution of low energy particles and plasma in the range of 0–20 keV/charge.	Physikalisches Institut der Universität Bern, Switzerl. and Max-Planck-Institut für extraterristrische Physik, Garching, Germany
S-310	Measurement of the pitch angle distribution as a function of energy in the range 0.2–20 keV for electrons and protons.	Kiruna Geophysical Observatory, Kiruna, Sweden
S-321	Electron-proton spectrometer for medium energies, (electrons: 30 keV to 200 keV; protons 40 keV to 1.4 MeV)	Max-Planck-Institut für Aeronomie, Lindau, Germany
S-325	Study of electromagnetic wavefields in the magnetosphere (3 components 0.1 Hz–3.4 kHz magnetic fields; 3 components 5 Hz–10 kHz for electric field)	Group de Recherches Ionosphériques, C.N.E.T., Issy-les-Moulineaux, France, and Danish Space Research Institute, Lyngby, Denmark
S-328	Measurement of DC, ELF and VLF electric fields in the magnetosphere; double sphere electric field probe	ESTEC, Noordwijk, Holland
S-329	Electron beam electric field experiment (DC electric field perpendicular to B)	Max-Planck-Institut, Garching, Germany
S-331	Measurement of DC and ELF magnetic field Fluxgate magnetometer (3 components: 0–5 Hz)	Laboratory for Space Research, C.N.R., Rome, Italy

to detect electromagnetic fields in the ULF, ELF and VLF range from 0.1 Hz to 10 kHz (S-325). Of the two active experiments, one transmits electromagnetic signals into the magnetospheric plasma. From the return signal, information of the ambient plasma is obtained (S-301). The other experiment employs a novel technique to measure the DC electric field (S-329). An electron beam is emitted from an electron gun, and particles returning to the satellite are detected after having traced a complete gyration circle.

The four remaining experiments measure particles from thermal energies up to medium energies. Thermal and subthermal electrons and ions in the energy range of

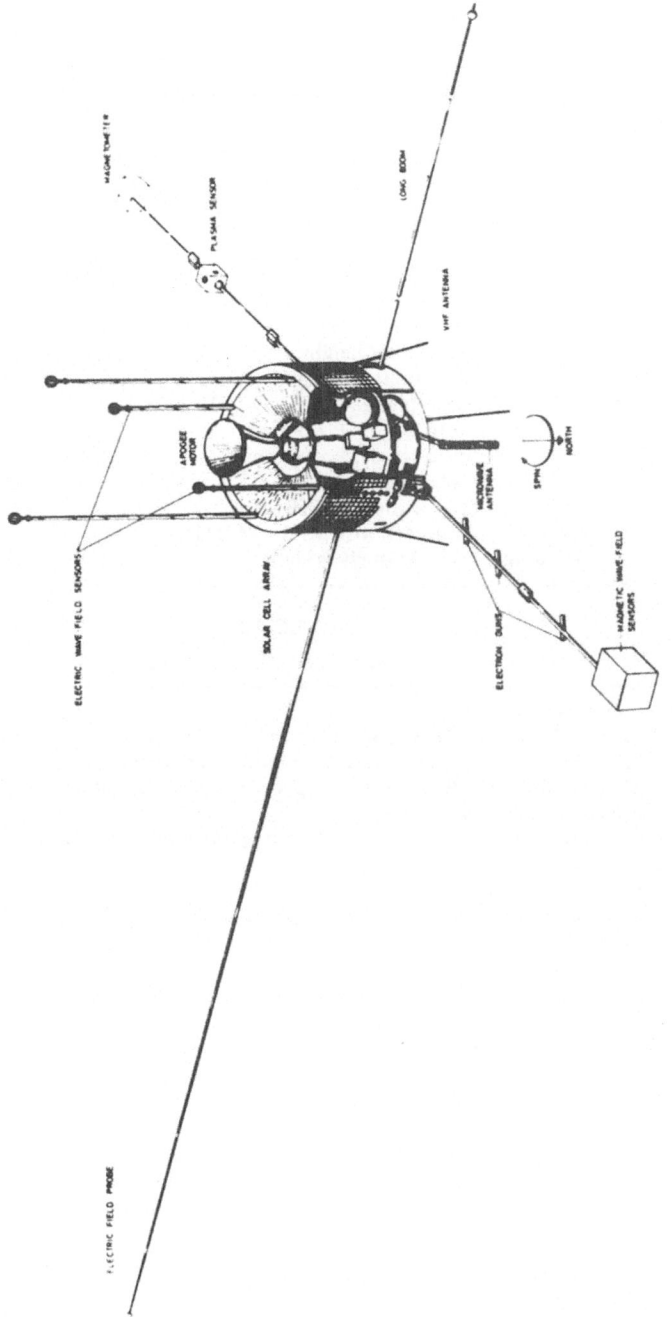

Fig. 1. Configuration of GEOS in orbital position.

2–500 eV are investigated by two hemispherical electrostatic analysers (S-302). A mass spectrometer is used to measure the ion composition, energy spectra and angular distribution of thermal plasma and low energy particles from thermal energies up to 20 keV (S-303). The energy spectrum of electrons and protons in the energy range of 0.2–20 keV as a function of pitch angle is obtained from 10 particle spectrometers consisting of curved plate analysers and channel multipliers (S-310). A magnetic deflection system and solid state detectors are used to measure the same parameters for medium energy particles up to 200 keV for electrons and 1.4 MeV for protons (S-321).

The overall spacecraft design, as it has been studied so far by ERNO Raumfahrt-

TABLE II

GEOS physical characteristics

Size:	1.0 m	main body height
	1.4 m	main body diameter
	6.1 m	overall height (E.F. to UHF antenna)
	21.9 m	tip to tip of $\pm y$ booms
	8.7 m	tip to tip of $\pm x$ booms
Weight:	30.0 kg	Experiments
	170.0 kg	Spacecraft (less ABM)
	160.0 kg	Apogee Boost Motor (ABM)
	360.0 kg	Total Launch Weight

TABLE III

GEOS subsystem characteristics

Telecommunication:
    VHF uplink at 154.2 MHz for command and tracking
    VHF downlink at 136–138 MHz to provide engineering data
    UHF downlink at 2.29–2.3 GHz to provide scientific and engineering data
Attitude measurement and control:
    Attitude reconstitution and control through ground processing
    Initial spin rate of 120 rpm
    Operational spin rate of 10 rpm
    Hydrazine propulsion system with minimum total impulse capability of
        34 000 Ns
Electric power supply:
    Solar array of 6000 cells provides max. 97 W at end of life
    Silver cadmium battery
    Decentralized dc/dc converter or centralized converter system
Boom:
    Two 10 m $\pm$ axial long booms
    Two 3 m $\pm$ axial short booms
    Four axial ($+z$) 1.5 and 3.0 long stacer construction booms
    One UHF antenna boom ($-z$)
Thermal control:
    Passive system ($+5$ °C to $+25$ °C)
Apogee motor:
    Epoxy glass filament or titan construction
    Provides 1470 m s^{-1} velocity increment
    Redundant safe and arm device

technik and the other members of the MESH Consortium, is strongly influenced by the stringent requirements of the experiments as well as by those of the launcher and the overall mission. Especially those requirements imposed for magnetic-, electric-, and chemical cleanliness, field of view, command and telemetry, satellite control accuracy, etc., have been successfully integrated in the configuration presented in Figure 1.

The spacecraft is characterized by 8 booms for scientific reasons and 1 boom for carrying the S-band antenna. It is further characterized by having an apogee motor system and uses a hydrazine propulsion system for station keeping as well as two telecommunication systems, i.e., an S-band system for telecommand, tracking and transmission of engineering data. The main characteristics are compiled in the following Tables II and III.

2. The Scientific Payload

Let us consider the relationship of scientific measurements in the geostationary environment and the different means used, in order to create a better understanding about this complex integrated payload.

The static and dynamic behaviour of the magnetosphere is not yet well understood, and it is necessary to measure particles and fields simultaneously to get a better knowledge about those phenomena.

The GEOS spacecraft is an ideal platform for those purposes for two reasons:

The first one is that due to the geostationary orbit at 6.6 R_E the temporal and spatial variations may be separated much easier than has been possible by earlier spacecraft missions.

The second reason is that the geostationary orbit is a very interesting one with respect to the geomagnetic position within the magnetosphere.

One interesting feature of this position is the plasmapause which defines the boundary of the ionosphere and which is marked by a sudden decrease of particle density when going outward from the ionosphere. The position of this boundary is a function of local time and magnetic activity and is normally located lower than 6.6 R_E. But during solar minimum, when GEOS will be launched, it is expected that due to the low magnetic activity for that period the magnetopause will reach 6.6 R_E at least for part of the time. It is supposed that at the plasma pause the separation of the part of plasma which is corotating with the Earth from the part which is not corotating is complete, but more detailed measurements have to confirm this.

The second interesting feature is the trapped radiation zone which ends at the outer radiation belt. This belt is situated outside the plasma-pause and involves large fluxes of low energy particles while the concentration of thermal plasma is relatively small.

The sources and the acceleration mechanisms for these particles are not well understood and it is not clear how far these particles help to extend ring currents which cause decrease of magnetic field at the Earth's surface at low latitudes during magnetic storms. The outer boundary of trapped radiation zone moves between $L=3$ and $L=8$ (magnetic shell) and may ideally be sampled by a geostationary satellite.

The third interesting feature of the environment of the GEOS satellite at 6.6 R_E is the magnetosphere just outside of the radiation belt. The lines of magnetic field at this region cease to be closed, and quite different possibilities of access of charged particles are allowed. These particles may fall into the atmosphere at the auroral oval which joins along the high latitude boundary of trapped electrons. The mechanisms for the injection and acceleration of those particles and their sources are not understood. It is therefore one of the prime objectives of the GEOS mission to measure low energy plasma concentration and bulk motion just outside the trapping zone together with simultaneous electric field measurements and to correlate these measurements with ground based observations at high latitudes.

The scientific payload of GEOS consists of two groups of instruments. One is related to the electric and magnetic fields in the frequency range from 0 to 60 kHz and the other to particles which are influenced by these fields on the one hand and which are influencing these fields by their own dynamics on the other hand.

The first striking features are the long booms which act as a dipole antenna for the S-301 experiment. The overall tip to tip length of this antenna will be 22 m, just the Debye-length of the environmental plasma. The experiment will transmit electromagnetic waves into the plasma over the frequency range from 300 Hz to 50 kHz. At certain frequencies such as plasma frequency, upper hybrid frequency and electron gyro frequency and its multiples, strong resonances will occur and may be detected by the same antenna for several msec after transmission.

A second part of the experiment consists of 4 spheres placed on long and short axial booms, acting as a quadrupole. A current will be emitted by two spheres on the axial booms and the induced voltage will be measured by the other two spheres on the long booms.

The ratio of current emitted to voltage induced is called the mutual impedance. Very characteristic changes of that mutual impedance occur at the hybrid frequencies and gyro frequency. From those measurements electron density will be determined to better than 1% and intensity of magnetic field will be derived. In addition information will be obtained on plasma temperature, drift velocity and on micro field distributions. Further information will be obtained because the resonance values of the total spectrum of received waves is transmitted by special 10 kHz wide band telemetry.

A second experiment taking advantage of the wide band telemetry is the S-325 electromagnetic wave field experiment which transmits three components of the electric field in the frequency range from 5 Hz to 10 kHz by means of spherical probes on the axial booms and further three components of the magnetic field in the range 0.1 Hz to 3.4 kHz by search coil sensors mounted on one of the short radial booms.

There are three basic directions for the planned investigations with that experiment:

(1) the condition of wave propagation when K is parallel to B (K=wave vector);
(2) the condition of generation of these waves and the related wave particle interaction mechanisms;
(3) the propagation and generation of waves with K vertical B.

The investigations concerning the first two items may be supported by simultaneous ground measurements. Assumptions on the path of propagation of the waves being necessary for the interpretation of simple ground based measurements up to now, are not necessary by use of an in situ receiver like GEOS, which readily separates spatial and temporal variations due to its stationary orbit. Measurements of this kind could not be performed until now, and it is unlikely that those measurements will be performed before the launch of GEOS.

Waves with $K \perp B$ are not observable at the Earth's surface. Measurements from OGO 3 and 5 and Injun 5 have discovered those waves but they have the disadvantage of non-stationary orbits. As these waves are important with respect to acceleration, diffusion and precipitation of particles, the study of their abundance, intensity, spectral shape, and polarization is fundamental for the processes within the magnetosphere.

A further experiment related to electric fields is the S-328 experiment which consists of two spherical probes mounted at the end of the long booms. This experiment differs from the former one in the method of coupling of the probes to the plasma. It measures one component of E-field from DC up to 12 kHz.

The DC part of the magnetic field up to 5 Hz is covered by a three axial fluxgate magnetometer S-331 which is mounted on the short radial boom opposite to the one carrying the S-325 AC magnetometer.

In order to measure DC magnetic fields, one has to select non-magnetic materials for all components of the spacecraft, and the unavoidable fields due to some special magnetic material which is needed, have to be shielded or compensated. Besides this, the DC magnetometer experiment must be mounted on the end of a boom of at least 3 m in order to get sufficient advantage from the dipole decrease character of the spacecraft fields.

Experiment S-329 measures electric fields by means of an electron beam which is injected into the plasma in a direction perpendicular to the magnetic field. The emitted electrons will return to the spacecraft after 1 gyration cycle and their drift path due to the electric field which is perpendicular to B will be measured. This technique of measuring electric fields in space is a new one and will be performed with GEOS for the first time. The sensitivity will be as high as $3 \times 10^{-5} \, \text{V} \cdot \text{m}^{-1}$ and the advantage of the method employed is that no spacecraft field will disturb that measurement because the information is gathered at distances as large as a gyration cycle diameter from the spacecraft.

Besides the electric field the magnetic field vector **B** and the gradient of B will be measured by this method as a byproduct and the same advantages with respect to the spacecraft cleanliness apply to these measurements as well.

The second group of experiments is related to the particles. Experiment S-302 measures concentration, energy distribution and angular distribution of thermal and supra-thermal electrons and ions in the range of 2–500 eV. From these measurements plasma temperature may be derived. Variations of the above parameters as a function of local time, solar activity, and geomagnetic disturbance in correlation with other GEOS measurements, ground based and near Earth observations may give a better knowledge of the dynamics of the magnetosphere.

Besides the plasma parameters the satellite potential and photo emission currents in the neighbourhood of the spacecraft may be determined by this experiment. Because of disturbance of the very low energy particles by spacecraft fields, the S-302 sensor has to be mounted on a boom.

An ideal complement to the S-302 experiment is the S-303 experiment which measures ion composition and the degree of ionization of particles in the energy range between 0 and 20 keV/charge. The measured quantities may be used as tracers for determining the sources of magnetospheric particles, their life time and genetic relationships. An ideal tracer for example is the ratio of $^4He/^3He$ which is 2×10^3 in the solar wind and 10^6 in the Earth's atmosphere.

An indicator for the 'age' of particles within the magnetosphere is the He^+/He^{2+} ratio.

An advantage of the ion plasma experiment against the electron plasma experiment is that drift velocity may be determined more readily because of the larger energies of the particles due to their higher masses. Simultaneous measurement of ion compositions in Aurorae will be very helpful when seeking the origin of precipitating particles.

One further very interesting measurement will be performed in the geostationary orbit in 1976 to 1977 by the S-310 experiment. It measures pitchangle distribution of electrons and protons in the range from 0.2 to 20 keV between angles of 0 to 100°.

The experiment is designed with particularly high resolution for the loss cone. The measurements will be completed by simultaneous ground based measurements at the conjugate regions. From this, far reaching conclusions may be drawn with respect to acceleration and precipitation mechanisms of particles. For this purpose it is imperative to measure within the loss cone, and with very high accuracy of instantaneous direction of the magnetic field by the S-331 sensor. In addition AC and DC magnetic and electric field measurements are necessary to treat the problem. At ground, sounding rockets will perform particle measurements and X-ray measurements will be performed by balloons simultaneously. In addition global networks of sky cameras, spectro-photometers, standard magnetometers, pulsation magnetometers, ionosounders, riometers and ELF, VLF emission recorders preferably should be available.

Another feature, which will be investigated by these complex measurements is the dynamic of substorms. Some useful answers to that question may be given by data from the HELIOS spacecraft which will be launched during the same time as GEOS.

Other questions which are related to the S-310 experiment are concerned with the generation processes of ring currents, the wave particle interaction mechanisms, and with the detailed magnetic conjugacy as a check of the existing model of the Earth's magnetosphere.

The last of the nine experiments described here is the S-321 pitchangle experiment which completes the S-310 measurements in the range of 30 to 200 keV for electrons and from 40 keV to 1.4 MeV for protons. The electrons are separated from the protons by 1000 G magnet which has to be taken carefully into account for spacecraft cleanliness. The scientific aims of this experiment correspond to those already described.

3. Requirements on Cleanliness

As can be seen from the experiment discussion this scientific payload has stringent demands in respect to cleanliness. This concerns electrostatic cleanliness, magnetic and electromagnetic cleanliness and chemical cleanliness. All these areas impose heavy requirements on the spacecraft design and will be discussed briefly.

The problems of electrostatic cleanliness encountered here, can be divided roughly into two classifications. The first is that of the environmentally imposed floating potential of the S/C. The second type of electrostatic problems is that of electric disturbances generated on the spacecraft which may be detected by the experiments. This problem is not nearly as straightforward as electromagnetic interference, nor is it as well investigated. The problem only begins with exposed electric potential surfaces and is complicated by modes of coupling between the electric noise sources to the plasma and from the plasma to those experiments sensitive to this type of disturbance. It is therefore important that attention is given to the equilibrium spacecraft potential to the order of some volts, as this will have a significant effect upon all experiments with detection thresholds below approximately 20 eV. The problem itself is less studied especially in synchronous equatorial orbit altitudes when photoelectric effects become the dominant charging mechanisms over the collection of electrons and ions from environment.

Magnetic cleanliness is the other area of interest. The magnetic fields generated by the spacecraft, consist of permanent and induced fields from materials and the stray fields from currents. They could therefore split in 'DC fields' and 'AC fields'. The 'DC fields' are composed of DC current fields, permanent fields and non-alternating induced fields, whereas 'AC fields' consist of the sum of AC current fields and alternating induced fields.

The requirements for magnetic DC-fields is confined to the S-331 DC magnetometer experiment, which possibly may allow 1 γ disturbance field parallel to the spin axis and 2 to 3 γ perpendicular to that. The realization of this requirement is controlled by a DC magnetic model, which sums algebraically the DC-magnetic fields at the S-331 sensor, arising from different components of the satellite. Table IV gives the DC magnetic characteristics of 'clean'-spacecraft.

TABLE IV

DC magnetic characteristics of 'clean' spacecraft

Spacecraft	Mass (kg)	R (m)	$B(\gamma)$ at R	M (gauss-cm^3)/kg	M normalize
GEOS (ESRO)	150	3.1	1.41	2.40	6.0
Pioneer F/G (spec. NASA)	205	6.0	0.11	0.61	1.5
Pioneer A/E (measured/NASA)	65	2.0	0.64	0.40	1.0
HEOS I (measured/ESRO)	53	2.0	0.45	0.35	0.9
OGO-C (measured/NASA)	450	7.3	0.55	2.46	6.2
Lunar P/F	30	1.65	1.20	0.90	2.3

The AC magnetic field requirements arise from the S-325 AC magnetometer experiment which allows different AC-field levels in different frequency ranges. The realization of these requirements is controlled by an AC-magnetic model, which also summarizes algebraically the magnetic noise of electronic boxes. The present AC requirements are summarized in Figure 2 and compared with these of other spacecrafts.

This satellite also imposes one of the most severe EMC requirements identified to date on any scientific satellite. The magnitude of the demands arising from the combination of the different proposed experiments, can be compared with current

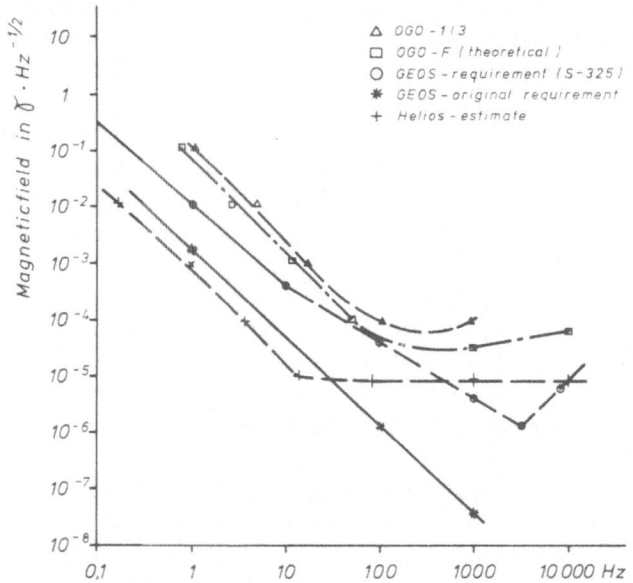

Fig. 2.   Magnetic noise spectral density requirements for different spacecrafts.

EMC specifications. This comparison makes it apparent that existing specifications will be inadequate for assuring GEOS compatibility for the frequency range below 54 kHz. The problem is further complicated by the requirements which will be imposed upon the test instrumentation (below 10 Hz) to verify the adequacy of the design. An adequate EMC control program must therefore be installed controlling generated interference from spacecraft below 54 kHz and the spacecraft design in the area of grounding power generation and data distribution have to take care of those requirements.

The last area for cleanliness is that of chemical outgassing or contamination effecting experiment sensors. Instruments of concern are channeltron multipliers inside analyser structures, solid state radiation detectors within the S-321 experiment, optical lenses on sensors and the spherical probes of S-328 made out of vitreous carbon.

Sources of chemical outgassing are the apogee motor during firing and afterwards, as well as dissociated byproducts from the attitude control thrusters using hydrazine (N_2H_4). Also other material used within the spacecraft are potential outgassing sources. Those certainly have to be investigated and a proper control program be initiated.

References

[1] *Feasibility Study of a Scientific Geostationary Satellite*, Vol. **2**, December 1968, ESRO.

[2] *Evaluation of the Recommended Payload for the Scientific Geostationary Satellite*, July 1970, ESRO.

[3] *MESH Proposal for the GEOS Satellite*, ERNO Raumfahrttechnik GmbH, No. 030/71/31.

SKYLAB
A Manned Scientific Space Laboratory

LELAND F. BELEW

NASA Geo. Marshall Space Flight Center, Huntsville, Ala. 35812, U.S.A.

Skylab will provide man with an observatory, unobstructed by the Earth's atmosphere, for conducting solar and stellar astronomy investigations. As a laboratory it will offer weightlessness and vacuum for exploring processes, technologies and biological experiments not practical on Earth. From this vantage point, Earth resource

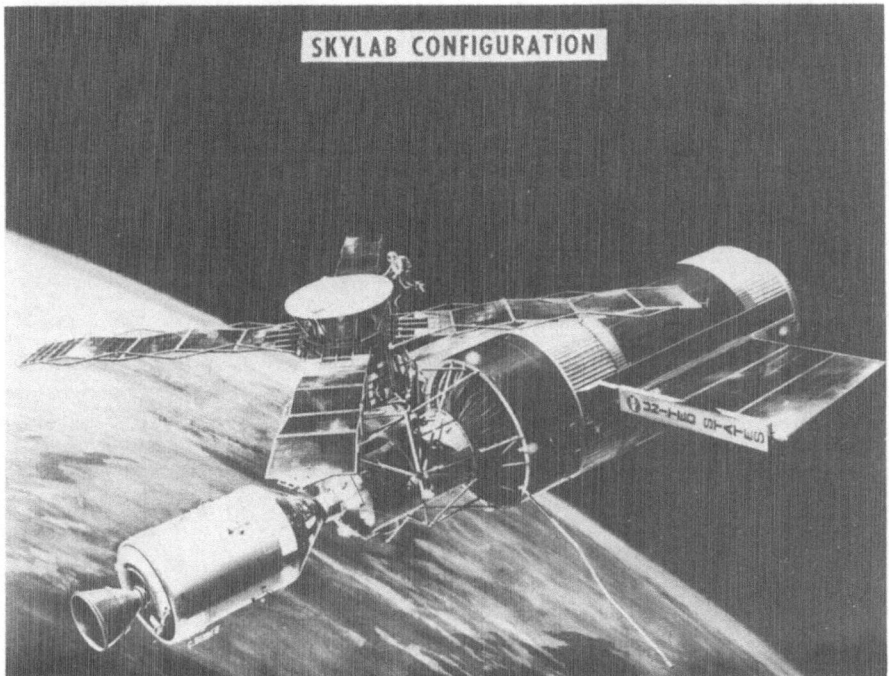

Fig. 1. Skylab in orbit. General characteristics: conditioned with volume 381 m³ (12 700 ft³); overall length 35.1 m (117 ft); weight including CSM 89 550 kg (199 000 lb); width-OWS including solar array 27 m (90 ft).

experiments designed to provide data related to the Earth's atmosphere and surface will be performed. This Earth resource experiments package is designed to give insights into solutions for many current problems effecting man and his environment (Figure 1).

Skylab will provide man with a unique environment, protecting and sustaining him, yet giving him freedom to move about, work and conduct experiments to determine the effects of long duration space flights on man and systems.

L. G. Napolitano et al. (eds.), Astronautical Research 1971, **455–469**. *All Rights Reserved.*
Copyright © 1973 by D. Reidel Publishing Company, Dordrecht-Holland.

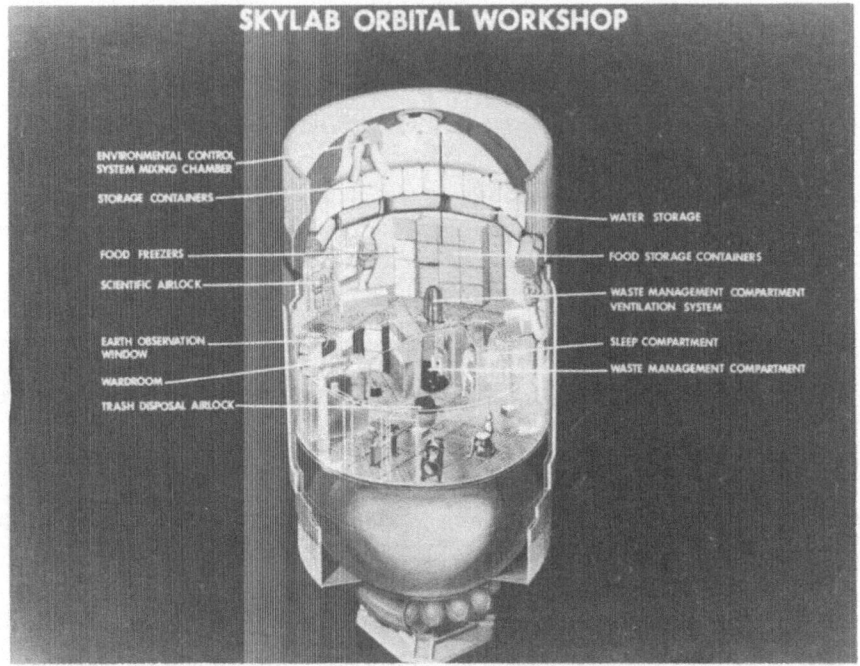

Fig. 2.   Skylab orbital workshop – internal configuration.

Fig. 3.   Airlock module.

The operational configuration of Skylab consists of: An orbital workshop (OWS), and air lock module (AM), a multiple docking adapter (MDA), an Apollo telescope mount (ATM) for conducting solar astronomy experiments and a command and service module (CSM).

The central base of activity is the workshop itself. This is where the crew sleeps, eats, attends to personal hygiene and where a large part of the experimental activity is carried on (Figure 2).

An airlock module was necessary to allow the crew to leave the laboratory for brief excursions into space. Due to its central location with respect to the other modules,

Fig. 4.   Multiple docking adapter.

and its proximity to systems support expendables/storage accommodations and exterior trusswork, the airlock module became the utility core. It contains the environmental control system, communications systems, data handling and recording systems, and is the electric power control and distribution station (Figure 3).

The multiple docking adapter provides docking ports for the command and service module. An axial port is provided for primary use and a radial port for backup and emergency purposes. The multiple docking adapter also serves as an experiment module for Skylab. It accommodates the Earth resource experiments and materials processing experiments and contains the control and display station for the solar astronomy experiments (Figure 4).

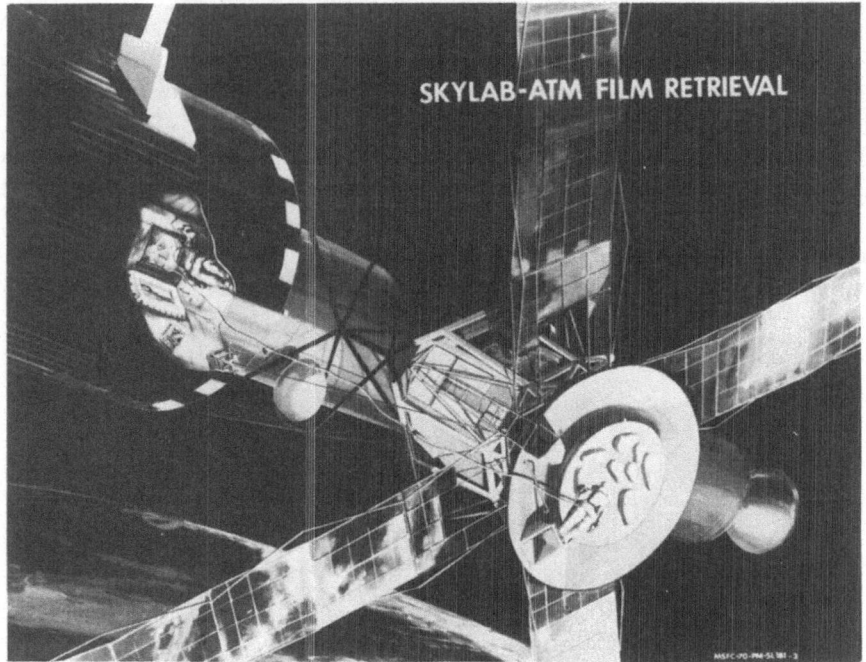

Fig. 5.   Skylab ATM film retrieval.

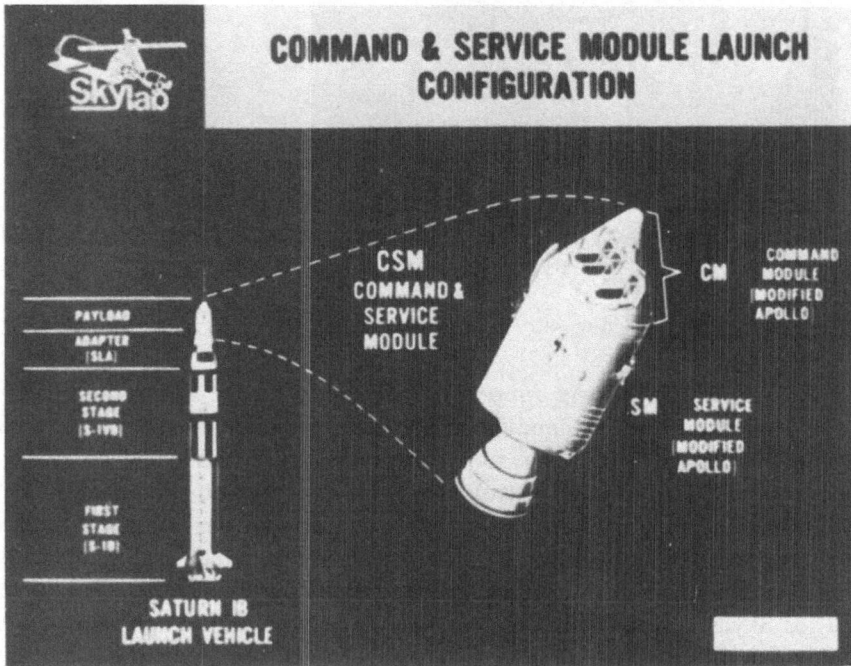

Fig. 6.   Command and service module – launch configuration.

The Apollo telescope mount completes the laboratory. This solar observatory is composed of a group of experiments integrated into a common facility (Figure 5).

The crew will be launched in the command and service module aboard a Saturn IB vehicle. Docking of the command and service module will complete the orbital assembly and Skylab will be operational (Figure 6).

The Skylab mission will begin with the launch of the unmanned Skylab from the Kennedy Space Center. The unmanned Skylab will be inserted into a near-circular orbit of about 234 NMI (434.75 km), with a nominal orbit inclination of 50°. Within $7\frac{1}{2}$ h, Skylab will be oriented to a solar-inertial (Sun-pointing) attitude mode. The

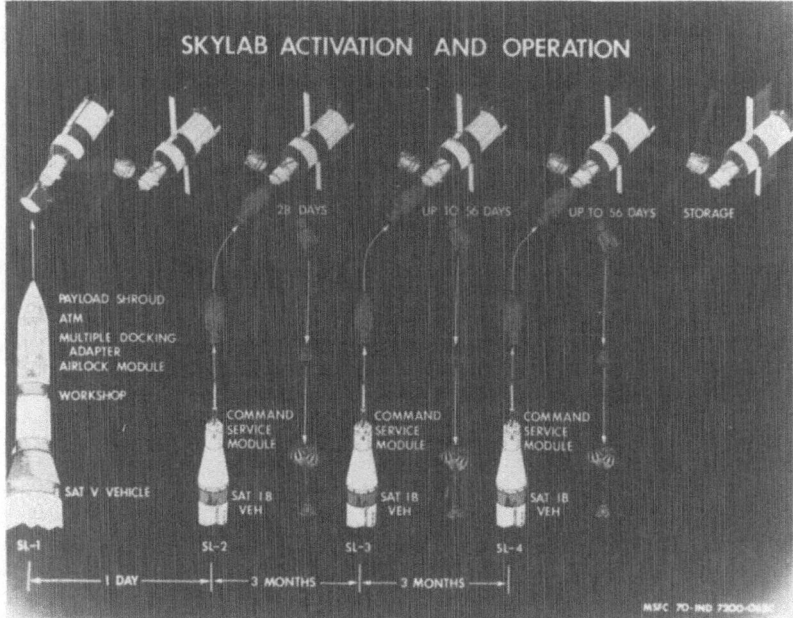

Fig. 7. Skylab mission activation and operation.

Skylab solar arrays will then be deployed. The pointing control system will be activated to maintain the solar-inertial attitude. The interior of the Skylab will be pressurized at 5 psia (0.5 kg m^{-2}) with an oxygen-nitrogen mixture, making it ready to accept docking of the command and service module and entry of the flight crew. A command and service module with a three man crew will be launched one day after the Skylab. The command and service module will rendezvous with the Skylab and will dock to the axial port of the multiple docking adapter, thus completing the manned Skylab. The habitability systems will be evaluated and the experiment program conducted placing emphasis on the medical, solar astronomy and Earth resources experiments. It is planned to obtain data from all experiments (54) during this mission. After the crew prepares the Skylab for the orbital unmanned operation, the command and service module will deorbit on the 28th manned mission day (Figure 7).

Fig. 8a. Skylab workshops crew quarters located on main floor.

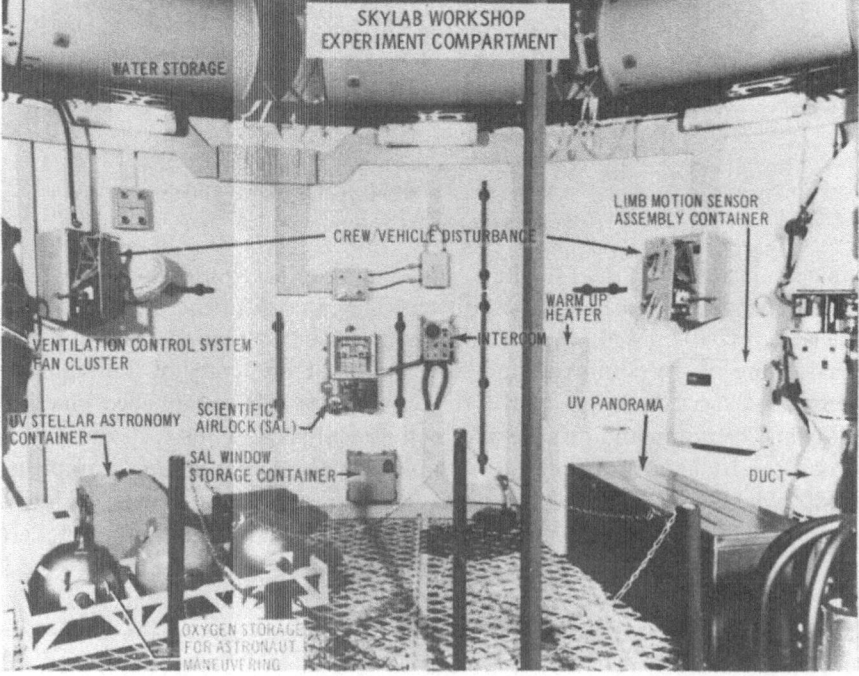

Fig. 8b. Skylab workshop experiment and stowage area above main floor.

The second mission will begin approximately 60 days after the first mission. The mission will be increased in duration up to approximately 56 days. More emphasis will be placed on the solar astronomy and Earth resource experiments.

The third mission will be launched approximately 30 days after the second mission. This mission will complete the planned experiment objectives, and will provide additional statistical data on the crew's adaptability and performance over the planned 56-day mission.

The crew is essential to the conduct of Skylab experimentation. Targets must be observed, cameras must be loaded and pointed; sensors must be positioned and activated; specimens must be processed, recovered and returned to Earth for examination.

Fig. 9. Skylab workshop crew quarters layout on main floor.

As a scientific laboratory, Skylab will be manned by crewmen trained in the appropriate scientific disciplines, qualified in the use of complex and sophisticated equipment and who will be possessed of unique ability to react to scientific observations. All equipment to be operated by the crew has been developed with active direct participation by crewmen over the full range of hardware design, manufacture and test. The Skylab has been designed to provide the crew with adequate space, pleasing surroundings, good food, privacy and acceptable hygiene facilities (Figures 8a and 8b).

The main 'floor' of the laboratory contains the wardroom of food preparation area, the waste management area containing hygiene facilities, individual sleeping compartments and an experiment work area (Figure 9).

The wardroom includes provisions for heating foods, including an assortment of frozen foods. Hot and cold beverages may also be prepared. Storage for food, utensils and napkins is provided in the wall cabinets. Food waste disposal is provided by an airlock for ejecting waste into the Aft tank which carried liquid oxygen for propulsive S-IVB stages. Such solid waste is retained in this tank. Food and other life support expendables are launched in sufficient quantity aboard the workshop on the first flight to support the three visiting crews. A limited quantity of selected items will be replenished aboard the manned command and service module. A window in the wardroom provides general viewing for the crew and will also accommodate several experiments (Figure 10).

Fig. 10.   Skylab workshop – ward room/galley.

Waste management facilities presented a unique challenge to designers of the spacecraft. In addition to collection of liquid and solid wastes, there is a medical requirement to dry all solid waste and return the residue to Earth for analysis. Similarly, liquid waste is sampled and the samples stabilized for return to Earth. In both cases the quantity is precisely measured. The waste management area contains the equivalent of a space lavatory. Bathing can be accomplished in the following way. The crewman wets a washcloth by placing it over the water discharge, applies soap to the cloth, bathes and rinses as he would at home. The wet cloth is discarded after use. A shower enclosure is under study which will afford total body bathing. Towels and tissues will be supplied in ample quantities, as well as antiseptic cleaning agents.

Shaving is afforded by either electric or safety razor. Shaving and grooming are assisted by the mirror above the lavatory (Figure 11).

Each crewman is assigned a separate space for sleeping. Sleeping can be accommodated quite comfortably in a 'bag' which restrains the body while enclosing it in a manner which is psychologically and physically pleasing (Figure 12).

Fig. 11.   Skylab workshop – waste management compartment located on main floor.

An experiment work area immediately adjacent to the wardroom, waste management and sleeping areas is where primary medical experiments will be conducted to assess the manner in which the crew is adjusting to spaceflight (Figure 13).

Provision of the life support systems, living accommodations and experiments associated with man's physiology represents a major step forward in planning for man in space and essential for eventual major space stations.

Facilities provided on Skylab that allow the performance of scientific investigations and exploration include a solar observatory, an Earth resource experiment facility, two scientific airlocks and a materials processing facility.

The solar observatory is comprised of ultraviolet, X-ray coronagraph and hydrogen alpha instruments and supporting systems. These instruments are a major step forward in performing solar research from an orbiting spacecraft. Special emphasis is placed on observations of those portions of the Sun's spectrum which are invisible to astronomers on the ground, because of absorption in the Earth's atmosphere. During the mission, ground based astronomy activities will be conducted in support of the orbital solar observatory operations (Figure 14).

Fig. 12. Skylab workshop – sleep compartment.

The Earth resources experiment experimental facility provides a flexible array of high-performance sensors to make a synoptic survey of selected areas on the Earth in the visible, infrared and microwave spectral wavelengths. Data for evaluation and development of remote sensors will be obtained in conjunction with data related to the Earth's surface and atmosphere that will be utilized in the fields of meteorology, agriculture/forestry, oceanography, hydrology, geology and geography. Insight will be obtained into global aspects of atmospheric pollution. The sensors and the sup-

Fig. 13. Skylab workshop – work area located on main floor.

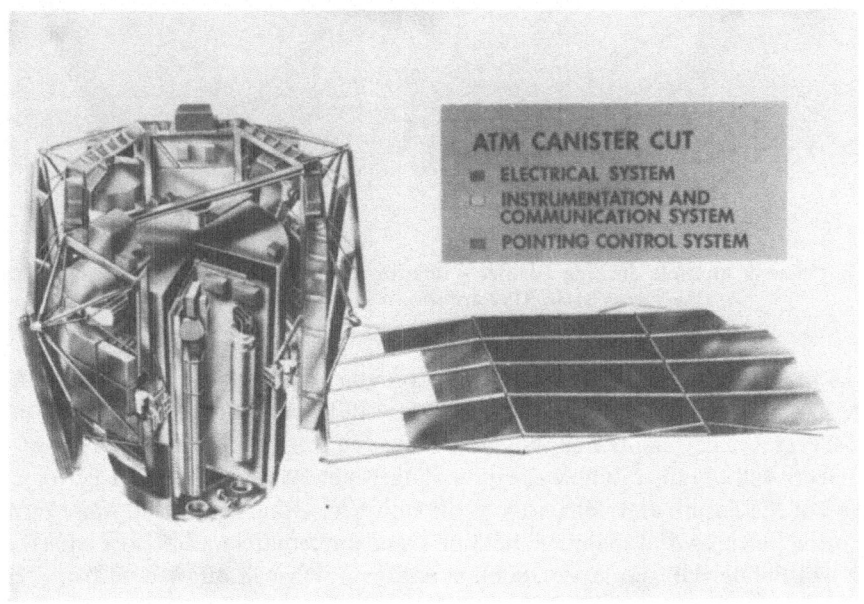

Fig. 14. Cutaway view of Skylab ATM.

port equipment are designed so that each experiment can be operated individually or in any combination. It is planned that many nations will participate in the use of the Earth resource experiment data (Figure 15).

Two scientific airlocks are available – one Sun oriented and one Earth oriented. This facility provides a means for attaching experiments internally, and deploying them through the airlock into space or looking into space with a variety of sensors. A universal extension mechanism is available on which sensors or cameras can be deployed up to 18 ft (5.4 m) from the workshop meteoroid shield. Shaft rotation and head elevation permit virtually unlimited visibility. Common services at the scientific

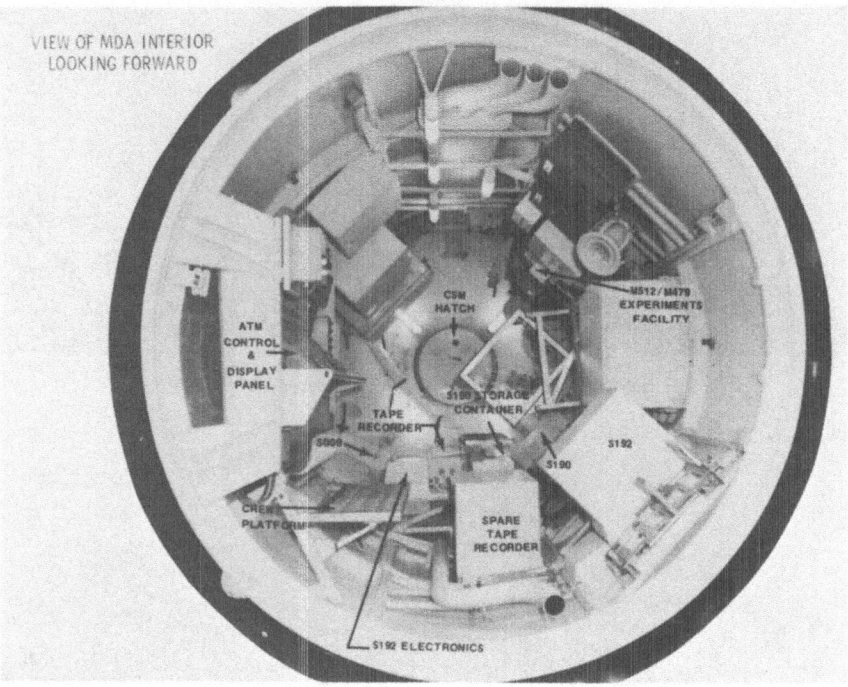

Fig. 15. View of multiple docking adapter – interior looking forward (Earth resources experiments S190, S192 are shown in lower centre).

airlocks include power and data outlets, a vacuum purge and repressurization connection. A window can also be installed for visibility and for photographic experiments. Several experiments are assigned to use the scientific airlocks. Some of these experiments will obtain valuable scientific data which will contribute to better understanding of the nature and behaviors of the universe. Others will yield more information on the particles and magnetic field of space immediately adjacent to Earth. Still others will obtain data on sensor contamination which will allow data from sensors of other experiments to be more accurately interpreted (Figure 16).

A materials processing facility will be used to investigate selected physical and

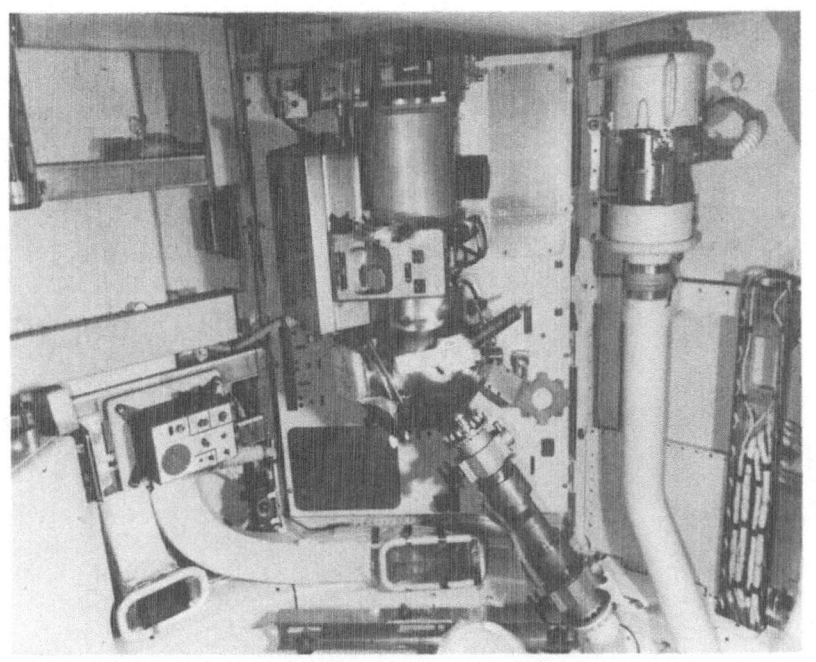

Fig. 17. Materials processing facility in multiple docking adapter.

Fig. 16. Scientific airlock with X-ray/UV solar photography experiment (SO 20) installed.

MEDICAL

M071	MINERAL BALANCE*
M073	BIOASSAY OF BODY FLUIDS
M074	SPECIMEN MASS MEASUREMENT
M092	IN-FLIGHT LOWER BODY NEGATIVE PRESSURE
M093	VECTORCARDIOGRAM
M131	HUMAN VESTIBULAR FUNCTION
M133	SLEEP MONITORING
M151	TIME & MOTION STUDY
M171	METABOLIC ACTIVITY
M172	BODY MASS MEASUREMENT
T003	IN-FLIGHT AEROSOL ANALYSIS
ESS	EXPERIMENT SUPPORT SYSTEM

TECHNOLOGY

D024	THERMAL CONTROL COATINGS
M415	THERMAL CONTROL COATINGS
M479	ZERO GRAVITY FLAMMABILITY (B)
M512	MATERIALS PROCESSING IN SPACE
M551	METALS MELTING
M552	EXOTHERMIC HEATING
M553	SPHERE FORMING
M554	COMPOSITE CASTING
M555	GALLIUM ARSENIDE CRYSTAL GROWTH
T002	MANUAL NAVIGATION SIGHTINGS (B)
T013	PRECISION OPTICAL TRACKING
T025	CORONAGRAPH CONTAMINATION MEASUREMENT
T027	ATM CONTAMINATION MEASUREMENT

SCIENCE

S009	NUCLEAR EMULSION
S019	UV STELLAR ASTRONOMY
S020	UV X-RAY SOLAR PHOTOGRAPHY
S063	UV AIRGLOW HORIZON PHOTOGRAPHY
S073	GEGENSCHEIN-ZODIACAL LIGHT
S149	PARTICLE COLLECTION
S150	GALACTIC X-RAY MAPPING (B)
S183	ULTRAVIOLET PANORAMA

APOLLO TELESCOPE MOUNT

S052	WHITE LIGHT CORONAGRAPH
S054	X-RAY SPECTROGRAPHIC TELESCOPE
S055A	UV SCANNING POLYCHROMATOR SPECTROHELIOMETER
S056	X-RAY TELESCOPE
S082A	XUV CORONAL SPECTROHELIOGRAPH
S082B	UV SPECTROGRAPH

OPERATIONS

M487	HABITABILITY & CREW QUARTERS
M509	ASTRONAUT MANEUVERING EQUIPMENT
T013	CREW VEHICLE DISTURBANCE
T020	FOOT CONTROLLED MANEUVERING UNIT

EARTH RESOURCES EXPERIMENT PACKAGES

S190A	MULTISPECTRAL PHOTOGRAPHIC CAMERAS
S190B	EARTH TERRAIN CAMERA (ETC)
S191	INFRARED SPECTROMETER
S192	MULTISPECTRAL SCANNER
S193	MICROWAVE RADIOMETER SCATTEROMETER & ALTIMETER
S194	L-BAND RADIOMETER
ESE	EREP SUPPORT SYSTEM

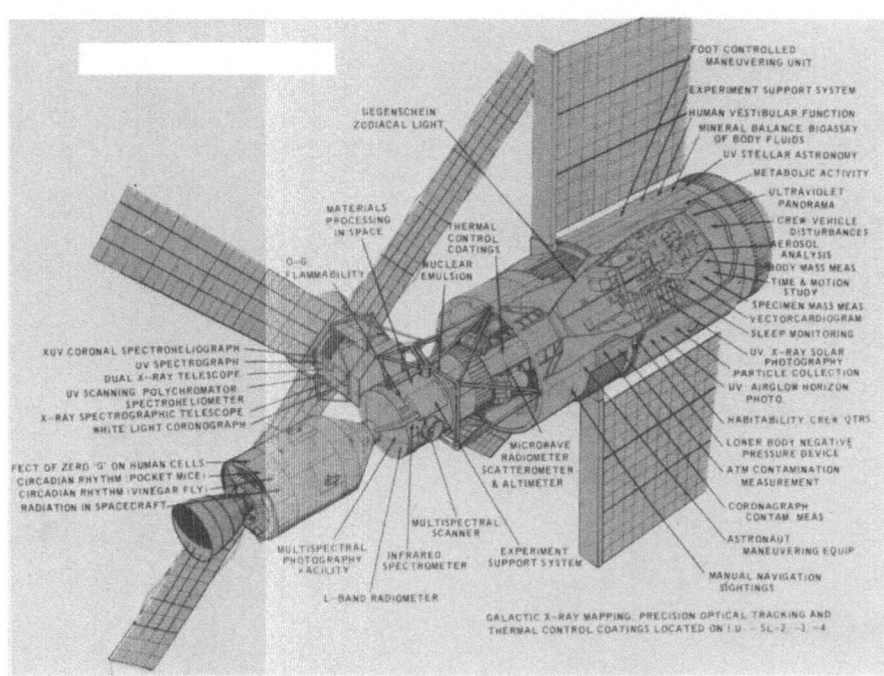

Fig. 18b.   Dispersion of Skylab experiments.

chemical aspects of metallurgical processes leading to the practical use of space for the processing of materials. The materials processing experiments utilize an electron-beam generating device and an electric furnace for melting, molding and welding materials. Provision is made for experimental crystal growth, for example. Crystals, widely used in the electronics industry, may be grown rapidly, virtually without defects and economically in weightlessness. Other experiments involve alloying of metals of great variation in density which is done poorly and expensively on Earth (Figure 17).

This represents a brief introduction to Skylab with a description of its general arrangement and some of its major features and accommodations. The brevity of this summary does not permit a complete discussion of the Skylab modules, supporting systems, and the experiments.

The laboratory support systems have been designed for maximum flexibility in adapting to the experiments. This design approach has provided a means of achieving compatibility with a wide variety of experiments. The complement of experiments will yield significant new knowledge in areas of the physical sciences and technologies which are critical in today's society. The Earth resources group of experiments will provide information on the condition of Earth's atmosphere, water and land areas, and provide direction in planning further development of resources. Some experiments will add to understanding the Sun and predicting its influence upon life on Earth. Others will exploit the unique environment of space for potentially producing improved products. Still others will study celestial bodies and may yield new knowledge of the development and behavior of the universe. Many of these experiments are being conducted with international cooperation. Many countries are participating in the Earth resources and science experimentation (Figures 18a and 18b).

The Skylab program is furnishing major opportunities for experimentation in many scientific disciplines. While the flight systems are within the state-of-the-art, they are the forerunners for further applications, and will establish design criteria for manned and/or man attended scientific spacecraft of the future.

THE UK 5 SPACECRAFT FOR EXPERIMENTS IN X-RAY ASTRONOMY

E. C. SEMPLE

Royal Aircraft Establishment, Farnborough, U.K.

Abstract. UK 5 is the fifth in the series of spacecraft built for the Science Research Council of the United Kingdom, and is part of the joint U.S./U.K. co-operative space programme. The spacecraft carries six experiments to investigate, with their own particular emphasis, the spatial distribution and energy spectra of emissions from X-ray sources in space, and the diffuse X-ray background. Studies will also be made of the polarization of X-ray emissions and of any pulsar periodicities.

Most of the experiments view areas of sky along the spin axis of the spacecraft. To be able to vary the pointing direction, the spin axis can be precessed by an on-board attitude control system.

This paper outlines the design of the spacecraft and the method by which it will be controlled in orbit.

1. Introduction

UK 5 is the fifth in the series of spacecraft built for the Science Research Council of the United Kingdom and is part of the joint United States/United Kingdom Co-operative Space Programme. The nature of the co-operation is that the U.S. provides the launches and the tracking facilities while the U.K. provides the spacecraft; the results of the scientific investigations are equally available to both parties. The first two in the series were built in the U.S. and contained British experiments; from UK 3 onwards the spacecraft have been made in the U.K. Under a contract from MOD Procurement Executive, UK 5 is being designed and built by Marconi Space and Defence Systems, part of GEC Ltd., who are prime contractors for the project. It is due to be launched in about 2 yr time.

2. Experiments

UK 5 carries six experiments to investigate, with their own particular emphasis, the position and energy spectra of X-ray sources in space, and the diffuse X-ray background. The variation of these X-ray emissions during the life of the spacecraft will be studied, and the pulsar features of any of the sources can be examined. A study of the polarization of X-ray emissions will also be carried out.

The scientific groups involved with the experiments are at University College London (experiments A and C), Leicester University (experiments B and D), Imperial College (experiment F) and NASA's Goddard Spaceflight Center (experiment G).

The experiments divide into two broad categories: those which exploit the spin of the spacecraft to do a sweep survey of the sky (experiments B and G), and those which view a limited area of sky along the direction of the spacecraft's spin axis (experiments A, C, F, G).

L. G. Napolitano et al. (eds.), Astronautical Research 1971, 471–481. All Rights Reserved.

Experiment A is designed to study X-ray emissions in the energy range 0.3 keV to 30 keV. Of all the experiments viewing along the spin axis it has the widest field of view – 35° FWHM – and uses a rotating collimator principle to discriminate the position of the various sources which may be in its field of view. The collecting window has an area of about 800 cm². Attached to the experiment is a star tracker, the output of which can be used to relate X-ray sources with sources of visible radiation. With this experiment it is hoped to position X-ray sources within 2–5′.

Experiment B views radially from the side of the spacecraft, and contains four large proportional counter tubes grouped together in pairs to form two detectors. The total sensitive area is about 700 cm². The field of view of the two detectors is different: one is $\frac{1}{2}° \times 5°$, the other $2° \times 15°$, with wider field of view dimension being parallel to the spacecraft spin axis. The primary purpose of this experiment is to survey the position and spectrum of X-ray sources in a band of sky at right angles to the spin axis of the spacecraft; however, by comparing the output of the two detectors the X-ray background can also be measured. The detector system is arranged to respond to radiation in the energy range 1.5 keV to 20 keV, and information on the direction of sources is obtained by dividing the spin period into a number of sectors, each sector corresponding to a direction in space relative to the Sun vector.

Experiment C has a fairly narrow field of view, about 3.5° FWHM and is set up to be at a slight offset from the spin axis of the spacecraft. The experiment is sensitive to X-ray energies in the range 2 keV to 30 keV, and uses a proportional counter with a window area of about 100 cm². The particular purpose of the experiment is to analyse the spectrum of X-ray sources in detail.

Experiment D contains two Bragg crystal spectrometers mounted to view along the spacecraft spin axis. The experiment may be used to conduct a search for emission lines in the range 2–8 keV, or alternatively may be used to investigate the presence of polarization in the radiation flux from certain X-ray sources. The reflected X-rays are detected by fixed counters, and the angle of the crystals can be altered by a stepper motor.

Experiment F views the sky in a direction slightly offset from the spin axis, and is designed to investigate the spectrum of selected X-ray sources in the energy region above 20 keV, and possibly up to 2 MeV; alternatively the experiment can be used to search for pulsar activity from certain X-ray sources. It is also hoped to be able to study the X-ray background. The experiment uses a Cs–I scintillator surrounded by an active collimator. The collimator controls the directional sensitivity of the experiment and reduces the effects of spurious background. The field of view is a 17° cone (full width) and the effective aperture is 8 cm².

Experiment G is designed to make a complete survey of the sky, and consists of a pair of one dimensional X-ray pinhole cameras. Each camera is in the form of a right angled isosceles triangle with the pinhole set across the 90° corner. The detector is in the form of a long proportional counter. The field of view of each camera is 90° × 4°, therefore, when the long field extension is arranged parallel to the spin axis, the two cameras can cover the full 180° of latitude; as the spacecraft rotates the whole celestial sphere will be continuously observed.

These descriptions of the experiments scarcely do them justice since nearly all include many more modes and sub-modes than are outlined here.

Most of the experiments include small radioactive sources for in-orbit calibration, and include a number of refinements to limit the effect of other sources of radiation to which the detectors may be susceptible. Further if the particle radiation in the vicinity of the spacecraft rises to an unacceptable level, the output from the experiments will be inhibited automatically by the data system until the radiation level falls below an acceptable set threshold: these 'stop/start' thresholds are variable on command. The experiments operate only in sunlight since the Sun is used as a reference.

3. Spacecraft Design Concept

The configuration of the spacecraft is shown in Figure 1, but basically the design of the spacecraft centres round the experiments and their relationship with the data system. How this is arranged is best described by an example:

As stated, Experiment B's primary objective is to perform a scan of the sky to observe the position and spectrum of X-ray sources in a band of sky at right angles to the spin axis of the spacecraft. Illustrated in Figure 2 is the way in which this is done for one particular sub-mode of this experiment. The experiment categorizes the incident X-ray photons by their energy and the direction from which they have come in space. To define direction, the sky is divided into 1024 equal segments around the axis of the spacecraft, each one thus being about $\frac{1}{4}°$ wide; for the case illustrated, only 128 of these sectors are used by the experiment and viewing is restricted to an arc of sky 45° wide. Within each sector the experiment categorizes the incident photons into 8 energy bands, thus in total the experiment can put an incident photon into one of 1024 categories – an array made up of 128 position categories, by 8 energy categories.

This method of data collection is made to suit the data storage system which centres on two core stores. 1024 core words are allocated to store the data from Experiment B, and when an X-ray is detected, the experiment, and the sector generator in the data system, assemble a 10 bit word which defines one of the above categories. This word is then used, as in a computer, to modify the address of the first core store word in the block allocated to storing data from Experiment B. This specific address in the core store is then incremented by one, i.e., if the first block address were 128 and the assembled word was 112 the address in the core store which would be incremented is 240. The sector generator is in effect a digital clock which is initialized on each complete revolution of the spacecraft by a pulse from a slit Sun sensor that triggers near the leading edge of the Sun; the direction of the Sun therefore acts as the basic position reference for defining the direction from which an X-ray photon has come.

By zeroing the contents of the core store at some stage, and allowing counts to build up in the various core store locations, a histogram of the occurrence of events in the different categories is built up, and it is relatively simple to interpret the data to deduce the position or spectrum of the sources observed. In practice, it is intended to allow data to accumulate in the core stores for about one orbit; the process of dumping the data also zeros the contents of the core store. On the ground, data

Fig. 1.   UK 5 spacecraft configuration.

accumulated in successive orbits will be processed together. A complete sampling period will last about one day.

The experiment is allowed considerable flexibility in how it uses the core store addresses which are allocated to it, and can trade spectral resolution for position resolution, and vice versa; at one extreme it can categorize the incident X-rays into 16 energy levels within an arc of sky $22\frac{1}{2}°$ wide; at the other it can have only one energy level, but do a sweep over the full 360° provided by the spin. Also when the experiment is viewing a limited arc of the sky, the first sector does not need to be adjacent to the spacecraft/Sun line as indicated for simplicity in Figure 2.

The various modes and sub-modes of the experiment are controlled by a register which contains 27 bits; this register is filled and altered by sending commands from

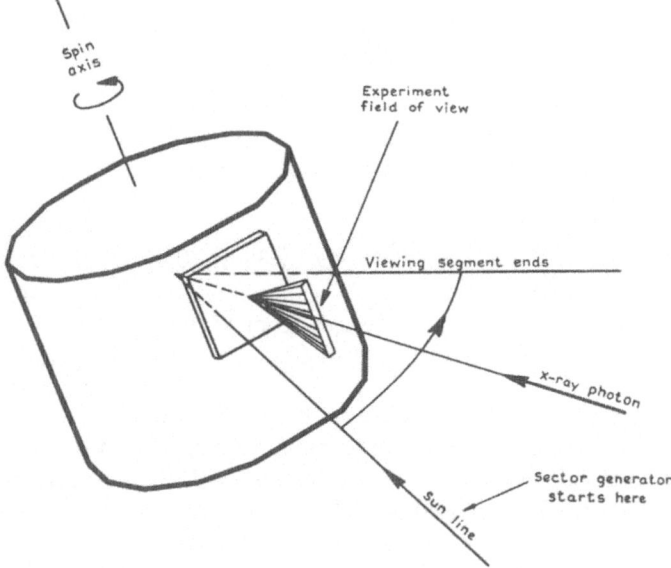

Fig. 2. Schematic of operation of experiment B in sub-mode 1/1.

the ground. Information in this word includes the mode of operation, the first sector in its viewing arc, etc.; this command word therefore sets up the electrical configuration of the experiment for the study to be performed on a particular occasion. It will be realized that this arrangement provides the experimenter with considerable scope to control the experiment in orbit.

The other experiments, and the other modes of Experiment B, use the core store and the data system in similar, but different, ways to that described above. All the experiments, except F, have a command register to control the electrical configuration of the experiment.

A particular variation of the above system is the pulsar mode used in experiments C, D and F. In this mode the experiments can look for X-ray pulsar features in sources of known pulsar frequency. A number related to the period of the pulsar can be transmitted to the spacecraft, and held in a register which at other times is used by the attitude control system. Using this information, counts from the experiments can be directed to various locations in the core store corresponding to pulsar phase. Broadly speaking, the data is interpreted by looking for dissimilarity in the counts recorded in the various phase locations: if the counts are all similar then there is no pulsar effect at the set period, if the counts are dissimilar then a X-ray pulsar may be taken to be present. Problems of phase change between the pulsar and the data system cycling period are limited by counting into one set of locations for only a few seconds; at the end of this time, a new set of core locations is used to continue counting.

4. The Spacecraft

4.1. GENERAL

The configuration of the spacecraft is shown in Figure 1, and it is designed to be launched on a Scout launcher with an Algol 3 first stage. A summary of the main parameters of the spacecraft and its orbit are given below.

Weight: 300 lb (135 kg)
Dimensions: Cylinder 34 in. (0.87 m) long by 38 in. (0.97 m) diameter
Orbit height: 450–500 km
Orbit inclination: 37° – Directly eastwards from Wallops Island
Spin rate: At launch 150 rpm, reduced to 10 ± 2 rpm during operation

The spacecraft is designed to operate over only a restricted range of solar aspects between 45° and 135°, i.e., with the Sun shining more or less on the side of the spacecraft. To enable various parts of the sky to be observed, the pointing direction of the spin axis can be altered by a pulsed gas jet system.

The position of all X-ray sources is related by the data system to the position of the Sun in the sky, and the experiments are dependent on the output of the sector generator to provide this reference. A consequence of this is that when the spacecraft is in the Earth's shadow the experiments will not work, and to save power the experiments are switched off during this period.

4.2. STRUCTURE

The main structural members are the honeycomb panels which in plan form the shape of a letter H with a double bar. These are augmented by webs and the total makes up an egg-box type construction.

A transition section is fitted into the bottom of the centre box section to make the structure interface with the circular separation ring on the E-section of the Scout launcher. There is a strong ring round the bottom of the spacecraft to support the antennae and the jet system manifold. The ends are closed with diaphragms supporting thermal blankets.

4.3. ATTITUDE SENSING SYSTEM

The mounting position of the attitude sensing system may be seen in Figure 1, and the details of the sensor plate are shown in Figure 3. Two sensing systems are used to provide enough information to define the pointing direction of the spin axis in space; one is a Sun sensor, the other is an albedo sensor. The Sun sensor consists of two slit sensors with a field of view $1° \times 150°$, and they are mounted adjacently on the sensor plate, one with its slit line parallel to the spin axis of the spacecraft, and the other one with its slit line inclined at 15° to the spin axis. As the spacecraft spins the Sun flashes into the detectors and the pulses are registered in the data system. Since the slits are inclined, the flash in each detector will occur at a different position in the spacecraft's spin phase, and for the inclined slit this position will vary depending

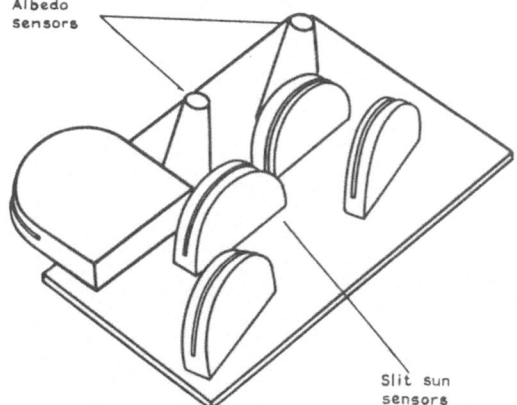

Fig. 3. UK 5 attitude sensor layout.

on the angle between the spin axis and the Sun line. A measurement of the time between the two flashes can therefore be directly related to Sun angle, at a given spin rate.

The albedo sensor has a conical field view of 2°, and triggers on crossing a sunlit horizon or the terminator.

Analysis of the data on the ground can distinguish, and reject, terminator crossings recorded by the albedo sensor, and can deduce the attitude of the spacecraft in celestial co-ordinates. The overall accuracy of the system is about $\pm 1°$.

For redundancy the sensor systems are duplicated. The fifth slit sensor shown on the plate, which is mounted at right angles to the others, is used only in conjunction with a back-up mode for operating the attitude control system.

The whole system is very similar to that developed by the Royal Aircraft Establishment for the X3 Technology Spacecraft which is now in orbit.

4.4. ATTITUDE CONTROL SYSTEM

The attitude control system uses gas jets, and is capable of increasing and decreasing the spin rate of the spacecraft over and above its primary function of altering the pointing direction of the spin axis. The jets are mounted on two manifolds which are on diametrically opposite sides of the spacecraft at the bottom (see Figure 4). There are 3 jets on each manifold – one for spin-up, one for de-spin, one for precession; the duplicate systems in general work together but can operate independently should there be a failure of any one in a pair of jets. The fuel used is propane which is stored in liquid form in a tank on the centre line of the structure.

The spin-up and de-spin jets are operated by direct command from the ground, and are switched off again by direct command, or automatically as the spacecraft enters the Earth's shadow. The spin rate can be changed at about 1 rpm min^{-1}. This system will be used to reduce the initial spin rate, and subsequently to control the spin rate at 10 ± 2 rpm.

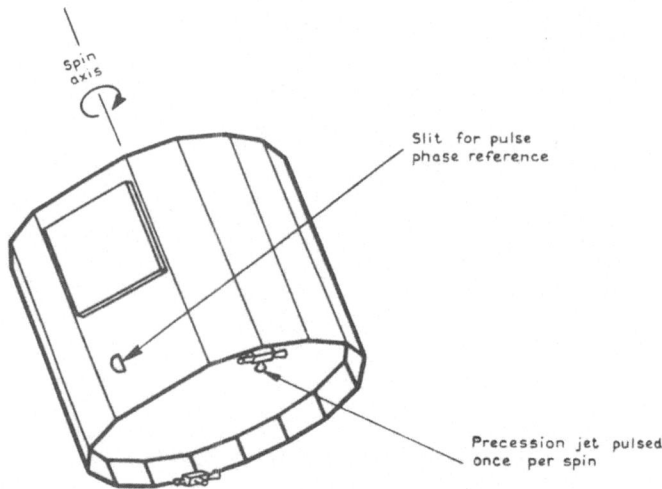

Fig. 4.   Schematic of UK 5 attitude control system.

Precession of the spin axis is achieved by pulsing the two precession jets which point radially towards the spin axis of the spacecraft. The direction in which the spacecraft is moved can be selected at will, and is altered by varying the point on the spin phase at which the jets are pulsed. The reference from which the triggering time is controlled is the pulse from one of the vertical Sun slit sensors on the attitude sensing system.

A command register similar to that used to control the experiments is used in conjunction with the attitude control system. Bits 1 to 9 give the sector number, produced by the sector generator, at which the jet is to be pulsed; bits 10–19 give the number of times the jets are to be pulsed to complete the manoeuvre. The maximum step length will be about 30°.

Bits 22–25 relate to the magnetotorquer which is carried by the spacecraft. This compensates for any remnant magnetic dipole component (which the spacecraft may have) parallel with the spin axis. By observing the drift rate of the spacecraft, it will be possible to calculate the level to which the magnetorquer should be magnetized to minimize the drift rate of the spin axis. The device can be set up to any of 16 discrete levels, positive or negative. This is controlled by the bits indicated.

4.5. POWER SYSTEM

The power system is an extensively modified version of the one used on the X 3 spacecraft and provides regulated lines at $+12$ V, -12 V and $+5.4$ V; there is also a $+11$ V line for driving stepper motors etc.

The system is driven from an array of solar cells distributed round the cylindrical surface of the spacecraft; the evenness of the output is, however, upset by the presence of the facet containing the experiment B and G detectors and the aspect sensors. At minimum the array output at 13 V is about 28 W. This minimum occurs at the end of the spacecraft's life, and when the solar aspect is at one of its allowable extremes, i.e., either 45° or 135°.

4.6. DATA SYSTEM

The method by which the data system stores data from an individual experiment has been outlined in Section 3. The essence of the system is that all X-ray photons detected by the experiments are categorized and for each category there is a location in the core store. When an event in a particular category is detected then the data system services the experiment and increments the contents of the appropriate core store location by one. Once every orbit the contents of the core stores are dumped, and then are set all to zero on completion of the dump.

The incrementing process carried out by the data system can only be done for one experiment at a time, and the problem of dealing with a number of events occurring virtually at the same time is dealt with by scanning the experiments in turn. When an experiment detects a photon, it sets a flag which causes the scanner to stop on the next occasion it interrogates that experiment; the output from the experiment is stored awaiting this interrogation. The time between successive interrogations is very short – about 50 μs – even when the system is busy, and the loss of data – due to a new photon arriving before the previous one has been processed – is very small; also such losses as there are can be taken into account statistically by the experimenters, when processing the data.

Other functions of the data system are as follows:

(a) An orbit clock is included which will count up to 8192 s before overflowing. This clock is reset after a command core store dump. The clock is also used in conjunction with the experiments, particularly when they are in pulsar mode.

(b) It integrates the output of particle radiation detectors associated with experiment C and D. When the level in either register exceeds a set threshold a pulse is sent to the experimenters which can be used to inhibit data collection; likewise when the level falls again, a pulse is sent to the experiments which can be used to switch them on again.

(c) The system generates two telemetry formats: one is transmitted almost continuously and includes all housekeeping data, the other is transmitted on command and gives the contents of the core stores – this data is transmitted three times.

(d) Periodic samples are taken of the housekeeping data – mainly attitude data, and background radiation data – and the system stores these in part of one of the core stores.

4.7. THERMAL SYSTEM

The thermal design of the spacecraft is arranged so that the main heat transfer path is through the solar cells on the side of the spacecraft. Blankets cover the top and bottom of the spacecraft, and thin 1 μ plastic sheets are likely to be put over the experiment detectors to cut down the heat losses where they protrude through the top blanket.

In-Orbit Operation

The final details of the method of in-orbit operation are still in the process of being worked out; the system, however, will be broadly as follows:

All data from the spacecraft will be received using the NASA STADAN network, and these stations will also be used to transmit commands to the spacecraft. A Post-launch Operations Control Centre will be set up in the U.K., and operated by the Science Research Council at the Radio and Space Research Station, Slough. UK 5 will be in a low inclination orbit – about 37° – and will not pass over the U.K., so this centre is connected by data links to the various tracking stations over which the spacecraft will pass. About three passes of data per day will need to be sent to the Control Centre for near real-time analysis. These will be used:

(a) to verify the health of the spacecraft;
(b) to check the contents of the experiment and attitude control command registers;
(c) to monitor the attitude of the spacecraft;
(d) to provide 'quick look' experiment data.

Consideration is also being given at present to the problem of converting the broad requirements of the experiments into an explicit, yet flexible, manoeuvre programme. It is anticipated that the sequence of events for doing this will be as follows:

(a) The experimenters will propose a list of 30 to 40 sources to be observed in the following month; this will be broadly compatible with the viewing restrictions of the spacecraft, and will take into account visibility of sources relative to the Earth, radiation effects, etc.

(b) With the assistance of the Operations Manager this list will be refined to 20 to 30 sources, and the optimum sequence in which the sources should be viewed to minimize fuel consumption will be computed using a 'travelling salesman' type linear programming routine.

(c) For each source, details of viewing time, pointing accuracy, command register contents, etc., will be agreed.

This programme would be valid for one month, but would be up-dated every two weeks for yet another month. It is intended to make the operating system sufficiently flexible so that it will be possible to change the programme, or repeat a source viewed a few days previously, at relatively short notice.

Acknowledgement

The author is indebted to the Science Research Council, the experimenters, and Marconi Space and Defence Systems Ltd. for their assistance in providing information for this lecture. In particular the author acknowledges the work of MSDS in providing illustrations for Figures 1 and 3.

C. EDUCATION

A SPACE BENEFITS EDUCATIONAL PROGRAM

KONRAD K. DANNENBERG

American Institute of Aeronautics and Astronautics, Huntsville, Ala., U.S.A.

and

FREDERICK I. ORDWAY

School of Graduate Studies and Research,
University of Alabama, Huntsville, Ala., U.S.A.

Abstract. A program developed to educate teachers, students, and the public on the diverse benefits of space research is described. It involved first a series of popular lectures given principally by local members of the Huntsville aerospace community. This was followed by a summer teacher workshop that emphasized the use of specially updated, space-oriented instructional materials, improved teaching methods, and increased student motivation through direct participation. The lectures and workshop were organized by a group of societies in the Huntsville area in cooperation with NASA's Marshall Space Flight Center, the University of Alabama in Huntsville, and the Alabama Space and Rocket Center. Preliminary studies of a space benefits source book for use by teachers and students in future space educational programs are also reviewed.

1. Introduction

After the excitement caused by unmanned and manned space flights that began in October 1957 and continued throughout the 1960 decade, a discouraging decline in public interest in astronautics has been observed. This is perhaps inevitable: no one program, nor any human endeavor for that matter, can be expected to enthrall or even interest society for an indefinite period. The adult, like the child, has a variable attention span that even the incredible feats of astronautics cannot attract forever.

In most societies, major government expenditures are closely scrutinized by the elected leaders and through them by the people themselves. In the long run, nations work and pay for what they want – or at least think they want. They tax themselves for roads, for education, for defense, for culture, for health and social services, for urban transportation, etc., to an extent dependant upon their tastes, desires, and resources, as well as the positions they occupy in world economic and power structures. Until recently, society looked upon astronautics partly in terms of sheer admiration for the capabilities of modern science, engineering and management; partly in response to an encoded sense of curiosity about the universe; and partly as a form of pure national prowess – the U.S. flag turned up on the Moon just as markers appeared along the coast of Africa as 15th century Portuguese navigators probed ever southward, and as Roald Amundsen's black flag appeared on a sledge marker as soon as he and his group first reached the South Pole in January 1912.*

---

* It is hard to not be moved by British explorer Robert Falcon Scott's keen disappointment at arriving at the pole shortly afterwards, only to learn that his party had been beaten in the conquest. 'The Norwegians have forestalled us and are first at the Pole', he wrote in his diary. 'It is a terrible disappointment, and I am very sorry for my loyal companions . . . All the daydreams must go; it will be a wearisome return'.

L. G. Napolitano et al. (eds.), *Astronautical Research 1971, 485–493. All Rights Reserved.*
Copyright © 1973 by D. Reidel Publishing Company, Dordrecht-Holland.

Today, some of the daydreams of space flight are over; others are still to come. During this transitory period, it is up to the International Astronautical Federation, its member organizations and the individuals and entities composing them to help inform the great world public not only how the space program came about and how it is conducted, but why mankind wants to explore beyond the confines of his terrestrial atmosphere and how what he learns benefits peoples from all echelons of society. To meet this objective, we must call upon the educational process, through formal channels (schools and universities) on the one hand and outside channels (adult education courses; popular lectures; teacher workshops; film, filmstrip and slide presentations) on the other.

One experimental approach to the objective of educating the non-specialist on the benefits of space was taken in 1971 by the Alabama Section of the American Institute of Aeronautics and Astronautics – better known as the AIAA. A study group was established first to try to determine why public interest in space activities had decreased and then to propose remedial measures. Among other things, the recommendations called for deeper involvement of the local community to help create broader public awareness of the many identified and potential benefits from space. To this end, other engineering and some non-engineering societies were invited to participate through the Huntsville Association of Technical Societies (HATS). Among the activities subsequently taken were the establishment of a space benefits lecture series, the organization of a national space benefits congress, and the creation of teacher workshops conducted jointly with local educational authorities.

This paper focusses on the workshop activity, which was structured around anticipated direct broadcasts from forthcoming Apollo lunar and Skylab orbital laboratory missions, as well as from possible later instructional programs involving space stations and research and application modules – the so-called RAMs.

Engineers, scientists and teachers are cooperating to define the requirements that such space educational broadcasts will impose on school systems in general and on teachers and students in particular at local, county and state levels. The Skylab mission, scheduled for spring 1973, may show that educational broadcasts should become a major element of later space station missions, placing education on the same level as such major disciplines as physics, astronomy and life sciences. It is presupposed that educational broadcasts from orbit will be preceded and followed by well-planned and well-coordinated classroom activities. Moreover, in order to insure that maximum benefits funnel into the educational process from Skylab and post-Skylab flights, source-book-type data will have to be prepared and distributed to teachers and to students. The value of science and technology to the human condition, the systems approach to problem solving, and the need for adaptability of any future system to changes were emphasized throughout the workshop.

2. Public Lectures on the Benefits From Space

The Alabama Section of the AIAA and the HATS organization decided that before the teacher workshop could be set up a series of experimental lectures should be given to the public, answering such questions as:

(1) What are the future goals of space research and exploration?

(2) How do these goals compare with those of the Apollo program?

(3) Why should a nation like the United States continue a major space program during a time of pressing domestic and foreign difficulties?

(4) How has space science and technology aided non-space industries and endeavors?

(5) What direct benefits have resulted, and predictably will result, from the space program?

Recognizing that these questions could best be answered initially through a series of lectures and directly coupled group discussions, deferring to later the teacher workshops, it was decided to embark upon a popular 'space benefits' program at the Alabama Space and Rocket Center – the largest museum in the United States devoted to rocketry and space. Ten subjects were covered (see Table I) in the lecture series to audiences numbering up to 100 persons each. Due to the fact that scientists and engineers were the speakers, the audience tended to be rather well informed to begin with – about half being involved with the aerospace field and the remainder public school science teachers, secondary school and university students, and a limited 'general public' which, however, grew as the series progressed. One of the problems faced was how to arrange for suitable publicity that would reach effectively into all levels of the local community and the area surrounding it. The fact that the series took place at the popular Alabama Space and Rocket Center helped draw in the public, which was – aside from the teachers and the students – the target audience. A result of the series was the establishment of a special exhibit at ASRC on the benefits and dividends from space research; it was dedicated at the beginning of the 1971 summer season. Another was the furnishing to the NASA-Marshall 'space mobile' operation new materials on space benefits especially oriented to younger pupils and their teachers.*

Following the lecture series, studies showed that a key audience to be reached by any space benefits awareness program is located in the schools, where a general cross-section of the general public is found. In the United States, education involves between 25 and 30% of the population. Youth, contrary to some opinions, is definitely interested in space exploration and often highly informed about it. Their teachers, however, may find themselves reluctant to enrich their courses with space-generated materials for the simple reason that they do not always have relevant and easily

* Space mobiles are closed vans that carry lecturers with their audio-visual aids and models to schools where presentations are made on a wide variety of subjects. A typical lecture program covers such subjects at the history of rocketry, rocket propulsion, bioastronautics, Earth resources, and space communications. It lasts about 50 min. In fiscal year 1970 the NASA space mobile program is estimated to have reached the following numbers of persons:

	Lectures	Audience
Live audiences	11 450	2 364 687
Radio and television	94	51 181 200
Teacher services	1 204	21 601

TABLE I

Space benefits lectures and speakers

1. Benefits from Space Stations	Dr K. Ehricke	North American Rockwell Corp.
	R. Holmen	McDonnell Douglas Astronautics Co.
2. Skylab Experiments and Objectives	C. De Sanctis	Marshall Space Flight Center
3. Dividends from Space Technology – An Overview	Prof. F. Ordway	University of Alabama, Huntsville
4. The Systems Approach – A Space Lesson	J. Aberg and others	Marshall Space Flight Center
5. Results from Lunar Exploration	G. Heller	Marshall Space Flight Center
6. Sound, Noise and the SST	I. Vatz	Teledyne-Brown Engineering
7. Nuclear Energy for Power	R. W. Hunt Dr J. B. F. Champlin	Westinghouse Electric
8. Weather Satellites and Meteorology	W. Vaughan and others	Marshall Space Flight Center
9. Space Exploration for World Peace	Dr Mercieca	Alabama A & M University
10. Why to Explore Space?	Dr Stuhlinger	Marshall Space Flight Center

TABLE II

Listing of seminar topics and speakers

1. Registration – Introduction	C. Hammett	U.S. Army Missile Com. Redstone Arsenal
2. NASA's Educational Program	E. Collins	NASA Headquarters
3. Dividends from Space Technology – An Overview	D. Christensen	University of Alabama, Huntsville
4. The Systems Approach – A Space Lesson	J. Aberg	Marshall Space Flight Center
5. Skylab Mission and Concept	L. Belew	Marshall Space Flight Center
6. Skylab Experiments and Objectives	C. De Sanctis	Marshall Space Flight Center
7. Earth Surveys	Dr McDonough	Marshall Space Flight Center
8. Weather Satellites and Meteorology	O. E. Smith	Marshall Space Flight Center
9. Space Manufacturing	H. Wuenscher	Marshall Space Flight Center
10. Application of Space Remote Sensing to Solution of Ecological Problems	A. Adelman	IBM, Huntsville, Al.
11. Results from Lunar Exploration	B. Jones	Marshall Space Flight Center
12. Information Management	Dr R. Vachon	Auburn University, Auburn, Al.
13. The Space Shuttle	T. O'Connell	Marshall Space Flight Center
14. The Lunar Roving Vehicle	S. Morea	Marshall Space Flight Center
15. Space Exploration for World Peace	K. Dannenberg	Marshall Space Flight Center
16. Why to Explore Space?	Dr E. Stuhlinger	Marshall Space Flight Center

understood information for their purposes. Although NASA and other elements of the aerospace community supply some very useful aids to the teachers,* not enough is available on the benefits and dividends from space. What is urgently needed is a space benefits source book (see Section 6, below).

The AIAA and HATS groups, and their associates at the NASA-Marshall center, the Alabama Space and Rocket Center, and the University of Alabama in Huntsville, concluded that education can probably benefit more and faster from space-derived knowledge and the application of advanced science and technology than any other discipline. By focussing on teachers, the multiplier effect can be expected as they impart space benefits information through the normal educational processes in the classroom. In a few years, today's pupils will become university students and then 'the general public'. Through the teachers, the entire future population can be reached at impressionable ages and made aware of the goals, the benefits, and the responsibilities of space science and technology.

3. The Summer Teacher Seminar on Space Benefits

To implement the recommendations of an ad hoc committee for a teacher seminar, the HATS organization announced that it would take place in August 1971. Held in Huntsville, Alabama, the one-week seminar was oriented, like the lectures that preceeded it, on the diverse benefits and dividends resulting from America's space program. The uses of new knowledge gained were discussed, and the plans for and the expected results from future astronautical activities were considered. The seminar was enhanced by visits to selected laboratories at the NASA-Marshall center, the ASRC, and local aerospace industrial concerns. The University of Alabama in Huntsville assumed the responsibility for the conduct of the seminar, with appropriate Marshall and ASRC support. The topics presented at the workshop are shown in Table II.

Of particular interest to attendees were the planned educational broadcasts from the remaining Apollo lunar flights and the 1973 Skylab orbital laboratory. The scientific and technological objectives of the latter were discussed in terms of potential benefits in such fields as solar and stellar astronomy, Earth resource surveys, space medicine, astrobiology, space physics and chemistry, demonstrations of the effects of zero gravity on materials and manufacturing processes, etc. Seminar attendees were stimulated to think of space research in terms of their own classroom curricula. It was hoped that this pilot seminar workshop would serve as a model for similar activities elsewhere.

---

* Typical resources are NASA's 'A Universe to Explore' covering space-oriented science experiments (e.g., solar cells as sensors; from energy – action and reaction; and simulating the space environment in the laboratory); the NASA/National Science Teachers Association 'Space' which has a sub-section on 'What Can Man Derive from Space Exploration?'; the Massachusetts Department of Education's 'Aerospace Curriculum Resource Guide', and NASA's 'Space Resources for Teachers: Biology'.

4. Seminar Extension by Teacher Workshop

Teachers desiring to spend an additional week for deeper study of the educational benefits from space were invited to participate in an exercise to use the workshop technique and the systems approach to prepare classroom programs tied in to later space flight activities. Visits to NASA-Marshall and other local facilities were made and discussions took place with interested engineers, scientists, systems analysists, and program managers.

5. Results of the Teacher Workshop Experiment

To the extent that this experimental teacher workshop responded to the investigations and recommendations of HATS, it differed from other aerospace workshops; however, far from replacing the techniques of earlier workshops, it prescribed a more precise role for them. The workshop described herein highlighted ways of transferring space knowledge, over a relatively direct path, to the educational system. Teachers were provided *first hand* with advanced scientific and technological information, which otherwise would probably not have reached them with such impact for months or even years – data normally would have entered the educational system via scholarly journals, learned monographs, then university textbooks, and finally, as text enrichment, into secondary school textbooks. It was hoped that the increased flow rate of information would help, in turn, to accelerate the speed at which space-generated knowledge would be placed into practical use in many sectors of activity here on Earth, helping create an alert, informed, and highly responsive public.

The experience gained from the pilot workshop, and from the lecture series that preceeded it, can be summarized as follows:

(1) The accelerated flow of space-generated knowledge into the educational system will help assure that such knowledge is rapidly and effectively applied to local, national, and global problems. Often major problems are tackled by persons unaware of the potential of space age techniques, be they technological, scientific, managerial or information-oriented.

(2) Rather than stressing what has been done in space, this workshop addressed itself to the 'why', helping lay the foundation for an appreciation of why, and how, space can affect the lives of all of us on Earth. Teachers learned why it is important to observe the Sun from beyond the Earth's atmosphere, and how the overview from orbit can affect studies of the lands, the oceans, the inland waters, pollution, and the atmosphere below; why the zero-gravity environment is so important to the conduct of certain technological, physical and biological experiments; and how space communications is already impacting on human ways of life.

(3) The workshop helped prepare teachers for forthcoming space missions, including the Skylab, space station, RAM, and Apollo missions already referred to; but, logically, including unmanned flights and experiments and activities carried out aboard space shuttles and space tugs. The workshop helped make teachers in particular, and

the educational system in general, more aware of the requirements for new equipment, new educational materials, and a well-trained and motivated teacher corps.

(4) A two-way channel of communications was initiated between the needs and the desires of the educational system and the aerospace community. On the one hand, aerospace experts learned at first hand the problems and requirements of educators, while the latter came to understand the capabilities, the constraints, and the potential of space research. It is hoped that a 'working group' relationship can be maintained for the mutual benefit of both communities of interest. In this regard, the Skylab mission and post-Skylab space station and RAM missions appeared to offer an excellent testing ground to cement this relationship.*

(5) This workshop also demonstrated the usefulness and applicability of a little understood by-product of the space program, a greatly enhanced system management capability. The workshop was organized in conformance with a sound systems approach, which was applied by the teachers to their analysis and implementation of a series of TV tapes produced to demonstrate the benefits from space. This first-hand experience underlined the value of systems management, and encouraged the teachers to apply system engineering to classroom education, to overcome existing hurdles, and to reconstructure educational methods and procedures. Moreover, a greatly increased necessity for continuing (adult) education would have to emphasize the new space-generated requirements. It was believed that only a thorough, system-oriented study and properly-devised total system management would provide an acceptable answer as to how these requirements should be fulfilled.

(6) The workshop also defined associated supporting activities needed in the schools. It was found that new technological inputs must be analyzed for their effect on classroom activities and the need for amplified teacher training. Moreover, additional teaching materials have to be prepared in the form of a space benefits source book. The manner in which this new information is presented requires that new types of equipment be obtained and installed, and operators for its use and repair trained.

6. A Source Book on Space Benefits

One of the most important needs identified as a result of the workshop is the systematic accumulation and organization of space benefits information for teachers and students. This information would appear in the form of a source book that would be used to enhance educational programs tied to actual space broadcasts, to provide a channel through which space-generated science and technology could be assimilated by teacher and student alike, and to supply a single fountain of data on the application of space knowledge to all fields of endeavor.

The feasibility of a space benefits source book has been studied by the University of

* As far as Skylab is concerned, inputs from educators into the 'scenario' of on-board operations are expected, though hardware modifications are practically impossible due to the advanced status of development of the program. A much more basic definition of educational broadcasts from space stations and RAMs is to be pursued. The University of Alabama in Huntsville is coordinating an effort with this goal in mind.

Alabama in Huntsville, and it was shown that extensive source book-type materials are already available for review, discussion and use at working level. Indeed, the pilot teacher workshop took advantage of some of them, while at the same time recommending follow-on improvements that would lead to a full-scale source book, supported by literature citations, teaching materials (charts, slides, filmstrips, films, filmloops, charts and posters), examples of benefits, and the like.*

7. Future Activities

While the space benefits lecture series and teacher workshops were being conducted, city, county, state, and private educational interests were queried regarding the program. Initial responses indicated their readiness to amplify space and space-related education in the schools in the form of selective courses. A recent questionnaire sent to schools in the state of Alabama by the State Educational System also indicated interest in implementing space instruction as an elective subject. It is believed that a similar situation exists throughout the country, and probably in many IAF member nations. To assure that instruction for elective subjects is valuable to the students, every effort must be made by the space community to generate up-to-date and interesting information tailored for direct use by school teachers.

The AIAA Alabama Section and HATS were requested to take the lead in the establishment of a Space Education Advisory Council that would advise the Board of Education while space education courses are being worked out. Moreover, they were asked to help coordinate assistance from the University of Alabama in Huntsville, the NASA-Marshall center, private industry, the Alabama Space and Rocket Center, state and city governmental organizations, and possibly other entities. The UAH assured full support in this endeavor, and made it possible to grant teacher credits and to continue regular workshop programs in support of space education.

The university also cooperated with NASA-Marshall and industry in making preliminary assessments of the impact of educational requirements on information management systems of down-the-road space station, RAM, and shuttle flights. Space stations and their associated attached and free-flying modules will follow orbits ideal for educational broadcasts to vast numbers of the world's peoples. Preliminary studies indicated that only minor investments in educational information systems would be necessary to produce very substantial results in the aim of carrying more and better educational material to more and more people.

In looking forward at the educational potential in tomorrow's space stations and RAMs, maximum advantage would be taken of unmanned geosynchronous satellite

---

* Efforts have already been made at bringing together into single volumes many of the benefits of space science and technology. See, for example, Ordway, Frederick I., Carsbie C. Adams, and Mitchell R. Sharpe, *Dividends from Space* (New York, 1971: Thos. Y. Crowell); Forbes, Fred W. and Paul Dergarabedian (eds.), 'Technology Utilization Ideas for the 70's and Beyond' (Tarzana, Calif., 1971: American Astronautical Society Proceedings); Ford, C. Quentin (ed.), 'Space Technology and Earth Problems' (Tarzana, Calif., 1970: American Astronautical Society Proceedings); and NASA's 'Ecological Surveys from Space' (Washington, NASA Publication SP-230, 1970).

educational program experience and of the Skylab broadcasts. It is contemplated that a UAH group will conduct a pilot study of the feasibility of conducting regular educational programs from manned space stations in the post-Skylab period. Attention will be focussed not only on the educational activities of the astronauts in their space-borne 'lecture room' (where they would typically demonstrate how physical, chemical, biological and other experiments are undertaken in the space environment), but on supporting ground efforts prior to, during, and following a given broadcast. Programs from orbit would range from explanations of the behaviour of living plants and animals in the zero-gravity state and its importance to medical research on Earth, to demonstrations of how Earth resources can better be exploited and studied from the overview; from how solar and stellar research is essential to a deeper understanding of the mechanisms at work in the physical universe, to how novel manufacturing processes may evolve in orbit yielding products that cannot be made in the presence of gravity. Through the teaching of fundamental and practical aspects of space research with the direct assistance of space-borne astronauts actually conducting that research, not only will far better informed students result but a more astronautically aware – and appreciative – public.

Appendix: Space Benefits Congress

Less than two months after the presentation of this paper in Brussels in September 1971, the Huntsville groups responsible for initiating the space benefits lecture series and pilot teacher workshop organized a third-phase effort in the form of a national space benefits congress. The proceedings of the conference are now available under the title *Space for Mankind's Benefit* (Washington, 1972, U.S. Government Printing Office).

A CASE FOR TOTAL CONCEPT DESIGN COURSES
IN EDUCATION

J. F. SLADKY, JR.

*Division of Engineering and Weapons, United States Naval Academy,
Annapolis, Md., U.S.A.*

The complexity of society is today an accepted fact. Let us examine a simplified analogy. From the study of thermodynamics we know that a property, or a number of properties, define the state of a system at any instant in time. Complexity can be considered a property of our social structure or the state at which our society finds itself in history.

The property, complexity, is a close indicator of the steps in the development of man's society. Distinct increments can be detected (Figure 1). At the earliest times

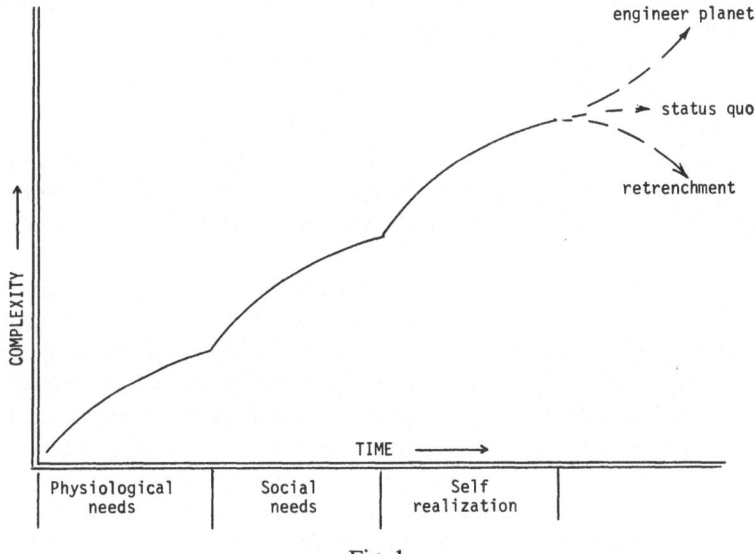

Fig. 1.

man was concerned with providing for his physiological needs; eating, breathing, etc. Existence was simple, usually life or death. As these factors became more secure man turned towards social interaction with his fellow man; existence was more complex, communication and transportation were needed, values defined and so on. Finally, once man attains social respect, he reaches for self realization. One step cannot be attained unless the preceeding is firmly secured, and each of these steps is characterized by an order of magnitude increase in complexity of man's society.

Recently, Dr Sinsheimer of the California Institute of Technology observed that man is leaving a world he never made and entering a life of human design. A passage from an era where man was subject to his environment to a time where he must control not only his surroundings, but himself as well.

Three alternatives exist; retrenchment, maintenance of status quo, or engineering a world to man's specification. The first two appear unacceptable for the processes of society are irreversible. This then leaves only one alternative; to engineer the world for the future of mankind: – a staggering thought. We no longer have unlimited supplies of resources, energy, water and space. Man must provide – design – for the existence of man. Designing the future planet must be a multisided and open ended process. That is; any meaningful solution or design to a complex question must take into consideration and place relative importance on the interplay of a large number of diverse disciplines. Further; the scope of these engineering projects is so vast, both in size and time frame, that very often a system design must be planned with technology that does not yet exist, but which is expected to exist when the system is fabricated. A designer of future systems will be a professional who is an expert in a particular area yet has a grasp of inputs from other disciplines, who is aware of technology in a number of fields, and who can operate within the constraints of numerous other factors such as time, economics, politics – all of which govern a system definition and solution.

This multidisciplinary capability in the design of large complex systems is more of a requirement than may at first appear. If we assume that there is an optimum design for a particular system and the design management does not have total concept capability then there is a chance that the end product could conceivably be considerably off-optimum.

The question then becomes how best to prepare individuals with the aforementioned characteristics, or more specifically, how can the *university* develop *total concept thinking capability* in its students. One approach is similar to the teaching of scientific or technical research; however, this has been found to be a sterile business. The only means found effective in developing competent researchers is to have students participate in a research project under the guidance of a competent and inspiring faculty. Similarly, it seems that an effective method of teaching multi-disciplinary design or total-concept thinking is to have students take part in a complex and creative design project. This is the purpose of multidisciplinary design courses.

Teams of students from all disciplines – engineering, humanities, and social sciences – collaborate to design a complex system which is technically feasible and socially desirable, but which has not been previously designed. It is the aim of the course to give a student an educational experience where he must consider the overall problems involved in carrying out the design phase of a typical complex system. The students are advised and guided by the faculty and assisted by invited lecturers from industry and government. The faculty provides lectures on relevant basic principles, while outside experts supply state-of-art information. The scope of these lectures is *not* to enable each student to solve problems in all disciplines, but to be able to discuss them intelligently and appreciate their significance.

The physical aspects of a multidisciplinary design course are as follows: The course is one semester in duration and has an academic value of three credit hours. The material or work consists of a preliminary design of one major complex system. The class meets two times a week for two two-hour working sessions. Initially, the first hour of each session is used for lectures given by the faculty on basic principles related to the design. The second hour is used for discussion and exploration of possible solutions. As alternative designs become evident, outside speakers from industry are brought in for one of the weekly periods to discuss the topic and provide current information. As early as possible, the students are allowed to manage the design and conduct classes themselves. The course concludes with a formal presentation of the final design and a publication of a design report.

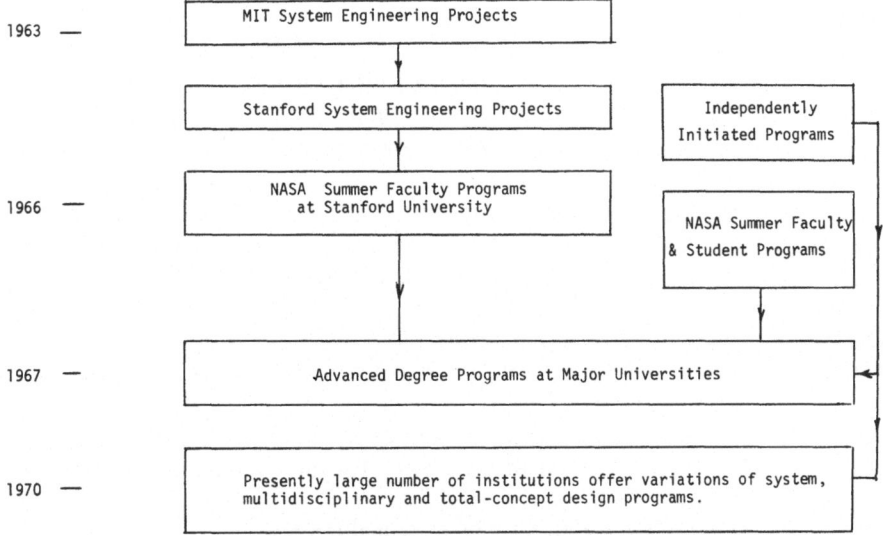

Fig. 2.

A total concept design course has a number of advantages that are not present in other courses.

(1) By selecting a design project which is current and of vital interest, a degree of relevance is introduced into the education process not matched by any other education technique. A problem with which a student identifies, assures and maintains his interest. Few interested students will fail to respond with their best effort.

(2) The design of a complex system forces the student to make decisions without the benefit of precise mathematical equations and presents him the opportunity of exercising his creativity. He is forced to employ imagineering and guestimating.

(3) Guest speakers, experts in their field, introduce the students to the state-of-the-art of technology. They provide, to use the slang expression, 'where its at' information. These lecturers give the students the opportunity to discuss their problems from the *real world* point of view and bring into the educational process a sense of vitality and relevance that goes beyond field trips and seminars.

(4) A total-concept design project broadens the student's views and gives them a depth of problem appreciation.

Very early in the U.S. space programs it was realized that a considerably more active interchange will have to take place among the engineering professions for massive space system designs to come to fruition. In 1963 Dr W. Bollay instituted, at the Massachusetts Institute of Technology, system design courses where students from different engineering disciplines collaborated to design relatively complex space engineering systems. He later introduced these courses at Stanford University (Figure 2).

These projects have been so successful that NASA supported a Summer Faculty Training Program at Stanford during 1966. The purpose was to give outside faculty experience in the course with the hope that they would initiate it at their own university. Until recently summer faculty programs, along with student multidisciplinary design summer courses, were offered at most major NASA centers and associated universities. As experience was gained and as space projects broadened in their scope participation was opened to students from fields other than engineering. Because of the success of these courses in space programs it was logical to extend them to other complex projects. At this time a number of universities in the U.S. are offering total concept design courses. Some of the projects undertaken are:

ICARUS   – One tenth A.U. solar probe.
SPINMAP   – Stanford proposal for an international network for meteorological
              analysis and predictions.
SWAMI   – Stanford worldwide acquistion of meteorological information.
SAMPLER   – Stanford advanced Mars project for life detection, exploration and
              research.
STAR   – Satellite television for the Appalachian region.
STRIDE   – Satellite television relay for India's development and education.
SENSOR   – Satellite for Earth natural and scientific observable resources.
USMM 1990   – United States merchant marine in 1990.
DART   – Downtown Atlanta rapid transit.

These are only a sample to give an idea of the range of problems examined. Studies have also been done on Moon colonies, airport designs, and rapid transit in cities.

Sufficient time has elapsed since the initiation of these courses in 1963 that at least a general evaluation can be made of their effectiveness. To achieve this, a survey was made of past students and faculty and included participants and organizers of many different total concept design projects. The results can be reduced to the following observations:

(1) The course was an enjoyable and interesting experience. This factor is probably due to the relevance of the design topic and to the relative freedom in creativity allowed to the students. (97%)

(2) Suitability of previous education. This is an interesting fact, for only half felt

that their previous education equipped them sufficiently to carry on multidisciplinary design. (55%)

(3) Change of opinion about design. Most students had misconceived ideas about design. (85%)

(4) Change opinion regarding other profession. (75%)

(5) Helped to relate education and reality. (80%)

(6) Caused major change in interest. Mostly research to design. (25%)

(7) Industrial orientation time has been shortened. (50%)

(8) Locate jobs after graduation. This factor is due to the industry – student contact through the guest speakers. (10%)

(9) Would take course again. (95%)

From consideration of this survey, several conclusions can be reached:

(1) Education, particularly design education, must be made more relative.

(2) Admissions to total-concept design programs must be closely scrutinized. Not all are suited for team design work. There still is, and will continue to be very great need for individual researchers.

(3) Total-concept design courses must be broadened to include as many disciplines as possible.

(4) It was found that, in the case of predominantly undergraduate participation, special attention must be paid to the selection of the design problem. The undergraduate generally does not have as deep a background as a graduate student. If the topic selected is too vague or general then too much time is spent gathering supporting information and the design itself becomes a series of evaluations and opinions.

From their inception in 1963, these courses have evolved in a number of steps. From the first system design courses with engineering participation, through multidisciplinary design of space systems, to total concept design and solutions of general social problems these courses have proved to be worthwhile efforts.

At present, we are witnessing another increase in complexity of design. We are going from programs and efforts of individual nations to massive designs on an international scale. These designs are not by any means restricted to space efforts. International cooperation in design will exist in transporation, ecology, environment, medicine, meteorology, and education. These systems are either too large for any one nation to undertake alone, or by their very nature, involve two or more countries. There indeed is a demand for total-concept design capability in the planning of future systems.

It may be to our advantage to plan now for the design of future systems and initiate international total concept design courses. A pressing problem of interest to a number of nations, would be identified and a solution sought by an international student design team from universities in the concerned countries. Benefits would be considerable at all levels for all concerned, however, greatest advantages would be derived by the teaming of universities and students from developing nations and technically established countries.

How the planet is *engineered* for man in the future depends on educators now. The university community has the responsibility of providing leaders for the society to come. There may be questions of the shape or configuration of this society, but its extreme complexity is not in doubt.

SAFETY IN YOUTH ROCKET EXPERIMENTS
(SYRE) SESSION

EDITOR:
A. INGEMAR SKOOG
Chairman of the 3rd SYRE Session, Dornier System, German Federal Republic

CO-EDITOR:
GEORGE S. JAMES
Chairman of the SYRE Study Group, Rocket Research Institute, Inc., U.S.A.

The 3rd SYRE session continued the international review of youth rocket activities in different countries. Two promising projects, one with a hybrid rocket and one with a monopropellant rocket, were also presented. It is very likely that they will increase the safety and at the same time raise student rocketry to a more advanced level.

The bilateral and international cooperation within the French ANCS/CNES youth rocket program is advancing, and the future plans include the forming of a European youth rocket organization.

As there is no possibility to reproduce all seven papers in their entirety, a review of each paper will be given.

PROJECT ASTREUS, A STUDENT HYBRID ROCKET PROGRAM
LAWRENCE R. HARVILL
University of Redlands, U.S.A.

During the spring semester of 1961, a group of six senior engineering students elected to devote their senior engineering design project to the design, development, and flight testing of a small hybrid rocket. This project was the first in a long series of similar student projects which have been undertaken as senior design projects for the purpose of providing an integrated engineering experience.

These projects have been centered about three separate series of vehicles which were named SYRUS, TAURUS and ASTREUS. The first two series were developed by students at the University of California, and project ASTREUS was carried out by students at the University of Redlands. In all instances the students were completely responsible for the decision as to what the project would be as well as the means by which it would be carried out.

The hybrid engine was selected by the students primarily on the basis of safety and relative simplicity. Also at the time the hybrid engine was a relative new development which supplied another point of interest for the students in that they were working with an engine which had not been extensively developed.

In order to deal with the complexities of such a project, the students very readily realized that they needed a formal structure in which to operate. In each project they organized themselves along the lines of a small engineering firm with specific areas of responsibilities assigned to subgroups.

1. History of Projects

The SYRUS project ran from Spring 1961 to Spring 1962 and was based on a 116 mm diam 1360 N thrust engine with a total impulse of 21 800 N s. The first engine was a static test motor which was fired November 1961. Based on the success of this firing, the students obtained permission from the officials at the Naval Ordnance Test Station China Lake, California to schedule a flight test in December 1961. Due to a leak in the liquid oxygen valve system the flight test failed. The second test in April of 1962 was successful except for the fact that approximately five seconds after engine ignition, the motor wall burned through and destroyed the stability of the vehicle thus causing a premature crash of the rocket. The SYRUS engines provided experimentation with both epoxy-polyurethane and polysulfide rubber base solid propellants. The oxidizer in all instances was liquid oxygen.

The TAURUS series of rockets began in the Spring of 1962 and ran through July of 1964. The Taurus engine was a 153 mm diameter hybrid with a 4500 N thrust and a total impulse of 45 000 N s. The combustion chamber was an annular section with fuel cast to the outer walls and supported on an inner core down the central axis of the chamber. Liquid oxygen was supplied by a symmetrical array of six injectors at the head end of the combustion section. TAURUS I was a static test motor with an epoxy base solid fuel tested in December of 1962. This test was a failure in that the motor exploded on the test stand due to several fractures in the epoxy fuel. A second static test engine was assembled with a polysulfide rubber base fuel and was successfully fired in April of 1963. TAURUS III flight vehicle had a two channel telemetering system. Parameters that were monitored by this system were axial acceleration and total pressure. This vehicle exploded on the launch stand as a result of a wiring error in the automatic countdown sequencer. TAURUS IV was identical to TAURUS III and was successfully flight tested on July 17, 1964. The rocket reached a maximum altitude of 4500 m.

In the fall of 1969 students at the University of Redlands elected to build another rocket vehicle, ASTREUS, in an attempt to improve upon the previous designs. To improve upon the TAURUS design, the oxygen pressurization and rupture disc system was deleted in lieu of using a positive explosive propellant valve. The entire engine design was refined by developing a computer model and seeking optimum values for various parameters. A six channel telemetry system was developed and a parachute recovery system was also included. The flight test was conducted in January of 1970 and was successful except for the failure of the recovery system. All things considered, project ASTREUS was the most demanding as well as rewarding project of this series.

2. Project Organization

After careful consideration, the students decided to establish five subgroups to improve the efficiency of the overall operation. Four of the five subgroups were formed about each of the major subsystems of the project: propulsion, telemetry, airframe and recovery, and launch and ground support. The fifth group was a system group which had the responsibility for coordinating all the subgroup activities via the functions of communications, scheduling, flight simlation, external communications, and budget control.

3. Safety Considerations

Due to the nature of these projects, safety was a prime factor. The hybrid engine system was selected primarily because of its greater safety combined with design simplicity. Although some additives were included in the fuel, their concentrations were low enough so that the material was essentially inert. Liquid oxygen was also chosen as the oxidizer because of its relative safety.

In all cases, the students developed detailed check out procedures for all steps of the preparation and countdown to minimize the chance of an accident happening.

All electrical firing circuits were doubly shorted with interlocks and special grounding to eliminate any possibility of induced potentials. Assembly and firing sequences were carefully checked out by holding dry runs.

The launch sites were selected so as to provide covered bunkers for all necessary personnel.

Because of the complexity involved in the design and construction of a hybrid rocket system, its relative safety, and the wide variety of problems encountered, it indeed is an excellent vehicle for engineering education.

A SIMPLE SOUNDING ROCKET FOR EDUCATIONAL PURPOSES

PHILIP RAMSDEN
Youngstown, U.S.A.

Any student rocket project which is initiated at this point in history faces very strong opposition from influential bodies which hold the opinion that rocket engineering is too dangerous for students. This opinion was formed in the years 1945–50 when amateur rocket experiments caused many accidents. All that was 20 years ago. It is now timely to consider changes that may be both desirable and possible in the light of the enormous advances made in rocket engineering in the meantime.

It is true that rocket engineering is dangerous, particularly at the development stage. Nevertheless, the danger is not itself sufficient reason for avoiding rocket development work. To do so would be to deny the practice of mechanical engineering which, in essence, involves the harnessing of energy and hence the risk of losing control of that energy. In everyday engineering, loss of control could mean a bursting pressure vessel, an overspeeding turbogenerator or a runaway reactor. All these events are very dangerous. The responsible professional engineer takes precautions against these dangers. In the same way he can take precautions against the special danger of rocket development.

Now the characteristics of a rocket vehicle which make it specially dangerous are its large store of energy and its large rate of energy release – to be more specific, its propellant and large propellant flow rate. Safety during development is a matter of taking precautions against loss of control of propellant containment or flow. In practical terms this means assessing the fire and explosion risk of all tests which use propellant and ensuring appropriate protection for test personnel and facilities.

The real problem is not that precautions are technically difficult but that they are expensive. Because money is short, there is a strong temptation to take risks. The correct approach is to estimate the cost of safety precautions along with the cost of design and development as an inescapable part of the total project cost. In planning the development program for minimum cost, this approach has the effect of minimizing the number of tests using propellant, particularly those using a large quantity of propellant. It also leads to a reduction in the different types of tests performed.

Up to this point, only the dangers caused by the energy of the propellant have been discussed. A great many more dangers exist if the propellant is also toxic. The important difference here is that whereas high energy is a necessary property of the propellant, toxicity is not. The only sensible approach to a student rocket project is to choose a non-toxic propellant. By doing so, the very large cost of safety precautions involved in storing and handling toxic propellants can be avoided.

Another way of reducing safety costs and at the same time benefitting the whole programme is to ensure high mechanical reliability of all components of the system before using them with propellant. Mechanical failures which can lead to fire or explosion are expensive and cause big delays in the program, all the more serious the later they occur in the course of development.

Summarizing them, for safety, personnel protection is needed in all tests where propellant is used. For minimum cost safety we need in addition:

(1) A non-toxic propellant.
(2) A development program arranged to require the minimum number of hazardous tests and the fewest possible types of hazardous tests.
(3) A proven high reliability of components prior to using them with propellant.

The rocket project proposed in this paper aims to satisfy these requirements.

1. Description of the Vehicle

When considering simple rocket vehicles, it is usual for thoughts to turn immediately to solid propellants. There is no doubt that solid propellant vehicles are simple to use, but they are neither simple to develop nor simple to make. Therefore, since this proposal is for a design and development exercise, solid propellants are not suitable.

Among the liquid propellants, all bipropellant oxidizers are in some way difficult to handle. Furthermore, having two propellants makes for a complex vehicle. It is by this reasoning that the liquid monopropellant iso-propyl nitrate has been chosen.

As a primary propulsion rocket propellant propyl nitrate has one disadvantage which will be discussed first so as to forestall the critics. It has only a middle-range specific impulse (approx. 1800 N s/kg for a 30:1 expansion ratio) which results in a rocket vehicle having a poor performance relative to its size. However, maximum performance for a given size is not an object of the program and all the other characteristics of the propellant make it by far the most attractive propellant for this application. Higher performance monopropellants do, of course, exist, but the better performance can only be had at the expense of detonation sensitivity. For an educational project there is nothing to gain from them. Iso-propyl nitrate itself is very insensitive. It cannot be accidentally detonated during normal handling. Even when a detonator is deliberately fired in a drum of propellant at temperatures up to 80°C, the detonation is not propagated.

It looks like water, has almost the same density and has very nearly the same boiling point. However, its freezing point is low. For a general feel of its properties it is best likened to common inflammable solvents and should be kept as in a paint store. It is bought and stored in sealed drums at a price of about $3.00 per U.S. gallon and can be simply poured into the vehicle from a can whilst wearing ordinary clothes.

The choice of vehicle size is a compromise between performance and cost. The biggest risk is that considerations of cost could push the scale so small that the vehicle becomes a mere toy, unrepresentative of the current generation of sounding rockets. A suitable compromise is to choose a vehicle which can just climb into real rocket domain, that is, above the operating height of aeroplanes. The major dimensions are shown in Figure 1.

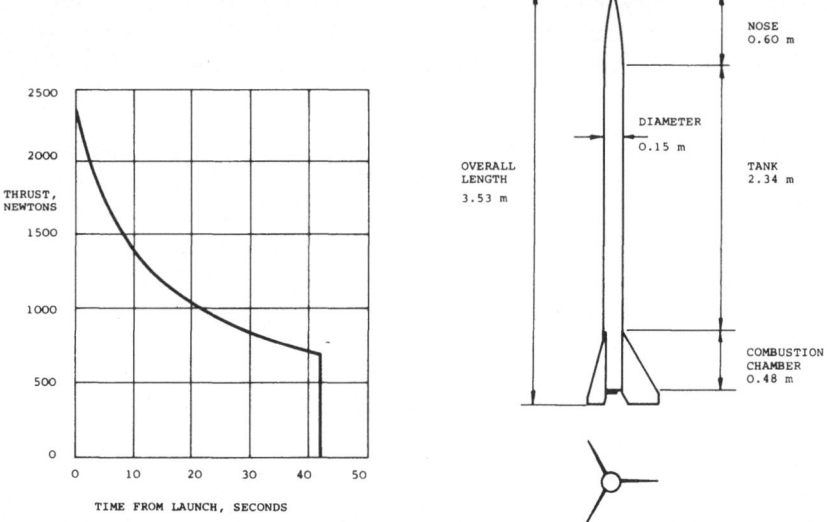

Fig. 1. Principal dimensions of the monopropellant vehicle and its thrust characteristics.

The aluminum alloy tube body is divided into tank and combustion chamber and closed at the ends by simple cylindrical plugs which carry O ring seals and are retained either by screws through the tube or by epoxy resin bonding. The nose cone is spigoted into the forward end of the tube and fixed by screws. The fins are carried by the aft closure, into which they can be aligned before fitting the closure in the vehicle.

The propellant is fed to the combustion chamber by pressurization with inert gas. For simplicity, 100% blowdown is proposed. In order to avoid carrying a high pressure gas bottle, stop valve and reducing valve it is proposed to fill the tank only 70% full of propellant, the remaining space being pressurized prior to flight with nitrogen at the maximum pressure which the tank will tolerate. During the engine run this gas will expel the propellant by expanding to fill the tank and the result will be a falling thrust characteristic as shown in Figure 1.

2. Performance

Allowing for 2 kg of instrumentation or range safety tracking equipment and for 26.5 kg of propellant, the take-off weight is estimated to be 48.5 kg. The effect of drag on a vehicle as small as this is large, so an approximate step-by-step calculation has been done using drag figures from similar vehicles to obtain a more realistic estimate of altitude capability. This calculation predicts an altitude of 43 km.

3. Development Program

The development program envisaged is one which will allow separate teams of students, working in successive years and separate establishments, to make their contributions to the project.

The project by its very nature integrates many different disciplines. An important part of the project, therefore, is to recognize these different disciplines, make sure that the capabilities are available at the right time and that the direction of their development efforts are mutually compatible.

4. Conclusion

It has been shown that a useful sounding rocket can be made safely, simply and cheaply and at the same time provide engineering students with a project typical of real life. The vital matter is to start as simple as possible so as to have some success to show as a foundation for continuing rocket and space research work.

YOUTH ROCKET ACTIVITIES IN ARGENTINA

JOSÉ F. ELASKAR
Universidad Nacional de Cordoba, Argentina

Youth rocket activities in Argentina appeared as spontaneous activities of private effort with no government aid whatsoever. Almost all the groups are maintained by the work and contributions of their members, with only occasional aid from other organizations in the form of materials, surplus equipment and technical advice.

Though individual youth rocket activities were held in this country well before 1963, this year could be taken as the starting-point of activities carried out by organized groups. The age of the group members goes from that of teen-agers to adults, the majority being between 18 and 25 years old, but with even older persons having an active part. To date it is difficult to determine how many groups in Argentina are dedicated to rocket activities. The main groups are shown in Figure 2.

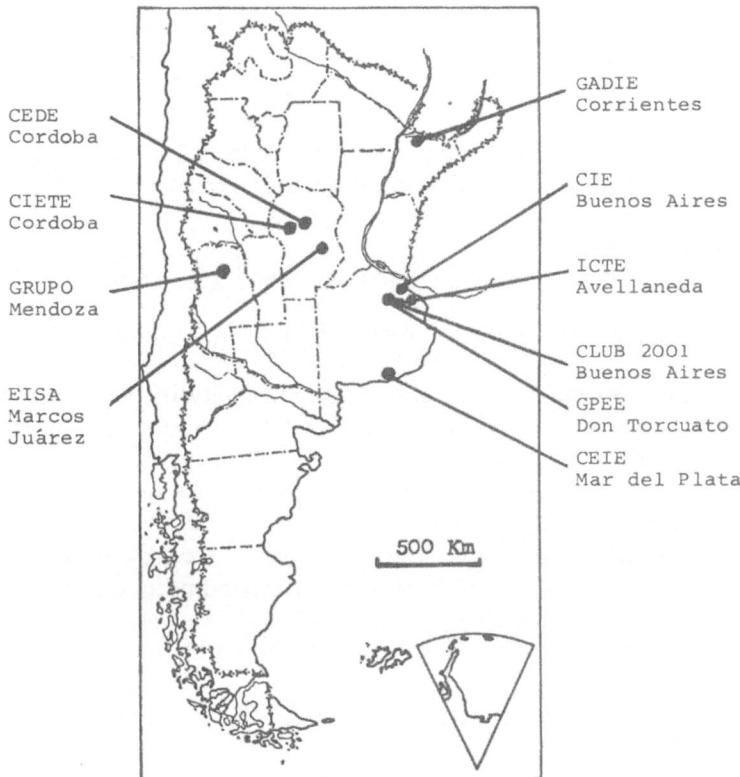

Fig. 2.   Geographical locations of various rocket groups in Argentina.

The more advanced groups have achieved interesting goals within youth rocketry. At present, several groups are working on multistage rockets intended for carrying small payloads to be used in meteorological and hail avoidance research. One of the most outstanding achievements made by the ICTE in Avellaneda was doubtlessly project 'Bio-Ensayo-6': on February 1, 1970 a 'Pantera X-1' rocket was launched carrying a 1.5 kg monkey to an altitude of 20 km. During the flight telemetric measurement of heart and respiratory rates were made.

Some groups dedicate part of their efforts to the divulgement of space knowledge, by means of lectures, seminars and public launchings for which purpose small rockets have been developed.

Safety and Legislation

Even without specific legislation, up to the present no serious accident related to youth rocket experiments is known to have happened in Argentina.

This is due mainly to the fact that all these groups have shown a marked inclination towards standards of safety, elaborated by them.

These standards include not only the safety of the group members, but also that of others during the launchings.

Usually before launchings, the groups contact the local police authorities and make a complete survey of the zone, arranging every necessary safety measure.

Notwithstanding these excellent records, the Argentine National Space Commission (CNIE) has elaborated and proposed to the Federal Government a draft of minimum regulations to cover this activity not only from the point of view of safety but also to promote their development.

The main points of these proposed regulations, which are not yet effective, are:

- Every private group working on rocketry must be registered in a National Registry.
- Damages caused to third parties are the responsibility of the owner or person authorized to operate the rocket or space vehicle, or in the case of minors, the person who exercises the tutorage.
- Design and materials of vehicles or space devices must be approved by the authorities as regards their safety.
- Construction and tests of vehicles must be made in sites previously approved.
- Every launching must be notified to the corresponding authority in order to obtain the necessary clearance.
- The CNIE is designated as acting authority in this matter and can be delegated in other appropriate institutions.
- Promotion of private rocket activities is the CNIE's responsibility.

SUPERVISED YOUTH ROCKET PROJECTS IN ISRAEL

Y. MANHEIMER-TIMNAT
Technion, Israel

Interest in rocketry and space research has been high among the youth in Israel ever since the launching of the first Sputnik in 1957. This generated a demand for lectures and courses, which were organized by the Israel Astronautical Society. Since 1965 youth groups divided into high school students and college undergraduates have been engaged in actual construction and launching of small rockets under professional guidance and supervision. The activities during 1965–70 were reviewed at the 2nd

Fig. 3. Meteorological sounding rocket, Technion, Israel, 1971.

SYRE session. The two-stage vehicle project has been continued, and in 1971 the operation was completely successful and shall be described in some more detail.

The overall length of the rocket (Figure 3) is 4.06 m and it weighs 73 kg at lift-off. The vehicle was launched from a railtype launcher (used, with modifications, since 1968) at an angle of 85°. The boost stage is a solid propellant rocket of 160 mm diameter, which burns for about 7.5 s, giving a total impulse of 75 000 N s. The second stage consists of a smaller solid propellant rocket (108 mm diam), burning for 2.8 s with a total impulse of about 20 000 N s. The casing of the second stage is made of epoxi-impregnated wound-fiberglass and has been developed for the sounding rocket student program in 1967–68 (as a one-stage rocket), giving valuable insight in problems connected with materials. Stabilization was achieved in both stages by fins.

The payload comprised a parachute, a four-channel telemetry transmitter connected to the output of the transducers (accelerometer, thermistor and pressure gauge) and a dry battery power supply. The separation systems (operating at 8 s after lift-off for the first stage and at 98.5 s-maximum height for the payload) were of mechanical type, triggered by pyrotechnic devices, with appropriate delays provided by electronic timers.

The rocket was tracked by cinetheodolites and for this purpose two flares were mounted on the second stage fins. These were ignited 5 s before lift-off. The first stage performed according to prediction and reached a height of 4.5 km, the maximum acceleration being 35 g about half a second after ignition. Separation occurred after 8 s and the second stage was ignited after 28 s, according to schedule. The theodolites lost the vehicle shortly after burnout of the second stage motor, at an altitude of about 20 km. From the transmitted acceleration trace it was, however, possible to establish that the unpowered coasting stage rose to about 30 km where the payload separated and the parachute opened.

In conclusion it can be stated that this type of project is very well suited for acquainting engineering students with the principles of rocketry and their application, provides them with very valuable engineering experience by going through all the stages of a rather complex problem and gives them a strong motivation for their work.

OPTIMALIZATION OF ROCKET ENGINES

BOHDAN E. WEGRZYN
Technical University of Warsaw, Poland

In the class of solid propellant model rocket engines characterized according to FAI by total impulse, engines with different combustion curves are available on the market. Hence arises the problem of defining an optimal characteristic curve of rocket propellant so that the rocket could achieve the highest altitude.

In this connection are considered eight different engine curves $P=f(t)$ for which the total impulse remains the same $I_t=80$ N s and the specific impulse is constant (Figure 4).

With the thrust curves (Figure 4) and known aerodynamic quantities of the Black

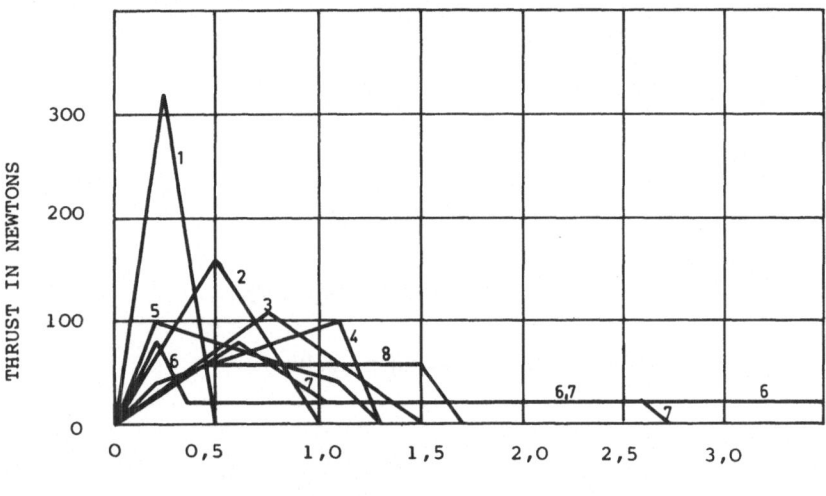

THRUST DURATION IN SECONDS

Fig. 4.   Thrust characteristics for various model rocket engines.

Brant rocket model the individual trajectories are calculated. Equations of rocket motion are solved employing the method of numerical integration.

1. Active Section of the Trajectory

For the active section of the trajectory acceleration components on axes x, z are:

$$a_x = \frac{g}{Q}(P - P_x)\cos\theta \tag{1}$$

$$a_z = \frac{g}{Q}(P - P_x)\sin\theta - g \tag{2}$$

in this section the following values vary: thrust force P, air drag force P_x, weight Q and acceleration of gravity g. That is why the rocket is not uniformly accelerated. If however the time of flight t_k is divided into sufficiently small time sections Δt, then it can be assumed that for a given section acceleration is constant. So the increase of velocity components is:

$$v_x = a_x \cdot \Delta t; \tag{3}$$

$$v_z = a_z \cdot \Delta t. \tag{4}$$

Velocity components on axes x, z corresponding to any time t_n are:

$$v_{xn} = v_{xn-1} + \Delta v_{xn}; \tag{5}$$

$$v_{zn} = v_{zn-1} + \Delta v_{zn}. \tag{6}$$

Velocity increases are equal:

$$\Delta v_{xn} = \left[\frac{g}{Q_n} (P_n - P_{xn-1}) \cos \theta_{n-1} \right] \Delta t \tag{7}$$

$$\Delta v_{zn} = \left[\frac{g}{Q_n} (P_n - P_{xn-1}) \sin \theta_{n-1} - g \right] \Delta t. \tag{8}$$

Change of weight on the active section of the trajectory is:

$$Q_n = Q_0 - \frac{\omega}{t_k} \cdot t_n \quad (\omega = \text{mass of propellant}). \tag{9}$$

Air drag force is:

$$P_{xn-1} = C_x \cdot \rho \ \frac{v_{n-1}^2}{2} \cdot S. \tag{10}$$

Rocket speed for any section is:

$$v_n = \sqrt{v_{xn}^2 + v_{zn}^2}. \tag{11}$$

Trajectory angle is calculated from the following dependence:

$$\theta = \text{arctg} \ \frac{v_{zn}}{v_{xn}}. \tag{12}$$

Coordinates of rocket location corresponding to any time t_n are equal to:

$$x_n = x_{n-1} + \Delta x_n; \tag{13}$$

$$z_n = z_{n-1} + \Delta z_n; \tag{14}$$

$$\Delta x_n = v_{xn} \cdot \Delta t; \tag{15}$$

$$\Delta z_n = v_{zn} \cdot \Delta t, \tag{16}$$

where Δx_n and Δz_n denote increases of the coordinates of rocket location.

2. Passive Section of the Trajectory

The following forces act on the rocket: head resistance of the air P_x, acceleration of gravity g, and the weight $Q_k = Q_0 - \omega$. Out of the projection of the forces on axes x, z

the acceleration components are received. Increase of velocity components for time section t_n is equal to:

$$a_x = -\frac{g}{Q_k} \, P_x \cos \theta \, ; \tag{17}$$

$$a_z = -\frac{g}{Q_k} \, P_x \sin \theta - g \, ; \tag{18}$$

$$\Delta v_{xn} = \left[-\frac{g}{Q_k} \, P_{xn-1} \cos \theta_{n-1} \right] \Delta t \, ; \tag{19}$$

$$\Delta v_{zn} = \left[-\frac{g}{Q_k} \, P_{xn-1} \sin \theta_{n-1} - g \right] \Delta t. \tag{20}$$

Besides formula (5), (6), (11), (12) are obligatory.

3. Conclusion

Out of the above presented rocket engine characteristics, curve No. 6 proved to be the most advantageous because it allows the rocket to achieve the highest altitude. Curves No. 7, 8, 3, 5, 2, 1, 4 are situated in further succession.

A PROGRAM FOR INTERNATIONAL STUDENT PARTICIPATION IN THE FRENCH ANCS/CNES AEROSPACE CLUB ACTIVITIES

H. MOULIN
ANCS, France

The 'Association Nationale des Clubs Aerospatiaux' (ANCS) contains all French youth clubs, which are aiming at the design, construction and launching of experimental rockets and balloon-probes. In a decree of August 7th, 1962 the conditions are laid down for the use of experimental launch vehicles among youth. This document serves as a regulation and it contains two main ideas:

- any experimentation with explosives and manufacturing of rocket engines shall be prevented;
- the youth shall be encouraged to develop aerospace activities by the use of standard propulsion units from 'Centre National d'Études Spatiales' (CNES).

Since 1964 the cooperation between CNES and ANCS, named 'Association Nationale des Clubs Scientifiques', has given most valuable experiences, and 88 experimental rocket launchings with standard propulsion units have been performed. In 1971 ANCS consists of 70 youth clubs with 1400 members.

1. The ANCS International Strategy

The ANCS's main task today is to present the principles and applications needed in order to improve the international youth rocket relations. This new policy of international cooperation started on September 1st, 1970 and is carried out at two levels:

- the bilateral level, which is mainly for short duration operations,
- the multilateral level, which is supposed to play an important role in the development of youth aerospace activities in the world.

A. BILATERAL COOPERATION

Bilateral cooperation seems to be the most realistic at the moment. It is initiated with the elaboration of the documents, which establish the responsibility and the mutual obligations. These documents are of two types.

'The general agreement' is usually for a one year period and contains a declaration of the intention to cooperate and specifies the targets to be met.

The second document, 'The convention', is issued for each separate project within the cooperation. The project is specified in detail together with the economical implications. Contained in the convention are the technical and cultural aspects as well as the exchange of components. The latter are subject to very strict regulations concerning the propulsion units and the balloon-probes.

Bilateral cooperation according to these rules has been carried out with Tunisia in 1971 leading to rocket campaigns in both France and Tunisia. Agreements also exist with Belgium and Canada and are in progress with Yugoslavia and Rumania.

B. MULTILATERAL COOPERATION

International meetings form the basis for this type of cooperation. This will make it possible to establish new ways to adopt common rules of safety and models for bilateral cooperation and to decide the level of responsibility to be shared between youth and adults.

2. Presentation of the Programs

The programs to be carried out within the framework of international cooperation are divided into three main groups.

Cultural programs: This is the traditional way to improve technical education by exchange of books, movies, discussions and exhibitions.

Technical assistance programs: The more advanced partner makes his knowledge, equipment and infrastructure available for an adaptation period.

Technical programs: This is the most advanced form of cooperation in which teams from various countries are developing and manufacturing payloads and putting the units into operation throughout launch campaigns.

ANCS has taken the initiative to establish a European Youth Rocket Organization, which will promote international youth rocket cooperation. The first step is the organization of a European Youth Rocket Conference in 1972.

THE BR-1 ROCKET FOR THE SAFE SUPERVISED LAUNCHING OF STUDENT PAYLOADS

GEORGE S. JAMES, CHARLES PIPER, JOHN BILLHEIMER and
JOHN CHRISTENSEN

Rocket Research Institute, Inc., Glendale, Calif., U.S.A.

This report describes the results to date of a college level student rocket safety science motivation program which is evaluating, through experimental rocket launching campaigns, the usefulness of a large, ready-made, solid propellant rocket motor called the BR-1. This motor is capable of carrying student experimental payloads weighing 10 kg to altitudes in excess of 10 000 m in theoretical vertical flight. Actual altitudes of 7000 m have been achieved.

The BR-1 motor is manufactured in small quantities by the Rocket Research Institute, Inc., at its Perkins Rocket Safety Center near Sacramento, California, and launched under the supervision of Institute personnel at the RRI, Inc., Smoke Creek Flight Range, located about 190 km north of Reno, Nevada. Since the start of the flight program in 1967, a total of fifteen propulsion units have been launched to test flight vehicle features and evaluate experimental payloads constructed by students, from 17 to 23 years old.

The characteristics of the BR-1 motor are as follows:

Length	(L)	0.80 m
Diameter	(D)	0.127 m
Weight loaded	(W_1)	30.0 kg
Weight empty	(W_0)	20.5 kg
Firing duration	(Tb)	0.7 s
Total impulse	(I_t)	16 700 N s
Average thrust	(F_{av})	21 000 N

The development of the BR-1 has been under the direction of Charles Piper of the Institute staff. The unit is fabricated by modifying MK 10 surplus military 12.7-cm aircraft rocket motors and their nozzles. A unique feature of the BR-1 design is the use of the long established GALCIT-61c castable propellant in star grain form. A remote control propellant mixer is located at the Perkins Center.

Payloads are designed to be attached directly to the front of the BR-1 motor. After the payload compartments have received their last ground checks, they are connected to the motors. The guidance fins are also designed to be attached directly to the motor. Thus the propulsion chamber becomes the rigid spine of the launch vehicle.

The nozzle is inserted into the motor just prior to installation of the vehicle on the launching rack. After the area is cleared, the igniter and other pyrotechnic devices on the vehicle are armed by Institute supervisors who then return to the control center to arm the firing circuits.

Unlike the small ready-made solid propellant rocket motors called 'model rocket engines', the approximately 17 000 N s total impulse of the BR-1 makes mandatory the same safety standards as those required for industry and government rocket

Fig. 5. BR-1 flight vehicle mounted for an 85° launch from a zero-length launcher, September 1971.

launches. The Smoke Creek Desert Flight Range is the only site in the United States currently approved by the Federal Aviation Administration of the Department of Transportation for the use of the BR-1. Institute range safety requirements require that all personnel be located at least 700 m from the launch point.

The primary emphasis of the BR-1 program to date has been to evaluate reliability of the propulsion unit, develop operational procedures, investigate methods of launching, and test payload recovery systems. The fifteen standard BR-1 motors launched since 1967 have all operated successfully and have attained altitudes of from 4200 to 7000 m carrying a variety of payloads. These flights have demonstrated the potential problems possible with recovery systems, staging mechanisms, ignition of second stage vehicles, and reliability of student-designed electronic payloads.

The BR-1 flight test program began in August 1967 with the successful launch of a two-stage vehicle, which consisted of BR-1 motor SN-7; an experimental SR-1

Fig. 6. BR-1 flight vehicle launch photographed by automatic camera, September 1971.

second stage (which produced 1500 N of thrust for 10 s); and an experimental parachute payload. The vehicle weighed 63.5 kg and reached an altitude of 4200 m.

In the years since this launch, the portable launching racks have evolved from the 5-m long rail launcher to the present zero-length launcher shown in Figure 5.

The use of the zero-length launcher was first investigated by Richard Bennett in the spring of 1969 as a result of the BR-1 trajectory computer program which indicated that a BR-1 propelled vehicle would be stable in flight immediately after lift-off. Examination of the motion pictures from launches indicated that the vehicle was completely stable upon leaving the launching-lug guide pins of the launcher. Variations of this zero-length launching rack have been used in the subsequent nine flight launches. The problem of erosion of the soil beneath the launcher was finally eliminated during the 1971 launching campaign by the use of a portable steel deflector plate firmly attached to the foundation of buried railroad ties.

The flight vehicle propelled by BR-1 motor SN-23, shown in Figure 5, launched in September 1971, carried a 20 kg payload consisting of a dual parachute compartment containing two Super-8 movie cameras and a radio transmitter to indicate operation of the cameras. The launch of this vehicle, recorded by a Zeiss-Ikon Contax, automatically triggered by the launch, is shown in Figure 6. At the peak altitude of 4200 m, the instrument compartment was separated from the rest of the vehicle by the detonation of a ring of JET CORD, a linear shaped-charge produced by Explosive Technology, Inc., Fairfield, California. The compartment successfully deployed its parachute. However, the BR-1 separated from its parachute and descended spinning rapidly making a loud whistling sound. When the BR-1 motor was recovered from the surface of the dry lake, it was found that the jagged edge of the parachute compartment, caused by the shaped-charge, had cut the parachute cord and allowed the parachute to drift away. However, this open ended tubular compartment with the BR-1 motor attached was unstable and spun down. This problem will be avoided in future launchings by using braided steel cable. The instrument compartment landed by its parachute approximately one and one half miles away, Figure 7.

This SN-23 flight demonstrated the best methods developed to date at Smoke Creek to avoid damage and to facilitate recovery of the BR-1 and payloads. By using both a parachute system and, as a back-up, a method of causing the vehicle to become unstable after accomplishing its mission, the parts of the rocket usually will not bury themselves too deeply in the semi-fluid mud beneath the surface of the dry lake bed.

Future projects using the BR-1 motor include the telemetry of motor chamber pressure and other parameters during flight; development of radio signal beacons for payload compartments; improved pre-launch work facilities; and study of the BR-1 ignition and initial combustion process through the use of a Contarex-Electronic camera system to be loaned the Institute by Zeiss-Ikon of Stuttgart, West Germany.

The BR-1 motor program has shown that students, who are engaged in advanced engineering or technical programs in school, have been able to supplement their formal training by constructing payloads to be used with this professionally designed and manufactured propulsion unit.

Fig. 7. BR-1 payload compartment after successful parachute deployment and descent from 4200 m, September 1971.

A staff member of the Institute is assigned to coordinate the vehicle fabrication and launching phases of each project and to work with the student organization in the design of each flight vehicle and launching system. Thus, maximum participation in the physics and aerodynamics of flight vehicles is obtained without the hazards and difficulties inherent in unsupervised experimental rocket motor development projects. Students design and construct the payloads, retrieval devices, and range sighting equipment while leaving the motor manufacture and safety aspects of the project to the RRI, Inc. program supervisors. The Institute also makes available flight-proven components such as nose cones, fin assemblies, launch clips, and even the launching rack if the student group decides it doesn't wish to construct these items.

Understandably, these BR-1 motors are available, at nominal cost, only for such activities as those described in this paper which are part of the supervised programs of the National Rocket Safety Registry program of the Rocket Research Institute, Inc. Contacts with RRI, Inc. can be made through the International Astronautical Federation, 250 Rue Saint-Jacques, Paris-5, France.

PART IV

FIRST I.A.F. STUDENT CONFERENCE

TRANSVERSE ACOUSTIC WAVE AMPLIFICATION DUE TO MASS INJECTION AROUND A SUBMERGED NOZZLE

A. CRAIG HANSEN

Dept. of Mechanical Engineering, University of Utah

Abstract. Submerged nozzles in solid propellant rocket motors are finding increased application due to size limitations and other factors. One little explored aspect of such nozzles is their effect on the growth of acoustic waves in the motor chamber. This paper shows that the flow from the annulus surrounding the nozzle to the chamber cylinder is a significant contributor to the wave growth rate. Theoretical growth rate calculations after F. E. C. Culick and G. A. Flandro show that the mean flow from the annulus to the chamber contributes to the growth under any conditions. The magnitude of the effect depends primarily upon the mean flow speed and direction at the annular exit. A parametric study is used to show simple motor configuration changes that will greatly enhance the acoustic stability.

1. Introduction

The use of submerged nozzles in solid propellant rocket motors has been found to have many advantages in current applications. The effect of submerged nozzles on the acoustic stability has not, however, been adequately investigated.

The annular slot around the nozzle will have considerably higher gas velocities than the normal burning surface because of the large burning surface area but small slot cross-section area. The effect of this high-speed mass injection on sound waves in the cavity is studied in this paper. Also, a parametric study is performed to determine the effect of minor grain configuration changes on the stability of acoustic disturbances.

Acoustic instability is a phenomenon that has been observed in solid propellant rocket motors since their conception. During the operation of the motor small amplitude sound waves are generated and propagated in the combustion cavity. These pressure waves are observed to propagate both along the axis of the chamber (axial waves) and in the plane normal to the axis (transverse waves).

Acoustic instability results, obviously, in vibration of the rocket motor. It also causes an increased burning rate that leads to an abnormally high chamber pressure. The vibration problem provides an undesirable environment for missile components and guidance control systems. The high chamber pressure can result in case failure.

It is now understood that there are two basic mechanisms that amplify any small acoustic disturbance occurring in the motor. The best known is called pressure coupling. The propellant burning rate depends primarily on the chamber pressure. Thus, as a pressure impulse reaches the burning walls of the chamber, it causes the burning rate to increase. This, in turn, causes a higher mass flux from the wall and an amplified

L. G. Napolitano et al. (eds.), Astronautical Research 1971, 523–534. All Rights Reserved.
Copyright © 1973 by D. Reidel Publishing Company, Dordrecht-Holland.

pressure impulse is sent back into the cavity. The net result is that energy from the combustion is added to the pressure wave. The pressure coupling driving is also observed in regions where a jet of gas enters the combustion chamber. Such jets occur in motors with slotted forward ends or, as in the case this paper considers, with an annular slot surrounding a submerged nozzle. The mass flux out of such a slot is affected by the downstream chamber pressure. Thus, oscillations in the flow velocity of the slot are caused by pressure oscillations in the chamber and driving results from the same pressure coupling seen at the propellant wall.

The response of a propellant wall or gas jet to a pressure oscillation is reflected through use of an admittance function. This function relates the pressure oscillation (at the wall or mouth of the jet) to the resultant velocity oscillation. That is

$$\hat{n} \cdot \tilde{\mathbf{u}} = -\frac{\nu}{\gamma} A \tilde{P}$$

where $\hat{n} \cdot \tilde{\mathbf{u}}$ is the normal component of velocity fluctuation, ν is the Mach number at the wall, $\tilde{P}$ is the pressure fluctuation, and A is the admittance function. The ratio of specific heats, γ, is inserted for later convenience. The admittance function can, in some cases, be determined analytically. It has been approximated with some success for the burning propellant. However, reliable values must normally be found experimentally using the T-burner technique for example. Use of the admittance function provides the great advantage that it separates the properties and response of the propellant from the geometry of the motor in the stability calculation. It will be shown that this reduces the single complex problem to two independent and simplified problems and thus allows much more work to be done analytically.

The second driving mechanism is less well known but equally important. Any mean flow into the motor chamber from secondary combustion regions such as the annulus surrounding a submerged nozzle can amplify acoustic waves. (By mean flow it is meant the steady flow that would occur if there were no pressure oscillations.)

It might appear that a mean flow into the combustion chamber would have no driving effect on acoustic oscillations. However, this mean flow can sustain and add energy to the oscillatory particle motion in much the same way turbulence is sustained and driven by a high Reynolds number mean flow. Thus, the mean flow represents a large reservoir of energy that is tapped by the oscillating particles through collision interactions. This same mean flow driving is caused by the flow away from a burning propellant wall. The magnitudes of the effects due to the mean flow from a slot and the mean flow from a burning wall will be compared in this paper.

The analysis that follows is directed at determining analytically the growth rate of small acoustic disturbances caused by the mean flow from the annulus around a submerged nozzle. The basic approach is that used by F. E. C. Culick in reference [1], the notation which follows is that of Flandro [2]. This paper demonstrates a new application of the general theory in that the calculations are reduced to a form amenable to parametric study of special or unusual flow situations.

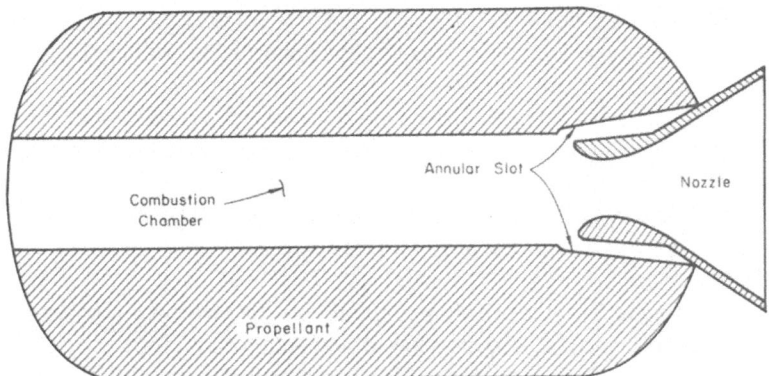

Fig. 1. Motor configuration.

2. Analysis

The geometry and coordinate system is shown in Figures 1 and 2. The case considered is a rigid hollow cylinder with inert walls. The effect of the burning at the walls is certainly a prime factor in the analysis of acoustic instability but it has been shown (2) that the various driving effects in the motor are additive to the first order. Thus, the driving of the mass injection from the annular slot can be computed independently of the other effects.

With no combustion in the chamber, it is obvious that the admittance function will be zero at all surfaces except the entrance to the annulus where it must be determined experimentally or estimated analytically.

Only transverse modes will be considered as the axial modes are complicated by the aft end geometry. Axial waves will be distorted by the interaction with the nozzle

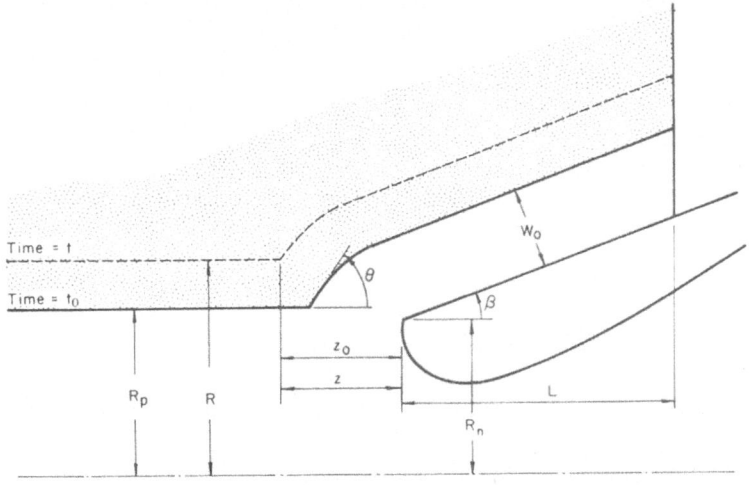

Fig. 2. Nozzle dimensions and coordinate system.

and annular slot ports. Unlike the other surfaces of the chamber, the aft end has no solid wall to reflect waves. In many cases this exclusion is insignificant because transverse modes are decidedly dominant.

Calculation of the detailed flow field would be extremely involved. Fortunately, this is not required to determine the tendency of the motor toward instability. The growth rate of a small acoustic disturbance can be calculated directly from general principles without precise knowledge of the flow field. It is shown that to the order of precision of the calculations, the growth rates depends only on conditions at the chamber surface.

The analysis hinges on perturbation expansions in two small parameters. The magnitude of the acoustic disturbance is assumed small and denoted by ε. The Mach number at the surface of the cylinder is also small and represented by the symbol ν. The mean flow is assumed inviscid, steady, and incompressible.

When the parameters are non-dimensionalized by a characteristic length R, velocity a_0, pressure P_0, temperature T_0 and time R/a_0, the conservation equations become

$$\frac{\partial \rho}{\partial t} + \nabla \cdot (\rho \bar{u}) = 0$$

$$\rho \frac{D\bar{u}}{Dt} + \frac{\nabla P}{\gamma} = 0 \tag{1}$$

$$\frac{DT}{Dt} - \frac{\gamma - 1}{\gamma} \frac{DP}{Dt} = 0$$

and the equation of state

$$P = \rho T. \tag{2}$$

Combining the energy and state equations yields the isentropic relationship

$$P = \rho^{\gamma}. \tag{3}$$

Expanding in integral powers of ε the dependent variables can be expressed in the form

$$P = 1 + \varepsilon P^{(1)} + O(\varepsilon^2)$$

$$\bar{u} = \nu \bar{U} + \varepsilon \bar{u}^{(1)} + O(\varepsilon^2). \tag{4}$$

A double expansion will ultimately be necessary, thus

$$P^{(1)} = P^{(10)} + \nu P^{(11)} + O(\nu^2)$$

$$\bar{u}^{(1)} = \bar{u}^{(10)} + \nu \bar{u}^{(11)} + O(\nu^2). \tag{5}$$

Using the relations (Equation (4)) in the governing equation (Equation (1)) gives the following result to order ε:

$$\frac{\partial P^{(1)}}{\partial t} + \gamma \nabla \cdot \bar{u}^{(1)} = -\nu \bar{U} \cdot \nabla P^{(1)} \tag{6}$$

$$\frac{\partial \bar{u}^{(1)}}{\partial t} + \frac{\nabla P^{(1)}}{\gamma} = -\nu [\bar{U} \cdot \nabla \bar{u}^{(1)} + \bar{u}^{(1)} \cdot \nabla \bar{U}]. \tag{7}$$

Differentiating Equation (6) with respect to time, taking the divergence of Equation (7), and combining the result yields

$$\frac{\partial^2 P^{(1)}}{\partial t^2} - \nabla^2 P^{(1)} = \nu \left[\gamma \nabla \cdot (\bar{U} \cdot \nabla \bar{u}^{(1)} + \bar{u}^{(1)} \cdot \nabla \bar{U} - \frac{\partial}{\partial t} \bar{U} \cdot \nabla P^{(1)} \right]. \tag{8}$$

In anticipation of oscillations it is convenient to express the dependent variables

$$P^{(1)} = \gamma P^{(1)} e^{ikt}$$
$$\bar{u}^{(1)} = \bar{q}^{(1)} e^{ikt} \tag{9}$$

where k is the complex frequency

$$k = \Omega + i\Lambda. \tag{10}$$

It can be seen that Ω represents the frequency of the oscillations and Λ is the growth rate. When Λ is positive, the wave amplitudes will decrease; a negative growth rate indicates increasing amplitude and, hence, instability.

When the values of Equation (9) are inserted into Equation (8), the result is

$$\nabla^2 P^{(1)} + k^2 P^{(1)} = \nu g^{(1)} \tag{11}$$

where

$$g^{(1)} = ik(\bar{U} \cdot \nabla P^{(1)}) - \nabla \cdot (\bar{U} \cdot \nabla \bar{q}^{(1)} + \bar{q}^{(1)} \cdot \nabla \bar{U}).$$

As previously discussed, an admittance function will be defined to relate the velocity fluctuations at the mouth of the annulus to the pressure fluctuations. Thus, define

$$\hat{n} \cdot \bar{u}^{(1)} = -\frac{\nu}{\gamma} A P^{(1)}. \tag{12}$$

With this admittance function and the momentum equation, the boundary condition at the slot is

$$\hat{n} \cdot \nabla P^{(1)} = -\nu h^{(1)} \tag{13}$$

where

$$h^{(1)} = -iKAP^{(1)} + \hat{n} \cdot [\bar{U} \cdot \nabla \bar{q}^{(1)} + \bar{q}^{(1)} \cdot \nabla \bar{U}].$$

One method of solution of this non-homogeneous Helmholtz equation is with Green's functions and is given in reference [1]. The Green's function satisfies the system

$$(\nabla^2 + K^2)G(\bar{r}|\bar{r}_0) = \delta(\bar{r} - \bar{r}_0)$$

$$\hat{n} \cdot \nabla G(\bar{r}|\bar{r}_0) = 0 \quad \text{at} \quad \bar{r} = \bar{r}^s \tag{14}$$

the superscript s denotes values on the surface of the cylinder and $\delta(\bar{r} - \bar{r}_0)$ is the Dirac delta function. Using the definition in Equation (11) and integrating over the volume yields

$$P^{(1)} = \nu \int_v G^*(r|r_0)g_0^{(1)} \, dV_0 - \int_v G^*(r|r_0)\nabla^2 P^{(1)} - P^{(1)}\nabla^2 G(r|r_0) \, dV_0.$$

Using Green's theorem and the boundary condition (Equation (13)) gives the pressure

$$P^{(1)} = \nu \int_v G^*(r|r_0)g_0^{(1)} \, dV_0 + \nu \int_s G^*(r|r_0)h_0^{(1)} \, dS_0. \tag{15}$$

The Green's function is assumed of the form

$$G^*(r|r_0) = \sum_\alpha A_\alpha \psi_\alpha(\bar{r}) \tag{16}$$

where the eigenfunctions satisfy

$$(\nabla^2 + k^2)\psi_\alpha = 0$$

$$\Delta \cdot \psi\alpha = 0 \quad \text{at the surface.} \tag{17}$$

The eigenfunctions of Equation (17) are, for cylindrical coordinates,

$$\psi_\alpha = \cos(K_l z) \cos(m\varphi) J_m(k_{mn}r) \tag{18}$$

where

$$m = 0, 1, 2, \ldots$$

$$k_l = \frac{l\pi R}{L}$$

K_{mn} are roots of

$$\left.\frac{dJ_m(K_{mn}r)}{dr}\right|_{r=1} = 0$$

$$K_N^2 = K_l^2 + K_{mn}^2.$$

The eigenfunctions of the wave equation must be orthogonal; that is

$$\int_v \psi_\alpha \psi_\beta \, dV = 0 \qquad \alpha \neq \beta$$

$$\int_v \psi_\alpha \psi_\beta \, dV = E_\alpha^2 \qquad \alpha = \beta$$

where E_α is the normalizing constant and α identifies the wave mode ($\alpha = (l, m, n)$). Using the definition in Equation (16) the constants A_α are shown to be

$$A_\alpha = \frac{\psi_\alpha(r_0)}{E_\alpha^2(K - K_\alpha^2)}. \tag{19}$$

Using the second expansion for $P^{(1)}$

$$P^{(1)} = P^{(10)} + \nu P^{(11)} + O(\nu^2)$$

it can be seen that

$$P^{(1)} = \psi_\alpha + O(\nu).$$

Using the result for Green's functions (Equations (16) and (19)) in the solution for $P^{(1)}$ (Equation (15)), it is shown that, for a particular mode, $\alpha = N$

$$K^2 = K_N^2 + \frac{\nu}{E_N^2} \left[\int_v g^{(1)} \psi_N \, dV + \int_s h^{(1)} \psi_N \, ds \right]. \tag{20}$$

Now recall

$$K = \Omega + i\Lambda.$$

Expanding Ω and Λ in a series in ν and expanding Equation (20) in a Taylor's series reveals for the indicated orders of ν:

$$O(1): \quad \Omega^{(10)} = K_N, \qquad \Lambda^{(10)} = 0$$

$$O(\nu): \quad \Omega^{(11)} = \frac{1}{2K_N E_N^2} \text{Re} \left[\int_v g^{(10)} \psi_N \, dV + \int_s h^{(10)} \psi_N \, ds \right] \tag{21}$$

$$O(\nu): \quad \Lambda^{(11)} = \frac{1}{2K_N E_N^2} \text{Im} \left[\int_v g^{(10)} \psi_N \, dV + \int_s h^{(10)} \psi_N \, ds \right].$$

Using the values of $g^{(10)}$ and $h^{(10)}$ previously given and performing the integration of Equation (21) yields

$$\Omega = K_N + O(\nu) \tag{22}$$

and to first order in ν

$$\Lambda = \frac{\nu}{E_\alpha^2} \int_s [\bar{U} \cdot \hat{n} - A^{(r)}] \frac{(P^{(10)})^2}{2} \, ds \tag{23}$$

where, since

$$P^{(1)} = \psi_\alpha + O(\nu)$$

$P^{(10)}$ is given by

$$P^{(10)} = \psi_\alpha$$

and ψ_α is given by Equation (18).

It is now clear that to order ν in the solution the amplification of sound waves depends only on conditions at the chamber surface. The first term under the surface integral represents the driving due to the mean flow across the surface as discussed earlier. The second term – the real part of the admittance function – represents the pressure coupling effect at the surface.

3. Calculation for a Submerged Nozzle

With the expression for the growth rate as given in Equation (23), it becomes a simple matter to determine the growth rate due to the submerged nozzle. A simplified geometry of a typical configuration as shown in Figure 2 will be analyzed. The walls of the annulus are straight and parallel, inclined at an angle β to the chamber axis. The angle θ represents the angle at which the flow enters the chamber. Inviscid 'slug' flow is assumed and, except for the change in flow direction, effects of the curvature at the annulus mouth are neglected. These simplifications are for convenience only. More exact calculation could be readily performed for a particular configuration.

The velocity of the flow at the annulus exit is calculated from the continuity relationships

$$m = \rho_s S\dot{r} = \rho_g A_c |\overline{U}| \tag{24}$$

where $\dot{m}$ is the mass flow rate; $\dot{r}$ the propellant burn rate; S the burning surface area; A_c the slot cross-section area; and ρ_g and ρ_s are densities of the gas and solid phases, respectively. Geometric considerations to find S and A_c and Equation (24) yield

$$|\overline{U}| = \frac{\rho_s}{\rho_g} \frac{S}{A_c} r = 2 \frac{\rho_s}{\rho_g} \dot{r} \; \frac{\left(r_p + \dot{r}t + \dfrac{L + z - z_0}{2} \tan \beta\right) \dfrac{L + z - z_0}{\cos \beta}}{\left(\dfrac{r_n}{\cos \beta} + w_0 + \dot{r}t\right)^2 - \dfrac{r_n^2}{\cos \beta}} \tag{25}$$

where t represents time after ignition. The dimensions, coordinates, and angles are shown in Figure 2. The mean flow velocity entering the chamber is

$$\overline{U} = -|\overline{U}| \sin \theta \, \hat{e}_r - |\overline{U}| \cos \theta \, \hat{e}_z.$$

The growth rate will be calculated in two parts. The first, Λ_r, is due to radial injection through the cylinder side wall and the second, Λ_z, is from axial flow through the aft end. Thus

$$\Lambda = \Lambda_r + \Lambda_z.$$

where

$$\Lambda_r = \frac{\nu}{E_N^2} \int_0^{z/R} \int_0^{2\pi} [\overline{U} \cdot \hat{e}_r - A^{(r)}] \frac{P^2}{2} \, d\varphi \, dz$$

and

$$\Lambda_z = \frac{\nu}{E_N^2} \int_{r_n/R}^1 \int_0^{2\pi} [\overline{U} \cdot \hat{e}_z - A^{(r)}] \frac{(P^{10})^2}{2} \, d\varphi \, dr$$

recall ν is the Mach number at the surface

$$\nu = \frac{|\bar{U}|}{a_0} \tag{26}$$

E_N^2 for transverse modes is given by

$$E_N^2 = \int_v (P^{(10)})^2 \, dV = \frac{L_c \pi}{2R} \left(1 - \frac{m^2}{K_{mn}^2}\right) J_m^2(K_{mn}). \tag{27}$$

where L_c is the combustion chamber length. Performing the integration yields

$$\Lambda_r = \frac{-\nu}{L_c} (1 + A^{(r)}) \sin \theta \, \frac{1}{1 - (m^2/K_{mn})} \, z \tag{28}$$

where

$$z = z_0 + \dot{r}t \tan \beta/2$$

$$\Lambda_z = \frac{-\nu R}{L_c} (1 + A^{(r)}) \cos \theta \, \frac{1}{[1 - (m^2/K_{mn}^2)]J_m^2(K_{mn})} \, \xi \tag{29}$$

where

$$\xi = \tfrac{1}{2} J_m^2(K_{mn}) \left(1 - \frac{K^2}{K_{mn}^2}\right) - \frac{1}{2K_{mn}^2} \left\{ \left[K_{mn}^2 \left(\frac{r_n}{R}\right)^2 - m^2 \right] J_m^2 \left(K_{mn} \frac{r_n}{R}\right) \right.$$

$$\left. + \left(\frac{r_n}{R}\right)^2 \left[- K_{mn} J_{m+1} \left(K_{mn} \frac{r_n}{R}\right) + \frac{mR}{r_n} J_m \left(K_{mn} \frac{r_n}{R}\right) \right]^2 \right\}.$$

The term ξ depends only on the geometry and is always positive. Then it can be seen from Equations (28) and (29) that the growth rate is directly proportional to the mean flow Mach number ν. Notice also that for values θ between 0° and 90° – the values of interest – the growth rate is always negative, indicating that the submerged nozzle will always drive the acoustic oscillations to higher amplitudes.

This analysis includes no damping effect such as viscosity or acoustic energy loss through the nozzle. These damping effects will certainly act to decrease the oscillation amplitudes and must be overcome by the driving forces before any net amplification can occur.

The dimensional form of the growth rate is found by dividing by the characteristic time R/a_0. Thus

$$\Lambda' = \frac{\Lambda a_0}{R}. \tag{30}$$

The growth rate for a typical motor was calculated for several slightly different configurations. All calculations used the following values for dimensions

$$r_p = 9.75 \text{ in.}, \quad r_n = 8.5 \text{ in.}, \quad z_0 = 1.7 \text{ in.}, \quad L = 15.5 \text{ in.}$$

$$\dot{r} = 0.4 \text{ ips} \quad P_{\text{chamber}} = 300 \text{ psi} \quad T = 2000 \text{ °R}$$

$$\beta = 20° \quad \rho_{\text{solid}} = 1.6 \text{ gm/cm}^3$$

θ and ω_0 are as noted.

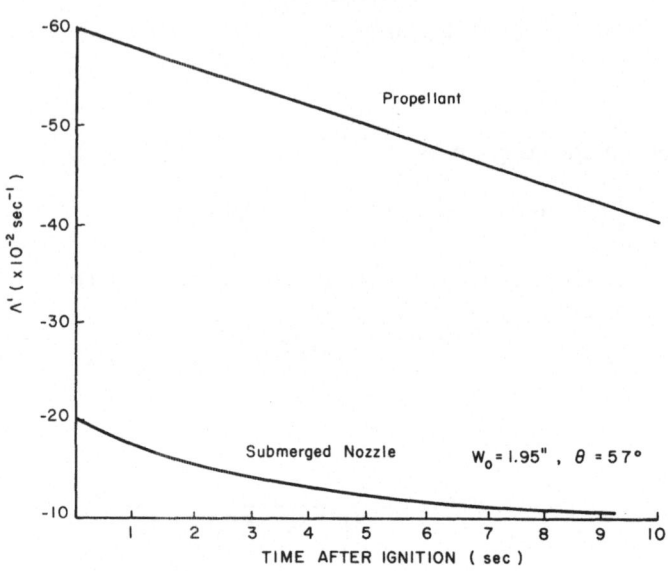

Fig. 3.   Comparison of nozzle and propellant induced growth rates.

To show the relative magnitude of the effect of the submerged nozzle, the growth rate due to the burning propellant alone was calculated for the same motor. The result, in dimensional form, is shown in Figure 3. It can be seen that the driving of the nozzle is as much as one-half the driving of the propellant. This is certainly not insignificant and any reduction in the nozzle induced growth rate will be very beneficial to the overall stability.

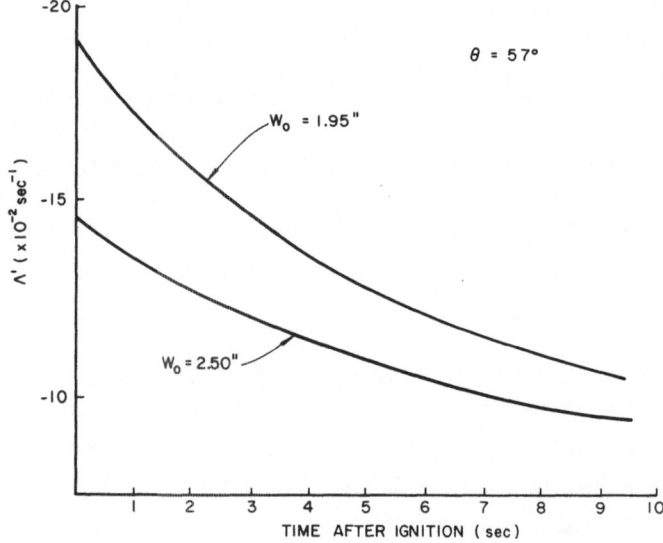

Fig. 4.   Growth rate vs. time for two initial slot widths.

The effects of changing the angle θ and the initial slot width were studied. Figures 4 and 5 show some of the results. The admittance function was not included in the calculations for lack of a numerical value. But it is not necessary to include this value to determine the relative magnitude of the effect and methods of decreasing the driving.

The velocity is the chief factor in the growth rate calculation. By decreasing the velocity, the growth rate will decrease proportionately. Calculation of Λ with a 25% greater initial slot width shows as much as a 30% decrease in the growth rate (see Figure 4). Thus, small configuration changes enhance markedly the acoustic stability.

The angle θ is seen to have a significant effect on the growth rate. It is obvious that a decrease in the angle θ results in a smaller maximum driving effect (see Figure 5).

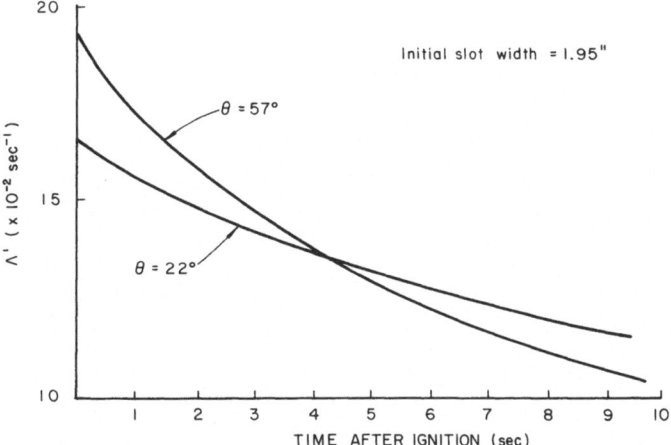

Fig. 5. Growth rate vs. time for different values of θ.

Also note, however, that a higher value of θ causes the growth rate to decrease more rapidly with time. If it is felt that a high maximum growth rate can be tolerated to obtain a lower value at a later time – at about ignition plus five seconds – a high value for θ may be desirable. The angle can be changed easily by rounding or smoothing the corner at the exit. By carefully considering the effect of this seemingly insignificant parameter, the designer can exercise important control over the acoustic instability.

4. Conclusion

The driving effect of the high-speed mean flow from the annulus around the nozzle into the combustion chamber is seen to be a significant factor in the stability of solid propellant rocket motors with submerged nozzles. The mean flow will drive acoustic oscillations to higher amplitude under any conditions at a rate proportional to the flow Mach number. Results from sample calculations show that the driving due to the submerged nozzle alone is comparable to the driving due to all of the burning propellant walls. On this basis it is recommended that the use of submerged nozzles in solid propellant rocket motors be discontinued wherever possible.

If other considerations, primarily motor length restrictions, require the use of submerged nozzles, several precautions must be taken in the nozzle design. It was shown that reduction of the mean flow speed by increasing the annular slot width could appreciably reduce the driving effect. Control over the angle θ by simple smoothing or rounding of the propellant grain at the slot exit can also improve the stability characteristics.

Though this paper does not consider axial modes because of computation difficulties, the same basic principles apply and the same driving mechanisms are important. Thus, axial modes will be driven by the submerged nozzle also.

Though the presence of the submerged nozzle necessarily increases the growth rate of acoustic waves, the increase can be kept at a minimum by the alert designer.

References

[1] Culick, F. E. C., *Astron. Acta*, **12** (1966), 113.
[2] Flandro, G. A. 'Acoustic Combustion Instability in Flight Application Solid Propellant Rocket Motors', Technical Publication, Hercules Inc., Magna, Utah.
[3] Dettman, J. W., *Mathematical Models in Physics and Engineering*, McGraw-Hill Book Company, Inc., New York, 1969.

AXISYMMETRIC BUCKLING OF AN
ANNULAR PLATE*

M. FEDER
Dept. of Aeronautical Engineering, Technion – Israel Institute of Technology, Haifa, Israel

1. Introduction and Historical Survey

The problem of the stability of annular plates is common, especially in aeronautical engineering where there is a need to create circular holes in compressed plates. The purpose of such holes may be to minimize the weight of the vehicle or to provide access, to pass cables and pipes and so on.

The practical loading conditions and the buckling shape are not necessarily of axial symmetry, but the results presented here for the axisymmetric problem may give a good estimation of the critical load of the real problem.

Plates with holes are usually reinforced by shell segments or rings attached to the edge of the hole. It is assumed here that the reinforcement is connected to the plate symmetrically, so that the plate does not bend at the pre-buckling stage. The reinforced plate is in self equilibrium; the reinforced edge is free to move axially, so that the axial force resultant is zero.

The plate isolated from the reinforcement can be described as an annular plate subjected to unequal uniform radial compressions along the inner and the outer edges, as is shown in Figure 1.

The equation for small axisymmetric deflections which takes into account the effect of forces in the plane of the plate is [1] [11]

$$\frac{D}{h} \nabla^4 w = \frac{1}{r} \frac{d}{dr} \left(r \sigma_r \frac{dw}{dr} \right),$$
(1)

where h is the thickness of the plate, w is the deflection (the displacement in axial direction), σ_r is the radial stress, ∇^2 is the operator of Laplace which is defined in the axisymmetric case as

$$\nabla^2 = \frac{d^2}{dr^2} + \frac{1}{r} \frac{d}{dr}$$
(2)

and can be written also in the form

$$\nabla^2 = \frac{1}{r} \frac{d}{dr} \left(r \frac{d}{dr} \right).$$
(3)

D is the flexural rigidity given by

$$D = \frac{Eh^3}{12(1 - \nu^2)}.$$
(4)

* This paper forms part of the author's M.Sc. thesis under the supervision of Professor A. Libai.

L. G. Napolitano et al. (eds.), Astronautical Research 1971, 535–546. All Rights Reserved.
Copyright © 1973 by D. Reidel Publishing Company, Dordrecht-Holland.

E is the modulus of elasticity and ν is Poisson's ratio.

Equation (1) can be easily integrated once to give

$$\frac{D}{h} \frac{d}{dr} \nabla^2 w = \sigma_r \frac{dw}{dr} + \frac{K}{r}. \tag{5}$$

The membrane state of stress for an annulus was solved by Lamé [2]. The solution is

$$\sigma_r = A + \frac{B}{r^2}$$

$$\sigma_\theta = A - \frac{B}{r^2}. \tag{6}$$

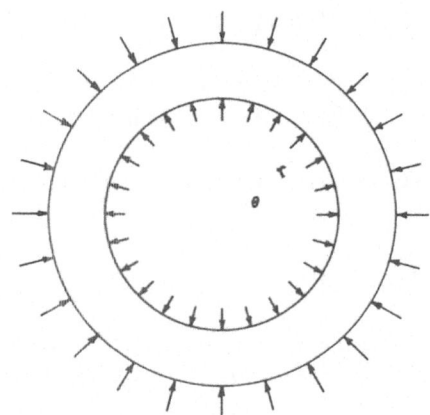

Fig. 1. Geometry of an annular plate and system of coordinates.

The constants A and B are such that the radial stress reaches the applied compressive stresses at the boundaries:

$$r = a: \quad \sigma_r = -P/h$$
$$r = b: \quad \sigma_r = -H/h. \tag{7}$$

The constants A and B are found to be

$$A = \frac{Hb^2 - Pa^2}{h(a^2 - b^2)},$$

$$B = \frac{a^2 b^2 (P - H)}{h(a^2 - b^2)} \tag{8}$$

After defining the ratio between the radii and the forces respectively as

$$\mu = b/a, \ \ 0 \le \mu \le 1 \tag{9}$$

$$\lambda = H/P \tag{10}$$

and introducing new constants, α and β

$$\alpha = -\frac{Ah}{D}$$

$$\beta = -\frac{Bh}{D} \tag{11}$$

which can be written in the form

$$\alpha = \frac{1 - \lambda\mu^2}{1 - \mu^2}\left(\frac{P}{D}\right)$$

$$\beta = -\frac{\mu^2(1 - \lambda)}{1 - \mu^2}\left(\frac{Pa^2}{D}\right) \tag{12}$$

All above equations lead to the equation for the deflection in the form

$$\frac{d^3w}{dr^3} + \frac{1}{r}\frac{d^2w}{dr^2} + \left(\alpha - \frac{1 - \beta}{r^2}\right)\frac{dw}{dr} = \frac{K}{r}. \tag{13}$$

This equation can be reduced into an equation of the second order only, using a new variable, the slope, which is

$$\varphi = \frac{dw}{dr}. \tag{14}$$

The equation for the slope is

$$\frac{d^2\varphi}{dr^2} + \frac{1}{r}\frac{d\varphi}{dr} + \left(\alpha - \frac{1 - \beta}{r^2}\right)\varphi = \frac{K}{r}. \tag{15}$$

Equation (15) is due to Meissner [1]. In order to simplify the equation Meissner has looked for a case in which the constant K would be zero. He has found that when the inner edge is free from stress, namely

$$\lambda = 0 \tag{16}$$

then K is zero and Equation (15) becomes homogeneous [3] [4] [11]

$$\frac{d^2\varphi}{dr^2} + \frac{1}{r}\frac{d\varphi}{dr} + \left(\alpha - \frac{1 - \beta}{r^2}\right)\varphi = 0. \tag{17}$$

The solution of Equation (17) is given by

$$\varphi = C_1 J_n(x) + C_2 Y_n(x), \tag{18}$$

where $J_n(x)$ and $Y_n(x)$ are Bessel functions of the first and second kinds, and of order n, with x as an argument. Both the order and the argument depend on the membrane state of stress by the constants α and β:

$$x = r\sqrt{\alpha} \tag{19}$$

$$n = \sqrt{1 - \beta}. \tag{20}$$

Two boundary conditions are needed, one for every edge. While the inner edge, in Meissner's case, is free, the radial moment is zero

$$r = b: \quad M_r = 0. \tag{21}$$

The radial moment can be expressed in terms of slope:

$$M_r = -D\left[\frac{d\varphi}{dr} + \frac{\nu}{r}\varphi\right]. \tag{22}$$

Introducing the solution (18) into Equation (21) leads to the boundary condition

$$C_1[x_2 J_n'(x_2) + \nu J_n(x_2)] + C_2[x_2 Y_n'(x_2) + \nu Y_n(x_2)] = 0, \tag{23}$$

where

$$J_n'(x) = \frac{dJ_n(x)}{dx} = \frac{n}{x}J_n(x) - J_{n+1}(x)$$

$$Y_n'(x) = \frac{dY_n(x)}{dx} = \frac{n}{x}Y_n(x) - Y_{n+1}(x) \tag{24}$$

$$x_2 = b\sqrt{\alpha}. \tag{25}$$

The second boundary condition refers to the outer edge:

$$x_1 = a\sqrt{\alpha}. \tag{26}$$

This condition may be zero radial moment again when the edge is simply supported, or may be zero slope when the edge is clamped. The corresponding equation for the first case is

$$C_1[x_1 J_n'(x_1) + \nu J_n(x_1)] + C_2[x_1 Y_n'(x_1) + \nu Y_n(x_1)] = 0 \tag{27}$$

and for the second case

$$C_1 J_n(x_1) + C_2 Y_n(x_1) = 0. \tag{28}$$

After eliminating C_1 and C_2 from Equations (23) and (27) or (23) and (28), the eigenvalue problem for each case is reduced to a transcendental equation, from which the critical load can be found. The critical load may be written in a dimensionless form

$$z = \frac{Pa^2}{D}. \tag{29}$$

It is difficult to find the critical load for a given annulus directly, because the order is yet unknown. The way to solve this problem is to choose several orders and to find to what problem they correspond. After plotting the curve of z as a function of μ, the critical load can be found graphically for any given μ.

Two graphs have been plotted by Meissner, one for the case of simply supported edge and one for clamped edge, but in both cases the inner edge is free from supports and forces.

The problem of the buckling load of an annulus has been considered after Meissner by other authors too, but for other loading conditions which have been found to be easier to solve. One of them is the case of equal uniform compressions along the inner and outer edges, namely

$$\lambda = 1. \tag{30}$$

In this case the state of stress is the same as in the case of a complete circular plate where $B=0$ and the order is well known

$$n = 1, \tag{31}$$

Olsson [5] and Schubert [6] have considered this case for several boundary conditions. An extension has been given by Yamaki [7] to include asymmetric buckling. The order is found to depend on the number of waves but not on the load.

Another loading condition is the one considered by Mansfield [8] in which the compression ratio is the inverse of the square of the radii ratio

$$\lambda = 1/\mu^2. \tag{32}$$

In this case

$$\alpha = 0 \tag{33}$$

and the equation for the slope is homogeneous in r. This case was considered by Mansfield for asymmetric buckling too. Here the exact solution is available in terms of elementary functions.

The problem of the buckling of an annulus compressed along the outer edge which is clamped and having the inner edge free was extended by Rózsa [9] and by Majumdar [10] to include asymmetric buckling, using iteration method and energy method respectively.

An extension to the studies of Meissner, Mansfield and Yamaki which includes an elastically restrained outer edge was done by Lizarev and Bareeva [11].

The problem of the stability of an annular plate submitted to distributed forces acting on the inner and outer edges in any given ratio has been considered by Buckens [12] using an approximate method.

In Buckens' case the boundary conditions are not very clear, and the results which are given for one value of radii ratio only are not very convenient to use. Buckens [13] extended his work later on for asymmetric buckling but without giving results.

The present work gives the exact solution for the axisymmetric buckling where the loading condition is of any given compression ratio, and the boundary conditions are of the usual types such as simple supports, free edge, clamped edge and movable edge. (A movable edge is an edge which can deflect freely but cannot rotate.)

2. Basic Equation and Method of Solution

It will first be shown now that taking $K=0$ in Equation (15) has a much wider use than Meissner's original assumption of a free edge.

Equation (5) can be written in the form

$$-Q_r = \sigma_r h\varphi + \frac{Kh}{r}, \tag{34}$$

where Q_r is the transverse shear force per unit length. The boundary condition for the inner edge which is free to move axially is

$$Q_r = -\sigma_r h\varphi. \tag{35}$$

Thus, it can be seen that K is zero for every loading condition, and the equation for the deflection is

$$\frac{D}{h}\frac{d}{dr}\nabla^2 w = \sigma_r\varphi. \tag{36}$$

This equation means that the axial force resultant is zero everywhere, and this is the common practical case of the reinforcement of a hole which is in self equilibrium axially.

TABLE I

Eight possibilities of different boundary conditions

Equation		Outer edge	Inner edge
(37)	$F_n(x_1)=F_n(x_2)$	Simply supported Free	Free Simply supported
(38)	$F_n(x_1)=G_n(x_2)$	Simply supported Free	Movable Clamped
(39)	$G_n(x_1)=F_n(x_2)$	Clamped Movable	Free Simply supported
(40)	$G_n(x_1)=G_n(x_2)$	Clamped Movable	Movable Clamped

where

$$F_n(x_i) = \frac{x_i J'_n(x_i) + \nu J_n(x_i)}{x_i Y'_n(x_i) + \nu Y_n(x_i)} \tag{41}$$

$$G_n(x_i) = \frac{J_n(x_i)}{Y_n(x_i)} \tag{42}$$

The solution for the slope is given by Equation (18). Applying this result to the various possible boundary conditions at hand, implicit eigenvalue problems are obtained. There are four possible eigenvalue equations due to eight cases of different boundary conditions as is shown in Table I.

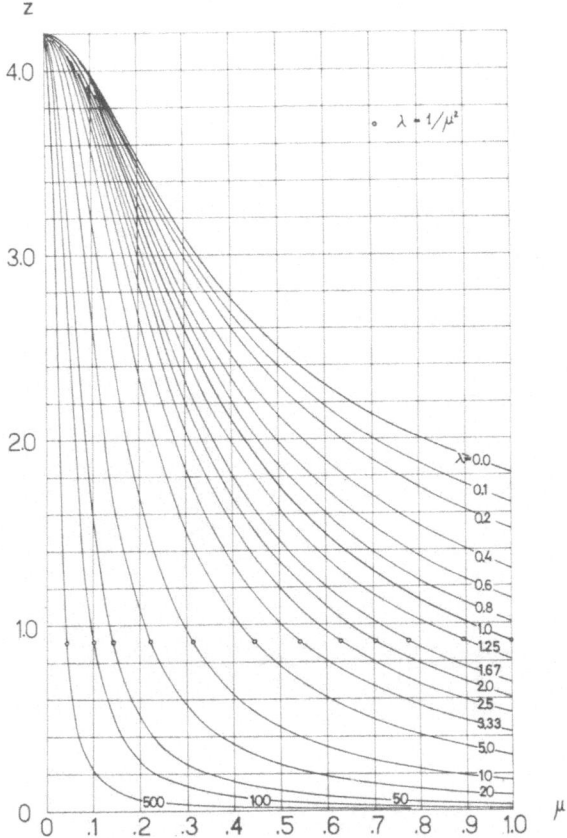

Fig. 2.   Critical load according to Equation (37):
$$F_n(x_1) = F_n(x_2) \quad \nu = 0.3.$$

An eigenvalue equation can also be obtained in a more general case, the case of elastically restrained edges, where there are torsional springs at the edges. In this case there is a linear connection between the radial moment and the slope. The boundary conditions are:

$$r = a: \quad M_r = k_1\varphi, \tag{43}$$

$$r = b: \quad M_r = -k_2\varphi, \tag{44}$$

where k_1 and k_2 are the rotational rigidities of the outer and inner springs, respectively. The change in sign between Equations (43) and (44) is due to the difference in definition of positive direction between M and φ. Introducing the solution (18) into the boundary conditions (43) and (44), using Equations (22), (24), (25) and (26), the eigenvalue equation for this case is found to be:

$$F_n(x_1, \nu_1) = F_n(x_2, \nu_2), \tag{45}$$

where

$$F_n(x_i, \nu_i) = \frac{x_i J_n'(x_i) + \nu_i J_n(x_i)}{x_i Y_n'(x_i) + \nu_1 Y_n(x_i)} \tag{46}$$

$$\nu_1 = \nu + \frac{k_1 a}{D} \tag{47}$$

$$\nu_2 = \nu - \frac{k_2 b}{D}. \tag{48}$$

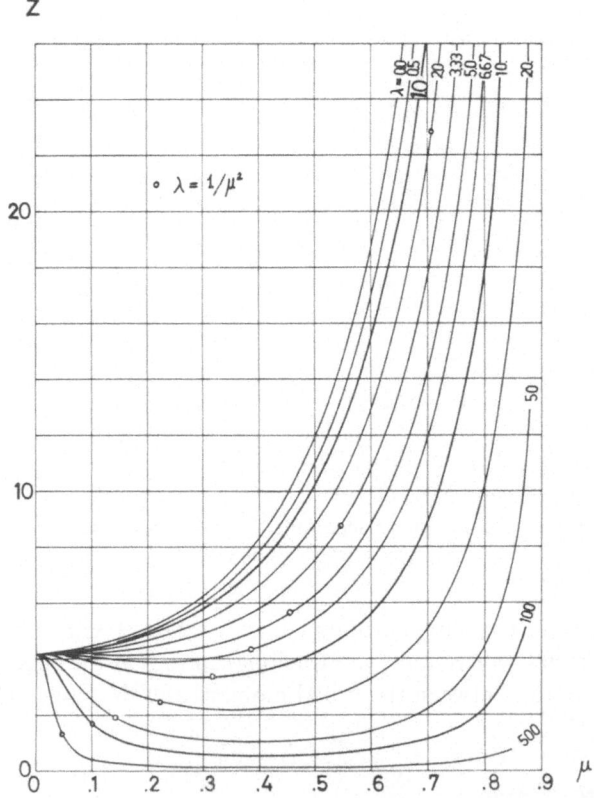

Fig. 3.　Critical load according to Equation (38):
$$F_n(x_1) = G_n(x_2) \quad \nu = 0.3.$$

It can be easily shown that when the rotational rigidity is zero, Equation (46) is reduced into Equation (41), and when the rotational rigidity is infinity, Equation (46) is reduced into Equation (42), so that Equations (37)–(40) are particular cases of Equation (45).

It can be shown from Equations (12), (20), (25) and (26) that the pair of roots x_1

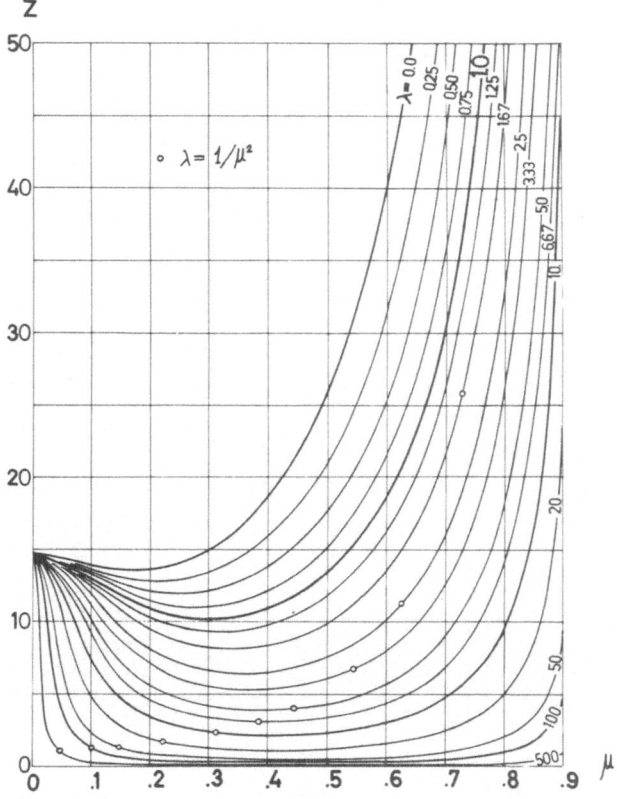

Fig. 4.   Critical load according to Equation (39):
$$G_n(x_1) = F_n(x_2) \quad \nu = 0.3.$$

and x_2 and the order n determine the geometry of the plate and the loading condition by the following equations:

$$\mu = x_2/x_1 \tag{49}$$

$$\lambda = \left(1 + \frac{1 - n^2}{x_2^2}\right) \Big/ \left(1 + \frac{1 - n^2}{x_1^2}\right) \tag{50}$$

The method of solving the problem here is the same as in Meissner's case: to choose several orders and to find to what problems they correspond. It can be seen from Equations (16) and (50) that Meissner's case is a particular case in which

$$x_2^2 = n^2 - 1. \tag{51}$$

For any given values of μ, λ, k_1 and k_2, there is an infinite number of pairs x_1 and x_2 that solve the eigenvalue equation, but, from the stability point of view, only one pair is important, the one that gives the lowest value of critical load which is given by

$$z = x_1^2 + 1 - n^2. \tag{52}$$

It should be noted here that the order n and the argument x may be either real or imaginary numbers.

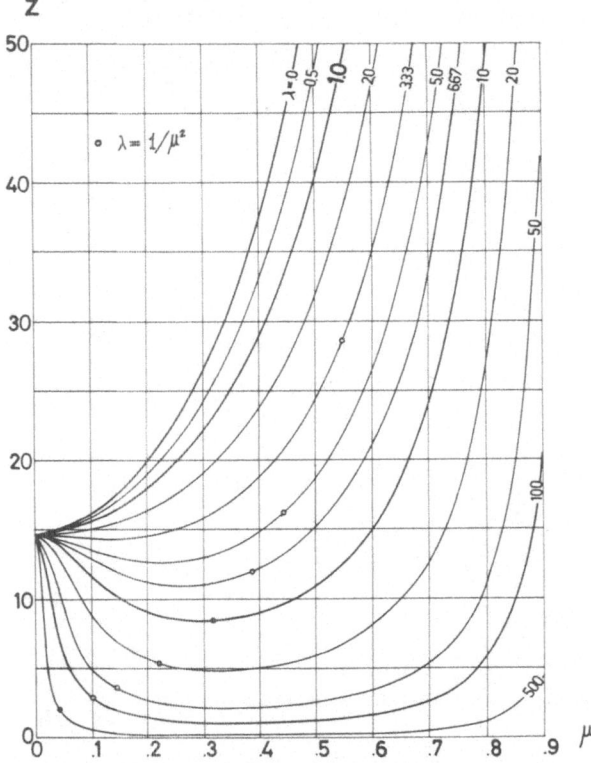

Fig. 5.   Critical load according to Equation (40):
$$G_n(x_1) = G_n(x_2).$$

Results are presented in the form of graphs. For any eigenvalue equation of Table I, Equations (37)–(40), z is plotted as a function of μ while λ is used as a parameter. (See Figures 2–5).

3. Conclusion

Using the graphs presented here, the critical load can be easily found for any given ratio of radii and forces. From these graphs it appears that the cases considered by Meissner, Yamaki and Mansfield are particular cases in which the loading conditions are: $\lambda = 0$, $\lambda = 1$ and $\lambda = 1/\mu^2$, respectively. These cases have no special importance from the engineering point of view, but it was believed that their mathematical complexity was less than that of the general case. The present work shows that adding any compression along the inner edge does not increase the mathematical complexity of the problem discussed by Meissner in which there is no inner force.

Starting with $\mu = 0$, all loading conditions give the solution of a complete circular plate with the proper boundary condition at the outer edge. By increasing μ, the loading condition becomes more and more important. As μ reaches the value 1, the critical load goes to infinity, except in the case where one edge is simply supported and the other edge is free, Equation (37). In this case the critical load reaches a finite value depends on the loading condition, that is

$$z = \frac{2}{1 + \lambda} (1 - \nu^2). \tag{53}$$

This equation is in full agreement with the assumption that the narrow annulus rotates as a ring without bending. The connection between the moment and the rotation of the 'ring' with a rectangular cross-section is [14]:

$$\varphi = \frac{12am}{Eh^3(1 - \mu)}. \tag{54}$$

Putting the moment as

$$m = \frac{1 + \lambda}{2} (1 - \mu)Pa\varphi \tag{55}$$

and using Equations (29), (4) leads to Equation (53).

The cases where the narrow plate is not fully free to rotate can be described as a strip subjected to two forces which have the magnitude of $[(1 + \lambda)/2]P$.

This assumption leads to the critical load

$$z(1 - \mu)^2 = \frac{2}{1 + \lambda} \pi^2 \tag{56}$$

in the case of Equation (40), and to

$$z(1 - \mu)^2 = \frac{2}{1 + \lambda} \left(\frac{\pi}{2}\right)^2 \tag{57}$$

in the case of Equations (38)–(39).

Acknowledgements

The author wishes to express his sincere appreciation to Professor A. Libai of the Department of Aeronautical Engineering, Technion – Israel Institute of Technology, for his helpful supervision, and Professor J. Singer for his helpful advice. Thanks are also due to Mr G. Meir, third year student, for his assistance with the programming.

References

[1] Meissner, E., *Schweizerische Bauzeitung*, **101** (1933), 87.
[2] Timoshenko, S. P. and Goodier, J. N., *Theory of Elasticity*, McGraw-Hill – Kōgakusha, Tokyo, 1970, p. 68.
[3] Timoshenko, S. P. and Gere, J. M., *Theory of Elastic Stability*, McGraw-Hill, New York, 1961, p. 389.
[4] Biezeno, C. B. and Grammel, R., *Engineering Dynamics*, Blackie & Sons, London, 1956, Vol. II, p. 474.
[5] Olsson, G. R., *Ingenieur Archiv* **8** (1937), 449.
[6] Schubert, A., *Z. Angew. Math. Mech.* **25/25** (1947), 123.
[7] Yamaki, N., *J. Appl. Mech.* **25** (1958), 267.
[8] Mansfield, E. M., *Quart. J. Mech. Appl. Math.* **13** (1960), 16.
[9] Rózsa, M., *Acta Techn. Acad. Sci. Hungaricae* **53** (1966), 359.
[10] Majumdar, S., 'Buckling of Thin Annular Plates Due to Radial Compressive Loading', Caltech. Engineer's thesis, California Institute of Technology, Pasadena, 1968.
[11] Lizarev, A. D. and Bareeva, G. N., *Soviet Eng. J.* **5** (1965), 401.
[12] Buckens, F., 'Relations caractéristiques entre les valeurs critiques des forces radiales agissant sur les bords intérieur et extérieur d'une plaque annulaire', *Bull. Soc. Francaise des Méc.* juin 1955.
[13] Buckens, F., *Ann. Soc. Sci. Brux.* **74** (1960), 120.
[14] Timoshenko, S. P., *Strength of Materials*, D. Van Nostrand, New Jersey, 1956, Vol. II, p. 138.

A MODULATION TECHNIQUE FOR MEASURING SMALL DISTURBANCES IN THE UPSTREAM FLOW FIELD OF A SHARP LEADING EDGE IN A RAREFIED HYPERSONIC FLOW

FERNAND DE GEYTER

Von Karman Institute for Fluid Dynamics, Rhode-Saint-Genese, Belgium

Abstract. The electron beam fluorescence probe was employed to measure the density disturbance in the vicinity of a sharp wedge in a hypersonic flow. A modulation technique was developed whereby the quantity of interest $\Delta\rho = \rho_{\text{disturbed}} - \rho_{\text{free stream}}$ is measured directly with an accuracy comparable to that obtained with an absolute density measurement. Details of the disturbance field are described quantitatively. As an example of the sensitivity of the method, the marked influence of the leading-edge bluntness has been conclusively demonstrated.

Nomenclature

SYMBOLS

M	Mach number
S	speed ratio
P	pressure
T	temperature
x	length coordinate parallel to free stream velocity
y	length coordinate perpendicular to free stream velocity
λ	mean free path
ρ	mass density
τ	leading edge thickness
ω	wedge angle
φ	beam diameter

SUBSCRIPTS

L.E.	leading edge
0	stagnation conditions
∞	free stream conditions
w	conditions of the body surface
bf	collision between a molecule reflected from the body and a free-stream molecule.

1. Introduction

During the past 15 yr considerable effort has been directed towards understanding the flow over a semi-infinite plate with sharp leading edge in a hypersonic flow (e.g., [1–5, 7, 9, 10]). The full spectrum of flow regimes may be identified, ranging from near-free-molecular to continuum (Figure 1). Although the geometrical arrangement is very simple, the wide variety of flow conditions presented makes the problem of considerable interest.

The purpose of the present study is to provide experimental data for the upstream flowfield and the early stages of the kinetic region. Conclusions will then be drawn

L. G. Napolitano et al. (eds.), Astronautical Research 1971, 547–556. All Rights Reserved.

regarding the manner in which such flows develop and the appropriate scaling parameters to be employed.

A novel measurement technique has been developed whereby a detailed map of the
density profile close to the leading-edge is obtained. The motivation for this particular
study is two-fold:

(a) Theoretical calculations of the flow properties above the plate downstream from
the leading edge require the upstream conditions as a starting point for the numerical
computations [4, 5, 14].

(b) Considerable scatter is evident in the few available experimental results [6–10].
This is certainly due to the fact that density measurements in the upstream region of
the flow field are difficult to perform accurately because the disturbance is very weak

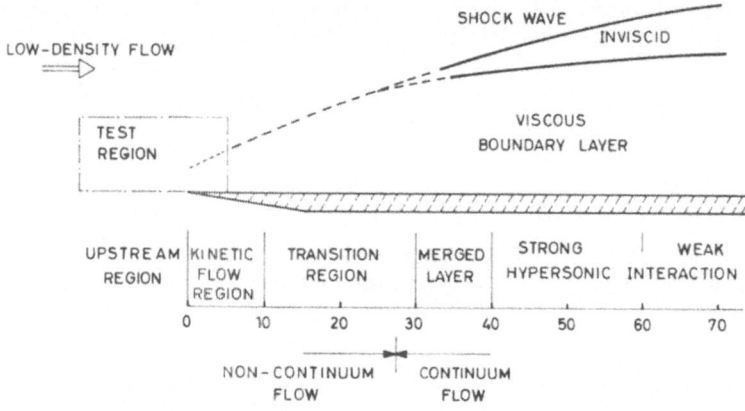

Fig. 1.   The semi-infinite flat plate problem in rarefied hypersonic flow.

(of order of 10%) resulting in a large relative error. It is also clear that the specific
shape of the leading edge has a marked influence on the flowfield both upstream and
perhaps several mean-free-paths downstream of the leading edge. This fact has not
always been sufficiently appreciated in earlier work and comparisons are thus not
always valid.

The purpose of this paper is to describe the philosophy behind the experimental
work and to present preliminary results.

2. Principle of the Measurement

The standard instrument to measure density in a rarefied flow is the electron-beam
fluorescence probe [11, 12]. In brief, the principle of operation is as follows: a narrow
beam ($\varnothing \approx 1$ mm) of high-energy (≈ 20 kV) electrons is passed through the flow region
of interest. A small fraction of the gas molecules is struck by the electrons. Each molecule is raised to an excited ionic state from which it relaxes to a lower state, emitting
light. The light intensity from a region of gas so excited is proportional to the local

gas density. The electron-beam-probe thus provides a local absolute measurement without disturbing the flow.

In the present problem, the basic quantity of interest at each point in the flow field relative to the model is

$$\frac{\Delta\rho}{\rho} = \frac{\rho_{\text{disturbed}} - \rho_{\text{free stream}}}{\rho_{\text{free stream}}} = \frac{\rho_{\text{with model}} - \rho_{\text{without model}}}{\rho_{\text{without model}}}$$

In all earlier experimental work [6–9], this quantity has been obtained by making two steady state measurements yielding individually $\rho_{\text{free stream}}$ and $\rho_{\text{with model}}$. Because of instabilities (low frequency drifts) and noise problems inherent in the operation of an electron-beam-probe, these two measurements each have a limited accuracy ($\sim 3\%$).

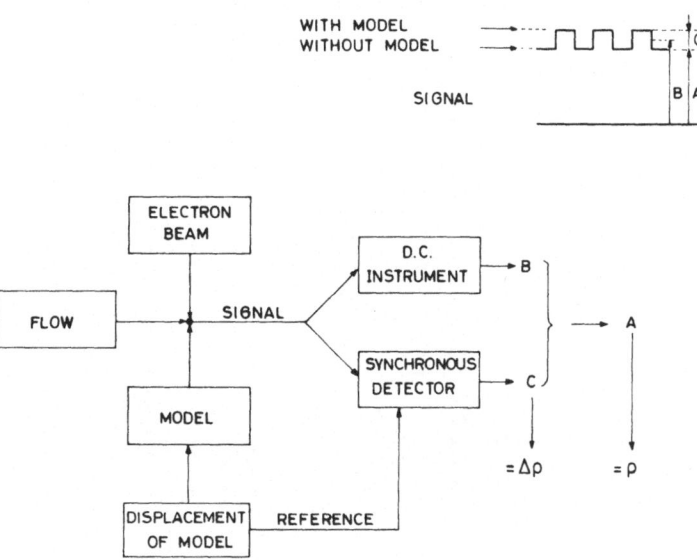

Fig. 2.   Principle of the modulation technique for density disturbance measurements.

As $\rho_{\text{disturbed}}$ and $\rho_{\text{free stream}}$ are only slightly different, the relative error in $\Delta\rho$ becomes so large as to mask any detailed flow variation.

A new measurement technique has been developed to overcome this basic limitation. The objectives of this method are, therefore,

(a) To bring out directly the magnitude of the disturbance $\Delta\rho$, with an accuracy comparable to that of $\rho_{\text{free stream}}$, and

(b) To obtain $\Delta\rho$ and $\rho_{\text{free stream}}$ simultaneously.

The principle of the measurement technique is illustrated in Figure 2. If the model is brought in and out of the flow periodically the disturbance is 'modulated'; i.e., the output signal has an ac component whose amplitude is directly proportional to the quantity $\Delta\rho$. At the same time the dc component is also recorded. From these two quantities $\rho_{\text{free stream}}$ is obtained. To measure the amplitude of the ac signal a lock-in amplifier with its high noise-rejection capability is used. The reference signal for the lock-in amplifier has the same frequency and is in phase with the displacement of the model. As $\Delta\rho$ and $\rho_{\text{free stream}}$ are obtained simultaneously and are both proportional to the instantaneous value of the beam current, the ratio $\Delta\rho/\rho$ is independent of beam current. This procedure eliminates one of the main difficulties encountered in the usual method: namely: the errors in ρ caused by an unstable beam.

3. Experimental Arrangement

To perform a systematic analysis of the influence of the leading edge geometry on the development of the flow, it was felt that a symmetrical geometry was appropriate. Therefore, the models used in the present study are wedges with various leading edge

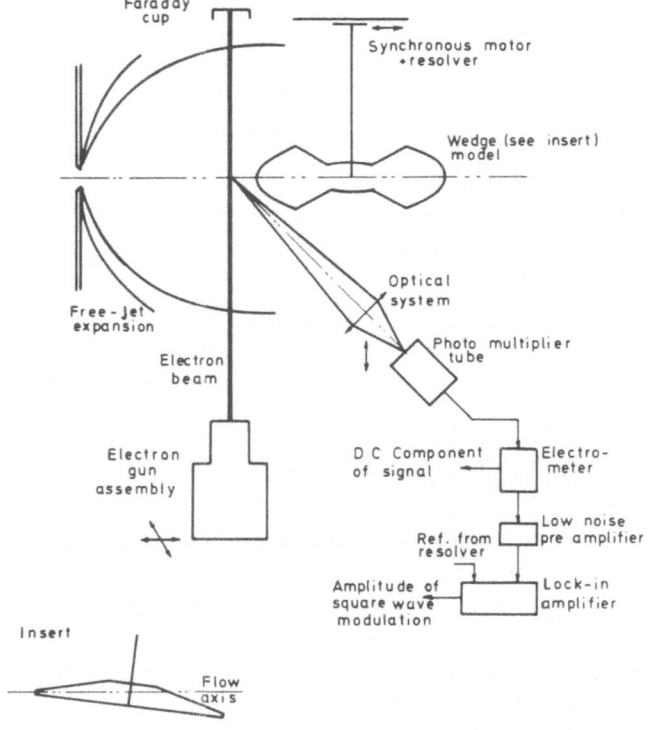

Fig. 3. Schematic for density disturbance measurements in leading edge region.

thicknesses and wedge angles. By extrapolating the results to zero wedge angle and zero leading edge thickness, one can arrive at the flowfield around a flat plate with an infinitely sharp leading edge.

The actual experimental set-up is shown in Figure 3. All experiments are performed in the low-density wind tunnel of the Von Karman Institute [13]. A free jet expansion is used to produce a hypersonic low density flow. The range of the relevant parameters describing the flow conditions, model geometry and electron-beam-probe are shown in Table I.

TABLE I

Characteristics of experimental study

Flow properties	
GAS = air	Maximum axial variation of free-stream flow properties
T_0 = room temperature	throughout the test region with respect to the conditions
P_0 = 5–25 Torr	at the leading edge
M_∞^* = 6–11	
λ_∞^* = 0.5 – 3.0 mm	
	M_∞: 5%
	λ_∞ : 15%
	ρ_∞ : 25%
Model characteristics	
$T_w \approx T_0$	
$\omega = 5°–20°$	
$\tau_{L.E.}$ = 0.1–1 mm	
Electron-beam-probe characteristics	
Beam current = 200 μA	
Beam energy = 17.5 ke V	
Beam diameter ≈ 1 mm	
Resolution of optics ≈ 0.5 mm³	

* Evaluated at leading edge.

The models are four-sector discs with a wedge-shaped cross-section having a diameter of approximately 30 cm. The discs are rotated by a synchronous motor while a resolver provides a reference signal in phase with the spinning disc. During one-quarter of a revolution the leading edge is near the measurement point (the location of the electron-beam), but during the next quarter-turn the edge is effectively removed to infinity. Thus it is clear that the intensity of fluorescence will vary periodically corresponding to a periodic change in density and that the amplitude will be directly proportional to $\Delta\rho$.

4. Results

Figures 4 and 5 illustrate two typical examples of the type of information obtained with the experimental arrangement described above. A map of the density variation ahead and above the wedge is plotted. Although the flow is hypersonic, the presence of the wedge is felt at least 10 free-stream mean-free-paths ahead of the body. Note also the formation of a shock wave (maximum in the density profile) in the down-stream part of the flowfield (Figure 4).

In Figure 5, λ_∞ is five times larger than in Figure 4. The disturbance at the same physical distance from the leading edge is therefore more pronounced. Downstream of the leading edge one observes a significant increase in density within one mean-free-path of the surface. From these and other similar runs a qualitative picture of the density variation around a wedge is obtained (Figure 6). Distances are normalized to λ_∞.

In Figure 7 radial profiles of the raw experimental data are shown. The disturbance maps given in the previous figures are taken from radial scannings similar to those shown here. The measured values of $\Delta\rho$ are indicated. The profiles are all symmetric with respect to the centerline of the flow with a single maximum on the axis. This

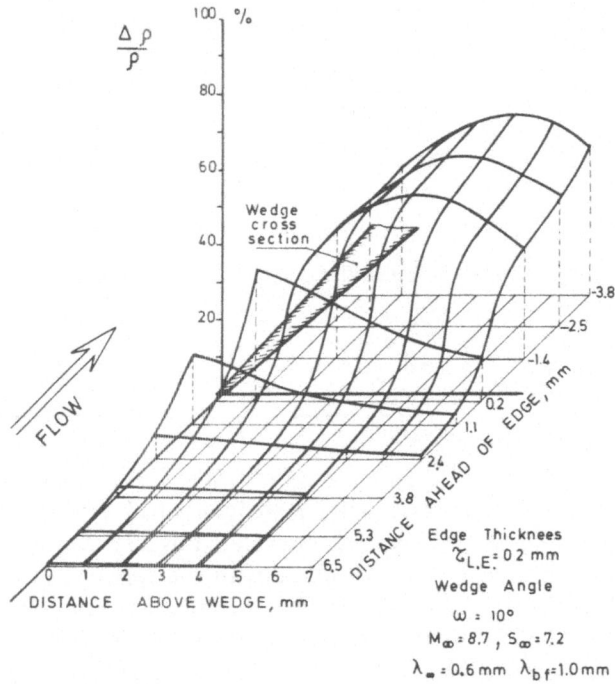

Fig. 4. Density disturbances in leading edge region.

indicates that the flow behavior is dominated by collisions everywhere. Scatter in the data-points remains small even for distances as far as 5 λ_∞ from the leading edge where $\Delta\rho/\rho$ is of the order of 2%.

Figure 8 indicates the on-axis variation of the upstream influence for two leading-edge thicknesses and a variety of flow conditions ($\lambda_0 = 0.5 \rightarrow 3.0$ mm). From these figures one can conclude that λ_{bf} (mean-free-path for a molecule reflected from the body and colliding with a free stream molecule) is a more appropriate scaling length for the density variation along the axis than λ_∞ because the slope of the on-axis variation is closer to the slope of $\exp(-x/\lambda_{bf})$ than to $\exp(-x/\lambda_\infty)$.

Finally in Figure 9 the influence of the parameter $\lambda_\infty/\tau_{L.E.}$ on the on-axis variation of the upstream influence is indicated. The extrapolation of $\lambda_\infty/\tau_{\mathrm{L.E.}}$ to infinity gives the value of $\Delta\rho/\rho$ for an infinitely sharp leading edge. In that limiting case, the disturbance is caused only by molecules reflected upstream from the top and bottom surface of the wedge. It should be noted that the data points were obtained by varying both λ_∞ and $\tau_{\mathrm{L.E.}}$. From these figures it can be concluded that $\lambda_\infty/\tau_{\mathrm{L.E.}}$ should be greater than at least 15 before the influence of the leading edge bluntness becomes negligible, i.e., less than $\sim5\%$.

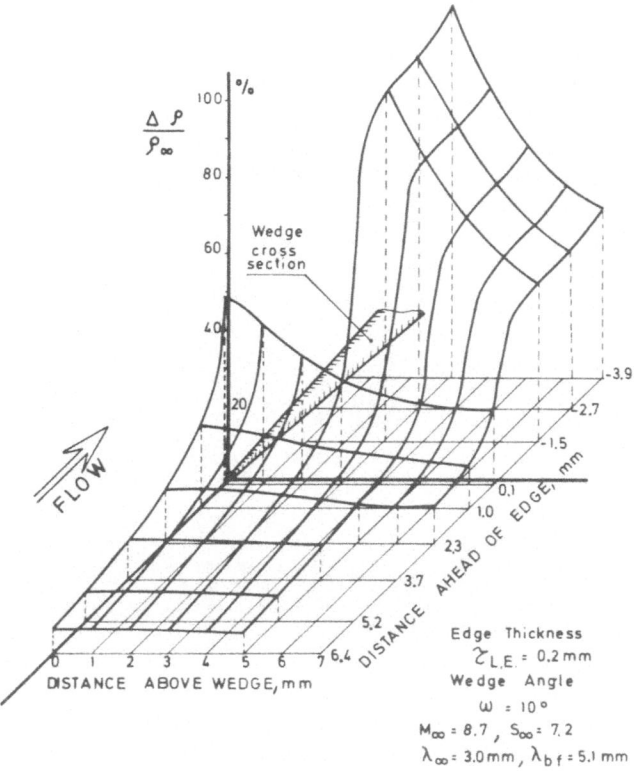

Fig. 5.   Density disturbance in leading edge region.

5. Conclusions

The 'modulation' technique has proved to be a valid method for obtaining a detailed picture of the density field around a wedge with a finite leading-edge thickness. Density disturbances as small as 2% can easily be measured with a good accuracy. Thus the details of the flow variation can be evaluated quantitatively. As an example, the marked influence of the leading-edge bluntness has been conclusively demonstrated.

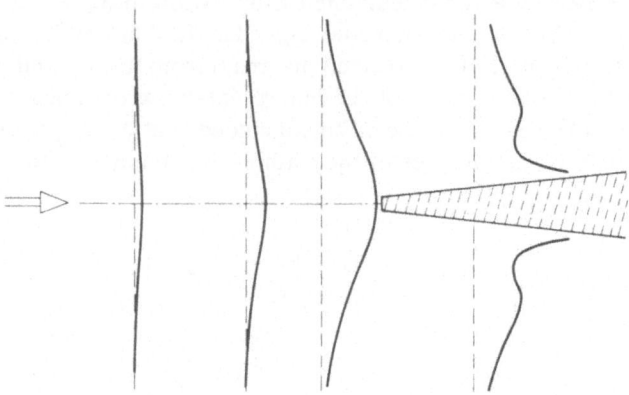

Fig. 6.   Qualitative density variation around a wedge.

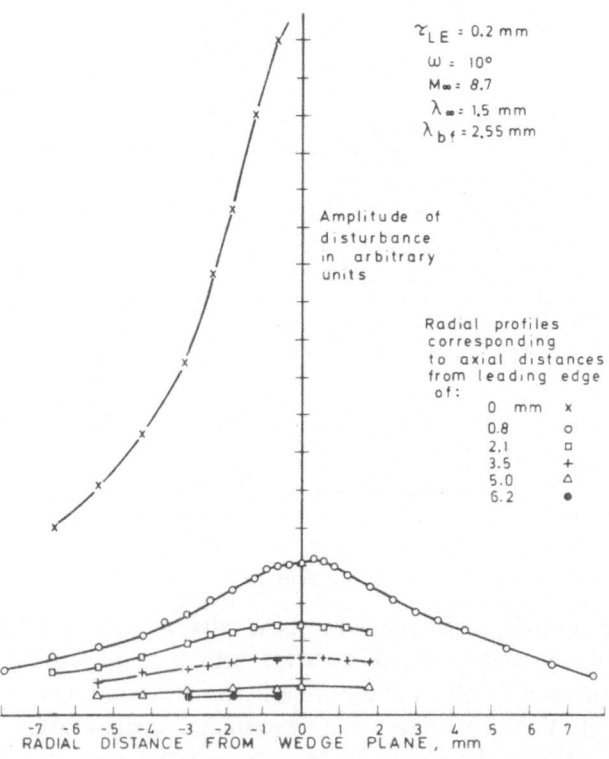

Fig. 7.   Radial profiles of experimental data proportional to density disturbance $\Delta\rho$.

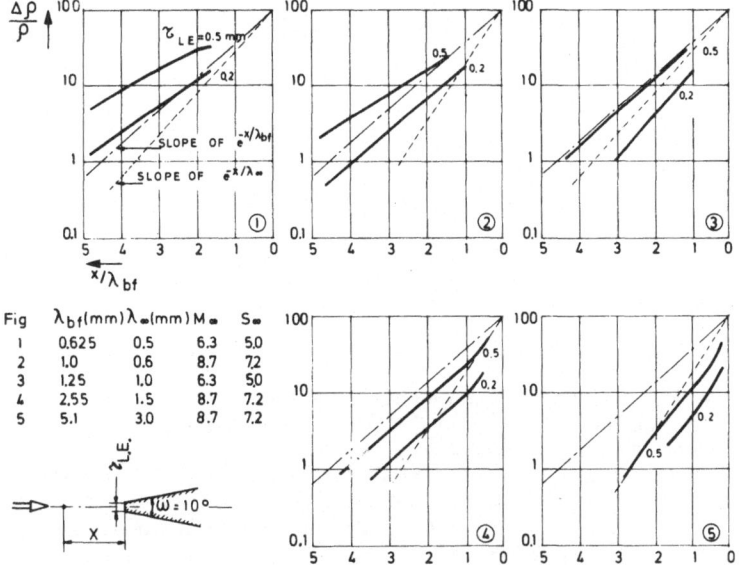

Fig. 8.   On-axis variation of upstream influence.

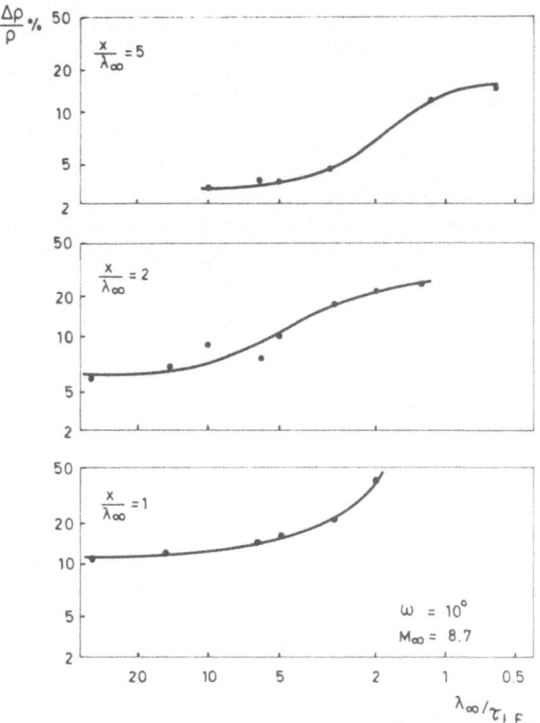

Fig. 9.   On-axis variation of upstream influence.

Acknowledgements

This research has been sponsored in part by the Air Force Office of Scientific Research through the European Office of Aerospace Research, O.A.R. United States Air Force under grant number EOOAR-70-0081.

A fellowship for specialization has been provided by I.W.O.N.L. (Belgium).

I would like to thank Dr J. J. Smolderen, Dr J. F. Wendt and M. A. Reynolds for their suggestions and assistance in the experimental work.

References

[1] Hayes, W. D. and Probstein, F. R., *Hypersonic Flow Theory*, Academic Press, New York, p. 341.
[2] Charwat, A. F., Rand. Corp. Memo RM-2553-PR 1963.
[3] Moulic, F. S. and Maslach, G. J., in C. L. Brundin (ed.), *Rarefied Gas Dynamics*, Vol. II, Academic Press, New York, 1966, p. 971.
[4] Huang, A. B. and Hwang, P. F., XIX IAF Congress, October 1968. Paper RE 63.
[5] Vogenitz, F. W., Broadwell, J. E. and Bird, G. A., *AIAA, J.* 8 (1970), 504.
[6] Harbour, P. J. and Lewis, J. H., in C. L. Brundin (ed.), *Rarefied Gas Dynamics*, Vol. II, Academic Press, New York, 1966, p. 1035.
[7] Joss, W. W. and Bogdonoff, S. M., in L. Trilling and H. Y. Wachman (eds.), *Rarefied Gas Dynamics*, Academic Press, New York, 1969, p. 403. – 'An Experimental Study of the Effects of Small Leading Edge Thickness on the Development of the Hypersonic Rarefied Flow over a Flat Plate', Presented at the 7th Int. Symp. on Rarefied Gas Dynamics at PISA, June 1970.
[8] Hickman, R. S., in L. Trilling and H. Y. Wachman (eds.), *Rarefied Gas Dynamics*, Vol. I, Academic Press, New York, 1969, p. 503. – *Phys. Fluids* 13 (1970), 3051.
[9] Lillicrap, D. C. and Berry, C. J., *Phys. Fluids* 13 (1970), 1146.
[10] Becker, M. and Boylan, D. E., in C. L. Brundin (ed.), *Rarefied Gas Dynamics*, Vol. II, Academic Press, New York, 1966, p. 993.
[11] Muntz, E. P., AGARDograph 132, 1968.
[12] De Geyter, F., V.K.I. PR 68-219, 1968.
[13] Smolderen, J. J. and Wendt, J. F., 'The Low Density Wind Tunnel and Associated Research Program at the von Karman Institute for Fluid Dynamics' Presented at Symposium 'Surface Vacuum Space' Liège, April 1971, to be published in *Vakuum Technik*.
[14] Garvine, R. W., in L. Trilling and H. Y. Wachman (eds.), *Rarefied Gas Dynamics*, Vol. I, Academic Press, 1969, p. 509.

DIAGNOSTICS OF AN ARGON FREE JET
EXPANDED FROM A HIGH PRESSURE
INDUCTIVE ARC SOURCE

W. F. PAYNE

Institute for Aerospace Studies, University of Toronto, Canada

Abstract. An experimental study has been made of the free jet flow field produced with argon heated by means of an inductive arc and expanded into a low density wind tunnel. The arc operates from 0.4 to 1 atm and thus the working gas is in thermodynamic equilibrium before expansion. The mass flow of gas was varied from 0.2 to 0.45 gm s^{-1}.

The following diagnostic methods were used:

– a total flow calorimeter, which measured the heat content of the exit gases;
– an electron beam densitometer which was modified to operate in plasma flows. This instrument was used to obtain the local density (ρ) at points in the free jet;
– a local mass flux probe, which was used to obtain a measure of the velocity times the density (ρu);
– impact probes, which may be shown to measure a quantity proportional to the density times the square of the velocity in these hypersonic flows (ρu^2).

The experimental density field was found to be described quite accurately by the same expressions used to calculate properties in cold free jets. Influences from background gas in the tunnel were observed in the overexpanded portion of the jet.

Consideration of the values for the density, mass flux and impact pressure allow one to calculate the velocity on the centerline of the free jet. As this velocity represents the terminal velocity attained when essentially all the random thermal motion of the gas atoms is converted to kinetic energy, the effective source stagnation temperature may also be calculated.

The flow is characterized by velocities from 1.8 to 2.4 km s^{-1}, neutral densities from 9×10^{13} to 3×10^{15}/cm^3 and ionization fractions of about 0.001. Effective centerline stagnation temperatures were measured to 5500 K.

1. Introduction

In the last decade it has become important to investigate the flight regime encountered at sub-orbital altitudes and velocities. In this regime some ionization and dissociation phenomena may occur. A number of aerodynamic facilities have been constructed to simulate aspects of flight in such rarefied high speed flows. The common method used in the past has been to pass the working fluid through a DC arc in such a manner that the gas is heated [1–5]. The hot gas is then expanded in some fashion and the resulting flow field used for aerodynamic testing.

Because of the electrodeless nature of the inductive discharge, no contamination can enter the flow from electrodes. The heat addition, while not uniform, is axially symmetric and is more extended in volume than that for a DC arc discharge. The discharge is also more constant in time. For these reasons, it was soon considered as a heater for supersonic wind tunnels [6–9].

The aim of the present study was to characterize the flow field produced by such a

L. G. Napolitano et al. (eds.), Astronautical Research 1971, 557–576. All Rights Reserved.

high enthalpy source. To this end, the density, impact pressure, and mass flux profiles have been measured.

2. Experimental Apparatus

2.1. THE INDUCTIVE ARC SOURCE

The inductively heated stagnation chamber is shown in cross section in Figure 1. It is a commercially made torch assembly (TAFA Model 56), that basically consists of a plastic body designed to hold a quartz discharge tube, a four turn exciting coil, the gas injection assembly and a mount for a copper orifice in such a manner that all the components may be water cooled.

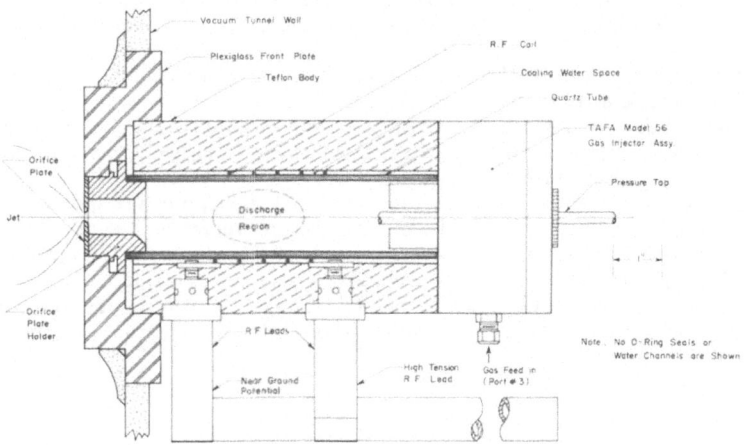

Fig. 1.   Modified inductive torch source (to scale).

The torch was driven by a radio frequency generator manufactured by Philips. It had a nominal 12 kW output and for these experiments it was run at 4 MHz. Meters were installed to measure the DC current and voltage supplied to the oscillator section.

Table I gives the range of operating points that were investigated by various flow probes. It was found that as the flow rate was lowered below 0.2 gm s^{-1}, the vortex stabilization in the discharge tube broke down rather easily and only a very narrow region of power inputs allowed stable operation of the arc. As the mass flow was increased, the arc became stable over a wider range of power input.

In the next section, proof is given that the stagnation pressure ratio hot/cold could be uniquely related to the total enthalpy in the flowing gas. This became the parameter used to compare the measurements of density, mass flow and impact pressure that were taken at different times.

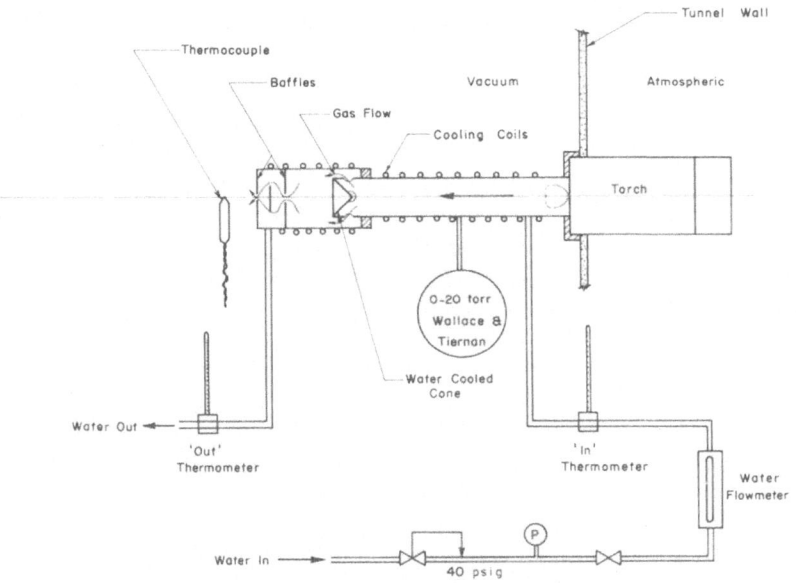

Fig. 2. Total flow calorimeter.

2.2. TOTAL CALORIMETER

One commonly made measurement in heated flow diagnostics is that of stagnation enthalpy.

The apparatus constructed to do this for this investigation is shown in Figure 2. It consists of a water cooled 2 in. copper tube which was blocked by a cooled copper cone. The hot gases were deflected by this cone out through a ring of holes and into a cooled 3 in. copper tube. The gases exited around a series of 3 baffles into the test section. The entire calorimeter was bolted onto the plexiglas orifice plate retainer in such a manner that the flow gases could not escape except by the rear exit. The temperature of the exit gas was measured by a thermocouple junction fixed in the exhaust.

The measurements were made as follows: The torch was operated in the normal manner by establishing the cold mass flow and allowing the source pressure to equilibrate. The arc was then struck and the power was varied over the range given in Table I.

TABLE I
Torch operating range

P_{0c}	P_{0h}	$\dot{m}$	DC power	P_{ts}
T	T	gm s^{-1}	kW	mT
100	267–273	0.194	2.9	17.32
125	297–378	0.240	1.4–3.9	20.61
150	335–478	0.288	1.4–5.1	23.76
175	367–586	0.335	1.4–6.5	27.80
200	475–694	0.380	2.2–8.2	30.65
225	511–777	0.430	2.2–9.9	33.94

For each power level, the DC oscillator power and stagnation pressure ratio hot to cold were calculated. After the water temperature had equilibrated at the exit from the calorimeter, the water flow rate and temperature difference (typically 20 °C) were recorded. The thermocouple output was converted to temperature and recorded as well as the internal pressure in the calorimeter.

The pressure downstream of the torch orifice was never more than 4 T. This means that the flow through the orifice was sonic at all times. Therefore the stagnation conditions would be identical with and without the calorimeter in place.

The total heat given up to the calorimeter was calculated from the temperature rise and flow-rate of calorimeter cooling water.

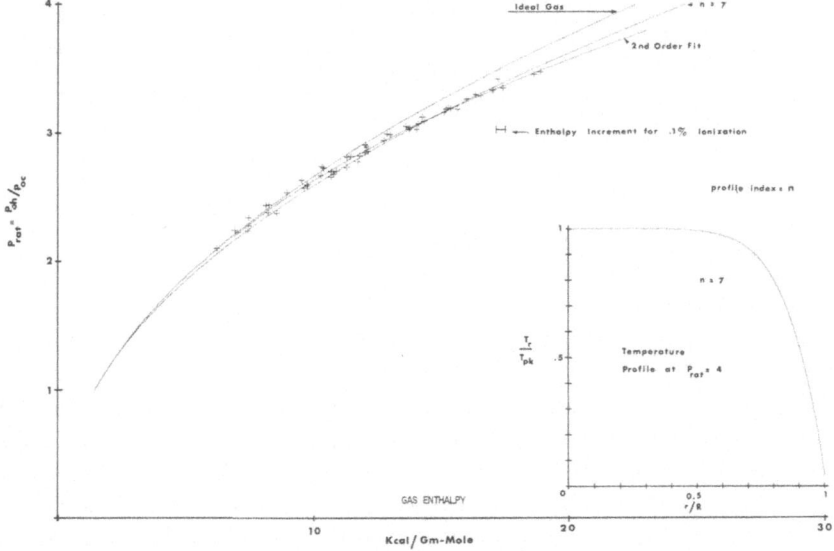

Fig. 3.   Pressure rise across choked orifice with non-uniform temperature profile-experiment.

The total heat carried by the gas was divided by the total mass flow of gas through the orifice. To this value was added the residual enthalpy of the cool exit gas, which was calculated from the thermocouple temperature (never higher than 70 °C) and the value of $C_p = 5/2$ R.

The resulting plot of pressure ratio versus total flow enthalpy (kcal gm^{-1} mole) is presented in Figure 3. The enthalpy values correspond to temperatures sufficiently low that negligible ionization would exist, if the flow were in equilibrium. Langmuir probe measurements [10] indicate an ionization fraction of about 10^{-3}. This would correspond then to about 0.36 kcal gm^{-1} mole additional heat given up to the calorimeter upon recombination. The magnitude of this additional heat is shown on the figure to scale. It is too small to account for the drop of the points (or the least squares fitted curve) below the calorically perfect equation.

The difference between the experimental points and the ideal case is postulated to be caused by a non-uniform temperature distribution. The details of calculations based on assumed temperature profiles in terms of a single shape parameter, n, are given in Appendix A. The pressure-ratio-enthalpy curve for $n = 7$ is shown on Figure 3, along with its associated temperature profile. It would appear that the explanation on the basis of non-uniform temperature is reasonable.

2.3. DENSITY MEASUREMENTS

2.3.1. *Theoretical Considerations for Density Measurement*

One fundamental property of a flow field is the local density. It is therefore an important parameter to attempt to measure in an aerodynamic testing facility. A method used in the past has been to observe the density dependent signal produced by an electron beam probe. This method does not introduce any perturbing objects into the flow field, and thus has obvious advantages.

In neutral (cold) gases the fluorescence produced by the electron beam has been so used to measure the density (Rothe [11]; Gadamer [12]). Also the same method has been used by Nakamura [13] and by Fraser [14] to get an indication of the density in arc heated argon flows. However, certain difficulties with the optical method in cold argon have been described by McMichael [15], where he has shown that the argon fluorescence signal contains a component that is velocity dependent.

It was thought that a better method and one that would not be sensitive to the already existing plasma would be the observation of elastically scattered relatively

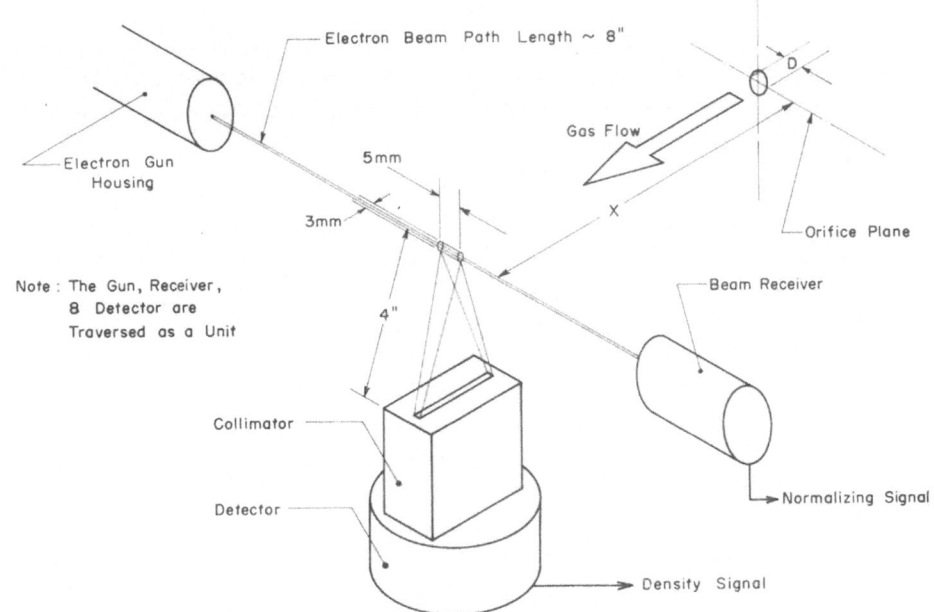

Fig. 4. Scattering experiment geometry.

high energy electrons recoiling off the nucleus of the flow particles (Camac [16]). This method was chosen over an X-ray method (Russell [17], Ziegler *et al.* [18]), because the theoretical photon yield was about 50 times higher. McMichael [15] presents a careful experimental comparison of the optical and elastic scattering methods in room temperature argon free jets. Figure 4 is a sketch of the geometry of the equipment used in this study. The scattered current density due to Rutherford (elastic) scattering of electrons off a nucleus of atomic number Z may be expressed by (Camac [16])

$$I_s = 7.8 \times 10^{-16} I_0 n x Z^2 \left[\frac{1 + V/(V + 10^6)}{V \sin^2 \theta/2} \right]^2 \bigg/ \text{ster} \qquad (1)$$

where n = number density/cm^3; V = beam energy in volts; x = gas sample thickness (cm); I_0 = incident beam flux; θ = angle through which the beam scatters; I_s = signal beam.

In the present case where 20 kV electrons are used, one gets:

$$n \propto \frac{I_s}{I_0} \frac{V^2 \sin^4 (\theta/2)}{Z^2 A}, \qquad (2)$$

where A is a constant that depends upon the beam-detector geometry.

Thus one may obtain a linear calibration curve of density against the scattered signal I_s, if the voltage and primary beam flux are held constant, and a detector is used that is accessible to electrons that have been scattered through a specific angle θ from the nominal observation point.

It is usual to obtain this curve by calibration in a static gas of known properties and then use the calibrated probe to measure the local density in flows. An important consideration with this type of measurement is the determination of the incident electron beam intensity.

2.3.2. *Experimental Apparatus*

The electron gun assembly used for this experiment was housed in a chamber which mounted on the front of a 4 in. glass cross assembly. This glass cross mounted on top of a water cooled baffle which in turn was placed upon an Edwards 4 in. diffusion pump. The rear and top arms of the glass cross were sealed by cans held at atmospheric pressure. It was necessary to use atmospheric pressure in these cans to avoid flash-over of the high voltage electron gun leads. From these cans, leads at atmospheric pressure were led to appropriate control panels. The diffusion pump backing line was led out to a separate mechanical pump beside the tunnel. The entire unit could be traversed both axially and radially to survey the flow field.

The 90° elastically scattered primary electrons were detected by means of a Calcium Fluoride (Europium doped) scintillator crystal. The front surface of this crystal was coated by vapor deposition with aluminum. (The thickness of this coating was not controlled but was made thick enough so that no obvious light leaks existed.)

The light output of the crystal was monitored by a water-cooled EMI 9502 photo multiplier operated in a pulse counting mode. Typical dark counts of 30 per sec were measured.

2.3.3. *Normalization of the Density Measurement*

As stated earlier, the accuracy of the density measurement is directly dependent on the accuracy with which the primary electron beam current can be measured by the commonly used Faraday cup.

Because of the difficulties involved in the direct method of beam measurement, a novel method was developed that is insensitive to the presence of background ionization, because it only responds to the high energy beam electrons. The primary beam was allowed to strike a heavy metal target, and the resulting radiation in the soft X-ray region was monitored as an indication of beam intensity.

To test the system, an extension was placed on the front of the monitor. With the extension in place, any beam entering the monitor could be measured by a micro-ammeter attached to the 2-in. copper tube housing the target. (This housing was insulated from ground.)

The monitor apparatus gave a linear calibration, with a sensitivity of about 180 counts per second per microamp of beam current. There was no change in this count rate when gas was admitted to the system.

2.3.4. *Density Measurement Results*

The density in room temperature free jets was found to agree with the theory to better than 5%. In general, the hot jet density was found to be about 5 times lower on the centerline than in the cold jet at the same axial distance and mass flow.

Figure 5 shows a series of typical heated jet centerline densities plotted against axial distance. It may be seen that the data follow a $1/x^2$ dependence until $x \sim 5$, as the data points are parallel to a line with a slope of -2 on the log–log plot. This distance of 5 in. is considerably closer to the orifice than the theoretical Mach disk location (~ 8.5 in.). The reason for this becomes obvious when the magnitude of the density is considered. The very low density hot jet is penetrated by the background gas at any point beyond the arrows (Muntz *et al.* [19]). Because the jet is highly overexpanded, the extent of the undisturbed jet is less than what one might expect from a calculation of the Mach disk location on the basis of the measured pressure ratio across the orifice.

The demonstrated $1/x^2$ dependence of the heavy species shows that the hot jet is described by the free jet theory. Thus it is possible to relate an equivalent source density to each case investigated.

As well, the ratio of the cold flow density to hot flow density may be obtained. This was done at each mass flow setting by placing the densitometer at a position of 25 orifice diameters downstream in the flow, recording the cold density at that point, then recording the hot density at a number of torch power settings. These results are

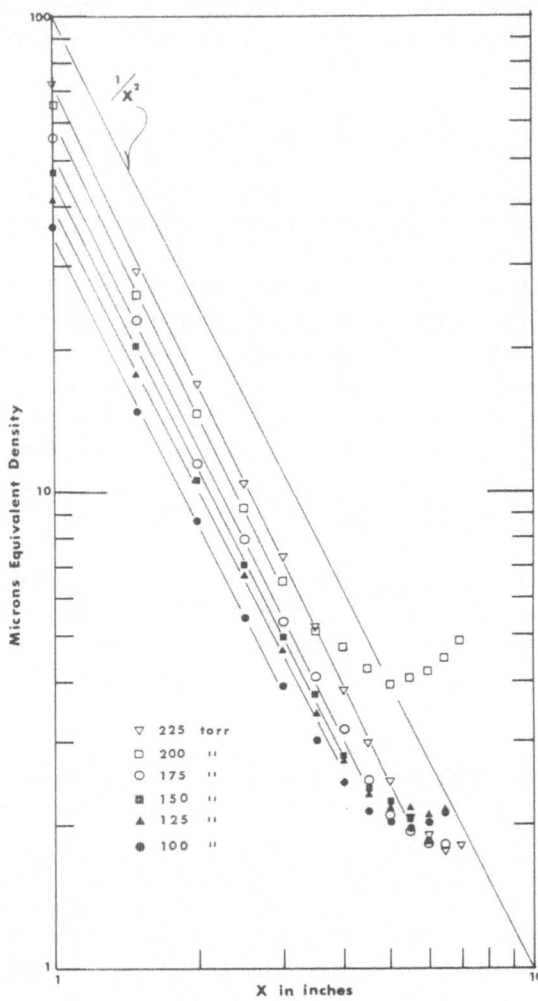

Fig. 5.   Density vs. axial distance.

presented and further data reduction is considered in section 3, where the mass flow and impact pressure results are considered in the same manner.

2.4. MASS FLOW

2.4.1. *Theory*

In the continuum limit, mass flow probes have been used in supersonic flows with success for a number of years (Reis [20], Potter *et al.* [21]). The criterion for the proper operation of such a probe is that the shock which stands in front of the probe is 'swallowed'. That is to say the shock wave produced by the probe is attached to the probe lip and does not spill flow around the entrance to the probe. When the flow is 'swallowed', the mass flow is given by:

$$\dot{m} = \rho u A_s \tag{3}$$

where A_s is the area of the probe orifice.

The sharp-edged skimmer used in the production of molecular beams has also been used to measure mass flow (O'Keefe [22], Bossel [23]). In these studies the probe is in a free molecule flow and the mass flow may be represented by:

$$\dot{m}_s = A_s \frac{mn}{\pi 4} \sqrt{\frac{8kT}{m}} \{\exp(-S^2) + \sqrt{\pi}\, S(1 + \mathrm{erf}\,(S)\} \tag{4}$$

where $S = \sqrt{\gamma/2}\, M$.

At the high speed ratios encountered in the present flows this expression simplifies to:

$$\dot{m}_s = m_a n u A_s \tag{5}$$

or the same expression as obtained in the continuum flow case. Thus experimentally one may find a number expressing the product of the local density and velocity:

$$\rho u = \dot{m}_s / A_s. \tag{6}$$

It was hypothesized that the skimmer will in the transition regime still approximately behave in such a manner that the measured mass flow will be the product of the local density and velocity.

The experimental data were reduced and are presented in a manner very similar to Bossel's.

A sketch of the experimental layout is shown in Figure 6. The rather low mass flow

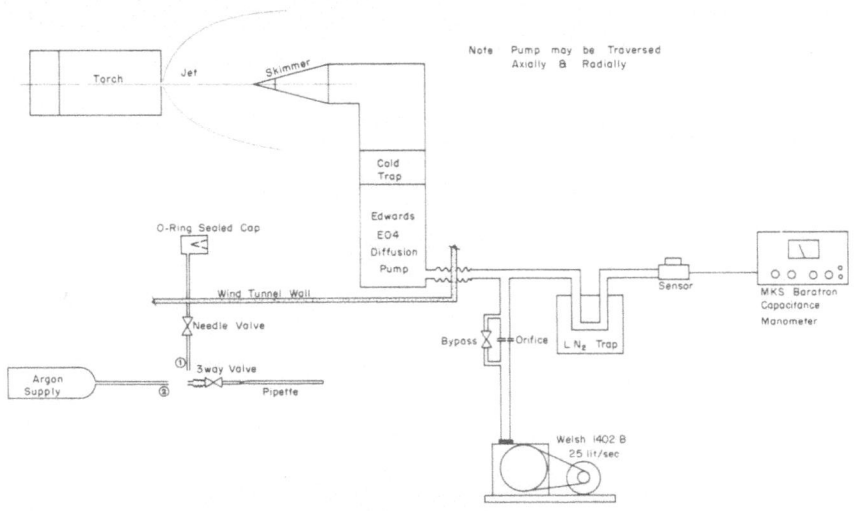

Fig. 6.   Layout of mass flux measuring apparatus.

rates were measured by monitoring the pressure upstream of a 0.068 in. orifice placed in the backing line of the diffusion pump. This pressure was measured by a capacitance manometer (Baratron Model H-77) isolated from the backing line by a liquid nitrogen cold trap. An upper limit on the mass flow that could be measured was a consequence of the limited back pressure tolerance of the diffusion pump. If the backing pressure rose above 180 mT the diffusion pump became unstable as evidenced by oscillations in the pressure measured in the glass cross by an ionization gauge.

The orifice assembly was calibrated by measuring the pressure upstream of the orifice for known mass flows of argon.

The skimmer mas flux results are presented as the ratio of

$$\frac{[\rho u]_{\text{exp}}}{[\rho u]_{\text{Theory}}} = \text{JREL} \tag{7}$$

versus x/D, the non-dimensional nozzle/skimmer distance.

2.4.2. *Experimental Apparatus*

The 4-in. diffusion pump assembly which was used to evacuate the electron gun chamber was used in a modified form for the mass flow measurements. This was done by removing the box containing the electron gun and installing a plate which served as a mount for a conical skimmer holder.

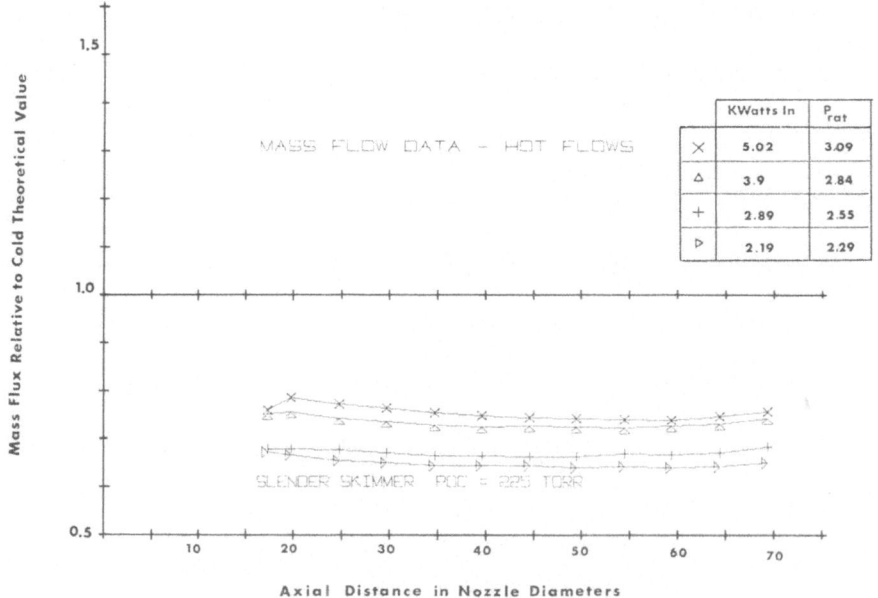

Fig. 7. Mass flux data hot flow Slender Skimmer.

The entire diffusion pump-skimmer assembly was mounted on the lower traversing gear which could be traversed both axially and radially with respect to the flow.

The skimmer geometry was made identical to Bossel's in an attempt to produce results in argon similar to those he obtained with nitrogen room temperature flows.

2.4.3. *Results of Mass Flow Measurements*

Some experimental data taken with this apparatus in room temperature argon free jets is given in [10].

Figure 7 shows a typical series of measurements for various power levels with the same mass flow. It can be noted that the flux increases with increasing power. When considered in conjunction with the density data, this is interpreted as an increase in velocity (or effective stagnation temperature).

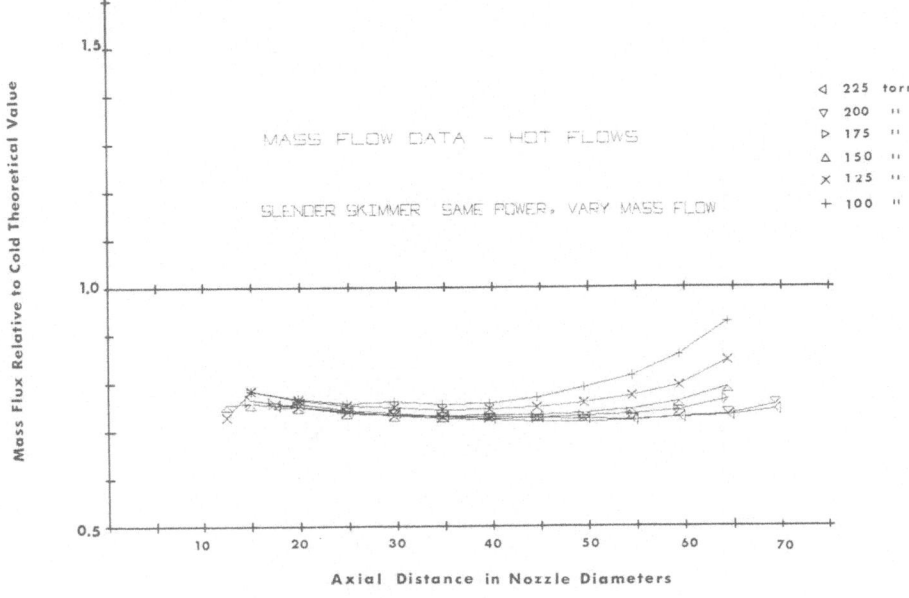

Fig. 8.   Mass flux data hot flow, $P_{rat} = 2.8$.

Figure 8 presents a series of curves of the same power level but different mass flow. At the lower mass flows an indication of background penetration in the jet may be discerned. Note however that the point at $x/D = 25$ is not affected by this penetration. This distance will be used later for source temperature determination. Figure 9 is a log–log plot of the mass flow data against distance downstream. Because the curve plots as a straight line parallel to the line with slope -2, one may conclude that the mass flow density is falling as $1/x^2$. Since the density data show the same fall-off this means the velocity has attained a constant value in the expansion.

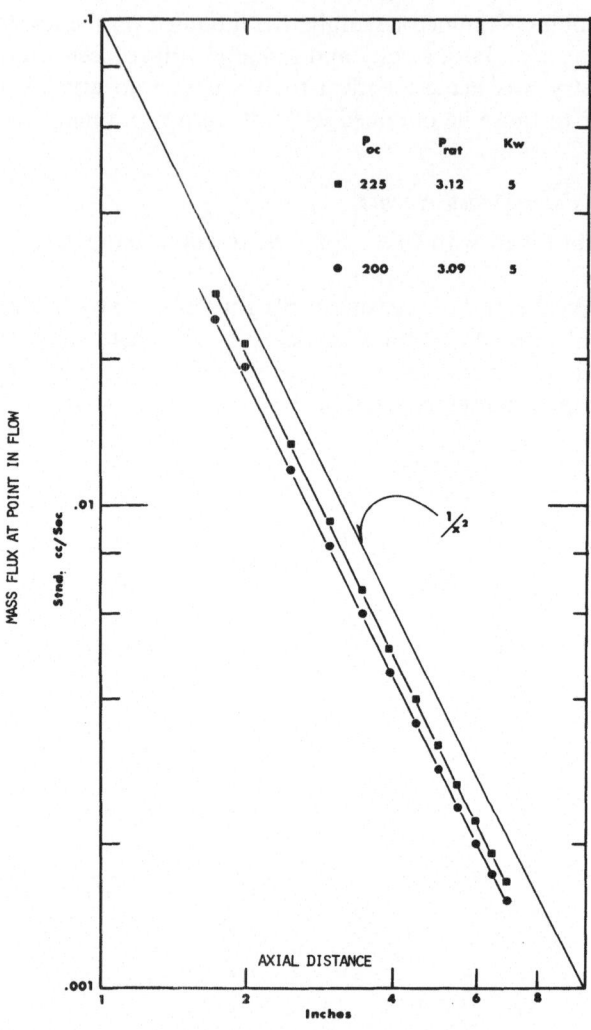

Fig. 9.   Mass flux vs. distance.

2.5. IMPACT PROBES

2.5.1. *Theory*

Ashkenas and Sherman [24] give a fitting formula for the impact pressure along the centerline of a free jet:

$$\frac{p_i}{p_0} = \left[\frac{\gamma + 1}{\gamma - 1}\right]^{\gamma/\gamma - 1} \left[\frac{\gamma + 1}{2\gamma}\right]^{1/\gamma - 1} A_m^{-2/\gamma - 1} \left[\frac{x - x_0'}{D}\right]^2, \qquad (8)$$

where P_i is the impact pressure; P_0 is the source (stagnation) pressure; and for argon:

$$\gamma = 1.67, \quad A_m = 3.26, \quad x_0'/D = 0.04$$

They also point out that in hypervelocity flows $P_i \propto \rho u^2$ quite accurately. Because in a free jet the velocity reaches an almost constant value beyond x/D values of 10 (Mach No. $= 15$), the impact pressure will follow the density profile further downstream. Thus in a free jet the impact pressure can be expected to fall as $1/x^2$.

The influence of the Mach disk which terminates the free jet will not be observed very clearly. This is because an impact probe measures a recovery pressure that has been processed by a normal shock. As the Mach disk is in itself a normal shock, the probe measures the same quantity both upstream and downstream of it.

2.5.2. *Experimental Apparatus*

In the present investigation, because of the importance of stand-off distance, a series of probes were constructed of 0.030 in. wall stainless steel tubing. The outside diameter was varied but the length-to-diameter ratio was fixed at 25. The probes were internally chamfered to best define the probe diameter for the shock stand-off correction. Probe dimensions are given in Table II.

TABLE II
Impact probe sizes

Outside diameter (in.)	Length (in.)
$\frac{1}{8}$	3.125
$\frac{1}{4}$	6.25
$\frac{3}{8}$	9.375
$\frac{1}{2}$	12.50

The raw data recorded (probe mouth location and bridge voltage) were put on punched cards for computer processing. The impact pressure was calculated by a 6th order polynomial fit to the calibration curve.

The theoretical impact pressure was calculated by subtracting the stand-off distance of the bow shock from the probe position, and using Equation (8). The stand-off distance used was calculated from the inviscid limit taken from Bailey and Sims, multiplied by the probe radius. Downstream in the jet, where the Reynolds number becomes small enough that additional corrections should be made for viscous effects, it was found that any more elaborate corrections for stand-off distance were fraught with uncertainty. This occurs because at x/D values greater than about 30, the shock thickness is greater than the calculated stand-off distance. As well, in the free molecule limit one cannot define a stand-off distance at all. An impact probe in free molecule flow indicates the flow properties that correspond to the probe mouth location. The alignment of the probe is critical in this regime, especially with high speed ratio flows. Fortunately, the correction becomes less important at larger nozzle-to-probe distances.

2.5.3. *Pitot Probe Results*

A plot on a log–log scale is presented in Figure 10 for a typical set of hot flow data. In the heated case, the impact pressure rather closely follows a $1/x^2$ dependence on distance. This would be expected from the previous density and mass flow results.

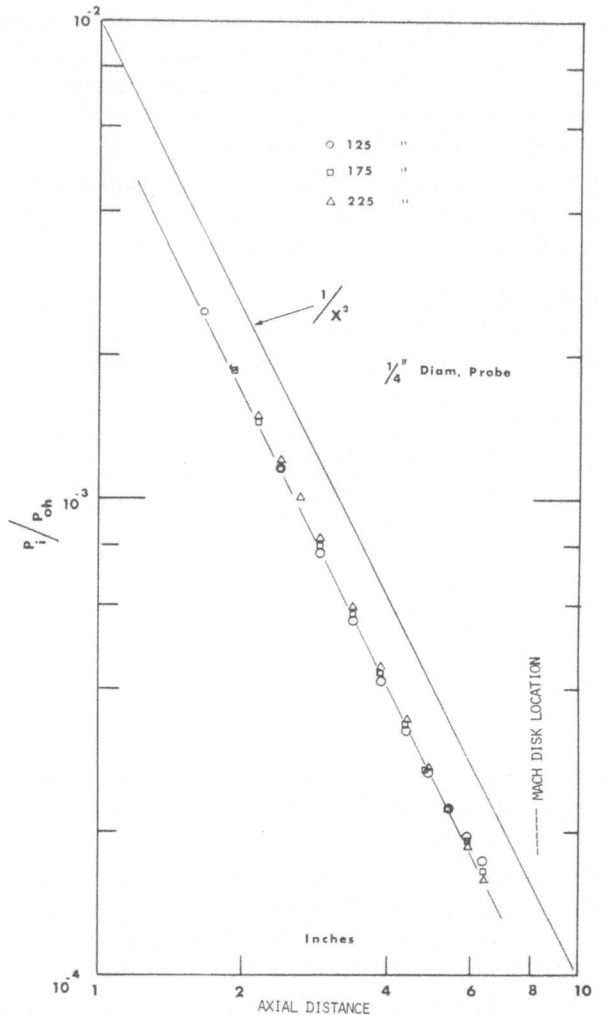

Fig. 10.   (Impact pressure)/(stagnation pressure) vs. distance.

The data follows the source flow pattern to a greater axial distance, thus the influence of the Mach disk is not readily detected for the shock will have become rather diffuse.

3. Discussion

3.1. VELOCITY CALCULATIONS

As mentioned in Sections 2.3, 2.4, 2.5; each specific flow property was measured at an x/D of 25 for the range of mass flows and power levels listed in Table I. Proof was presented in each case that the position used in the jet was well-behaved gasdynamically. In addition, each property was shown to follow a source-like expansion.

The pressure ratio P_{0h}/P_{0c} was measured in each case, and as was shown in Section 2.2 this can be related to the average enthalpy in the gas flow. Thus for each separate power and mass flow setting, the ratio of the appropriate gasdynamic parameter (density, mass flux, and impact pressure) was related to the theoretical value of that same parameter in the cold case. It was also shown that the measured cold flow value agreed reasonably well with the theoretical cold value in the case of the density and impact pressure.

If we consider

$$\text{the density ratio} \quad = \frac{1}{A} = \frac{\rho_{\text{COLD}}}{\rho_{\text{HOT}}} \tag{9}$$

$$\text{the mass flux ratio} \quad = B = \frac{(\rho u)_{\text{HOT}}}{(\rho u)_{\text{COLD}}} \tag{10}$$

$$\text{and} \quad \text{the impact pressure ratio} = C = \frac{(\rho u^2)_{\text{HOT}}}{(\rho u^2)_{\text{COLD}}} \tag{11}$$

where the 'cold' value in each case is the theoretical cold value, then it is obvious that the velocity ratio hot/cold may be calculated by any of the ratios:

$$\frac{v_{\text{HOT}}}{v_{\text{COLD}}} = u_1 = B/A$$

$$= u_2 = C/B \tag{12}$$

$$= u_2 = \sqrt{C/A}.$$

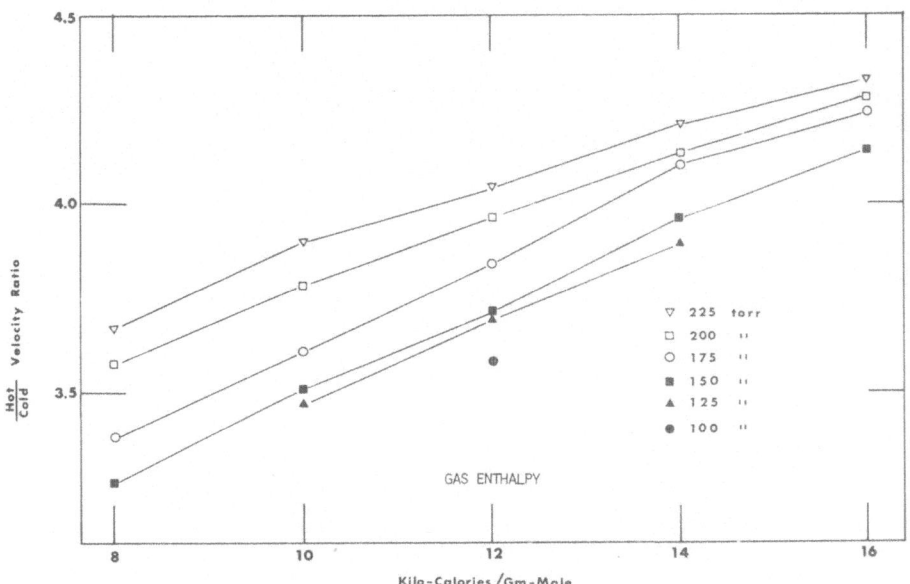

Fig. 11.   Velocity ratio, hot (measured) to cold (theory) vs. source enthalpy.

As the three methods had about the same level of confidence, a simple average was taken for the velocity ratio at each source condition.

Figure 11 is a plot of the average velocity ratio $\bar{u}$ versus source enthalpy.

$$\bar{u} = \tfrac{1}{3}(u_1 + u_2 + u_3). \tag{13}$$

In addition, recalling that $a = \sqrt{\gamma RT}$, the velocity at any axial distance may be calculated. Above a Mach number of about 5 (which corresponds to an x/D of 2.5) the flow has already attained 95% of the terminal velocity given by:

$$u = \sqrt{\frac{2\gamma}{\gamma - 1}\, RT_0}. \tag{14}$$

Thus the flow at any distance greater than say an x/D of 5, will have attained an almost constant velocity which is determined by the source temperature.

$$T_{\text{RAT}} \propto (\bar{u})^2. \tag{15}$$

The plot of effective centerline temperature ratio versus source enthalpy is given as Figure 12 and typical temperatures are drawn in assuming $T_{0c} = 290$ K. The horizontal axis of this graph is the measured enthalpy of the exit gas, and thus corresponds to $C_p T_{\text{AVE}}$, where T_{AVE} is the average temperature across the orifice. At a given enthalpy,

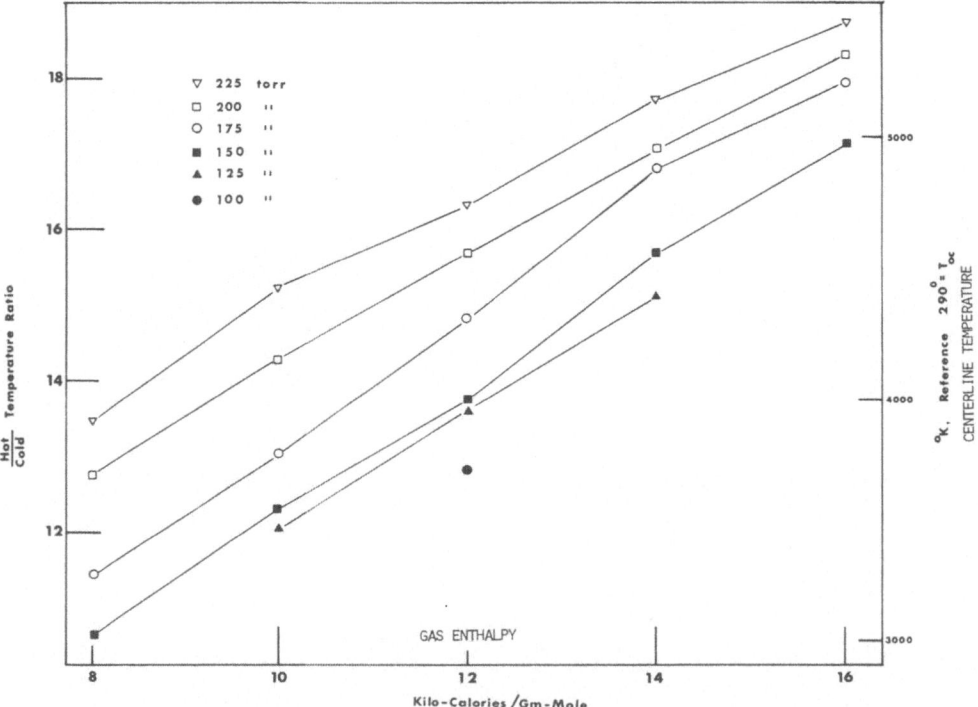

Fig. 12.   Temperature ratio, hot (measured) to cold (theory) vs. source enthalpy.

the temperature of the exit gas is seen to be higher for the higher mass flow cases. This has been explained earlier as a consequence of greater cooling in the lower mass flow cases. The temperatures calculated in this manner are centerline values and thus will be the maximum values, because the temperature profile is expected to be parabolic. For a specific case of mass flow and power setting then, this temperature will correspond to that called T_p in Appendix B, where T_p is the peak (centerline) temperature used in the calculation. Because we may calculate the average source temperature from the pressure ratio, and we know the wall temperature and the heat and mass flow transported out of the orifice, sufficient information exists to calculate a temperature profile. (The details are given in Appendix A.)

4. Conclusion

(1) An inductively coupled arc has been used to produce a high enthalpy source of argon. This arc was studied over a stagnation pressure range, from 300 to 800 T at mass flows between 0.2 and 0.5 gm s^{-1}. The enthalpy of the exit gas was measured to be between 8 and 20 kcal gm^{-1} mole.

(2) The following items of instrumentation were developed or modified for these flow diagnostics.

- A flow total calorimeter was constructed.
- An electron beam densitometer was constructed specifically for ionized flows.
- A mass flux probe for use in low density flows was tested.
- A thermister gauge and controller compatible for the use near the electrically noisy RF arc was constructed and calibrated.

(3) A study has been made of the free jet flow field produced from this source by expanding the hot gases through a 2.54 mm orifice.

The flow field has the following properties:

Dimensions – Physically, the usable portion of the jet is about 6 in. long and has about 2 in. of almost uniform core region radially. The pumping speed of the tunnel calculated from a consideration of this free jet size was 6500 l s^{-1} which is in agreement with the speed measured from mass flow consideration.

Densities – The heavy species number density on the centerline could be varied from 3×10^{15}/cm^3 at an x/D of 10 in the most dense jet to 9×10^{13}/cm^3 at an x/D of 45 in the least dense jet. The percent ionization was about 0.1% and depended on the power input to the generator.

Velocities – The velocities inferred from the experimental measurements ranged from 3.25 to 4.35 times the escape velocity of the gas at 290 K. This corresponds to 1.8 to 2.4 km s^{-1}.

The present investigation has shown that it is possible to produce uncontaminated high velocity argon flows, and that the flow properties may be measured with devices similar to those used in arc jet studies. Because the energy addition is virtually an equilibrium one, the stagnation conditions may be determined and used to calculate the flow field properties for an aerodynamic investigation.

References

[1] Stursburg, K., 'Lichtbogen-Brennkammern niedriger Leistung für Hypergeschwindigkeits-Vakuum-Windkanale'. (Low Power Arc Heaters for Low Density Hypervelocity Tunnels.) Deutsche Luft and Raumfahrt. Communication 67-10, June 1967.

[2] Brundin, C. L., Talbot, L., and Sherman, F. S., 'Flow Studies in an Arc Heated Low Density Supersonic Wind Tunnel', University of California (Berkeley), Institute of Eng. Research Report HE-150-181.

[3] Potter, J. L. and Kinslow, M., 'Description and Preliminary Calibration of a Low Density, Hypervelocity Wind Tunnel', AEDC TN-61-83.

[4] Christensen, D., Chen, C. J., and Price, R., 'A Feasibility Study and Preliminary Design of Instrumentation Suitable for Measuring the Gas Flow Parameters of an Arc Heated Hypersonic Gas Dynamic Facility', ASD-TDR-62-154.

[5] Becker, M. and Heyser, A., 'Der DVL-Plasmawind Kanal, eine Anlage für hypersonische Geschwindigkeiten, geringe Dichten und hohe Temperaturen', Deutsche Versuchsanstalt fur Luft- und Raumfahrt, FB 67-66.

[6] Reed, T. B., J. Appl. Phys. 32 (1961), No. 5.

[7] Barger, R. L., Brooks, J. D., and Beasley, W. D., 'The Design and Operation of a Continuous Flow Electrodeless Plasma Accelerator', NASA TN D-1004.

[8] Chuan, R. L., 'Plasma Heating of Hypersonic Gas Flow', University of Southern California Engineering Center, Report 56-202, December 1957.

[9] Eckert, H. U., J. Aerospace Sci. 26 (1959).

[10] Payne, W. F., 'Characteristics of a Free Jet Expanded from a High Pressure Inductive Arc Source', UTIAS Report 165, July 1971.

[11] Rothe, D., 'Electron Beam Studies of the Diffusive Separation of Helium Argon Mixtures in Free Jets and Shock Waves', UTIAS Report No. 114.

[12] Gadamer, E. O., 'Measurement of the Density Distribution in a Rarefied Gas Flow Using the Fluorescence Induced by a Thin Electron Beam', UTIA Report No. 83.

[13] Nakamura, Y., 'Temperature and Density Measurements of Induction Heated Free Jets of Nitrogen Seeded Argon', Univ. Southern Calif. Engr. ASCAE 116, February 1970.

[14] Fraser, R. B., 'Diagnostic Studies in an Induction Heated Supersonic Argon Plasma', Univ. of California, Berkeley, Report AS-70-5.

[15] McMichael, G. E., 'Investigation of Diffusive Separation in Helium Argon Flows with an Electron Beam Probe', UTIAS Report No. 167.

[16] Camac, M., 'Argon and Nitrogen Shock Thickness', AIAA preprint 64-35.

[17] Russell, D. A., Phys. Fluids 11 (1968), 1679.

[18] Ziegler, C. A., Bird, L. L., Olson, K. H., Hull, J. A., and Morreal, J. A., Rev. Sci. Inst. 35 (1964), 450.

[19] Muntz, E. P., Hamel, B. B., and Maquire, B. L., AIAA J. 8 (1970), 1651.

[20] Reis, V. H., 'Free Expansion of Pure and Mixed Gases from Small Sonic Nozzles', Dept. of Science and Technology, Princeton University, Report FLD-7, May 1964.

[21] Potter, J. L., Arney, G. D., Kinslow, M., and Carden, W. H., 'Gasdynamic Diagnostics of High Speed Flows Expanded from Plasma States', AEDC-TDR-63-241, Nov. 1963.

[22] O'Keefe, D. R., 'Initial Performance Study of the UTIAS High Energy Molecular Beam Facility', UTIAS Tech. Note 75.

[23] Bossel, U., 'Investigation of Skimmer Interaction Influences on the Production of Aerodynamically Intensified Molecular Beams', Univ. of California, Berkeley, Report AS-68-6.

[24] Ashkenas, H. and Sherman, F. S., 'The Structure and Utilization of Supersonic Free Jets in Low Density Wind Tunnels', in J. H. deLeeuw (ed.), Rarefied Gas Dynamics, Suppl. III.

Appendix A

PRESSURE RISE ACROSS A CHOKED ORIFICE WITH NON-UNIFORM TEMPERATURE

A flow through a circular sonic orifice is postulated to be formed by expansion to $M = 1$ from a stagnation region with a radial temperature profile given by:

$$T(r) = (T_{0c} - T_p)\xi^n + T_p \tag{A-1}$$

where $T_{0c} = 285$ K (wall temperature); $T_p =$ peak centerline temperature; $\xi = r/R$ non-dimensional radial coordinate.

For any stream-tube from the stagnation region the flow will then be sonic, with a local sound speed given by:

$$\sqrt{\gamma R T(r)^*} = a^*(r).$$

The mass flow in any stream tube is then:

$$\mathrm{d}(\dot{m}) = R^*(r)a^*(r)\,\mathrm{d}A$$

where ρ is the density and an asterisk denotes sonic conditions. Expressing $\mathrm{d}(A)$ as $r\,\mathrm{d}r\,\mathrm{d}\theta$ and integrating over the orifice one obtains

$$\dot{m}_{\mathrm{AVE}} = \int_0^{2\pi} \int_0^1 R^*(r)a^*(r)\,r\mathrm{d}r\mathrm{d}\theta, \tag{A-2}$$

Substituting

$$r = R\xi$$
$$\mathrm{d}r = R\,\mathrm{d}\xi$$

and using the isentropic relations

$$\dot{m}_{\mathrm{AVE}} = \frac{\Omega P_0 A}{\sqrt{T_{\mathrm{AVE}}}} = \int_0^{2\pi} \int_0^1 \frac{\Omega P_0 \xi\,\mathrm{d}\xi\,\mathrm{d}\theta}{\sqrt{(T_{0c} - T_p)\xi^n + T_p}} \tag{A-3}$$

where

$$\Omega = \left(\frac{2}{\gamma + 1}\right)^{\gamma + 1/2(\gamma - 1)} \sqrt{\gamma}$$

and T_{AVE} is the temperature one would calculate from the measured pressure ratio if the temperature profile were uniform.

This expression is an implicit relation between the average temperature T_{AVE} and the peak temperature T_p.

Equation (A-3) may be rewritten as:

$$\sqrt{T_p} = \sqrt{T_{\mathrm{AVE}}}\, 2 \int_0^1 \frac{\xi\,\mathrm{d}\xi}{\sqrt{(B - 1)\xi^n + 1}} = \sqrt{T_{\mathrm{AVE}}} \cdot Y1. \tag{A-4}$$

with $B = T_{0c}/T_p$.

Equation (A-4) may be solved by numerical integration for a given value of n by an iterative technique. A maximum of four iterations were required when the initial value of $B = B_1$ was chosen as

$$\frac{nT_{0c}}{(n + 1)T_{AVE} - T_{0c}}.$$

Thus a value $Y1$ could be obtained such that:

$$T_p = T_{AVE} \cdot (Y1)^2 \tag{A-5}$$

for any given profile.

This expresses the peak centerline temperature in terms of the average temperature (inferred from the stagnation pressure and mass flow) for a given exponent, n. Thus the effective temperature profile may be calculated.

The incremental heat carried out of the torch $d(Q)$, may be written

$$d(Q) = d(\dot{m}C_p T(r)). \tag{A-6}$$

Thus

$$Q_{AVE} = \int_0^{2\pi} \int_0^R \frac{\Omega P_0[(T_{0c} - T_D)\xi^n + T_p]C_p r \, dr \, d\theta}{\sqrt{(T_{0c} - T_p)\xi^n + T_p}} \tag{A-7}$$

$$Q_{AVE} = 2\pi R^2 \Omega P_0 \sqrt{T_p C_p} \int_0^1 (T_{0c} - T_p)\xi^n + T_p \rho \, d\rho. \tag{A-8}$$

Writing

$$H_{AVE} = \frac{Q_{AVE}}{\dot{m}_{AVE}} \qquad A = \pi R^2$$

$$Y2 = 2 \int_0^1 \sqrt{(T_{0c} - T_p)\xi^n + T_p} \, \xi \, d\xi$$

one gets for a given n:

$$H_{AVE} = C_p \sqrt{T_p} \sqrt{T_{AVE}} \cdot Y2. \tag{A-9}$$

Substituting (A-5) into (A-9), one gets:

$$H_{AVE} = c_p T_{AVE} \cdot Y1 \cdot Y2 \tag{A-10}$$

which is an expression for the average enthalpy transported through the orifice as a function of the profile parameter n.

As T_{AVE} may be given by the following expression,

$$T_{AVE} = T_{0c} \left(\frac{P_{OH}}{P_{0c}}\right) \tag{A-11}$$

then for any given pressure ratio hot/cold, H_{AVE} may be calculated. This was done for $T_{0c} = 285$ K, and is shown in Figure 3.

AN EXPERIMENTAL INVESTIGATION OF AIR FLOW
AT AN INLET TO A CENTRIFUGAL COMPRESSOR

ANANT Y. BARODEKAR and JOHN H. BLACKBURN

The University of Liverpool, Dept. of Mechanical Engineering, Liverpool, England

Abstract. The first stage of a Rolls Royce Dart centrifugal compressor provided a flow with suitable Mach number and compressibility for testing a quasi-three-dimensional flow model. As successful operation of the flow model depended on supplying the program with the correct inlet conditions an accurate experimental survey was required to determine the values of pressure temperature and velocity just upstream of the rotating guide vanes.

It was the aim of this project to establish whether the flow was axisymmetric and to obtain the required parameters by traversing with suitable probes. Because of the number of inaccessible positions around the circumference the proof of axisymmetry was dependent on being able to relate the results from a small number of traversing planes with the rest of the intake.

Nomenclature

SYMBOL		UNITS
N_c	compressor speed	rpm
P_T	total (stagnation) pressure	lb(in^2 abs.)$^{-1}$
P_S	static pressure	lb(in^2 abs.)$^{-1}$
r ·	radius	in.
ρ	density	lb ft^{-3}
T	total (stagnation) temperature	K
t	static temperature	K
U	blade speed	ft s^{-1}
V	fluid velocity	ft s^{-1}

CONSTANTS
C_p	specific heat at constant pressure
J	mechanical equivalent of heat
g	gravitational constant
R	gas constant

SUBSCRIPTS
1	intake inner surface
2	intake outer surface

1. Introduction

According to Ferguson [1], for ideal conditions at entry to a centrifugal compressor, the flow should be both uniform and axisymmetric. Any deviation by the air flow, from these ideal conditions, had to be investigated at the experimental rotational speed (7500 rpm – half design speed), which had been imposed by power considerations.

In order to investigate the flow at inlet to the rotating guide vanes some form of traverse within the flow was required. A circumferential traverse at varying radii

L. G. Napolitano et al. (eds.), *Astronautical Research 1971*, 577–586. *All Rights Reserved.*

would lead to a complete representation of the flow, but because of the complex construction of the intake casing, this form of investigation was impractical and impossible in the time available. Hence, radial traversing at selected points was adopted. Essentially, the choice of traverses is a compromise between those which were necessary to describe the flow adequately and those which were accessible.

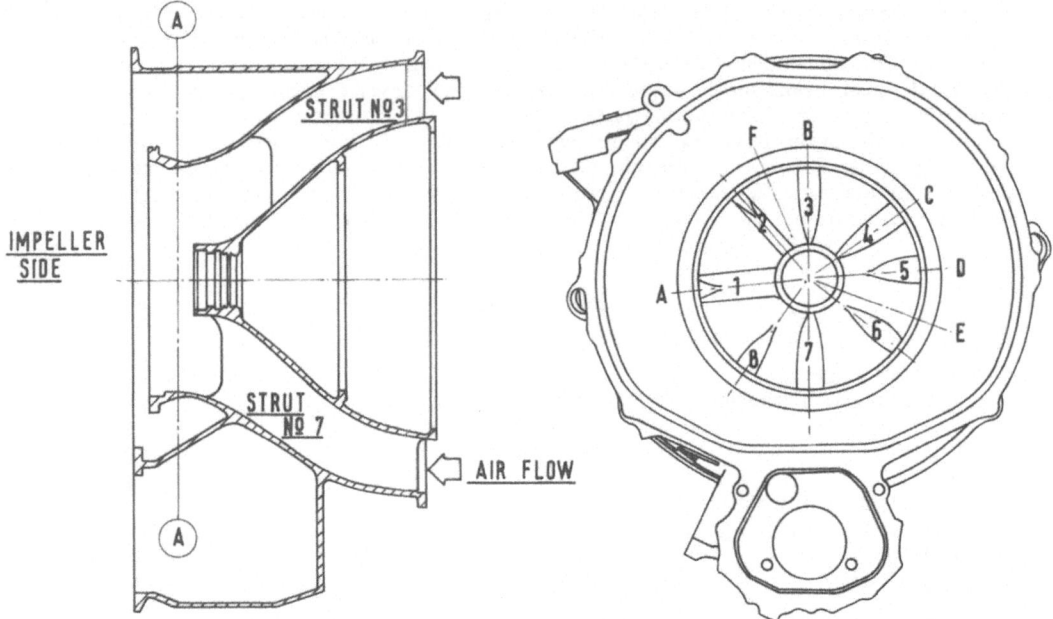

Fig. 1.   Left: sectional view of intake casing. Right: view on intake casing from impeller showing the positioning of struts 1 to 8 and traverse points A to F.

Reference to Figure 1 shows that the constructional detail presents no opportunity to install instrumentation on the lower half of the intake casing. By traversing at accessible points and correlating with those at similar geometrical positions, effective use was made of the minimum number of traverses.

Measurement of total and static pressure together with temperature and yaw angle (angle between flow and axial direction), provided sufficient data for the input of the starting boundary conditions of the flow model which was based on the Marsh [2] through-flow program.

2. Description of the Rig and Instrumentation

2.1. THE DART COMPRESSOR INTAKE

In view of the complex geometry of the intake, the following description is provided to enable the reader to visualize the position of the traverse plane and the restrictions offered to instrumentation necessary to carry out a flow study in the intake.

Basically the inlet casing consists of a converging annular duct containing eight struts of approximate aerofoil shape (Figure 1). Although the intake is designed to provide the annular passage for air to reach the impeller, it also includes an integral oil tank and service ducts for the supply of oil to the innermost section of the intake casing which forms a gearbox for the reduction gearing. Oil is fed and returned from this gearbox across the air passage by means of tubes incorporated in the struts. The struts also serve to remove the prewhirl induced in the air flow by the propeller. Chord lengths of these struts vary and since the leading edges are coplanar, the distances of the trailing edges from the impeller are dissimilar. Struts labelled 1 and 2 (Figure 1) are the least efficient aerofoils in fact, their trailing edges are almost cylindrical.

For the purpose of laboratory experimentation, the second stage of the compressor has been removed, resulting in power saving. The compressor is driven by a 650 H.P. AC motor through a gearbox having a step up ratio of 6:1. In order to prevent choking at exit from the second stage, a new exhaust plate with an increased exit area has been fitted.

The layout of the test rig can be seen in Figure 2.

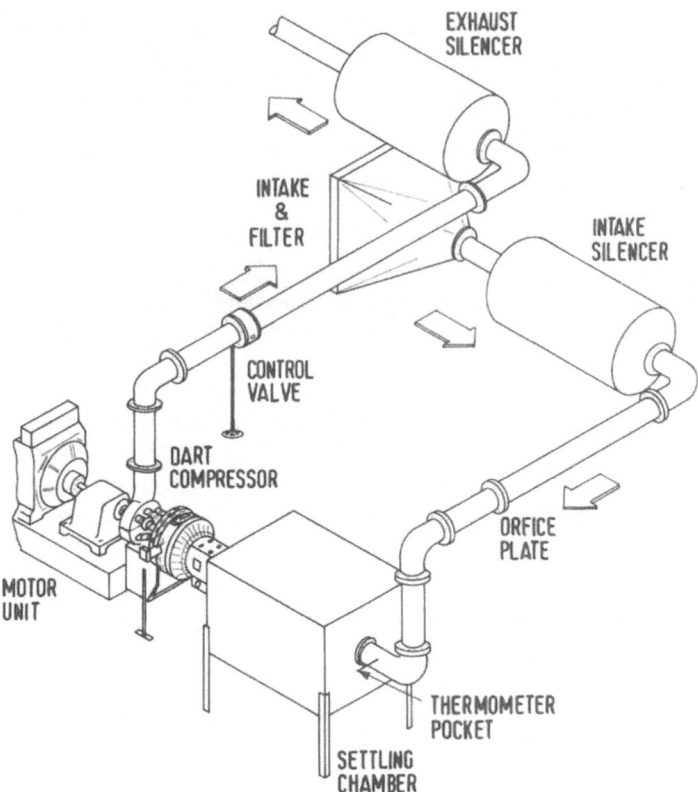

Fig. 2. Diagrammatic arrangement of the compressor system.

During service as a turboprop engine, various auxilliaries are positioned around the Dart intake casing (oil coolers, starter motor, etc.) and in some instances the opportunity arose to utilize the redundant brackets, studs, etc. in order to locate the probe traversing unit. To minimize the manufacture of new equipment, a range of mounting plates was used to position the probes for the measurement of fluid properties previously listed. The traversing unit consists of a friction drive for advancing and retracting the probes, and incorporates a protractor for angle measurement.

2.2. PRESSURE MEASUREMENT

An instrument capable of accurate air velocity measurement was required, but because of the multi-wall intake casing (see Figure 1), inherent in the Dart compressor, a limited number of instruments was suitable. The final type chosen had a measuring wedge smaller than the constant diameter shaft and was based on a Napier [3] design, capable of measuring total and static pressure, together with flow angle. All conventional hook-type pitot-static tubes were impractical since the probe had to be introduced into the inlet annulus through three separate walls. Over a number of years, valuable information has been acquired in the Department concerning the manufacture and use of Napier-type wedge probes of various dimensions. As an experimental tool it is relatively robust, simple to use and has a very good response to changes in flow direction.

2.3. TEMPERATURE MEASUREMENT

Selection of temperature probes was again influenced by the complicated intake casing and also the range of temperature expected. The latter was anticipated as 5 °C to 20 °C.

3. Experimental Procedure

3.1. INITIAL OPERATING CONDITIONS

In every test performed, the compressor was allowed to run for at least thirty minutes to ensure that thermal equilibrium was reached before any readings were taken. All tests were carried out at the same mass flow rate (11.5 lb s^{-1}) and compressor speed (7500 rpm). Motor speed was monitored on a 'RACAL' electronic counter-timer and the mass flow rate was measured using a previously calibrated orifice plate. To act as a reference, the wall static pressures round the annular passage were noted for each test.

3.2. STAGNATION TEMPERATURE MEASUREMENT

Copper-constantan thermocouples mounted in stainless steel stagnation shields of constant diameter were used to measure total temperatures. The final design of the probe complies with the constraints suggested by Dimmock [4], viz.:

(i) It must provide adequate support for the element.
(ii) It must bring the fluid almost to rest at the thermocouple junction and yet,
(iii) Pass a flow sufficient to ensure a good heat transfer.

(iv) This bleed air flow should pass along the wires for a short distance to counteract heat conduction in the wires from the external environment.

(v) The shield must protect the junction from radiation effects.

This form of probe does not satisfy (v) but in these tests the effects of radiation were negligible.

Temperatures were measured directly in degrees centigrade on a 'Comark' electronic thermometer which has an accuracy of $\pm \frac{1}{2}$ °C with the internal reference junction. The accuracy was further improved by the use of an external ice junction to $\pm \frac{1}{4}$ °C.

3.3. WALL STATIC PRESSURE MEASUREMENT

Thirty-three holes of $\frac{1}{16}$ in. diameter and 1 in. depth were used throughout the intake and were connected to a mercury manometer bank (reservoir type). These gave a direct visual representation of the static pressure distribution around the annulus. Holes 1–15 were in the same plane as the traverses and could be used as a check on the static pressure recorded by the wedge probe. Holes 16–33 were positioned round the annulus at the inlet end of the intake, and thus static pressure losses through the intake could be evaluated if required.

4. Derived Results

4.1. COMPUTATION OF VELOCITY AND DENSITY FROM STAGNATION AND STATIC PRESSURE AND STAGNATION TEMPERATURE

In compressible flow the following equation relates total and static pressures with density and velocity.

$$P_T = P_S + \tfrac{1}{2}\rho V^2(1 + M^2/4 + (2 - k)M^4/24 + \text{Higher order terms}).$$

For the purposes of this analysis, the Mach number effect was ignored, since the omission of terms involving M results in a difference of less than 2 ft s^{-1} in the final velocity of approximately 200 ft s^{-1}. Hence

$$V = \sqrt{\left(\frac{2(P_T - P_S)}{\rho}\right)}. \tag{1}$$

Values of P_T and P_S were measured at $\frac{1}{4}$ in. intervals between the impeller hub and the shroud, the total radial distance being 4 in. Total temperature was obtained at discreet points in a similar manner and hence with a knowledge of these properties at individual points the velocity V was calculated.

The method of calculating V involved first making an estimate of the density in the intake based on external ambient conditions. Then as Equation (1) above

$$P_S/\rho_1 = Rt \tag{2}$$

and

$$V = (T - t)2C_p Jg \tag{3}$$

have to be satisfied, the initial estimate of ρ from Equation (1) leads to a velocity V which can be inserted into Equation (3) to obtain the static temperature t. This can be used in Equation (2) to obtain an improved estimate of density. This iterative procedure was continued until all three equations were satisfied.

4.2. NON-DIMENSIONALIZING OF VELOCITY AND TEMPERATURE

Since the mass flow rate was sensibly constant for all tests, the velocities were non-dimensionalized with a velocity computed from the expression

$$\bar{V} = \frac{\int_{r1}^{r2} 2\pi r V \rho \, dr}{\int_{r1}^{r2} 2\pi r \rho \, dr}.$$

Evaluation of $\bar{V}$ was simplified when the variation of density across the passage width was found to be less than 2% of the total and was therefore assumed constant. The expression thus becomes

$$\bar{V} = \frac{\int_{r1}^{r2} V r \, dr}{\int_{r1}^{r2} r \, dr}$$

which was evaluated using a graphical technique.

4.3. ANGLE OF ATTACK AT RGV INLET

Velocity triangles were estimated at impeller inlet from the velocity and flow angle measurements. It was found that a considerable angle of attack existed in the entry flow and was almost constant along the entire leading edge.

Angle of attack is defined as the difference between the angle of the relative fluid velocity and the rotating guide vane inlet angle.

5. Discussion

Initially, flow predictions were to be based on results obtained at stations A to E, but two traverses produced some unexpectedly high values of temperature and consequently further investigation was undertaken. During a rig rebuild the opportunity was taken to establish another traverse position F, to further the investigation of the cause of these high temperature readings.

An examination of the flow environment produced several possible reasons for the temperature behaviour:

(i) Eddy flows caused by recirculation from the impeller (probable in view of excessive angles of attack);

(ii) Local hot spots caused by air flow from the front bearing seal;

(iii) Faulty apparatus, i.e., the stagnation probes or associated instrumentation.

A further series of tests showed that the temperature profiles at all stations were similar and corresponded to the lower values originally recorded.

No conclusive evidence was found to substantiate any particular one of the reasons suggested previously. The possibility of local hot spots could not be eliminated as a rig rebuild took place after the traverses at B and E. An alteration in the position of bearing seal, however small, could have influenced the flow of air from it.

Near to the outer wall of the duct the immersion length of the probe in the flow is short and the bleed air flow holes (in the probe) lie outside the flow passage but between the casing walls in a region of hot air. Thus this type of probe is unsuitable for measurement in this region due to the effect of conduction. Temperature readings near the wall were discarded as being unreliable.

The wedge probe cannot be used near to the wall with any degree of accuracy. In this position, static pressure readings become unreliable owing to wall interference and air leakage from the radial traverse holes into the air flow. Predictably a distortion of the pressure readings also occurs near to the shaft. A detailed study of the boundary layer close to the wall would require sophisticated equipment and considerable time, in view of the complicated casing construction.

Separation is anticipated from the outer wall firstly because of the curvature of the duct and secondly the high angle of attack suggest a partly stalled operating condition with the possibility of some reverse flow along the shroud in the region of the eye. Further investigation in this region is desirable.

Static pressure is always a difficult quantity to measure. The problems increase with turbulent flow conditions and the presence of swirl. According to Rayle's [5] data, by using the design of static tappings chosen, an error of $+1\%$ of the dynamic pressure can be expected with Mach numbers less than 0.4. Since the maximum Mach number encountered is 0.25 (based on results obtained, this amounts to 0.01 in. of mercury) the error introduced is negligible. The static pressure distribution shows no major discontinuities, when plotted for the whole of the circumference.

6. Conclusions

From the experimental results, the following conclusions have been reached.

(1) Some variation in total pressure distribution occurs downstream of strut No. 1. This has been attributed to the thick rounded edge and causes no significant change in the velocity profile, Figure 3.

(2) The presence of the struts within the air flow has little effect on either the profile or the values of velocity with the exception of strut No. 1 as mentioned above. A maximum difference in velocity of some 15 ft s^{-1} (6% of the total) occurred between traverses directly downstream of struts and those in the free stream. From these results it would appear that in service the major effect of these struts would be the removal of any prewhirl induced by the contra-rotating propeller.

(3) The velocity profiles from all the traverse positions were non-uniform (Figure 3). They were anticipated however as this would be the shape produced if an irrotational

flow was imposed on the intake geometry, i.e., higher velocity along the outer wall which has a smaller radius of curvature in the through-flow plane.

Of the two ideal conditions, i.e., symmetry and uniformity, the former has been

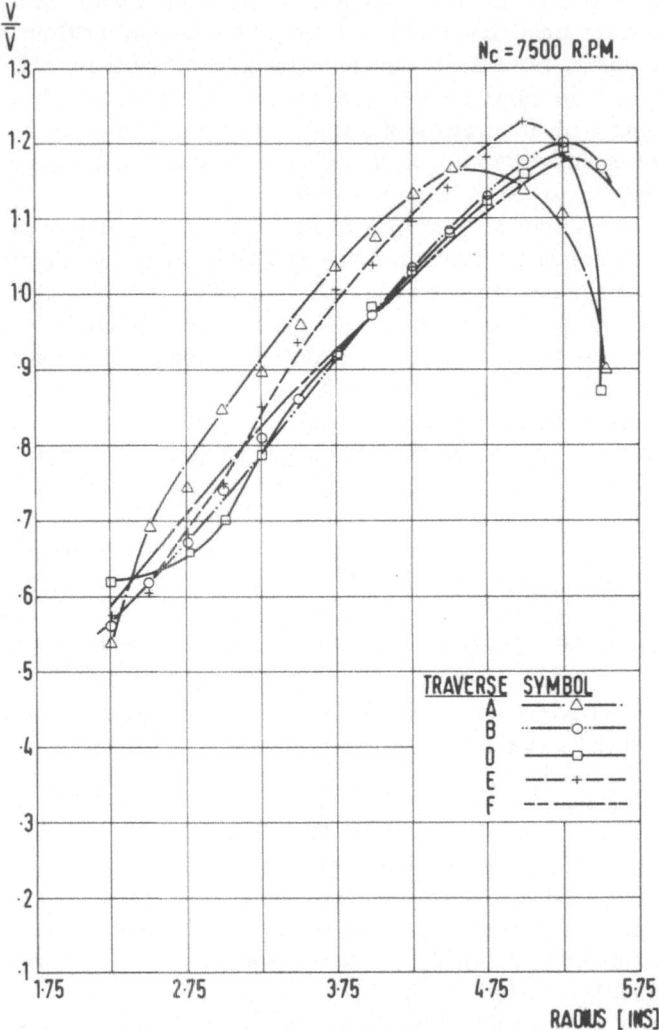

Fig. 3.   Radial distribution of velocity.

attained while the latter has had to be discarded due to other aspects of the overall engine design.

(4) Relative air angles calculated using velocity triangles show that there is a marked deviation from the blade angles. This difference (angle of attack) is almost constant from root to tip.

Under normal conditions such large deviations are unacceptable, but the compressor is working in an environment far removed from normal operating conditions. A reduction in density within the inlet would, for the same relative mass flow rate, cause an increase in axial velocity V and a consequent reduction in the angle of attack.

Appendix

The input data required by the quasi-three-dimensional computer program, apart from the geometrical positions of the hub and shroud boundaries and the impeller rotational speed, was the thickness of the blading and the angle of the mean-stream

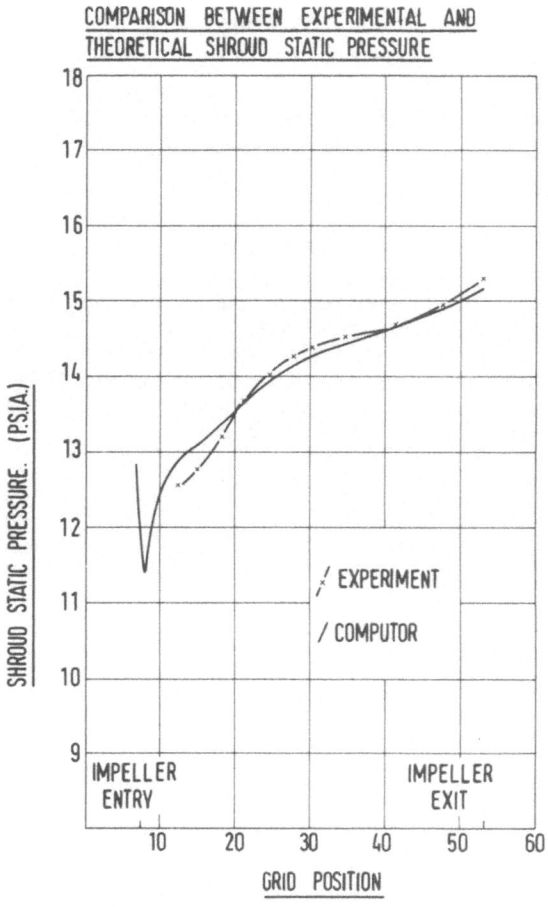

Fig. 4. Comparison between experimental and theoretical shroud static pressure.

surface. The starting grid line has to be situated in a duct region and has also to be sufficiently far upstream so as not to impose inflexible parameters and thus nullify the upstream influence of the rotating impeller. Therefore the asymmetric blading in the intake made it difficult to position the starting grid line in a duct and also choose suitable values for the following grid lines to represent the axisymmetric mean stream surface.

By obtaining the impeller inlet conditions it was possible to ignore the asymmetric blading in the intake and treat the annular passage as an unrestricted duct. Then by the choice of an appropriate velocity distribution on the starting grid line the program could produce a velocity profile just upstream of the impeller identical to the experimental profile, Figure 3. Also the extremely high positive angles of attack indicated that separation was likely from the rotating guide vane suction surface, and this would have to be allowed for in the theoretical solution.

When the experimental information regarding the inlet is fed into the computer program a realistic solution is obtained, which gives a similar impeller isentropic efficiency and also shows good agreement with the experimental static pressures from the shroud, Figure 4.

Further testing of the diffuser and the rotating impeller is in progress and will provide additional experimental results for comparison with the theoretical solution.

References

[1] Ferguson, T. B., *The Centrifugal Compressor Stage.*
[2] Marsh, H., *A Digital Computer Program for the Through-Flow Mechanics in an Arbitrary Turbomachine*, A.R.C. R & M No. 3509, 1968.
[3] *Manufacture of ¼" Wedge Probe*, REM/MW/1040, 1958, D. Napier & Sons.
[4] Dimmock, N. A., *A Compressor Routine Test Code*, A.R.C. R & M No. 3337.
[5] Rayle, R. E., 'An Investigation of the Influence of Orifice Geometry on Static Pressure Measurements', S.M. Thesis, M.I.T. Dept. of Mech. Eng.